Neuroanatomy for the Neuroscientist

Stanley Jacobson • Elliott M. Marcus

Neuroanatomy
for the Neuroscientist

 Springer

Stanley Jacobson
Tufts University Health Science Schools
Boston, MA
USA

Elliott M. Marcus
University of Massachusetts
School of Medicine
Worcester, MA
USA

ISBN 978-0-387-70970-3   e-ISBN 978-0-387-70971-0
DOI: 10.1007/978-0-387-70971-0

Library of Congress Control Number: 2007934277

© 2008 Springer Science+Business Media, LLC
All rights reserved. This work may not be translated or copied in whole or in part without the written permission of the publisher (Springer Science+Business Media, LLC, 233 Spring Street, New York, NY 10013, USA), except for brief excerpts in connection with reviews or scholarly analysis. Use in connection with any form of information storage and retrieval, electronic adaptation, computer software, or by similar or dissimilar methodology now known or hereafter developed is forbidden.
The use in this publication of trade names, trademarks, service marks, and similar terms, even if they are not identified as such, is not to be taken as an expression of opinion as to whether or not they are subject to proprietary rights.

Printed on acid-free paper

9 8 7 6 5 4 3 2 1

springer.com

To our families who showed infinite patience:
To our wives Avis Jacobson and Nuran Turksoy
To our children Arthur Jacobson and
Robin Seidman
Erin Marcus and David Letson
To our grandchildren Ross Jacobson
Zachary Letson and Amelia Letson

Preface

The purpose of this textbook is to enable a neuroscientist to discuss the structure and functions of the brain at a level appropriate for students at many levels of study, including undergraduate, graduate, dental, or medical school level. It is truer in neurology than in any other system of medicine that a firm knowledge of basic science material (i.e., the anatomy, physiology, and pathology of the nervous system) enables one to readily arrive at the diagnosis of where the disease process is located and to apply their knowledge at solving problems in clinical situations.

The two authors have a long experience in teaching neuroscience courses at the first- or second-year level to medical and dental students in which clinical information and clinical problem-solving are integral to the course. In addition, the first author has taught for many years an upper-level biology course on the central nervous system to undergraduates at Tufts University in Medford, MA, utilizing many of Dr. Marcus' cases to help engage the students. The second author has developed a case history problem-solving sessions in the book *Integrated Neurosciences* by E.M. Marcus and S. Jacobson (Kluwer, 2003) and he also conducts a problem-solving seminar in which all medical students at the University of Massachusetts participate during their clinical neurology clerkship rotation. This provides the students an opportunity to refresh their problem-solving skills and to review and update that basic science material essential for clinical neurology. At both levels, we have observed that this inclusion of case history materials reinforces the subject matter learned by markedly increasing the interest of the students in both basic and clinical science material. This text is a modified version of *Integrated Neurosciences*. This book is also an updated version of an earlier integrated textbook originally developed by the authors along with Dr. Brian Curtis and published by W. B. Saunders in 1972 as *An Introduction to the Neurosciences*. The text provides an updated approach to lesion localization in neurology, utilizing the techniques of computerized axial tomography (CT scanning), magnetic resonance imaging (MRI), and magnetic resonance angiography (MRA). Multiple illustrations demonstrating the value of these techniques in clinical neurology and neuroanatomical localization has been provided. The clinical cases illustrations have been utilized in the body of the text. An anatomical atlas, including MRI images, is provided on the accompanying CD and they are referred to as Atlas CD.

Decisions had to be made so that the size of the textbook remained within limits that could be managed in most of today's neuroscience courses. The printed book contains the core topics concerned with the central nervous system. We have divided this book into three sections: I: Introduction to the Central Nervous System (Chapters 1–8), II: The Systems (Chapters 9–15), and III: The Non-Nervous Elements (Chapter 16–18). Section III includes Chapter 16 (Meninges, Ventricular System and Vascular System). Chapter 17 (An Overview of Vascular Disease) and Chapter 18 (Movies on the Brain-a review of the movies that feature diseases of the nervous system). We have used several of these movies as an adjunct to a course (*Young Dr. Frankenstein* directed by Mel Brooks has a wonderful scene introducing the central nervous system and *Little Shop of Horrors* directed by Frank Oz features Steve Martin as a dentist and this a great introduction to the trigeminal nerve). There are many movies in the Science Fiction genre that are also useful for discussion and *Star Trek* and its many episodes and with its Medical Manual are at the top of our list! A number of other topics, including cell biology, cell physiology, embryology, nerve, and muscle are usually covered in other courses and the student should examine these topics in those courses. We have included a discussion on the olfactory system and the eighth nerve on the CD. The anatomy of the peripheral nervous system and autonomic nervous system should be reviewed in one of the standard gross anatomy texts.

A webpage has been established by the publisher (www.springer.com). This will provide a means for sending information to our readers, including errata and additions.

Most of the case histories utilized in the chapters have been drawn from the files of Dr. Marcus. For a number of the cases, our associates at the New England Medical Center, St. Vincent Hospital, Fallon Clinic, and the University of Massachusetts School of Medicine either requested our opinion or brought the case to our attention, and they provided information from their case files. These individual neurologists and neurosurgeons are identified in the specific case histories. We are also indebted to the many referring physicians of those institutions. Medical house officers at St. Vincent Hospital presented some of the cases to Dr. Marcus during morning report. In particular, our thanks are due to our associates in Worcester: Dr. Bernard Stone, Dr. Alex Danylevich, Dr. Robin Davidson, Dr. Harold Wilkinson, and Dr. Gerry McGillicuddy. Dr. Sandra Horowitz, Dr. Tom Mullins, Dr. Steve Donhowe, Dr. Martha Fehr, and Dr. Carl Rosenberg provided clinical information from their files for some of the case histories. Our associates at the New England Medical Center, Dr. John Sullivan, Dr. Sam Brendler, Dr. Peter Carney, Dr. John Hills, Dr. Huntington Porter, Dr. Thomas Sabin, Dr. Bertram Selverstone, Dr. Thomas Twitchell, Dr. C. W. Watson, and Dr. Robert Yuan, likewise provided some of the clinical material. Dr. Milton Weiner at St. Vincent Hospital was particularly helpful in providing many of the modern neuroradiological images. Dr. Sam Wolpert and Dr. Bertram Selverstone provided this material for the earlier version of the text. Dr. Val Runge from the Imaging Center at Texas A&M provided the normal MRIs. Dr. Anja Bergman (left-handed) had the patience to be our normal case and the images from her brain form the normal MRIs in the basic science chapters and atlas. Dr. Tom Smith and his associates in pathology provided much of the recent neuropathological material. Dr. John Hills and Dr. Jose Segaraa provided access to neuropathological material for the earlier version of the text. Dr. Sandra Horowitz and Dr. David Chad provided the critic of particular chapters. Dr. Brian Curtis contributed material for inclusion in the spinal cord chapter and on the physiology of the Visual System.

Dr. Mary Gauthier Delaplane provided many of the new anatomical drawings while a medical student at Boston University School of Medicine. Anne Que, Paul Ning, Tiffany Mellott, Elizabeth Haskins, and Tal Delman aided Dr. Delaplane. Dr. Marc Bard provided drawings for the earlier version of this text while a student at Tufts University School of Medicine. Dr. Brian curtis kindly provided much of the discussion in the spinal cord on its. We have continued to utilize or have modified some of the illustrations that were borrowed with permission from other published sources for the earlier version of this text. We have attempted to contact these original sources for continued permissions. We will acknowledge subsequently any sources that have been inadvertently overlooked. In many of the clinical chapters, various medications are recorded. Before utilizing these medications, the reader should check dosage and indications with other sources. It is with great pleasure that we extend our thanks to our publishers and particularly our editor Marcia Kidston and Joseph Burns. Any faults or errors are those of the authors and we would therefore appreciate any suggestions or comments from our colleagues.

Stanley Jacobson
Elliott M. Marcus

Contents

Part I Introduction to the Central Nervous System

Chapter 1 Introduction to the Central Nervous System 3

    I. The Neuron 3
        A. The Senses 5
        B. Muscles 6
    II. The Nervous System 7
    III. Central Nervous System 7
        A. Spinal Cord 8
        B. Brain 9
    IV. Glands Associated with the Brain 22

Chapter 2 Neurocytology: Cells of the CNS 23

    I. The Neuron 23
        A. Dendrites 23
        B. Soma 24
        C. Golgi Type I and II Neurons 24
        D. Dendritic Spines 24
        E. Cytoplasmic Organelles 26
        F. Nucleus 26
        G. Rough Endoplasmic Reticulum: Nissl Body 27
        H. Mitochondria 28
        I. Neurosecretory Granules 30
        J. Neuronal Cytoskeleton 30
        K. Microtubules and Axoplasmic Flow 31
        L. Neurofibrillar Tangles 32
        M. Axon and Axon Origin (Axon Hillock) 33
        N. Myelin Sheath 33
        O. Myelination 34
        P. Central Nervous System Pathways 35

	II. Synapse	35
	A. Synaptic Structure	36
	B. Synaptic Types	36
	C. Synaptic Vesicles	37
	D. Synaptic Transmission	38
	III. Supporting Cells of the Central Nervous System	40
	A. Astrocytes	40
	B. Oligodendrocytes	42
	C. Endothelial Cells	42
	D. Mononuclear Cells	43
	E. Ependymal Cells	45
	IV. Supporting Cells in the Peripheral Nervous System	46
	A. Satellite Cells	46
	B. Schwann Cells	46
	C. Neural Crest Cells	47
	V. Response of Nervous System to Injury	47
	A. Degeneration	47
	B. Regeneration	49
	VI. Blood-Brain Barrier	52
	A. Blood-Brain Barrier	52
	B. Extracellular Space and the CSF	54
Chapter 3	**Spinal Cord**	55
	I. Anatomy of the Spinal Cord	55
	A. Spinal Cord: Structure and Function	56
	B. Laminar Organization of Central Gray	61
	C. Segmental Function	64
	II. Nociception and Pain	70
	A. Nociceptive Stimulus	70
	B. Pain Receptors	70
	C. Projection Fibers	71
	D. Modulation of Pain Transmission	71
	E. White Matter Tracts	73
	F. Motor and Sensory Pathways	75
	III. Upper and Lower Motor Neuron Lesions	77
	A. Upper Motor Neuron Lesion	77
	B. Lower Motor Neuron Lesion	79
	IV. Other Spinal Pathways	83
Chapter 4	**Brain Stem**	85
	I. Gross Anatomical Divisions	85
	II. Functional Localization in Coronal Sections of the Brain Stem	87

Contents   xiii

                III. Differences between the Spinal Cord and Brain Stem   89
                     A. Medulla   89
                     B. PONS...   97
                     C. Midbrain 103
                IV. Functional Centers in the Brain Stem................... 110
                     A. Reticular Formation.............................. 110
                     B. Respiration Centers.............................. 114
                     C. Cardiovascular Centers 114
                     D. Deglutition...................................... 115
                     E. Vomiting 115
                     F. Emetic Center 116
                     G. Coughing.. 116
                     H. Taste.. 116
                V. Guidelines for Localizing Disease to and within
                     the Brain Stem 118

Chapter 5   The Cranial Nerves 121
                I. How the Cranial Nerves Got Their Numbers 121
                     A. Anterior Cranial Fossa (CN I and II).............. 121
                     B. Middle Cranial Fossa (CN III, IV, V, and VI) 122
                     C. Posterior Cranial Fossa (CN VIII–XII)............... 122
                II. Functional Organization of Cranial Nerves 122
                III. Embryological Considerations 124
                IV. The Individual Cranial Nerves 126
                     A. Cranial Nerve I: Olfactory........................ 127
                     B. Cranial Nerve II, Optic 128
                     C. Cranial Nerve III, Oculomotor 129
                     D. Cranial Nerve IV: Trochlear, Pure Motor 131
                     E. Cranial Nerve VI: Abducens, Pure Motor............ 131
                     F. Cranial Nerve V: Trigeminal, Mixed Nerve (Sensory
                         and Motor But No Parasympathetic)................ 132
                     G. Cranial Nerve VII, Facial, Mixed Nerve
                         (Sensory, Motor, Parasympathetic) 134
                     H. Cranial Nerve VIII, Vestibulo-cochlear,
                         Special Somatic Sensory, Receptive Organs in Petrous
                         Temporal Bone, Nerve Exits via Internal Acoustic.
                         Nerve in Posterior Cranial Fossa (Exits Internal
                         Acoustic Meatus in Petrous Temporal).............. 136
                     I. Cranial Nerve IX, glossopharyngeal,
                         Mixed (Sensory, Motor, Parasympathetic). Nerve to
                         third pharyngeal arch. Nerve in posterior cranial fossa
                         (exits via jugular foramen) 139
                     J. Cranial Nerve X. vagus, Mixed (Sensory, Motor,
                         Parasympathetic), and Longest Cranial Nerve. Nerve
                         to Fourth and Sixth Pharyngeal Arch............... 140

K. Cranial Nerve XI, Spinal Accessory, Pure Motor 141
L. Cranial Nerve XII, Hypoglossal, Pure Motor.......... 141
V. Cranial Nerve Dysfunction........................... 141
   A. Motor Cranial Nerve Lesion 141
   B. Sensory Cranial Nerve Lesion.................... 143
VI. Cranial Nerve Case Histories 144

Chapter 6 Diencephalon 147

I. Nuclei of the Thalamus............................. 148
II. Functional Organization of Thalamic Nuclei 149
   A. Sensory and Motor Relay Nuclei: The Ventrobasal
      Complex and Lateral Nucleus..................... 149
   B. Limbic Nuclei: Anterior, Medial, Lateral Dorsal,
      Midline, and Intralaminar Nuclei 153
   C. Specific Associational Polymodal/Somatic Nuclei:
      The Pulvinar Nuclei.............................. 155
   D. Special Somatic Sensory Nuclei: Vision and Audition.
      Lateral Geniculate and Medial Geniculate 155
   E. Nonspecific Associational....................... 156
III. White Matter of the Diencephalon 158
   A. Internal Capsule............................... 158
   B. Anterior Limb of the Internal Capsule 158
   C. Genu of the Internal Capsule 158
   D. Posterior Limb of the Internal Capsule 159
IV. Relationship between the Thalamus and the Cerebral
    Cortex 160
   A. Thalamic Input onto the Cortical Layers............ 160
   B. Thalamic Radiations and the Internal Capsule 160
   C. Other Possible Inputs to the Thalamus 161
V. Subthalamus 163

Chapter 7 Hypothalamus, Neuroendocrine System, and Autonomic Nervous System 165

I. Hypothalamus...................................... 165
   A. Hypothalamic Nuclei............................ 165
   B. Afferent Pathways.............................. 168
II. Neuroendocrine System: The Hypothalamus and Its
    Relation to Hypophysis............................ 171
   A. Hypophysis Cerebri............................. 173
   B. Hypothalamic–Hypophyseal Portal System.......... 173
   C. Hypophysiotrophic Area......................... 174
   D. Hormones Produced by Hypothalamus.............. 174
   E. Hormones Produced in Adenohypophysis 175
   F. Hypothalamus and the Autonomic Nervous System ... 179
   G. Functional Localization in Hypothalamus 180

Contents   xv

          III. Autonomic Nervous System........................ 183
               A. Enteric Nervous System 185
               B. Parasympathetic System (Cranio-sacral)............. 185
               C. Sympathetic System 186

Chapter 8 Cerebral Cortex Functional Localization 189

           I. Anatomical Considerations......................... 189
              A. Cytology 191
              B. Basic Design and Functional Organization
                 of Cerebral Cortex: 194
              C. Correlation of Neocortical Cytoarchitecture
                 and Function 197
          II. Methods for Study of Functional Localization........... 207
              A. How Do We Study Function? 207
              B. How Do We Confirm the Location
                 of the Pathology?............................. 209
              C. Neurophysiology Correlates of Cortical
                 Cytoarchitecture and the Basis of the EEG............ 209
          III. Subcortical White Matter Afferents and Efferents 211
              A. Projection Fibers (Fig. 8.8) 211
              B. Commissural Fibers........................... 212
              C. Associational Fibers 215
              D. Afferent Inputs and Efferent Projections
                 of Neocortex 215
              E. Nonthalamic Sources of Input 216
              F. Efferent Projections........................... 216
          IV. Development of the Cerebral Cortex 216
              A. Primary Sulci............................... 216
              B. Myelination................................ 217

Part II   The Systems within the Central Nervous System

Chapter 9 Motor System I: Movement and Motor Pathways 221

          I. Cerebral Cortical Motor Functions.................... 221
              A. Concept of Central Pattern Generators 221
              B. Effects of Spinal, Brain Stem, and Cerebral Lesions
                 on the Motor System........................... 223
          II. Postnatal Development of Motor Reflexes.............. 229
          III. Relationship of Primary Motor, Premotor
             and Prefrontal Cortex 229
              A. Functional Overview.......................... 229
              B. Primary Motor Cortex Area 4 231
              C. Areas 6 and 8: Premotor Cortex 235
              D. Area 8: Premotor............................. 238

            E. Suppressor Areas for Motor Activity
                (Negative Motor Response)...................... 238
            F. Prefrontal Cortex (Areas 9.14 and 46) 239
      IV. Disorders of Motor Development...................... 240
       V. Studies of Recovery of Motor Function in the Human..... 240
      VI. Cortical Control of Eye Movements.................... 241
            A. Saccadic Eye Movements 242
            B. Central Control of Saccades 242
            C. Smooth Pursuit in Contrast to Saccade............. 243
            D. Fixation System............................... 243
            E. Vergence Movements 243
            F. Vestibulo-ocular Movements..................... 243
            G. Opticokinetic Movements....................... 244
     VII. Major Voluntary Motor Pathways...................... 244
            A. Basic Principles of Voluntary Motor System.......... 244
            B. Corticospinal Tract: Voluntary Control of the
                Limbs, Thorax, and Abdomen.................... 245
            C. Corticonuclear/Corticobulbar System:
                Voluntary Control of the Muscles Controlled
                by Cranial nerves V, VII, and IX–XII 245
            D. Corticomesencephalic System: Voluntary Control
                of Muscles Associated with Eye Movements
                (Cranial Nerves III, IV, and VI).................. 248

Chapter 10   Motor System II: Basal Ganglia...................... 249

         I. Anatomy....................................... 249
            A. Connections................................ 250
            B. Microanatomy of the Striatum 253
            C. Overview of the Dopaminergic Systems............ 254
            D. Overlap with the Cerebellar System 254
        II. Clinical Symptoms and Signs of Dysfunction 255
            A. General Overview 255
            B. Parkinson's Disease and the Parkinsonian
                Syndrome 256
            C. Differential Diagnosis of Parkinson's Disease 265
            D. Chorea, Hemichorea, Hemiballismus and
                Other Dyskinesias 265

Chapter 11   Motor Systems III: Cerebellum and Movement.......... 273

         I. Anatomy....................................... 273
            A. Longitudinal Divisions 273
            B. Transverse Divisions.......................... 273
            C. Cytoarchitecture of the Cerebellum 275
            D. Cerebellar Fibers............................ 276

Contents

    II. Functions of the Cerebellum Topographic Patterns
of Representation in Cerebellar Cortex 277
   III. Effects of Disease on the Cerebellum.................. 278
       A. Overview 278
       B. Major Cerebellar Syndromes...................... 280

Chapter 12 Somatosensory Function and the Parietal Lobe 293

     I. Postcentral Gyrus: Somatic Sensory Cortex
[Primary Sensory S-I] 293
       A. Organization of the Postcentral Gyrus 293
       B. Postcentral Gyrus Stimulation..................... 294
       C. Postcentral Gyrus Lesions........................ 295
    II. Superior and Inferior Parietal Lobules 300
       A. Stimulation 300
       B. Lesions 301
       C. Parietal Lobules in the Dominant Hemisphere 301
       D. Parietal Lobules in the Nondominant Hemisphere 302
   III. Parietal Lobe and Tactile Sensation from the Body 306
       A. Basic Principle of Sensory System 306
       B. Tactile Sensation from the Body – Medial
Lemnsicus 306
       C. Tactile Sensation from the Head 308

Chapter 13 Visual System and Occipital Lobe 311

     I. Structure of the Eye 311
       A. Anatomy of the Eye............................ 311
       B. Optic Nerve 316
       C. Blind Spot 317
    II. Visual Pathway 317
       A. Retina and Visual Fields 317
       B. Visual Pathway: Overview 319
   III. Occipital Lobe 320
       A. Areas in Occipital Lobe-17, 18, 19 (VI–V5)......... 320
       B. Parallel Processing in the Visual Cortex 321
       C. Effects of Stimulation of Areas 17, 18, and 19 324
       D. Effects of Lesions in the Occipital Visual Areas....... 325
       E. Occipital Lobe and Eye Movements
(See Also Chapter 9) 326
   IV. Visual Field Deficits Produced by Lesions
in the Optic Pathway................................ 327
       A. Overview of Localized Lesions in the Visual System... 327
       B. Case Histories from a lesion on the Visual System:
Optic Nerve • Optic Chiasm • Visual
Radiations • Striate Cortex 327

Chapter 14 Limbic System, the Temporal Lobe, and Prefrontal Cortex 337

    I. Limbic System 337
        A. Subcortical Structures 339
        B. Cortical Structures in the Limbic System 341
   II. Principal Pathways of the Limbic System 350
        A. Fornix 350
        B. Circuits in Limbic Emotional Brain 351
  III. Temporal Lobe 353
        A. Auditory and Auditory Association 354
        B. Visual Perceptions................................ 354
        C. Symptoms of Disease Involving the Temporal Lobe ... 354
  IV. Role of the Limbic System in Memory 359
        A. Anatomical Substrate of Learning in Humans 359
        B. Disorders of Recent Memory; the Amnestic Confabulatory Syndrome of Diencephalic Origin; Wernicke–Korsakoff's. 361
        C. Other Lesions of the Diencephalon and Adjacent Regions Producing the Amnestic Confabulatory Disorder Seen in the Korsakoff Syndrome 363
        D. The Amnestic Confabulatory Syndrome Following Lesions of the Hippocampus and Related Structures........................... 363
        E. Progressive Dementing Processes................... 364
   V. Prefrontal Granular Areas and Emotions 367
        A. Anatomy and Functional Localization 367
        B. Connections of the Prefrontal Cortex................ 368
        C. The Case of Phineas P. Gage...................... 368
        D. Studies of Jacobsen and Nissen.................... 369
        E. Functional Neurosurgery......................... 370
        F. Role of the Limbic System in Psychiatric Disorders ... 371
  VI. The Limbic Brain as a Functional System 372
        A. Hierarchy of Function 372
        B. Reticular Formation.............................. 372
        C. Hypothalamus 372
        D. Pleasure/Punishment Areas....................... 372
        E. Limbic Cortical Regions 373

Chapter 15 Higher Cortical Functions........................... 375

    I. Cerebral Cortex and Disturbances of Verbal Expression ... 375
        A. Cerebral Dominance 376
        B. Development Aspects 377

Contents   xix

         II. Aphasia: Dominant Hemispheric Functions............. 377
             A. Cortical Areas of the Dominant Hemisphere
                 of Major Importance in Language Disturbances....... 378
             B. Types of Aphasia................................. 378
             C. Nonfluent Aphasia: Anatomical Correlation
                 of Specific Syndromes Involving Broca's area........ 381
             D. Fluent Aphasia: Anatomical Correlation
                 of Specific Syndromes (Wernicke's Aphasia and
                 Wernicke's Area).................................. 386
       III. Language Functions in the Nondominant
           Parietal Hemisphere.................................. 394
       IV. Role of Corpus Callosum in Transfer of Information...... 395

Part III   The Non-Nervous Elements within the Central Nervous System

Chapter 16   Meninges, Ventricular System and Vascular System....... 399

         I. Meninges: Coverings of the Brain..................... 399
             A. Dura Mater...................................... 399
             B. Arachnoid....................................... 400
             C. Pia Mater....................................... 401
         II. Ventricular System.................................. 401
       III. Blood Supply to the Brain........................... 403
             A. Arterial Supply to the Brain...................... 403
             B. Venous Circulation of the Brain.................. 406

Chapter 17   Cerebral Vascular Disease........................... 409

         I. Overview... 409
             A. Definitions...................................... 409
             B. Demographics................................... 409
         II. Ischemic–Occlusive Cerebrovascular Disease........... 410
             A. Definitions...................................... 410
             B. Role of Anastomoses............................. 411
             C. Major Types of Ischemic–Occlusive Disease......... 411
       III. Clinical Correlates of Vascular Territories: Syndromes.... 412
             A. Internal Carotid Artery........................... 412
             B. Middle Cerebral Artery Syndromes................ 414
             C. Anterior Cerebral Artery Syndrome................ 418
             D. Posterior Cerebral Artery Syndromes.............. 418
             E. Vertebral and Basilar Artery Syndromes
                 of the Brain Stem................................. 423
             F. Ischemic Occlusive Disease Involving
                 the Cerebellum................................... 427

		G. Ischemic Occlusive Disease of the Spinal Cord (the Anterior Spinal Artery Syndrome).............	428
	IV.	Primary Intracerebral Hemorrhage....................	428
		A. Demographics and Risk Factors	429
		B. Location.......................................	429
		C. Diagnostic Studies in Intracerebral Hemorrhage	429
		D. Clinical Correlates of Intracerebral Hemorrhage	430
	V.	Subarachnoid Hemorrhage...........................	430
		A. Demographics	430
		B. Major Clinical Features..........................	432
		C. Complications	432
		D. Management and Treatment	432

Chapter 18 Movies on the Brain................................. 435

	I.	Developmental Disorders	435
	II.	Spinal Cord/Brain Stem Disorders	436
	III.	Disorders of Motor Systems and Motor Control	437
	IV.	Limbic System	438
	V.	Cerebrovascular Disease............................	439
	VI.	Brain Trauma	440
	VII.	Brain Tumors and Increased Intracranial Pressure........	441
	VIII.	Infections	442
	IX.	Toxic and Metabolic Disorders......................	443
	X.	Disorders of Myelin	445
	XI.	Memory..	445
	XII.	Seizures and Epilepsy..............................	446
	XIII.	Coma..	447

Bibliography 449

Index... 487

List of Figures

Chapter 1 **Fig. 1** Types of neurons in the central nervous system (From EM Marcus and S Jacobson, Integrated neuroscience, Kluwer, 2003).. 4

Fig. 2 Hairy skin showing receptors (From EM Marcus and S Jacobson, Integrated neuroscience, Kluwer, 2003)............... 6

Fig. 3 The central nervous system *in situ* (Modified from Curtis, Jacobson, and Marcus. An introduction to neurosciences, Saunders, 1972)... 8

Fig. 4 Spinal cord. (From EM Marcus and S Jacobson, Integrated neuroscience, Kluwer, 2003).................................... 9

Fig. 5 The brain: MRI sagittal plain T1 (From EM Marcus and S Jacobson, Integrated neuroscience, Kluwer, 2003).............. 10

Fig. 6 Lateral surface of the cerebrum. The sensory (precentral) and motor strips (precentral) are marked. The somatotopic organization of this region is labeled: Head, Arm, Thorax Abdomen, and Foot. The foot region of the sensory–motor strip extends onto the medial surface of the hemisphere as the paracentral lobule. (Modified from Curtis, Jacobson, and Marcus, An introduction to the neurosciences, Saunders, 1972)............ 14

Fig. 7 Medial surface of a cerebral hemisphere, with brain stem removed. Gyri in frontal, parietal, occipital and temporal lobes identified and regions in the corpus callosum noted: (1) rostrum, (2) genu, (3) body, and (4) splenium. (Modified from Curtis Jacobson, and Marcus, An introduction to the neurosciences, Saunders, 1972)... 16

Fig. 8 Coronal section showing basal nuclei and their relationship to the internal capsule and thalamus (From EM Marcus and S Jacobson, Integrated neuroscience, Kluwer, 2003)......... 17

Fig. 9 Meningioma in the left cerebral hemisphere (From EM Marcus and J Jacobson, Integrated neuroscience, Kluwer, 2003).. 21

Chapter 2 **Fig. 1** Golgi type I cells (cells with long axons) in the motor cortex of a rat. (**A**) Soma, axon, and dendrite (×150); (**B**) dendritic spines (×1100) (A and B: Golgi rapid stain). (**C, D**) Electron micrographs of dendritic spines with excitatory synapses (×30,000) (From EM Marcus and J Jacobson, Integrated neuroscience, Kluwer, 2003).. 25

Fig. 2 Golgi type II cells (stellate cells, neurons with short axons in the motor cortex of the rat (Golgi-Cox stain; ×450.) (From EM Marcus and J Jacobson, Integrated neuroscience, Kluwer, 2003).. 25

Fig. 3 Ventral horn cell of a female squirrel monkey. Note the nucleus, nucleolus, and the accessory body of Barr (arrow); 1-μm epoxy section (×1400) (From EM Marcus and S Jacobson, Integrated neuroscience, Kluwer, 2003)................ 27

Fig. 4 Electron micrograph of the cerebral cortex showing the principal cell types in the central nervous system: neuron astrocytes (ASTRO), oligodendrocyte (OLIGO), and a blood vessel (BV) (From EM Marcus and S Jacobson, Integrated neuroscience, Kluwer, 2003)... 27

Fig. 5 Cytoskeleton: neurofibrillary stain of a ventral horn cell in the cat spinal cord, showing neurofibrillary network in soma and dendrites (**A**) and in the axons (**B**) (×400) 28

Fig. 6 Electron micrographs of a pyramidal neuron in the rat cerebral cortex: (**A**) soma and nucleus; (**B**) dendrite. Note the large amount of RER/Nissl substance in the soma and mitochondria. Dendrites have less Nissl substance and many microtubules. (×33,000). (From EM Marcus and S Jacobson, Integrated neuroscience, Kluwer, 2003).................................... 29

Fig. 7 Appearance of the axon hillock, axon origin: (**A**) in a Nissl stain (× 400), (**B**) in an electron micrograph (×15,000), and (**C**) in a Golg–rapid stain (×350) (From EM Marcus and S Jacobson, Integrated neuroscience, Kluwer, 2003) 30

Fig. 8 Myelin sheath. Electron micrograph of myelin sheath from the optic nerve of the mouse demonstrating repeating units of the myelin sheath, consisting of a series of light and dark lines. The dark line, called the major dense line (MDL), represents the apposition of the inner surface of the unit membranes. The less dense line, called the interperiod line

(IPL), represents the approximation of the outer surfaces of adjacent myelin membranes (×67,000) (Courtesy of Alan Peters, Department of Anatomy, Boston University School of Medicine)... 34

Fig. 9 Silver stain of a 1-μm plastic embedded section. **(A)** Synaptic boutons on neurons in the reticular formation; **(B)** boutons on ventral horn cells (×400) (From EM Marcus and J Jacobson, Integrated neuroscience, Kluwer, 2003) 36

Fig. 10 Synapse in the sensory cortex of the rat demonstrating agranular synaptic vesicles (300–400 Å in the presynaptic axonal side). Note the electron-dense synaptic membranes and the intersynaptic filaments in the synaptic cleft. Electron micrograph (65,000). (From EM Marcus and J Jacobson, Integrated neuroscience, Kluwer, 2003).................................... 37

Fig. 11 Electron micrograph of a human cerebral cortex demonstrating differences in the density of the DNA in the nuclei of oligodendrocytes (OLIGO) and microglia (×30,000) (From EM Marcus and S Jacobson, Integrated neuroscience, Kluwer, 2003).. 43

Fig. 12 Electron micrograph of a reactive astrocytes, gitter cell, in the cerebral cortex of a patient with Jakob–Creutfeld disease. Note the prominent digestion vacuoles in higher power A. (**B**, ×8000; **A**, ×35,000). (From EM Marcus and S Jacobson, Integrated neuroscience, Kluwer, 2003) 45

Fig. 13 Ependymal lining cell in the third ventricle of a rat. Note prominent cilia extending into the ventricle (arrow) in this 1-μm plastic section. (From EM Marcus and J Jacobson, Integrated neuroscience, Kluwer, 2003).................................... 46

Fig. 14 Ventral horn cells in the human lumbar spinal cord: **(A)** normal and (**B** to **D**) Wallerian retrograde chromatolytic changes in ventral horn cells following injury to the peripheral nerve. **(B)** Chromatolytic neuron with eccentric nucleus and some dissolution of the Nissl substance; **(C)** chromatolytic neurons, showing a peripheral ring of Nissl substance (peripheral chromatolysis); **(D)** chromatolytic neuron, showing eccentric nucleus and only a peripheral ring of Nissl substance (Nissl stain, ×400) (From EM Marcus and J Jacobson, Integrated neuroscience, Kluwer, 2003).. 48

Fig. 15 Wallerian degeneration in the medullary pyramid in a human many months after an infarct in the contralateral motor–sensory strip. Left side is normal; note the absence of myelin on right side. (Weigert myelin stain; ×80). (From EM Marcus and J Jacobson, Integrated neuroscience, Kluwer, 2003).. 49

Chapter 3 **Fig. 1** The posterior and anterior roots in relation to the gray and white matter of the spinal cord (From EM Marcus and S Jacobson, Integrated neuroscience, Kluwer, 2003)......... 56

Fig. 2 The relationships of the vertebral column, the meninges, and spinal cord. (From EM Marcus and S Jacobson, Integrated neuroscience, Kluwer, 2003).. 57

Fig. 3 The lateral aspect of the spinal cord exposed within the vertebral column. The spinous process and the laminae of the vertebrae have been removed and the dura mater has been opened longitudinally. (From C. Clemente (ed.), Gray's anatomy, Lea & Febiger, 1988) 57

Fig. 4 Typical spinal cord cross sections (From Gross, C.M. (ed.): Gray's anatomy. Lea and Febiger. From EM Marcus and S Jacobson, Integrated neuroscience, Kluwer, 2003) 59

Fig. 5 The technique of lumbar puncture (Modified from House and Pansky: A functional approach to neuroanatomy, McGraw-Hill, 1960). (From EM Marcus and S Jacobson, Integrated neuroscience, Kluwer, 2003).................................... 60

Fig. 6 Rexed's lamination pattern of the spinal gray 1954 matter on the right and the location of ventral horn cells on the left, lumbar section. myelin stain from 61

Fig. 7 Key dermatome boundaries in man: **A** Anatomical position. **B**-anterior surface, and **C** posterior surface (From J Zimmerman and S Jacobson, Gross anatomy, Little Brown, 1990)... 63

Fig. 8 Functional localization within the anterior horns (From Bossy, Atlas of neuroanatomy, WB Saunders 1974) (From EM Marcus and S Jacobson, Integrated neuroscience, Kluwer, 2003) 65

Fig. 9 The pathway for the monosynaptic stretch reflex (From EM Marcus and S Jacobson, Integrated neuroscience, Kluwer, 2003)... 67

Fig. 10 A muscle spindle (From Gardner, Fundamentals of neurology. Saunders, 1968 (From EM Marcus and S Jacobson, Integrated neuroscience, Kluwer, 2003) In contrast to this type of movement, which requires constant attention if the desired position is to be achieved, many of our movements, such as walking, require little attention beyond the decision to walk along the sidewalk. Most of us can even chew gum at the same time! These movements are probably carried out through mediation of the *gamma system*, which allows the brain to set a desired position and then forget about it. Compare then a corticospinal system, which produces a force, and a second system, the gamma system, which produces a new position. The gamma anterior horn cells are innervated by descending motor tracts other than the corticospinal. This system is controlled by neurons in the brain stem: the reticular formation, the vestibular nucleus, and the red nucleus.................................. 68

Fig. 11 The lamination pattern of the major tracts of the spinal cord (From Walker, Arch Neurol. Psychiat (Chicago), 43:284, 1940) (From EM Marcus and S Jacobson, Integrated neuroscience, Kluwer, 2003)... 73

Fig. 12 The location of the corticospinal tracts as shown by degeneration caused by a lesion in the internal capsule (From Is Wechsler,: Clinical neurology, Saunders, 1963) (From EM Marcus and S Jacobson, Integrated neuroscience, Kluwer, 2003)... 74

Fig. 13 The Babinski response. Upper: The normal adult response to stimulation of the lateral plantar surface of the foot; lower: the normal infant and upper motor neuron lesion effects in an adult (From EM Marcus and S Jacobson, Integrated neuroscience, Kluwer, 2003) 75

Fig. 14 Diagram illustrating a chordotomy. The cross section of the spinal cord shows the lamination of the spinothalamic tract, the position of the pyramidal tract in relation to it, and the presence of other tracts in the lower quadrant. A piece of bone wax is mounted 4.5 mm. from the tip of the knife as a depth gauge. Heavy curved lines in the ventral quadrant indicate the sweep of the knife. Note that a desire to spare the lateral corticospinal tract would result in sparing the sacral dermatomes. (From A. Kahn and S. Rand, J. Neurosurg 9:611–619, 1952) (From EM Marcus and S Jacobson, Integrated neuroscience, Kluwer, 2003).. 76

Fig. 15 Anterolateral pathway. The lateral and anterior spinothalamic tract Incoming fibers that originate in the dorsal root ganglion are activated by tissue-damaging stimuli. The axonal endings rise ipsilaterally from up to three spinal cord segments (as the fiber entering on the left; the fiber entering on the right synapses at the level of entry, one axon crosses the neuro-axis and rises and the other axon enters the anterior horn to participate in local reflexes, such as withdrawal from a hot surface after synapsing in the dorsal horn on the second-order neuron). The axons from the second-order sensory neuron cross and enter the contralateral spinothalamic tract. The major projections of the spinothalamic tract are the following: midline medulla, periaqueductal gray of the midbrain, and thalamus. In the thalamus, these end in either the ventral posterior lateral (VPL) nucleus or the intralaminar or dorsomedial thalamic nucleus. The information from VPL projects onto the postcentral gyrus and as a result, one knows where in the body the pain is originating. The information that reaches the prefrontal cortex via the DM nucleus permits one to decide how to respond to a painful stimulus. (From EM Marcus and S Jacobson, Integrated neuroscience, Kluwer, 2003) 78

Fig. 16 Effects of ALS on the corticospsinal treat and nentral horn cells. Courtesy of Klüver-Berrera Stain 83

Chapter 4 **Fig. 1** Gross view of the brain stem in sagittal MRI-T1. (From EM Marcus and S Jacobson, Integrated neuroscience, Kluwer, 2003) 86

Fig. 2 A Regions in the tegmentum of the brain stem. Coronal sections: Zones: d = ventricular, l = lateral zone, c = central zone, m = medial, v = basilar zone (From EM Marcus and S Jacobson, Integrated neuroscience, Kluwer, 2003) 86

Fig. 2 B Tegmentum of the brain stem with regions labeled (From EM Marcus and S Jacobson, Integrated neuroscience, Kluwer, 2003) 87

Fig. 3 Brain Stem Level 1 at transition level between cervical spinal cord and low medulla, with motor and sensory decussation. Coronal. (From EM Marcus and S Jacobson, Integrated neuroscience, Kluwer, 2003) 90

Fig. 4 A Brain Stem Level 2 at lower medullary level. Coronal (From EM Marcus and S Jacobson, Integrated neuroscience, Kluwer, 2003) 93

Fig. 4 B MRI, T_2 Brain stem at lower medullary level MRI-T1. (From EM Marcus and S Jacobson, Integrated neuroscience, Kluwer, 2003) 94

Fig. 5 A Brain stem level 3 at midmedullary level. (From EM Marcus and S Jacobson, Integrated neuroscience, Kluwer, 2003) 96

Fig. 5 B Brain stem at medullary level, inferior olive. MRI, T1 97

	Fig. 6 Brain stem level 4 at pontine level of cranial nerve VI and VII. Coronal lateral zone (From EM Marcus and S Jacobson, Integrated neuroscience, Kluwer, 2003)	99
	Fig. 7 A Brain stem at pontine level of cranial nerves V and VII. MRI, T1	101
	Fig. 7 B Brain stem at pontine level of cranial nerve V. Coronal (From EM Marcus and S Jacobson, Integrated neuroscience, Kluwer, 2003)	102
	Fig. 8 A Brain stem level 6 at midbrain inferior collicular level, Coronal (From EM Marcus and S Jacobson, Integrated neuroscience, Kluwer, 2003)	104
	Fig. 8 B Brain stem at inferior collicular level. Coronal MRI, T1	105
	Fig. 9 A Brain stem level 7; midbrain: superior collicular level	107
	Fig. 9 B Brain stem at superior collicular level. Coronal, MRI, T1	108
	Fig. 10 Brain stem at level of cerebral peduncles and III cranial nerve. Weil myelin stain. Coronal section. (From EM Marcus and S Jacobson, Integrated neuroscience, Kluwer, 2003)	110
Chapter 5	**Fig. 1** Locations of the cranial nerve nuclei in the gross brain stem. Motor nuclei of CN III–VII and CN IX–XII on the left; sensory nuclei of CN V and CN VII–X on the right. See Figure 5.3 for a demonstration of the functions of each of the 12 cranial (From EM Marcus and S Jacobson, Integrated neuroscience, Kluwer, 2003)	124
	Fig. 2 The 12 cranial nerves (From EM Marcus and S Jacobson, Integrated neuroscience, Kluwer, 2003)	125
	Fig. 3 Cranial nerve I, the olfactory nerve (From EM Marcus and S Jacobson, Integrated neuroscience, Kluwer, 2003)	128
	Fig. 4 Cranial nerve II, the optic nerve (From EM Marcus and S Jacobson, Integrated neuroscience, Kluwer, 2003)	129
	Fig. 5 Cranial nerves III, IV, and VI (From EM Marcus and S Jacobson, Integrated neuroscience, Kluwer, 2003)	130
	Fig. 6 Cranial nerve V, trigeminal nerve (From EM Marcus and S Jacobson, Integrated neuroscience, Kluwer, 2003)	132
	Fig. 7 Cranial nerve VII, facial nerve (From EM Marcus and S Jacobson, Integrated neuroscience, Kluwer, 2003)	134

Fig. 8 Parasympathetic innervations: Cranial from cranial nerves III, VII, IX, and X and sacral from spinal cord levels S2–S4 (From Curtis, Jacobson, & Marcus, 1974) 135

Fig. 9 Cranial nerve VIII, vestibulo-acoustic nerve, (From EM Marcus and S Jacobson, Integrated neuroscience, Kluwer, 2003).. 137

Fig. 10 Cranial nerve IX, glossopharyngeal (From EM Marcus and S Jacobson, Integrated neuroscience, Kluwer, 2003)......... 139

Fig. 11 Cranial nerve X, vagus. Motor nuclei distribute to branchial motor muscles in larynx and pharynx (ambiguous of CN X) and parasympathetic smooth muscles in gastrointestinal system (dorsal motor of CN X. (From EM Marcus and S Jacobson, Integrated neuroscience, Kluwer, 2003).................. 140

Fig. 12 Cranial nerve XI, accessory (From EM Marcus and S Jacobson, Integrated neuroscience, Kluwer, 2003)................ 142

Fig. 13 Cranial nerve XII, hypoglossal (From EM Marcus and S Jacobson, Integrated neuroscience, Kluwer, 2003)................ 142

Chapter 6 **Fig. 1** Brain in sagittal plane demonstrating the relationship between brain stem, diencephalons, and cerebrum (MRI T1.) Location of the thalamus. MRI. Diencephalon about a third of the way up from the floor of the third ventricle. (From EM Marcus and S Jacobson, Integrated neuroscience, Kluwer, 2003).. 148

Fig. 2 Dorsal view of gross specimen of brain stem, diencephalon, and basal ganglia. Medulla, pons, midbrain, thalamic regions, and internal capsule are labeled. (From EM Marcus and S Jacobson, Integrated neuroscience, Kluwer, 2003)................ 149

Fig. 3 Midthalamic level. **A** Schematic coronal section showing nuclei of the diencephalon: thalamus, epithalamus, hypothalamus, and suthalamus, and internal capsule and basal ganglia; **B** MRI, T2 weighted (From EM Marcus and S Jacobson, Integrated neuroscience, Kluwer, 2003) 150

Fig. 4 Three-dimensional reconstruction of the thalamic nuclei with the upper portion showing the three major nuclear masses and the lower portion showing individual nuclei (Modified from Carpenter, Core text, Williams & Wilkins, 1992) (From EM Marcus and S Jacobson, Integrated neuroscience, Kluwer, 2003).. 154

Fig. 5 Horizontal section demonstrating the anterior limb, genu, and posterior limb of the internal capsule. Note that the putamen

List of Figures   xxix

and globus pallidus are external to the internal capsule. MRI, T2 weighted (From EM Marcus and S Jacobson, Integrated neuroscience, Kluwer, 2003).................................... 156

Fig. 6 Internal capsule (right). Horizontal section with the major tracts and radiations labeled. Note the relationship between the lenticular nuclei (globus pallidus, putamen) and caudate to the capsule. (Modified after Carpenter, Core text, Williams & Wilkins, 1992) (From EM Marcus and S Jacobson, Integrated neuroscience, Kluwer, 2003).................................... 159

Fig. 7 Representation of the thalamic nuclei: **A** Connections of ventrobasal, midline, anterior, and lateral nuclei and geniculate nuclei; **B** connections of lateral dorsal, lateral posterior, and pulvinar nuclei. (Modified from RC Truex and MB Carpenter, Human neuroanatomy, Williams and Wilkins, 1970)................. 161

Fig. 8 Major thalamic projections onto the cerebral cortex. **A** Projections onto the lateral surface of the hemisphere; **B** thalamic projections onto medial surface (From EM Marcus and S Jacobson, Integrated neuroscience, Kluwer, 2003) 162

Chapter 7 **Fig. 1** The zones and nuclei of the hypothalamus (From EM Marcus and S Jacobson, Integrated neuroscience, Kluwer, 2003).. 166

Fig. 2 Myelin stain of diencephalon at the level of the anterior hypothalamus and optic chiasm showing corpus callosum, precommissural anterior commissure, and corpus striatum (From EM Marcus and S Jacobson, Integrated neuroscience, Kluwer, 2003).. 166

Fig. 3 A Electron micrograph of the neurohypophysis of an albino rat showing neurosecretory granules in the axoplasm of fibers in the hypothalamic–hypophyseal tract. (×30,000); **B** a Herring body, a storage site of neurosecretory material (×8000) (From EM Marcus and S Jacobson, Integrated neuroscience, Kluwer, 2003).. 168

Fig. 4 Myelin and cell stain (Kluver–Berrera) of diencephalon at midhypothalamus, demonstrating tuberal nuclei and mammillary bodies (From EM Marcus and S Jacobson, Integrated neuroscience, Kluwer, 2003).. 169

Fig. 5 Myelin and cell stain (Kluver–Berrera) of diencephalon at posterior hypothalamus (corpus callosum removed) (From EM Marcus and S Jacobson, Integrated neuroscience, Kluwer, 2003).. 169

Fig. 6 Afferent pathways into the hypothalamus (From EM Marcus and S Jacobson, Integrated neuroscience, Kluwer, 2003) 170

Fig. 7 Hypothalamic efferent pathways (From EM Marcus and S Jacobson, Integrated neuroscience, Kluwer, 2003) 171

Fig. 8 The hypothalamic–hypophyseal tract (neurosecretory system). This tract provides a direct connection between the hypothalamus and the neurohypophysis. The neurosecretory granules pass down the axons of this tract and are stored in the neural lobe until they are released into the bloodstream. (From EM Marcus and S Jacobson, Integrated neuroscience, Kluwer, 2003) 172

Fig. 9 The hypothalamic–hypophyseal venous portal system. These venous channels provide vascular continuity between the hypothalamus and the adenohypophysis, with releasing factors from the hypothalamus draining into the veins of the hypothalamus, which connect to the capillary bed in the infundibular stalk and median eminence. The factors are then carried to the adenohypophysis. (From EM Marcus and S Jacobson, Integrated neuroscience, Kluwer, 2003) 174

Fig. 10 Information flow between the brain and endocrine system 175

Fig. 11 Target glands for adneohypophyseal homones in an androgynous human (From EM Marcus and S Jacobson, Integrated neuroscience, Kluwer, 2003) 176

Fig. 12 Case 7.1 Adenomas: **A** MRI, T1-weighted midline sagittal sections; B control MRI, T1, weighted (From EM Marcus and S Jacobson, Integrated neuroscience, Kluwer, 2003) 178

Fig. 13 Tumor of the pineal. This is an enlarged nonmalignant cyst arising from the pineal gland (arrow). (Courtesy of Dr. John Hills and Dr. Jose Segarra) (From EM Marcus and S Jacobson, Integrated neuroscience, Kluwer, 2003) 183

Fig. 14 Autonomic nervous system: left, sympathetic (thoracolumbar); right, parasympathetic (craniosacral) (From EM Marcus and S Jacobson, Integrated neuroscience, Kluwer, 2003) 184

Chapter 8   **Fig. 1** Structure of the cerebral cortex. The results obtained with (left) a Golgi stain, (middle) a Nissl stain or other cellular stain, and (right) a myelin stain are contrasted. I = molecular layer; II = external granular layer; III = external pyramidal layer; IV = internal granular layer; V = large or giant. Pyramidal layer (ganglionic layer): VI = fusiform layer. The following features should be noted in the myelin stain: 1a = molecular layer; 3a1 = band of Kaes Bechterew; 4 = outer band of Baillarger; 5b = inner band of Baillarger. [After Brodmann (1909) from SW Ranson and SL Clark, The anatomy of the nervous system, Saunders, 1959, p. 350].. 190

Fig. 2 Pyramidal and stellate cells as demonstrated in the Golgi stain of cat cerebral cortex. PC = pyramidal cells; arrow = basket type of stellate cell; AD = apical dendrite of pyramidal cell; BD = basal dendrite of pyramidal cells; AX = axons of pyramS Jacobson, Integrated neuroscience, Kluwer, 2003) 191

Fig. 3 The five fundamental types of neocortical areas. Type 1 = agranular motor cortex; type 2 = frontal granular; type 3 = parietal homotypical–associational type of cortex; type 4 = polar cortex area 18; type 5 = granular koniocortex of striate cortex (From C Von Economo, The cytoarchitectonics of the human cerebral cortex, Oxford University Press, 1929, p. 16)............. 194

Fig. 4 Cytoarchitectural areas of cerebral cortex as designated by Brodmann. (From. Curtis BA, S. Jacobson, and EM Marcus, An introduction to the neurosciences: Saunders, 1972. Modified from Ranson S and Clarks S. The Anatomy of the Nervous System. W.B. Saunders 1959).. 195

Fig. 5 (From EM Marcus and S Jacobson, Integrated neuroscience, Kluwer, 2003). **A** Gross brain, **B** lobes labeled, **C** Major sulci labeled, **D** Major gyri labeled. Lateral surface................. 198

Fig. 6 Medial surface (From EM Marcus and S Jacobson, Integrated neuroscience, Kluwer, 2003). **A** Gross brain, **B** lobes labeled, **C** sulci labeled, and **D** Gyri labeled 200

Fig. 7. Map of somatic motor and sensory areas. Note that the primary motor and sensory areas extend into the depth of the rolandic fissure/central sulcus and extend onto the medial surface of the hemisphere. Note that the premotor cortex, area 6 also extends onto the medial surface of the hemisphere as the supplementary motor cortex. Note the location of the adversive gaze field in relation to motor cortex. (From W Penfield and H Jasper, Epilepsy and the functional anatomy of the human brain, Little, Brown and Company, 1954, p.103) 204

Fig. 8 Projection fiber systems passing through the internal capsule. The superior thalamic radiation includes fibers projecting from ventral thalamus to the sensory and motor cortex. (From EM Marcus and S Jacobson, Integrated neuroscience, Kluwer, 2003).. 211

Fig. 9 **A** Development of the Lateral surface of the brain during fetal life 18–41 weeks (From J Larroche, The development of the CNS during inuterine life. In F Falkner (ed.) Human development, Saunders 1966). **B** Development of the Medial surface of the brain during fetal life 18–41 weeks (From J Larroche, The development of the CNS during inuterine life. In F Falkner (ed.) Human development, Saunders 1966)......... 212

Fig. 10 The International Ten Twenty Electrode Placement System. The patient is awake but in a resting recumbent state with eyes closed. 8BF = frontal, T = temporal, C = central and O = occipital. Bottom: Effects of eye opening and closure. (From EM Marcus and S Jacobson, Integrated neuroscience, Kluwer, 2003). The International 10-20 Electrode Placement System is described in Jasper, 1958. 214

Chapter 9   **Fig. 1** Evolution of the automatic grasping responses of infants (From TE Twitchell, Neuropsychologia 3:251, 1965) 228

Fig. 2 Somatic figurines. The primary motor and sensory representation and the effects produced by supplementary motor and second sensory stimulation are demonstrated. (From W Penfield and H Jasper, Epilepsy and the functional anatomy of the human brain, Little Brown and Co., 1954) 232

Fig. 3 Parasagittal meningioma. Case 9.1. Arteriograms demonstrate the characteristic features of a circumscribed vascular tumor in the parasagittal central sulcal area, supplied by the right anterior cerebral artery and meningeal branches of the external carotid artery. **A** Arterial phase lateral common carotid injection; **B** arterial phase view, common carotid injection. See text for details... 234

Fig. 4 The motor cortical fields of the human as determined by Foerster employing cortical stimulation. Stimulation of the area indicated in black produced discrete movements at low threshold and was designated pyramidal. Lines or cross-hatching indicates other areas producing movements and were designated in a broad sense as extrapyramidal. These movements were more complex synergistic and adversive movements and included the adversive responses obtained by stimulation of area 19; areas extend to the midline is continued on to the medial surface. (Modified from O Foerster, Brain 59:137, 1936) (From EM Marcus and S Jacobson, Integrated neuroscience, Kluwer, 2003) 236

Fig. 5 Corticospinal pathway. voluntary movement of the muscles in the extremities, thorax and abdomen (From EM Marcus and S Jacobson, Integrated neuroscience, Kluwer, 2003)........... 246

List of Figures   xxxiii

Fig. 6 Corticonuclear system: voluntary control of the muscles controlled by cranial nerves V, VII, and IX–XII (From EM Marcus and S Jacobson, Integrated neuroscience, Kluwer, 2003) 247

Chapter 10   **Fig. 1** Diagram of the connections of the basal ganglia with transmitters and major transmitter action (+) excitatory or (−) inhibitory. GABA = gamma-aminobutyric acid (−); GLU = glutamate (+); dopamine = Dopa (+). The action of acetylcholine within the striatum has been omitted. See text for details. (From EM Marcus and S Jacobson, Integrated neuroscience, Kluwer, 2003) 251

Fig. 2 Major connections of the basal ganglia with sites for surgical lesions or implantation of stimulators in Parkinson's disease. Lesion I: globus pallidus; lesion II: ventral lateral (VL) nucleus of the thalamus (the sector of VL involvement is termed VIM = nuc. ventral inferior medial); lesion III: subthalamic nucleus (Modified from FH Lin, S Okumura, and IS Cooper, Electroenceph. Clin. Neurophysiol. 13:633, 1961) (From EM Marcus and S Jacobson, Integrated neuroscience, Kluwer, 2003) 252

Fig. 3 Parkinson's disease. The substantia nigra in Parkinson's disease. Left: normal substantia nigra; right: similar region in the case of idiopathic Parkinson's disease. A marked loss of pigmentation is evident. (Courtesy of Dr. Thomas Smith, University of Massachusetts) (From EM Marcus and S Jacobson, Integrated neuroscience, Kluwer, 2003) 258

Fig. 4 Lewy body in a pigmented neuron of the substantia nigra. H&E stain, original at ×125. (Courtesy of Dr. Thomas Smith) (From EM Marcus and S Jacobson, Integrated neuroscience, Kluwer, 2003) 258

Fig. 5 Parkinson's disease. Case 10.1. See text. On 05/03/83 before therapy with L-dopa/carbidopa and on 05/10/83, 1 week after starting therapy. In each example, the larger circles have been drawn by the examiner. (From EM Marcus and S Jacobson, Integrated neuroscience, Kluwer, 2003) 264

Fig. 6 Hemiballismus. Myelin stain of basal ganglia demonstrating a discrete hemorrhage into the right subthalamic nucleus (arrow points to the subthalmic nucleus on the unaffected side) (From LA Luhan, Neurology, Williams and Wilkins, 1968.) (From EM Marcus and S Jacobson, Integrated neuroscience, Kluwer, 2003) 266

Fig. 7 Huntington's disease. Marked atrophy of the caudate and putamen with secondary dilatation of the lateral ventricles is evident. Cortical atrophy was less prominent in this case. (Courtesy of Dr. Emanuel Ross, Chicago) (From EM Marcus and S Jacobs Integrated neuroscience, Kluwer, 2003).. 268

Fig. 8 Huntington's disease. Case 10.2. Marked atrophy of cerebral cortex and of caudate nucleus in this patient with a familial history of the disorder and a significant increase in the CAG trinucleotide repeats MRI, T_2. (From EM Marcus and S Jacobso Integrated neuroscience, Kluwer, 2003) 270

Chapter 11 **Fig. 1** Major transverse and longitudinal subdivisions of the cerebellum. The surface has been unfolded and laid out flat. (From CR Noback. et al,. The human nervous system. 4th ed, Lea & Febiger, 1991, p. 282) 274

Fig. 2 The most significant cells, connections, and the afferent and efferent fibers in the cerebellar cortex. Arrows show direction of axonal conduction. (+) = excitatory;(−) = inhibitory. Cf = climbing fibers; pf = parallel fiber; mf = mossy fiber. Refer to text. (From EM Marcus and S Jacobson, Integrated neuro science, Kluwer, 2003) 278

Fig. 3 Somat-topographical localization in the cerebellum of the monkey: **A** summary of projections from sensory area of the motor cortex; **B** summary of corticocerebellar projections. Note that there is a unilateral representation on the dosal surface and a representation within each paramedial lobule. (From RS Snider, Arch. Neurol. Psych. 64:204, 1950) 279

Fig. 4 Metastatic midline cerebellar tumor. Case History 11.2 (presented below. From EM Marcus and S. Jacobson, Integrated Neuroscience, Kluwer 2003)................................... 283

Fig. 5 Alcoholic cerebellar degeneration. The anterior lobe syndrome. There is a loss of Purkinje cells, atrophy of the cerebellar folia with relatively selective involvement of the anterior lobe. **A** Schematic representation of the neuronal loss; **B** atrophy of the anterior superior vermis in the sagittal sac (From M Victor, RD Adams, and EL Mancall, Arch. Neurol. 1:599–688, 1959)... 284

Fig. 6 Cerebellar atrophy most prominent in anterior superior vermis with predominant anterior lobe syndrome. MRI. This 70-year-old female 23 years previously had prolonged marked elevated temperature (104 °F, possibly as high as 110 °F) related to meningitis, coma, and convulsions with a residual ataxia of stance and gait. She was unable to walk a tandem gait but had no impairment on finger-tonose test. The patient also had episodes of tinnitus, vertigo, olfactory hallucinations, déjà vu, and musical hallucinations followed by loss of contact (temporal lobe seizures) (From EM Marcus and S Jacobson, Integrated neuroscience, Kluwer, 2003).................................. 286

Fig. 7 Syndrome of the lateral cerebellar hemisphere (hemiangioblastoma). This 72-year-old had a 2-year history of vertex headaches (worse on coughing), blurring of vison, diploplia, papliodema, and a lateral ipsilateral tremor on finger-to-nose testing (Courtesy of Dr. Jose Segarra) (From EM Marcus and S Jacobson, Integrated neuroscience, Kluwer, 2003).............. 287

Fig. 8 PICA, Cerebellar infarct. This 64-year-old right-handed male woke at 3 AM with a severe posterior headache, vertigo with projectile vomiting, and slurred speech. Shortly thereafter, he had left facial paraesthesias, clumsiness of the left extremity, and diplopia. Two weeks previously he had been seen for a transient episode of difficulty in controlling the right arm and impairment of speech. Past history was significant for hypertension, coronary bypass, 6 years previously episodes of transient weakness, and numbness of the right arm with transient aphasia. He had been placed on long-term anticoagulation. After intensive rehabilitation, 6 months later he showed only slowness of alternating hand movements. T2-weighted axial sections at level of upper medulla. (From EM Marcus and S Jacobson, Integrated neuroscience, Kluwer, 2003).............. 290

Chapter 12 **Fig. 1** Sensory representation as determined by stimulation studies on the human cerebral cortex at surgery. Note the relatively large area devoted to lips, thumb, and fingers. (From W Penfield and T. Rasmussen, The cerebral cortex of man, Macmillan, 195 p. 214).. 295

Fig. 2 Case 12.1. Glioblastoma, somatosensory cortex; focal sensory left facial seizures. MRI scans demonstrating the more extensive involvement of the cortex and white matter: horizontal (T2) (From EM Marcus and S Jacobson, Integrated neuroscience, Kluwer, 2003)... 297

Fig. 3 Case 12.3. Nondominant parietal constructional apraxia. The patient's attempts to draw a house are shown on the upper half of the page. Her attempts to copy a drawing of a railroad engine are shown on the lower half of the page (B and C). The examiner's (Dr. Leon Menzer) original is designated (A). (From EM Marcus and S Jacobson, Integrated neuroscience, Kluwer, 2003).. 304

Fig. 4 Nondominant parietal lobe syndrome. CT scans. This 68-year-old left-handed male with diabetes mellitus and hypertension had the sudden onset of left arm paralysis, loss of speech, and ability to read and write, all of which recovered rapidly. On examination 9 months after the acute episode, he continued to have the following selective deficits: (1) he was vague in recalling his left hemiparesis; (2) in dressing, he reversed trousers and failed to cover himself on the left side, (3) he had extinction in the left visual field on bilateral simultaneous visual stimulation and over the left arm and leg on bilateral simultaneous tactile stimulation; (4) a left Babinski sign was present. CT scan now demonstrated an old cystic area of infarction in the right posterior temporal-parietal (territory of the inferior division of the right middle cerebral artery). (From EM Marcus and S Jacobson, Integrated neuroscience, Kluwer, 2003) 305

Fig. 5 Tactile sensation from the upper extremity, fasciculus cuneatus. Posterior columns → medial lemniscus →VPL → postcentral gyrus.. 307

Fig. 6 Trigeminal: tactile sensation from the head: trigeminal lemniscus → medial lemniscus → VPM → postcentral gyrus 309

Chapter 13 **Fig. 1** Horizontal meridional section of the eye. (From Leeson and Lesson, Histology, Saunders................................ 312

Fig. 2 Schematic diagram of the ultrastructural organization of the retina. Rods and cones are composed of outer (OS) and inner segments (IS), cell bodies, and synaptic bases. Photo pigments are present in laminated discs in the outer segments. The synaptic base of a rod is called a spherule; the synaptic base of a cone is called a pedicle. Abbreviations: RB = bipolar cells, MB = midget bipolar cell, FB = flat bipolar cell, AM = amacrine cell, MG = midget ganglion cell, DG = diffuse ganglion cell. (Modified from M.B. Carpenter, Core text of neuroanatomy, Williams and Wilkins)................................. 315

Fig. 3 Visual pathway. Light from the upper half of the visual field falls on the inferior half of the retina. Image from the temporal half of the visual field falls on the nasal half of the retina; image from the nasal half of the visual field falls on the temporal half of the retina. The visual pathways from the retina to the striate cortex are shown. The plane of the visual fields has been rotated 90° toward the reader (From EM Marcus and S Jacobson Integrated Neuroscience, Kluwer, 2003)... 318

Fig. 4 Common lesions of the optic tract: (1) normal; (2) optic nerve; (3) optic chiasm; (4) optic tract or complete geniculate, complete geniculocalcarine, radiation or complete cortical; (5) if lesion in optic radiation temporal segment (Meyer's loop) it produces a superior quadrantanopia;- if it is in the parietal segment of the radiation, it produces an inferior quadrantanopia; (6) calcarine or posterior cerebral artery occlusion producing a homonymous hemianopia with macular sparing. (From EM Marcus and S Jacobson, Integrated neuroscience, Kluwer, 2003)... 328

Fig. 5 Case History 13.1. Sphenoid-wing meningioma producing compression of the right optic and olfactory nerves. Refer to text. Right carotid arteriogram, venous phase, demonstrating tumor blush in the subfrontal area of the anterior fossa, extending into the middle foesa. (Courtesy of Dr. Samuel Wolpert. New England Medical Center Hospitals.) (From EM Marcus and S Jacobson, Integrated neuroscience, Kluwer, 2003).. 329

Fig. 6 Pituitary adenoma. Case History 13.2. Visual fields demonstrating a bitemporal hemianopia. S. = left eye; O.D. = right eye. This 51-year-old obese male with large puffy hands and a prominent jaw, with declining sexual interest for 18 years, an 8-year history of progressive loss of visual acuity, a 6–7-month history of diplopia due to a bilateral medial rectus palsy, and headaches, had stopped driving because he was unable to see the sides of the road. Urinary adrenal and gonadal steroids and thyroid functions were low, with no follicle-stimulating hormone. Imaging studies demonstrated a large pituitary adenoma with significant suprasellar extent. (From EM Marcus and S Jacobson, Integrated neuroscience, Kluwer, 2003).. 330

Fig. 7 Case 13.2. A large pituitary adenoma in a 44-year old man with a bitemproal hemianopsia. Significant extrasellar extension compresses the optci chiasm. MRI scans: **(A)** midline sagittal section; **(B)** coronal section. (From EM Marcus and S Jacobson, Integrated neuroscience, Kluwer, 2003) 331

Fig. 8 Case History 13.3. Brain abscess, right occipital area. Perimetric examination of visual fields. Top row: Initial examination demonstrated a somewhat asymmetrical (noncongruous) *left homonymous hemianopsia*, less in the left eye than the right. The fields are shown from the patient's point of view. Middle row: 16 days later; with antibiotic therapy, an improvement had occurred. A *noncongruous quadrantanopia* is now present. Bottom row: Fields after an additional 24 days. A relatively *complete homonymous hemianopia*, present at the time of admission, persisted following surgery. (From EM Marcus and S Jacobson, Integrated neuroscience, Kluwer, 2003).. 333

Fig. 9 Case History 13.4. Focal visual seizures characterized by "flashing lights" and the sensation of movement of the visual field with secondary generalization beginning at age 14 due to metastatic (thyroid) tumor of the left occipital lobe. MRI, T2-weighted, nonenhanced demonstrated a small tumor at the left occipital pole with surrounding edema. (From EM Marcus and S Jacobson, Integrated neuroscience, Kluwer, 2003)......... 334

Fig. 10 Occlusion of right posterior cerebral artery. Tangent screen examination of visual fields 3 months after the acute event. This 75-year-old woman had the acute onset of headache, bilateral blindness confusion, vomiting, mild ataxia, and bilateral Babinski signs. All findings cleared except a homonymous hemianopia with macular sparing. O.S. = left eye; O.D. = right eye. (From EM Marcus and S Jacobson, Integrated neuroscience, Kluwer, 2003).................................. 335

Fig. 11 CT scan of a total infarct, presumably embolic in the right PCA cortical territory (occipital and posterior temporal lobes) obtained 5 days after the acute onset of confusion and possible visual hallucinations in an 86-year-old right-handed male. Dense left homonymous hemianopia with no evidence of macular sparing. In addition, a left field deficit is present As confusion cleared, examination indicated the patient would look to the right and to the midline but would not follow objects to the left. (From EM Marcus and S Jacobson Interpreted Neuroscience Kluwer, 2003) 336

Chapter 14   **Fig. 1 A** Medial surface of a cerebral hemisphere, including the entire brain stem and cerebellum (From EM Marcus and S Jacobson, Integrated neuroscience, Kluwer, 2003) **B** Medial surface with the brain stem and cerebellum now removed (From EM Marcus and S Jacobson, Integrated neuroscience, Kluwer, 2003) **C** Medial surface of the cerebrum with thalamus removed to demonstrate relationship between fornix and hippocampus (From EM Marcus and S Jacobson, Integrated neuroscience, Kluwer, 2003)................................. 338

Fig. 2 Coronal section through diencephalon showing hippocampus in medial temporal lobe and mammillothalamic tract leaving the mammillary body. (From EM Marcus and S Jacobson, Integrated neuroscience, Kluwer, 2003).............. 342

Fig. 3 Coronal section through mammillary bodies and amygdala. MRI-T2 (From EM Marcus and S Jacobson, Integrated neuroscience, Kluwer, 2003).. 343

Fig. 4 The sectors in the hippocampus of the human. The dentate gyrus, subiculum, and related structures are demonstrated. Nissl stain. (From EM Marcus and S Jacobson, Integrated neuroscience, Kluwer, 2003)................................. 343

Fig. 5 Sagittal section. Temporal lobe with amygdala and hippocampus. Myelin stain (From EM Marcus and S Jacobson, Integrated neuroscience, Kluwer, 2003)................................. 346

Fig. 6 Sagittal section showing the medial surface of the cerebrum demonstrating the limbic structures that surround the brain stem and are on the medial surface of the cerebrum, with portions of the Papez circuit labeled: fornix, mammillary bodies, and anterior thalamic nucleus cingulate cortex 351

Fig. 7 Perforant pathway. (Modified from M.B. Carpenter (1970), Core text of neuroanatomy, William and Wilkins) (From EM Marcus and S Jacobson, Integrated neuroscience, Kluwer, 2003).. 352

Fig. 8 Glioblastoma in left temporal lobe. **Case History 14.2.** Complex partial seizures with secondary generalization in a 47-year-old. MRI.T2 nonenhanced. Horizontal (From EM Marcus and S Jacobson, Integrated neuroscience, Kluwer, 2003).. 359

Fig. 9 Hippocampus. **A** Hippocampus of a normal male patient, age 91; **B** hippocampus of a patient with Alzheimer's disease, age 72. Note the smaller hippocampus and larger ventricles in the patient with Alzheimer MRI, T2 weighted. (From EM Marcus and S Jacobson, Integrated neuroscience, Kluwer, 2003) Courtesy of Dr. Daniel Sax 365

Fig. 10 Prefrontal lobotomy. A surgical section has separated the prefrontal connections with the thalamus. (Courtesy of Dr. Thomas Sabin and Dr. Thomas Kemper) (From EM Marcus and S Jacobson, Integrated neuroscience, Kluwer, 2003).. 370

Chapter 15 **Fig. 1** Speech areas of the dominant hemisphere, summary diagram combining the data of pathological lesions and stimulation studies. Precise sharp borders are not implied. The designations of Penfield and Roberts (1959) the terms "anterior speech area", and "superior speech cortex. The basal temporal area is not labeled. Stimulation of these regions produces arrest of speech." 379

Fig. 2 Case History 15.2. Anterior aphasia embolic infarct secondary to rheumatic atrial fibrillation. MRI T2-weighted, horizontal section (From EM Marcus and S Jacobson, Integrated neuroscience, Kluwer, 2003) 385

Fig. 3 The arcuate fasciculus: dissection of the long-fiber system, the superior longitudinal fasciculus and the arcuate fasciculus, passing beneath the cortex of inferior parietal, and posterior temporal areas. A small bundle of U fibers are also seen interconnecting adjacent gyri in the occipital and temporal lobes (From EM Marcus and S Jacobson, Integrated neuroscience, Kluwer, 2003) 388

Fig. 4 Case History 15.5. Posterior aphasia, infarction left posterior-temporal and parietal areas, and inferior division of left MCA (probably embolic following left carotid occlusion). CT scan 391

Chapter 16 **Fig. 1** Diagram showing relationship among the skull, meninges (dural falx, pia, arachnoid) subarachnoid space, arachnoid granulations, and venous sinus. (From BA Curtis, S Jacobson, and EM Marcus, Introduction to the neurosciences, Saunders, 1972) 400

Fig. 2 The ventricular system (From EM Marcus and S Jacobson, Integrated neuroscience, Kluwer, 2003) 402

Fig. 3 The blood supply to the brain. MRA. ACA = anterior cerebral artery; MCA = middle cerebral artery; PCA = posterior cerebral artery, ICA = Internal carotid artery; VA = vertebral artery. (From EM Marcus and S Jacobson, Integrated neuroscience, Kluwer, 2003) 403

Fig. 4 The arterial circle of Willis (From BA Curtis, S Jacobson, and EM Marcus, Introduction to the neurosciences, Saunders, 1972) 405

Fig. 5 Venous circulation from the brain, head, and neck (From BA Curtis, S Jacobson, and EM Marcus, Introduction to the neurosciences, Saunders, 1972) 407

List of Figures xli

Chapter 17 **Fig. 1** Left carotid stenosis with old infarction predominantly within the border zone between anterior and middle cerebral arteries. This 72-year-old male had a persistent paralysis of the right upper extremity with defective position sense at fingers. (Courtesy Dr. John Hills and Dr. Jose Segarra) (From EM Marcus and S Jacobson, Integrated neuroscience, Kluwer, 2003).. 413

Fig. 2 Lenticulostriate penetrating branches of the MCA. This 73-year-old hypertensive, at age 53, had sustained a right hemiplegia with persistent and dense motor deficits involving the right central face, arm, and leg. **A** Coronal section; **B** close-up of the cavity. (Courtesy of Dr. John Hills and Dr. Jose Segarra) (From EM Marcus and S Jacobson, Integrated Neuroscience, Kluwer, 2003).. 415

Fig. 3 Acute hemorrhagic infarction of the central cortical territory of the LT-MCA: calcified embolus occluded MCA following atheromatous occlusion of the carotid artery at the bifurcation. This 74-year-old, right-handed, white male, 5 days prior to death, had the sudden onset of pain in the left supraorbital region and lost consciousness. Upon regaining consciousness, the patient had a right central facial weakness and a right hemiparesis and was mute but was transiently able to follow commands. (Courtesy of Drs. John Hills and Dr. Jose Segarra) (From EM Marcus and S Jacobson, Integrated neuroscience, Kluwer, 2003).. 416

Fig. 4 Anterior cerebral arteries: old bilateral infarcts due to severe bilateral stenosis ACA (complete occlusion on right). **(A)** Left hemispheric involvement was marked with **(B)** a lesser involvement of the right hemisphere. This 61-year-old, right-handed female 16 months prior to death developed difficulty in speech, right-sided weakness, apraxia of hand movements, and progressed 2 days later to an akinetic mute state with urinary incontinence. (Courtesy of Dr. John Hills and Dr. Jose Segarra) (From EM Marcus and S Jacobson, Integrated neuroscience, Kluwer, 2003).............. 419

Fig. 5 Weber's syndrome. An area of infarction is noted in the right cerebral peduncle (arrow), so located as to involve the fibers of the right third cranial nerve. This 61-year-old patient with rheumatic heart disease, endocarditis, and auricular fibrillation had the sudden onset of paralysis of the left face, arm, and leg, plus partial right third nerve palsy. (Courtesy of Dr. John Hills and Dr. Jose Segarra) (From EM Marcus and S Jacobson, Integrated neuroscience, Kluwer, 2003).. 420

Fig. 6 Posterior cerebral artery syndrome. Left: MRA. Embolic occlusion of distal right posterior cerebral artery (arrow) with filling of more distal branches by anastamotic flow. Right: MRI. Infarct right occipital area, calcarine artery branch territory. This 55-year-old white female had the sudden onset of a left visual field defect followed by tingling paresthesias of the left face and arm. On exam, she was hypertensive and obese, with a small ecchymosis under the left second digit toenail. A noncongruous left homonymous hemianopsia was present greater in the left temporal field than the right nasal field. The left plantar response was possibly extensor. Pain and graphesthesia were decreased over the left foot. Reevaluation of the patient 22 hours after the onset of symptoms demonstrated clearing of all neurological findings. Holter monitoring demonstrated multiple brief episodes of paroxysmal atrial fibrillation, a cause of emboli. Echoencephalogram demonstrated a patent foramen ovale, so that shunting of blood occurred from the right to left atrium, which is also a cause of emboli. She received long-term anticoagulation and treatment for hypertension. (From EM Marcus and S Jacobson, Integrated neuroscience, Kluwer, 2003).. 422

Fig. 7 Paramedian infarct upper pons. MRI T2. This 70-year-old, right-handed male had the acute onset of weakness of the right face (central), arm, and leg plus slurring of speech. A similar episode had occurred 2 months previously but had cleared over 2 weeks. His findings indicated a relatively pure motor syndrome. (From EM Marcus and S Jacobson, Integrated neuroscience, Kluwer, 2003).. 424

Fig. 8 Dorsal–lateral medullary infarct (arrows) due to occlusion of the posterior inferior cerebellar artery (arrow) in a 72-year-old diabetic with the correlated clinical syndrome of acute onset nystagmus on left lateral gaze, decreased pain sensation on the left side of face and right side of body, with difficulty swallowing, slurring of speech, and deviation of the uvula to the right. (Courtesy of Dr. Jose Segarra) (From EM Marcus and S Jacobson, Integrated neuroscience, Kluwer, 2003).. 426

List of Tables

Chapter 1	Table 1	Types of neurons in the nervous system	4
	Table 2	Sensory receptors	5
	Table 3	Cranial nerve localization in brain stem	11
	Table 4	Divisions of the diencephalon	12
	Table 5	Overview of cerebral functions (Figs. 1.6 and 1.7)	13
	Table 6	Comparison of upper motor neuron to lower motor neuron lesion	20
Chapter 2	Table 1	Parts of a neuron	26
	Table 2	Rate of axonal transport of cellular structures	32
	Table 3	Location and functions of neurotransmiters	39
	Table 4	Mechanicoreceptor	40
	Table 5	Functions of central nervous system supporting cells	41
	Table 6	Role of astrocytes in the central nervous system	41
	Table 7	Types of microglia cell	44
Chapter 3	Table 1	Sensory Receptors	66
	Table 2	Major pathways in the spinal cord	82
Chapter 4	Table 1	Regions in the brain stem	88
	Table 2	The five zones in the tegmentum	88
	Table 3	Gray and white matter equivalents in spinal and brain stem structures	89
	Table 4	Major tracts in the tegmental zone	90

	Table 5	Functional groupings of nuclei in the reticular formation	111
	Table 6	Major pathways of the reticular formation	112
Chapter 5	Table 1	Components of the cranial nerves	123
	Table 2	Cranial nerve components (see Fig. 5.3 for a visual summary)	126
	Table 3	Dysfunction in the eye due to lesions in the brain stem	130
Chapter 6	Table 1	Nuclei in the thalamus	151
	Table 2	Functional groups of thalamic nuclei	157
	Table 3	Intrinsic fiber bundles of the diencephalon	158
	Table 4	Thalamic terminations of major ascending pathways of the central nervous system	164
Chapter 7	Table 1	Functional localization in hypothalamic nuclei	172
	Table 2	Releasing factors and hormones produced in the hypothalamus	175
	Table 3	Hormones produced in adenohypophysis	176
	Table 4	Effects of hypothalamic stimulation on the autonomic responses	179
Chapter 8	Table 1	Major gyri in the frontal lobe	201
	Table 2	Major gyri in the parietal lobe	203
	Table 3	Gyri in the temporal lobe	205
	Table 4	Gyri of the occipital lobe	206
Chapter 9	Table 1	Definition of a reflex	222
	Table 2	Developmental changes in motor responses	230
Chapter 10	Table 1	Anatomical divisions of corpus striatum	250
	Table 2	Clinical symptoms and signs of basal ganglia disease	255
	Table 3	Clincal signs in Parkinson's disease	257
Chapter 11	Table 1	Longitudinal subdivisions of the cerebellum	274
	Table 2	Transverse divisions of the cerebellum	275
	Table 3	Cerebellar fibers	276

	Table 4	Regional functional correlations	280
Chapter 12	**Table 1**	Modalities of sensation in postcentral gyrus	296
	Table 2	Mechanicoreceptor	306
Chapter 14	**Table 1**	Regions of the limbic system	340
	Table 2	Connections of hippocampal formation (Fig. 14.6) 349	
	Table 3	Anatomical localization of symptoms from temporal lobe seizures or stimulation of the temporal lobe	355
Chapter 15	**Table 1**	Milestones of normal development of speech in the child there is considerable variability	377
	Table 2	Evaluation of language functions	379
	Table 3	Fluent aphasias	381
Chapter 16	**Table 1**	Components of plasma and CSF	402
	Table 2	Venous sinus	407
Chapter 17	**Table 1**	Intracerebral hemorrhage	430
	Table 2	Summary of the causes of SAH, aneurysm location, and manifestations	431

Part I
Introduction to the Central Nervous System

Chapter 1
Introduction to the Central Nervous System

The Central Nervous System consists of the spinal cord and brain. The spinal cord is found in the vertebral column and the brain is housed in the cranium/skull. The spinal cord has 32 segments and the brain consists of the brain stem, diencephalon, cerebellum, and cerebrum. At the foramen magnum, the highest cervical segment of the spinal cord is continuous with the lowest level of the medulla of the brain stem. The 12 cranial nerves attached to the brain form the upper part of the part of the peripheral nervous system and record general sensations of pain, temperature, touch, and pressur; in addition, we now find the presence of the special senses of smell, vision, hearing, balance, and taste. The blood supply to the brain originates from the first major arterial branches from the heart, ensuring that over 20% of the entire supply of oxygenated blood flows directly into the brain.

I. The Neuron

Human beings enter the world naked but equipped with a nervous system that, with experience, is ready to function in almost any environment. One word summarizes the function of the nervous system: "protection." The central nervous system (brain and spinal cord) monitors and controls the entire body by its peripheral divisions, which are distributed to all the muscles, organs, and tissues. The brain has an advantageous site in the head and above the neck, which can move in about a 140° arc. Close to the brain are all of the specialized sense organs, which permit us to see, smell, taste, and hear our world. The central nervous system is protected by fluid-filled membranes, the meninges, and surrounded by the bony skull and vertebrae.

The basic conducting element in the nervous system is the nerve cell, or neuron (Fig. 1.1). A neuron has a cell body, dendrite, and axon. The cell body contains many of the organelles vital to maintain the cells structure and function, including the nucleus and nucleolus, and is considered the tropic center of the nerve cell. The dendrites extend from the cell body and increase the receptive surface of the neuron. The axon leaves the cell body and connects to other cells. Axons are covered by a lipoproteinaceous membrane called *myelin* that insulates the axons from the

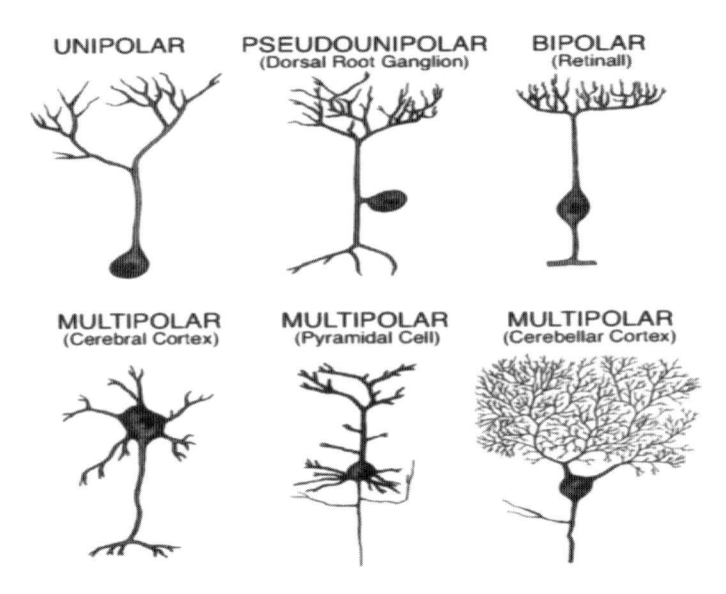

Fig. 1.1 Types of neurons in the central nervous system (From EM Marcus and S Jacobson, Integrated neuroscience, Kluwer, 2003)

Table 1.1 Types of neurons in the nervous system

Neuronal type	% of neurons	Location
Unipolar 0.05	0.5	Dorsal root ganglia of spinal cord
		Cranial nerve ganglia of brain stem
		Mesencephalic nucleus of CN V in midbrain
Bipolar	0.05	Retina, inner ear, taste buds
Multipolar:		
Peripheral	0.1	Autonomic ganglia
Central	99.8	Brain and spinal cord

fluids in the central nervous system. The site of contact between the axon of one nerve cell and the dendrites and cell body of another neuron is the *synapse* (see Chapter 2). The cells in the nervous system are classified based on their shapes: unipolar, bipolar, and multipolar (Table 1.1). In the central nervous system, the nerve cells are supported by glia and blood vessels; in the peripheral nervous system, they are supported by satellite cells, fibroblasts, Schwann cells, and blood vessels.

There are three basic categories of neurons: (1) receptors, the ganglia of the spinal dorsal roots and of the cranial nerves with general sensory components; (2) effectors, the ventral horn cells, motor cranial nerve nuclei, and motor division of the autonomic nervous system; and (3) interneurons, the vast majority of the neurons in the central nervous system. The areas in the central nervous system that contain high numbers of neuronal cell bodies are called *gray matter* and the regions that contain primarily myelinated axons are called *white matter*. Neurons are organized into ganglia, nuclei, or layered cortices.

Ganglia. *Sensory ganglia* are found outside the central nervous system and contain the first-order neurons in the sensory systems and they are the dorsal root ganglia on the 32 segments of the spinal cord and the sensory ganglia on cranial nerves CN V, CN VII, CN VIII, CN IX, and CN X. *Motor/autonomic ganglia* are found throughout the body and they are either sympathetic or parasympathetic ganglia.

Nuclei. Throughout the brain and spinal cord there are groupings of neurons with a common functions; these are the nuclei. They are found throughout the spinal cord (ventral and dorsal horn), brain stem (cranial nerve nuclei, reticular formation) and diencephalon (nuclei of thalamus, hypothalamus, subthalamus and metathalamus), basal ganglia (caudate, putamen, globus pallidus, substantia nigra), and in the cerebral cortex (amygdaloid nuclei).

Lamina. In the cerebral cortex, cerebellar cortex, and superior colliculus, the gray matter is on the surface and organized anatomically into horizontal columns and physiologically into vertical columns permitting a nearly infinite number of interconnections.

A. *The Senses*

Aristotle distinguished five senses: hearing, sight, smell, taste, and touch. Modern neuroscience, however, includes the *five special senses* (balance, vision, hearing, taste, and smell) and the *four general senses* (pain, temperature, touch, and pressure). Humans have evolved a series of specialized receptors for each of these different sensory functions (Table 1.2). The special sensory apparatuses are found in the head: the eye and its protective coverings and muscles, the membranous labyrinth in the temporal bone for hearing and balance, the nose with olfactory receptors, and the tongue with taste buds The *receptors for general sensation* (mechanicoreceptors, nociceptors, and thermoreceptors) are located primarily in

Table 1.2 Sensory receptors

Class of receptor	Function	Location
Chemoreceptors	Taste	Taste buds on tongue
	Smell	Olfactory mucosa in nose
Mechanicoreceptors	Balance	Inner ear, semicircular canals
	Sound	Inner ear, cochlea
	Tactile discrimination and pressure	Skin, muscle, tendons, joints
Nociceptor	Pain	Free nerve endings in skin and organs
Thermoreceptor	Temperature	Skin, tissues, and organs

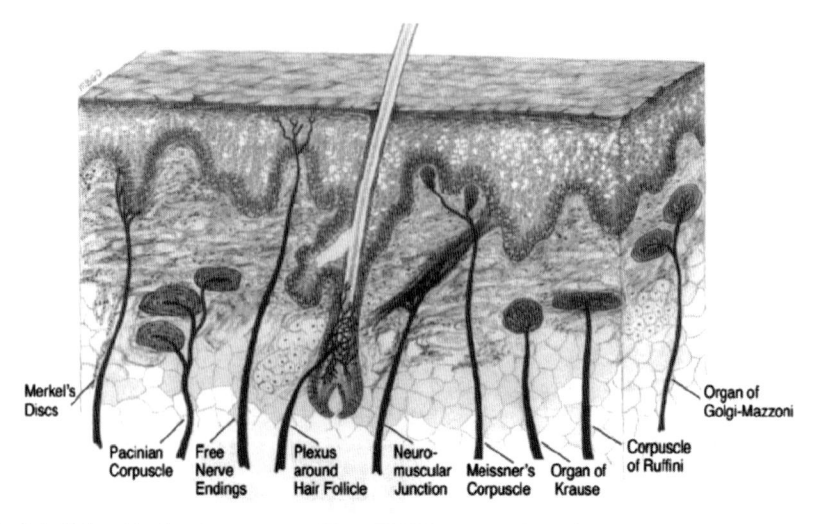

Fig. 1.2 Hairy skin showing receptors (From EM Marcus and S Jacobson, Integrated neuroscience, Kluwer, 2003)

the bodies largest organ, the skin.. Certain areas (e.g., the lips, fingers, feet, and genitalia) have a proliferation of the tactile mechanicoreceptors. We have hair everywhere, except on the soles and palms, which is an important tactile receptor but is continually being depleted by our concern for grooming. (See Fig. 1.2.) The pain receptors, or free nerve endings in the skin, are located throughout the body, but probably more receptors are in the skin over the face, lips, hands, and feet than over the rest of the body. As you review the receptors in Table 1.2, sense on your own body how the soles are especially good for feeling pressure and placing the body safely in light or darkness and the fingers and face are sensitive to touch and temperature. Remember that we have only discussed the skin receptors so far, which respond to external stimuli. However, there are also similar receptors within the respiratory, cardiovascular, endocrine, gastrointestinal, and urogenital systems that monitor our internal milieu.

B. Muscles

The approximately 240 muscles in the body form the bulk of the body and consist of three different functional and histological entities: *skeletal, smooth,* and *cardiac*. Skeletal muscles are found in the head, neck, arms, legs, and trunk and permit us to undertake voluntary movements. Smooth, or unstriated, muscles are found in the viscera, blood vessels, and hair follicles. Cardiac muscles form auricles and ventricles of the heart.

Each muscle group has a specialized nerve ending that permits the impulse carried down the motor nerve via a peripheral nerve to stimulate the muscle through release of a specific chemical. Contraction of the three muscle groups in response to sensory information originates from the central nervous system via the efferent/motor peripheral nerves.

The general and special sensory receptors in the skin provide the *afferent* nerves that carry sensory information to the spinal cord and brain. The brain often analyzes the sensory input before the muscles, which are controlled by the efferent nerves carrying information from the brain or spinal cord, make a response. These integrative functions of the central nervous system form the bulk of the discussion in this volume.

II. The Nervous System

The nervous system consists of a peripheral and central division. The central nervous system (brain and spinal cord) is surrounded by fluid-filled membranes (meninges; see Chapter 16) and housed in either the bony skull or vertebrae. In contrast, the peripheral nervous system that brings information from and to the central nervous system lacks a bony covering but is protected by the fascia, skin, muscles, and organs where it distributes. Sensory information enters the central nervous system through the afferent divisions of the peripheral nerves.

Peripheral nerves are found everywhere in the body: skin, muscles, organs, and glands. Peripheral nerves originate from either the spinal cord or brain. The peripheral nervous system is divided into a somatic and a visceral division. The *somatic division* innervates the skin and skeletal muscles in the body. The visceral, or *autonomic division*, innervates the cardiac muscles of the heart and the smooth muscles and receptors in the blood vessels and gastrointestinal, respiratory, urogenital, and endocrine organs. The details of the peripheral nervous system are usually taught as part of Gross Anatomy, so the student might want to review an anatomy text.

III. Central Nervous System (Fig. 1.3)

The central nervous system consists of the spinal cord and brain (brain stem, cerebellum, diencephalons, and cerebrum). The organization of the gray matter varies in each of these regions. Attached to all of the 32 segments of the spinal cord and the brain stem are sensory ganglia that form the first link in the sensory system and bring the sensory information into the central nervous system. Motor axons exit from each of the 32 segments of the spinal cord and all levels of the brain stem and connect the central nervous system to all muscles and organs in the body. In the spinal cord, much of the brain stem, and diencephalon, the neurons are organized into nuclei; in the superior colliculus of the brain stem, cerebellum, and cerebrum, the neurons are organized anatomically into layers and functionally into vertical columns.

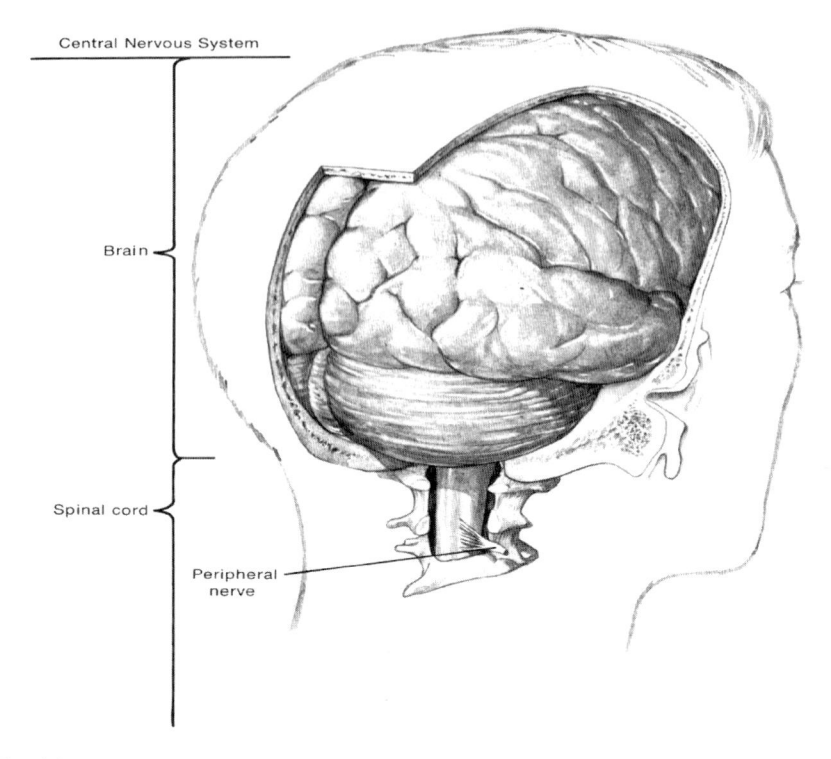

Fig. 1.3 The central nervous system *in situ* (Modified from Curtis, Jacobson, and Marcus. An introduction to neurosciences, Saunders, 1972)

A. *Spinal Cord*

The spinal cord is that portion of the central nervous system that lies in the vertebral canal from the upper border of the atlas (first cervical vertebrae) to the lower border of the first lumbar vertebrae in the adult (or third lumbar vertebrae in the neonate). The spinal cord has 32 segments divided into five regions (cervical, thoracic, lumbar, sacral, and coccygeal) and these regions innervates specific regions in the neck and upper extremity (cervical segments), thorax and abdomen (thoracic levels), anterior leg and thigh (lumbar segments), and buttock and posterior leg and thigh (lumbar segments). This ordered relationship between the spinal cord and body produces a somatotopic organization throughout the central nervous system.

In the spinal cord the parenchyma is organized into columns of gray and white matter, with the gray matter centrally placed and surrounded by the white matter. This organization is not evident as one looks at isolated cross sections, but when these sections are reconstructed serially, this columnar organization in the gray and white matter is apparent. The columns of gray matter in the spinal cord appear in

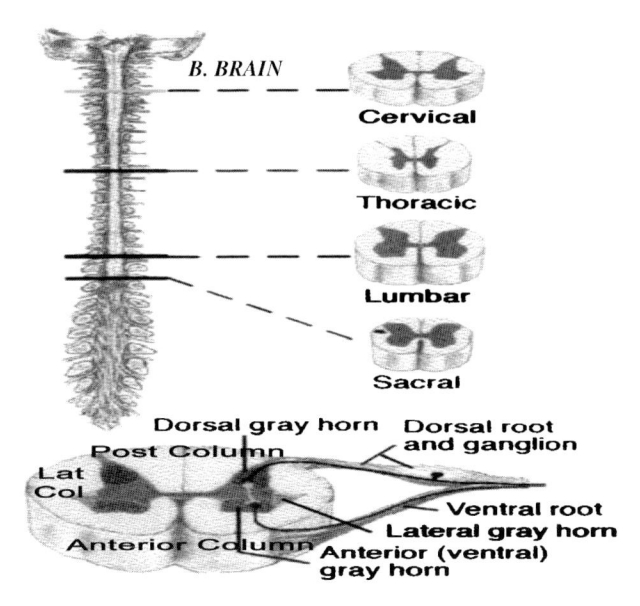

Fig. 1.4 Spinal cord. (From EM Marcus and S Jacobson, Integrated neuroscience, Kluwer, 2003)

the shape of a butterfly and are called horns and are divided into a dorsal sensory horn, a ventral motor horn, intermediate zone, and commissural region. The largest neuronal cell bodies are found in the ventral horn (ventral horn cells), whose axons form the efferent division of the peripheral nervous system and innervate the skeletal muscles (Fig. 1.4). The white matter of the spinal cord is divided into three columns: anterior, posterior, and lateral. The pathways interconnecting the spinal cord and brain are found in these columns.

The spinal cord has a tubular shape and has two regions of enlargement: the lower cervical that controls the upper extremity and the lumbosacral enlargement that controls the lower extremity.

B. Brain

1. Brain Stem (Fig. 1.5)

The columnar organization seen in the gray and white matter of the spinal cord is modified in the brain stem by the development of the ventricular system and the presence of the cranial nerves. The brain stem (Fig. 1.5) consists of three regions from inferior to superior: *medulla, pons*, and *midbrain*. The brain stem is often the most difficult region of the central nervous system for the student to learn because

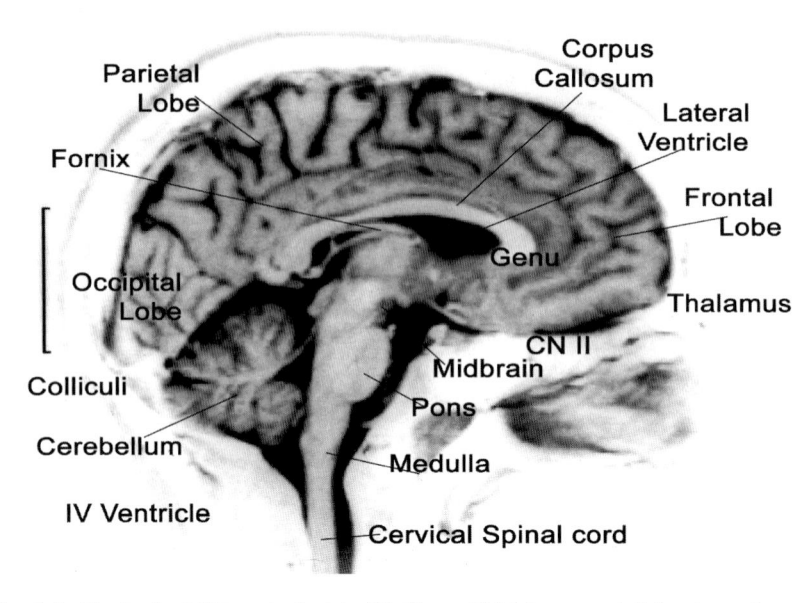

Fig. 1.5 The brain: MRI sagittal plain T1 (From EM Marcus and S Jacobson, Integrated neuroscience, Kluwer, 2003)

of the presence of the cranial nerves and associated nuclei. You might initially feel overwhelmed by its intricacy, but be patient. Approach neuroanatomy as you would a foreign language: first master the vocabulary and grammar before becoming fluent in conversation.

Reorganization of Gray and White Matter from Spinal Cord Gray to Tegmentum of Brain Stem

In the center of the brain stem, the narrow spinal canal enlarges, forming the wide fourth ventricle that divides each level of the brain stem into a region forming the floor of the ventricle, the *tegmentum*, and a region forming the roof of the ventricle, the *tectum* (consisting of the cerebellum and the inferior and superior colliculi). The region that lies on the most anterior surface of the tegmentum is called the *basilar zone* and it contains the descending fibers from the cerebral cortex. The columnar arrangement in the gray and white matter in the spinal cord slowly reorganizes in the lower medulla, resulting in the ventral and dorsal horn equivalent regions now lying on the floor of the ventricle with the intermediate zone forming the center of the tegmentum of the brain stem, the reticular formation. With this reorganization, the *ventral horn equivalent* contains cranial nerve motor nuclei of CN X and CN XII in the medulla, CN VI in the pons, and CN III and CN IV in the midbrain. The dorsal horn equivalent, general somatic sensory, and visceral sensory X lie lateral to the motor cranial nerve nuclei and are separated by the sulcus limitans. One also now finds the presence of the special sensory nucleus of CN VIII on the floor of the ventricle. The intermediate zone of the spinal cord evolves into the reticular formation in the center of the brain stem containing the motor and sensory cranial nerve nuclei innervating skin and muscles of pharyngeal arch origin and nuclei associated with the cerebellum (inferior olive) and basal ganglia (red nuclei). The medial surface of each half of the brain stem contains sensory and motor tracts, with the basilar surface containing the descending corticospinal and corticobulbar pathways. The lateral margin of the brain stem contains the anteriolateral system.

Cranial Nerves (Chapter 5)

In order to understand the functions of the brain stem, the student must first appreciate the distribution of the 12 cranial nerves, with nerve I attached to the olfactory bulbs on the base of the frontal lobe, nerve II entering the diencephalon, and nerves III–XII originating in the brain stem and distribute in the head, trunk, and abdomen. Additionally, the key to identifying dysfunction in the brain stem is to know which cranial nerve nucleus lies in each level.

The nerve rootlets of CN III and CN V–CN XII are found on the anterior surface of the brain stem and the rootlets only of CN IV exit on the posterior surface of the brain stem. Table 1.3 lists the location of the cranial nerve rootlets and nuclei in the appropriate brain stem level. Chapter 6 discusses the nuclei and tracts of the brain stem in greater detail and includes many examples of the consequences of disease in the brain stem. Because the tegmentum, where the cranial nerve nuclei are located, is an especially important region for the student to master, we have divided this region bilaterally into five zones like an apple with the core, the center, being the reticular formation and the pulp being the surrounding four zones (Chapter 4, Table 4.2).

Table 1.3 Cranial nerve localization in brain stem

Level	Cranial nerve nuclei
Medulla: rootlets of VII–XII	CN V: spinal nucleus and tract (pain)
	CN VIII: cochlear and vestibular
	CNs IX, X, XI: ambiguous nucleus
	CN IX
	CN X: preganglionic parasympathetic
	CN XII: tongue musculature
Pons: rootlets of V–VII	CN V: motor nucleus(muscles of mastication), chief sensory nucleus (tactile), descending nucleus (pain), mesencephalic (proprioception)
	CN VI: lateral rectus
	CN VII: muscles of facial expression
	CN VII: taste and parasympathtics to face
	CN VIII: vestibular and cochlear
Midbrain: rootlets of III and IV	CN IV and CN III
	CN V: mesencephalic nucleus

2. Cerebellum (Fig. 1.5)

The cerebellum, similar to the cerebrum, consists of gray matter on the surface and white matter inside. It attaches to the tegmentum of the medulla and pons by three peduncles. The cerebellum is divided into two lateral hemispheres and the midline vermis. From the phylogenetic (evolutionary development) standpoint, there are three regions: archicerebellum, paleocerebellum, and neocerebellum. The cerebellum functions at an unconscious level to permit voluntary motor functions to occur smoothly and accurately. Alcohol consumption and many drugs affect this region and temporarily impair, for example, gait and coordination.

In the white matter of the cerebellum are found four deep cerebellar nuclei: *fastigius, dentatus, emboliformus*, and *globosus*. The cerebellar cortex projects onto these nuclei, and then these nuclei project to other areas of the brain stem or diencephalon. The functions of the cerebellum are detailed in Chapter 11.

3. Diencephalon (Fig. 1.5)

This region, identifiable in the gross brain after a sagittal section separates the cerebral hemispheres, lies at the upper end of the brain stem. The third ventricle separates the right and left diencephalic masses. Their lateral margin is the posterior limb of the internal capsule, and their superior surface is the body of the lateral ventricle and the corpus callosum. The diencephalon stands as the great waystation between the brain stem and cerebral cortex, as all the major ascending pathways

Table 1.4 Divisions of the diencephalon

Division	Subnuclei	Functions
Thalamus with metathalamus	Anterior, medial, lateral, and metathalamic nuclei	Termination of most ascending general and special sensory and motor pathways project onto sensory, motor, and associational areas in cerebral cortex
Epithalamus	Habenula, stria medullaris	Similar to hypothalamus but with connections to the pineal gland
Hypothalamus	Anterior, medial, and posterior	Center for homeostatic mechanisms and highest subcortical center for emotions, as it controls autonomics and the periventricular pineal and pituitary glands.
Subthalamus	Subthalamus, zona incerta	Subcortical unconscious center for movement.

terminate here. The diencephalon consists of the following divisions: thalamus, hypothalamus, epithalamus, and subthalamus. The functional organization in the diencephalon is the basis for much of the function seen in the cerebrum and also contains the multimodal associations that make the cerebrum of the human such a potent analyzer (Table 1.4).

The *thalamus* forms the largest division of the diencephalon and it is divided into several groupings of nuclei. All of the ascending pathways terminate in thalamic nuclei and then their information is projected onto their respective region of the cerebral cortex. A detailed discussion of the thalamus is found in Chapter 6.

The *epithalamus* is a small zone on the mediodorsal portion of the diencephalon consisting of habenular nuclei, with its associative pathway, the stria medullaris, and the pineal gland; its functions are similar to the hypothalamus.

The *hypothalamus* is the smallest subdivision of the diencephalon and is found inferiorly in the third ventricle. However, because it controls the pituitary gland and functions as the "head ganglion" in the autonomic nervous system, it might well be the most important portion and emotional center of the diencephalon. The hypothalamus and the autonomic nervous system are discussed in Chapter 7.

The *subthalamus*, found below the thalamus, consists of the subthalamic nucleus and the zona incerta with its fiber pathways. It is an important subcortical region in the basal ganglia (Chapter 10)

The functions of this region are to integrate sensory and motor information and to begin to interpret these data according to the perceptions of the emotional areas in the brain.

4. Cerebrum: Cerebral Hemispheres (Fig. 1.6)

The cerebrum, forming the bulk of the brain and thus of the central nervous system, consists of a left and a right hemisphere containing the cortical gray matter, white matter, and basal nuclei (Table 1.5). In each hemisphere, we find four lobes: the rostral frontal lobe, the middle parietal lobe, and the posterior occipital lobe with the inferiorly located temporal lobe. The cerebral cortex consists of a corrugated surface, the cortical gray matter, which is laminated and has six layers, is broken up into numerous gyri, separated by narrow spaces or grooves, the sulci (Fig. 1.6). The 12 to 15 billion cortical neurons are found in the gray cortical mantle.

Table 1.5 Overview of cerebral functions (Figs. 1.6 and 1.7)

Frontal lobe	Volitional movement of the muscles of the body, including limbs, face, and voice; judgmental center
Parietal lobe	Tactile discrimination from the body, body image, taste, and speech
Occipital lobe	Vision (sight and interpretation)
Temporal lobe	Emotions and new memories, hearing, language, olfaction, and visual recognition
Cingulate gyrus	Emotions
Insular gyri	Emotions and language

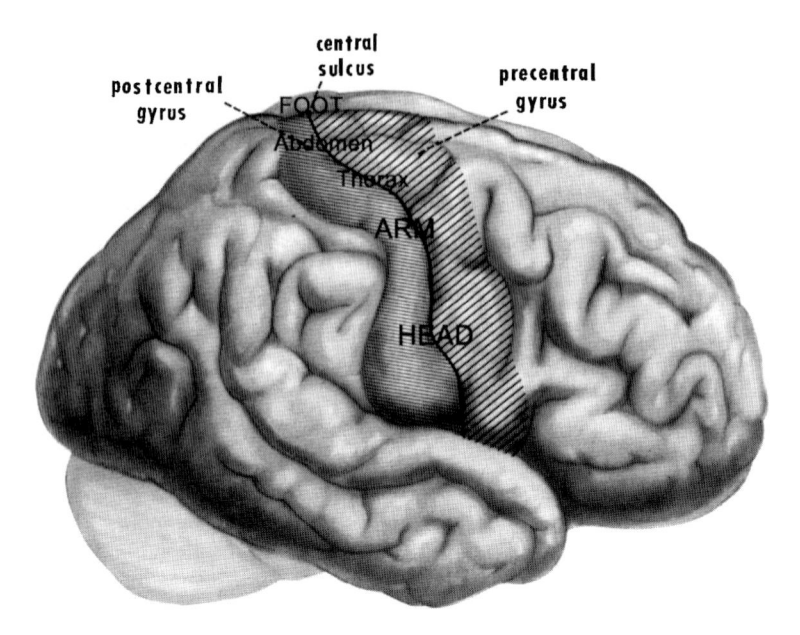

Fig. 1.6 Lateral surface of the cerebrum. The sensory (precentral) and motor strips (precentral) are marked. The somatotopic organization of this region is labeled: Head, Arm, Thorax Abdomen, and Foot. The foot region of the sensory–motor strip extends onto the medial surface of the hemisphere as the paracentral lobule. (Modified from Curtis, Jacobson, and Marcus, An introduction to the neurosciences, Saunders, 1972)

Students handling a preserved brain are often initially struck by the number of the gyri and sulci in the cerebral hemispheres and surprised by its weight (about 1500 g). If you have a series of brains to observe, note that there is a great variation in size. Remember that the size of the brain is related to the size of the skeleton, muscles, and viscera (but not adipose tissue), so a person of small stature has a proportionally smaller brain. Because women tend to be smaller than men, a woman's brain is usually smaller, although this in no way reflects intelligence or abilities. One of the first clues about the different functions in the two cerebral hemispheres was observed in stroke victims. Because the *left cerebral hemisphere* is dominant for language in right-handed people (93% of the population), strokes in that hemisphere usually affect speech. In addition to language, dominant functions include initiation of movement, emotions, and artistic abilities.

Note that each hemisphere provides motor controls to the opposite side of the body. This is because the motor pathway from each cerebral cortex crosses in the transition between the spinal cord and medulla; consequently, the sensory fibers also cross over to the opposite side of the body. In over 90 % of the human population, movement initiates from the left hemisphere, making it dominant for initiation of movement.

General Features

The cerebral cortex is divided into four lobes, the frontal, parietal, occipital, and temporal lobes (Fig. 1.6). Two other important regions in the cerebrum are the insula deep within the lateral sulcus and the cingulate gyrus on the medial surface of each hemisphere. Three of the lobes also have poles: the frontal, temporal, and occipital pole. The frontal lobe is separated from the parietal lobe by the central sulcus and from the temporal lobe by the lateral sulcus. The other lobes can be identified (Fig. 1.6) by first drawing a line from the parieto-occipital sulcus on the medial surface of the hemisphere to the preoccipital notch on the inferior lateral surface of the hemisphere. A line is then extended posteriorly from the lateral sulcus until it intersects the first line. The parietal lobe lies above the lateral sulcus and behind the central sulcus, anterior to the occipital lobe. The temporal lobe lies below the lateral sulcus anterior to the occipital lobe. The occipital lobe is most posteriorly placed.

Functional Localization

The cerebral cortex includes motor, sensory, auditory, and visual regions. In addition, broad areas are involved with multimodal integration, which combines sensory and motor with an emotional content to determine how to respond in any situation. These emotional or limbic areas occupy much of the temporal and frontal lobes. (The *limbic* region denotes the cortical and subcortical brain areas with tracts primarily involved with emotions, including the cingulate gyrus, hippocampal formation, parahippocampal gyrus, and nuclei in the diencephalon; see Chapter 14.) Human beings also use language extensively and much of the frontal–parietal–occipital–temporal regions that abut the lateral sulcus in the left hemisphere undertake these functions. Similar areas of the right hemisphere are devoted to visual–spatial integration. You might wonder just how we have such detailed information on the organization and functions of the human nervous system. Much of it comes from research on primates or from defects observed in patients after strokes, trauma, tumors, or degenerative neurological diseases like Alzheimer's disease.

Cortical White Matter

The axons entering or leaving the cerebral hemispheres form three distinctive groups of fibers: associational, commissural, and subcortical.

1. **Associational Fibers.** These two types of fiber (short U and long associational) provide the integrative circuitry for movement, language, memory, and emotions. Short associational fibers (the U fibers) form the bulk of local connections within a hemisphere, whereas the long associational fibers interconnect diverse areas in a hemisphere, providing the multimodal associations essential one for cortical and cingulum.

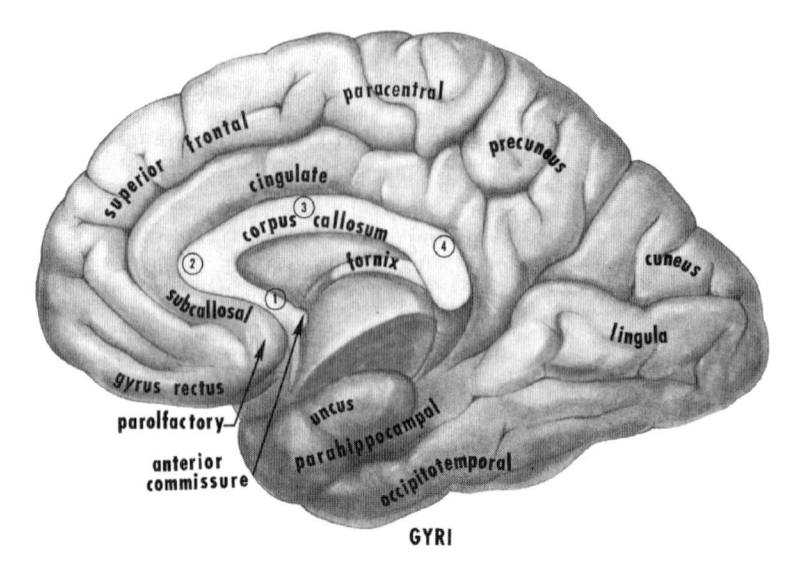

Fig. 1.7 Medial surface of a cerebral hemisphere, with brain stem removed. Gyri in frontal, parietal, occipital and temporal lobes identified and regions in the corpus callosum noted: (1) rostrum, (2) genu, (3) body, and (4) splenium. (Modified from Curtis, Jacobson, and Marcus, An introduction to the neurosciences, Saunders, 1972)

2. **Commissural Fibers.** Commissural fibers interconnect areas in the contralateral hemispheres and permit learning and memory in one hemisphere to be shared with the other. The bulk of the frontal, parietal, occipital, and temporal lobes are interconnected by the *corpus callosum*, best seen in the medial surface of the hemisphere. The corpus callosum consists of a rostrum, genu, body, and splenium (Fig. 1.7). The most rostral part of the temporal lobe and the olfactory cortex of the uncus are interconnected by the *anterior commissure*. The hippocampus, an important area for memory, is also connected by a commissure associated with the fornix.
3. **Subcortical Fibers.** This category of fibers includes fiber bundles reaching the cortex from subcortical areas, corticopetal (to cortex), and axons leaving the cortex and connecting to subcortical nuclei, corticofugal (from the cortex). *Corticopetal fibers* are the subcortical afferents to each cerebral hemisphere and are primarily from the thalamus of the diencephalon. *Corticofugal fibers* from the cerebral hemisphere project onto subcortical structures, including the basal nuclei, diencephalon, brain stem, cerebellum, and spinal cord. The major subcortical fiber tracts leaving the cerebrum are the corticospinal, corticonuclear, corticopontine, corticomesencephalic, and the fornix.

The *internal capsule* contains the major grouping of corticofugal and corticopetal fibers of the cerebral cortex and consists of an anterior limb, genu, and posterior limb (Fig. 1.7). The anterior limb provides fibers to and from the frontal lobe. The genu provides fibers to and from the lower part of the frontal and parietal lobes (corticonuclear). The posterior limb is the largest portion of the internal capsule and includes the auditory radiations (auditory fibers to the auditory cortex), visual radiations, as well as projections from the sensorimotor cortex to the spinal cord and brain stem (corticospinal, corticonuclear, and corticopontine fibers)

Functions in the Lobes of the Cerebrum

In addition to all of the functions already mentioned, the cerebral hemispheres also have the memory stores, which are the foundation for most conscious and unconscious thought, cultural activity, sexual behavior, and all of the other positive or negative traits that make the human being so distinctive. Each cortical region has a memory store that permits the function of that region (Table 1.5). A detailed analysis of the functions of each cortical region is presented in Chapters 8 to 15.

5. Basal Nuclei (Chapter 10)

In a coronal section through a midpoint of the cerebral hemispheres we can identify laterally the cerebral cortex, cortical white matter, and basal nuclei (Fig. 1.8). Three important nuclear groups deep within the hemispheres are the corpus striatum,

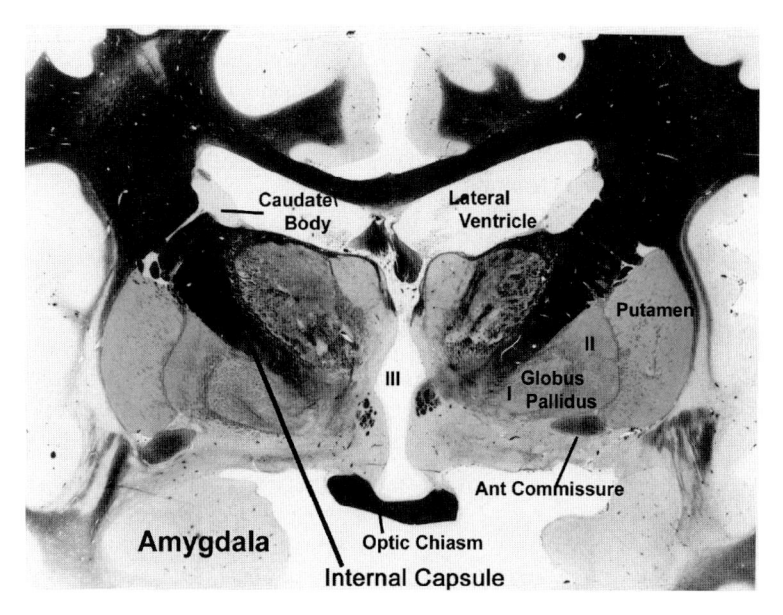

Fig. 1.8 Coronal section showing basal nuclei and their relationship to the internal capsule and thalamus (From EM Marcus and S Jacobson, Integrated neuroscience, Kluwer, 2003)

claustrum, and amygdala. This subsection focuses on the corpus striatum: the caudate nucleus, putamen, and globus pallidus They are separated from the cerebral cortex by white matter but are functionally linked to the cerebrum. These nuclei are motor associative in function. They carry on at a subconscious level until a dysfunction, such as in Parkinson's disease or Huntington's chorea, occurs. The caudate nucleus and putamen are more laterally placed surrounding the white matter, which connects the cortex with the subcortical areas, the internal capsule. The caudate nucleus lies medial and dorsal to the thalamus. The functions of these nuclei will be discussed in further detail in Chapter 11.

6. Central Nervous System Pathways

Up to this point, the diverse functional gray regions in the brain and spinal cord have been identified. In order for the nervous system to work, an extensive circuitry has been established to interconnect areas throughout the central nervous system. We have just described the intrinsic white matter circuitry within the cerebrum: the associational, commissural, and subcortical fibers. The subcortical fibers consist of a multitude of axonal pathways that provide afferents to the cerebrum and efferents from the cerebrum.

Many names are used for pathways within the central nervous system, including fasciculus (bundle), lemniscus (ribbon), peduncle (stalk), and tract (trail). Many of the pathways also have a name that indicates its function (e.g., the optic tract connects the eyes and visual regions, the corticospinal (Chapter 9, Fig. 9.6) tract runs from the cortex to the spinal cord, and the corticonuclear (Chapter 9, Fig 9.7) tract connects the cerebral cortex to the cranial nerve nuclei). Unfortunately, the majority of the pathways have names that give no clue to function (e.g., the medial longitudinal fasciculus, lateral lemniscus, and cerebral peduncle). The major tracts and their role in the central nervous system are discussed in Chapter 6. For each pathway, the student needs to learn the following:

1. Cells of origin
2. Location of the tract in the brain
3. Site of termination of pathway
4. Function.

As an example of function within the central nervous system we will now use a functionally very significant region that provides volitional motor control to the muscles of the body: the motor–sensory cortex.

7. Motor–Sensory Cortex (Fig. 1.6)

This cortex is found around the central sulcus that separates the frontal from the parietal lobe. It has six layers. Figure 1.6 identifies the vertically oriented central sulcus, which is surrounded by the precentral/motor strip and postcentral/sensory

III Central Nervous System

cortex. This region controls the skeletal muscles in the body that permit us to undertake so many tasks. In order to perform a skilled motor task, one is dependent on precise sensory input, so we will first describe the sensory portion of this circuit.

Sensory Information from the Foot to the SensorimotorCortex

In the sensory system, three different neuronal groupings are traversed before the sensory information reaches the cerebral cortex from the periphery: the primary, secondary, and tertiary neurons. The sensory information of importance for precise movement originates from stretch receptors in the muscles, tendons, and joints of the hand and is encoded as electrical impulses that detail the contraction status of the muscles, tendons, and joints. The cells of origin, the primary cell bodies, are in the large sensory ganglia attached to the lumbar and sacral segments of the spinal cord. The information is carried in by the dendrite of the primary sensory cell and its axon enters the central nervous system and ascends uncrossed and ipsilaterally in the posterior column of the cord until the medullospinal junction. In the medullospinal junction *second-order nerve cells* appear in the posterior column, the gracile nucleus. The axons from the primary neurons synapse on this nucleus. The axons of the second-order axons cross the midline and ascend contralaterally in one of the major ascending white matter pathways (the medial lemniscus) into the thalamus of the diencephalon where they synapse on the *third-order neurons in the ventral posterior lateral nucleus*, which then send fibers ipsilaterally onto the foot region on the upper end and medial surface of the postcentral gyrus.

Motor Control of the Foot from the Motor–Sensory Cortex

In the motor system, two distinct neuronal groups are necessary for a movement: one in the motor cortex of the cerebrum (*the upper motor neuron*) and the other in the spinal cord or brain stem (*the lower motor neuron*). Motor control to the foot is carried centrifugally by the corticospinal pathway. The *corticospinal pathway* to the foot originates from layer V on the medial surface of the motor cortex of the precentral gyrus and to a lesser degree in the postcentral gyrus. Before exiting the gray matter, the fibers are covered with myelin and descend on the same side through the internal capsule, cerebral peduncle, pons, and medullary pyramid. The corticospinal fibers cross (decussate) to the other side of the brain in the medullospinal junction and enter the lateral column of the spinal cord. Fibers to the musculature of the leg descend to lower lumbar and sacral levels and synapse in either the intermediate region of the gray matter of the spinal cord or directly on the ventral horn cells. The axons from the ventral horn cell leave the central nervous system form the peripheral nerves that synapse on the motor endplates of the muscles in the leg, which produce the contraction. The actual initiation of the movement is from the prefrontal and premotor region of the frontal lobe (see Chapter 9).

Table 1.6 Comparison of upper motor neuron to lower motor neuron lesion

Upper motor neuron lesion (UMN)	Lower motor neuron lesion LMN
Lesion in Motor Cortex or Corticospinal tract	Lesion in ventral horns or ventral rootlets
Reflexes: Increase (2+ normal), 3+, or 4+	Reflexes; Decrease or are absent (+1 or 0)
Babinski (extensor plantar response)	
Muscles spastic (sign of injury to corticospinal)	Muscles: Limb Flacid Fasciculations, lead to atrophy

Interruption of the corticospinal pathway or destruction of the motor cortex produces *upper motor neuron paralysis* with increased reflex tone and spasticity, whereas injury to the spinal cord ventral horn cells produces *lower motor neuron paralysis* and atrophy of the muscle and decreased reflex tone (Table 1.6).

In summary, it takes three orders of sensory neurons to provide the information to the cerebral cortex and two levels of motor neurons to produce a hand movement in response to the sensory information. Note that the sensory information ascends and crosses over, whereas the motor information descends and crosses so that the same side is represented in the cerebral cortex. Of course, the hand movement also needs a conscious decision to be made, which is what the cerebral cortex working as a unit does. The following Case History illustrates the effects of a lesion in the leg region of the motor strip on movement.

Case History 1.1 (Fig. 1.9)

This 45-year-old right-handed, married, white female, mother of four children was referred for evaluation of progressive weakness of the right lower extremity of 2 years' duration. This weakness was primarily in the foot so that she would stub the toes and stumble. Because she had experienced some minor nonspecific back pain for a number of years, the question of a ruptured lumbar disc had been raised as a possible etiology. The patient labeled her back pain as "sciatica" but denied any radiation of the back pain into the leg. She had no sensory symptoms, and no bladder symptoms. Six months before the consultation, the patient had transient mild weakness of the right leg that lasted a week. The patient attributed her symptoms to "menopause" and depression, and her depression had improved with replacement estrogen.

Neurological examination: Neurological examination revealed that the patient's mental status was normal and the cranial nerves were all intact. There was a minimal drift down the outstretched right arm, and there was significant weakness in the right lower extremity with ankle dorsiflexion and plantar flexion was less than 30% of normal. Inversion at the ankle was only 10–20% of normal. Toe extension was 50% of normal and there was minor weakness present in the flexors and extensors at the knee. In walking, the patient had a foot drop gait and had to overlift the foot to clear the floor. Examination of the shoes showed greater wear at the toes of the

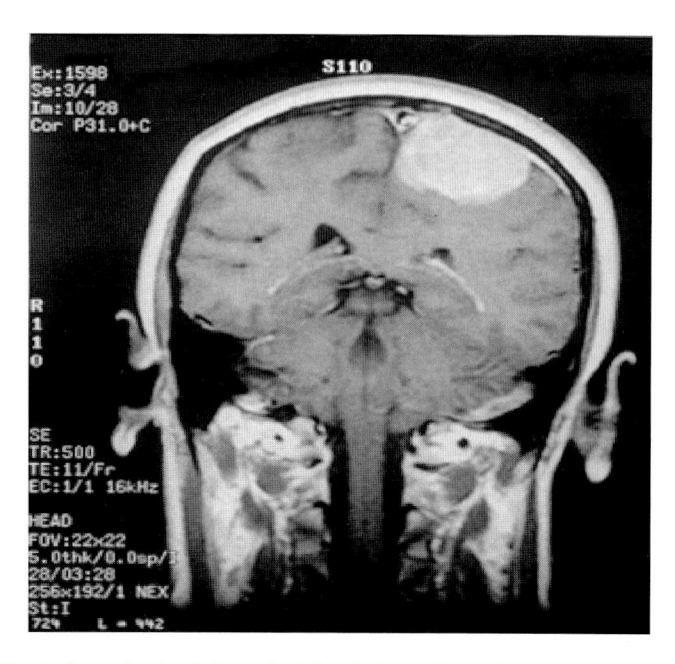

Fig. 1.9 Meningioma in the left cerebral hemisphere (From EM Marcus and J Jacobson, Integrated neuroscience, Kluwer, 2003)

right shoe, with evidence of scuffing of the toe. The patient's deep tendon stretch reflexes were increased at the patellar and Achilles tendons on the right. In addition, reflexes in the right arm at the biceps, triceps, and radial periosteal were also slightly increased. The plantar response on the right was extensor (sign of Babinski) and the left was equivocal. All sensory modalities were intact. Scoliosis was present with local tenderness over the lumbar, thoracic, and cervical vertebrae, but no tenderness was present over the sciatic and femoral nerves.

Comments: It was clear that the symptoms and signs were not related to compression of the lumbar nerve roots by a ruptured disc. Such a compression (lumbar radiculopathy) would have produced a lower motor neuron lesion, a depression of deep tendon stretch reflexes, lumbar radicular pain in the distribution of the sciatic nerve and tenderness over the sciatic nerve, no increase in reflexes in the upper or lower extremities, and no sign of Babinski. The fact that reflexes were more active in both upper and lower extremities on the right, with the sign of Babinski on the right, suggested a process involving the corticospinal tract. The fact that the weakness was greatest in the foot suggested a meningioma involving the parasagittal motor cortex where the foot area is represented.

Clinical Diagnosis: With these clinical considerations in mind, an MRI of the patient's head was obtained and the MRI confirmed the clinical impression of a meningioma in the right cerebral hemisphere. This tumor was successfully removed by the neurosurgeon, with an essentially complete restoration of function.

IV. Glands Associated with the Brain

The brain has two glands (the *pituitary* and the *pineal*), both of which are attached to the diencephalon and associated with the hypothalamus. The *pituitary* or *hypophysis* cerebri is attached to the base of the hypothalamus. Its functions are described in the hypothalamus (Chapter 7). The *pineal* or *epiphysis* cerebri is found in the epithalamus (Chapter 7). By its location its functions in relationship to light levels and diurnal cycles would not be apparent.

The coverings of the brain (meninges), the ventricular system, and vascular system are described in Chapter 16.

Chapter 2
Neurocytology: Cells of the CNS

There are two major cell types that form the nervous system: supporting cells and conducting cells. The supporting cells of the peripheral nervous system consist of Schwann cells, fibroblasts, and satellite cells; the supporting cells in the central nervous system consist of the glia and the lining cells of the ventricles the ependyma (the meningeal coverings of the brain, the circulating blood cells, and the endothelial lining cells of the blood vessels). The conducting cells, or neurons, form the circuitry within the brain and spinal cord and their axons can be as short as a few microns or as long as 1 meter. The supporting cells are constantly being replaced, but the majority of conducting cells/neurons, once formed, remain throughout our life.

I. The Neuron

The basic functional unit of the nervous system is the *neuron*. The neuron doctrine (postulated by Waldeyer in 1891) described the neuron as having one axon, which is efferent, and one or more dendrites, which are afferent. It was also noted that nerve cells are contiguous, not continuous, and all other elements of the nervous system are there to feed, protect, and support the neurons.

Although muscle cells can also conduct electric impulses, only neurons, when arranged in networks and provided with adequate informational input, can respond in many ways to a stimulus. Probably the neuron's most important feature is that each is unique. If one is damaged or destroyed, no other nerve cell can provide a precise or complete replacement. Fortunately, the nervous system was designed with considerable redundancy; consequently, it takes a significant injury to incapacitate the individual (as in Alzheimer's or Parkinson's disease).

A. Dendrites

The dendritic zone receives input from many different sources. The action potential originates at the site of origin of the axon and is transmitted down the axon in an all-or-nothing fashion to the synapse, where the impulse is transmitted to the dendritic zone of the next neuron on the chain.

Dendrites have numerous processes that increase the neuron's receptive area. The majority of synapses on a nerve cell are located on the dendrite surface. With the electron microscope, the largest dendrites can be identified by the presence of parallel rows of neurotubules, which might help in the passive transport of the action potential. The dendrites in many neurons are also studded with small membrane extensions: the dendritic spines.

B. Soma

The soma (perikaryon or cell body) of the neuron varies greatly in form and size. Unipolar cells have circular cell bodies; bipolar cells have ovoid cell bodies; and multipolar cells have polygonal cell bodies.

C. Golgi Type I and II Neurons

Neurons can also be grouped by axon length: Those with long axons are called Golgi type I (or pyramidal) cells; those with short axons are called Golgi type II (or satellite) cells (Gray, 1959). The Golgi type I cell (Fig. 2.1) has an apical and basal dendrite, each of which has secondary, tertiary, and quaternary branches, with smaller branches arising from each of these branches that extend into all planes. They form the basic long circuitry within the central nervous system. The axons of pyramidal neurons run long distances within the cortex, but they might also exit from the cortex.

D. Dendritic Spines

In the cerebral cortex. spines are numerous on Golgi type I neurons, are small knob-shaped structures approximately 1–3 microns (μm) in diameter (Fig. 2.1). Dendritic spines are bulbous, with a long neck connecting to the dendrite. Many axon terminals are located on the spines. Their importance stems from the fact that they greatly expand the dendrite's receptive synaptic surface. Spines are absent from the initial segment of the apical and basal dendrite of pyramidal neurons, but they become numerous farther along the dendritic branches. The transmitter in type I cells is glutamate. The Golgi type II cell has a short axonal field and dendrites (Fig. 2.2). The axon usually extends only a short distance within the cerebral cortex (0.3–5 mm). Golgi type II cells have few, if any, dendritic spines.

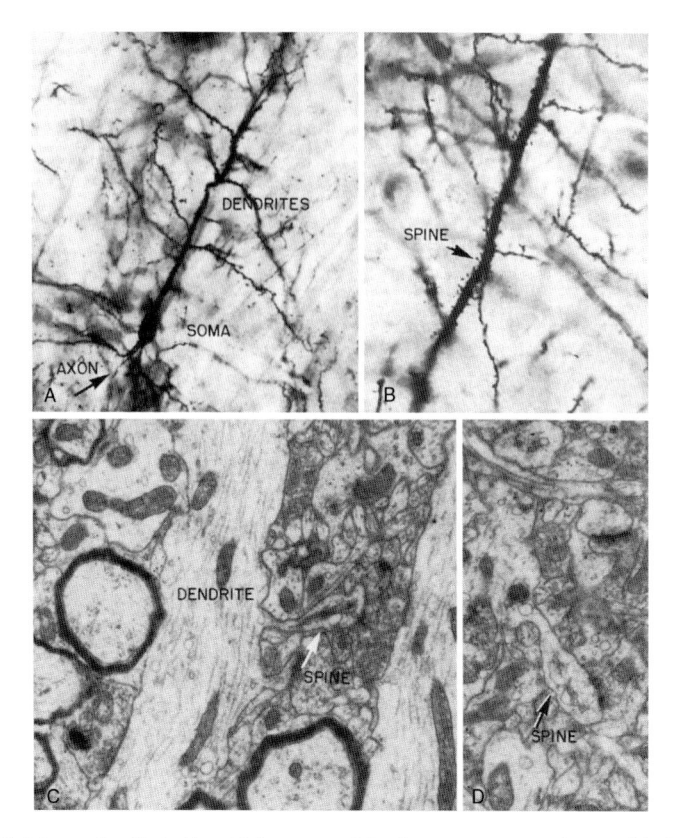

Fig. 2.1 Golgi type I cells (cells with long axons) in the motor cortex of a rat. **(A)** Soma, axon, and dendrite (×150); **(B)** dendritic spines (×1100) (A and B: Golgi rapid stain). **(C, D)** Electron micrographs of dendritic spines with excitatory synapses (×30,000) (From EM Marcus and J Jacobson, Integrated neuroscience, Kluwer, 2003)

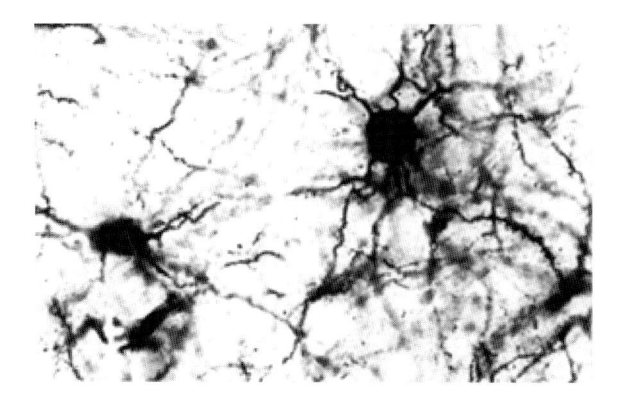

Fig. 2.2 Golgi type II cells (stellate cells, neurons with short axons in the motor cortex of the rat (Golgi-Cox stain; ×450.) (From EM Marcus and J Jacobson, Integrated neuroscience, Kluwer, 2003)

E. Cytoplasmic Organelles

These structures are seen in most cell and in all neurons and glial cells types and they permit each neuron to function (Table 2.1). In these eukaryotic cells, the organelles tend to be compartmentalized and include the nucleus, polyribosomes, rough endoplasmic reticulum, smooth endoplasmic reticulum, mitochondria, and inclusions (Fig. 2.4). Most neuronal cytoplasm is formed in the organelles of the soma and flows into the other processes. Newly synthesized macromolecules are transported to other parts of the nerve cell, either in membrane-bound vesicles or as protein particles. As long as the soma with a majority of its organelles are intact, the nerve cell can live. Thus, it is the *trophic center* of the neuron. Separation of a process from the soma produces death of that process.

F. Nucleus

The large ovoid nucleus is found in the center of the cell body (Figs. 2.3 and 2.4). Chromatin is dispersed, as nerve cells are very active metabolically. Within the nucleus there is usually only a single spherical nucleolus, which stains strongly for RNA. In females, the nucleus also contains a perinuclear accessory body, called the Barr body (Fig. 2.3). The Barr body is an example of the inactivation and condensation of one of the two female sex, or X, chromosomes (Barr and Bertram, 1949). The process of inactivation of one of the X chromosomes is often called lyonization, after the cytogenetist who discovered it—Mary Lyons. Recently, much progress has been made in the localization of genes associated with neurologic diseases, including trisomies, Down's, Alzheimer's, Huntington's, and Parkinson's.

Table 2.1 Parts of a neuron

Region	Contents
Soma: majority of inhibitory synapses are found on its surface	The neuron's trophic center, containing the nucleus, nucleolus, and many organelles.
Dendrite: continuation of the soma and covered with synapses and Type I neurons have dendritic spines (Fig. 2.1)	Has many branches forming a large surface area; contains neurotubules and the majority of synapses on its surface;.
Axon myelin: an insulator, covers axon.	Conducts action potentials to other neurons via the synapse. Length: a few millimeters to a meter.
Synapse: site where an axon connects to another neurons' dendrites, soma or axon	Consists of presynaptic part containing neurotransmitters and postsynaptic portion with membrane receptors separated by a narrow cleft.

I The Neuron

Fig. 2.3 Ventral horn cell of a female squirrel monkey. Note the nucleus, nucleolus, and the accessory body of Barr (arrow); 1-μm epoxy section (×1400) (From EM Marcus and S Jacobson, Integrated neuroscience, Kluwer, 2003)

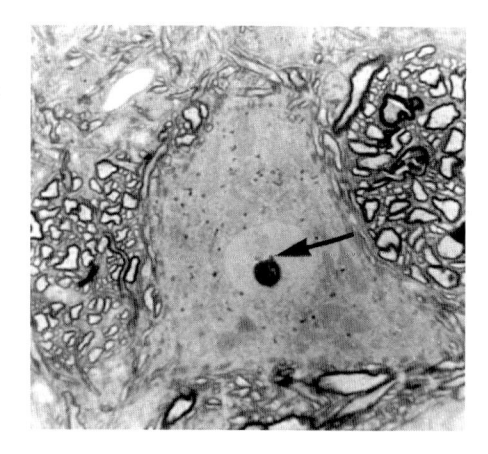

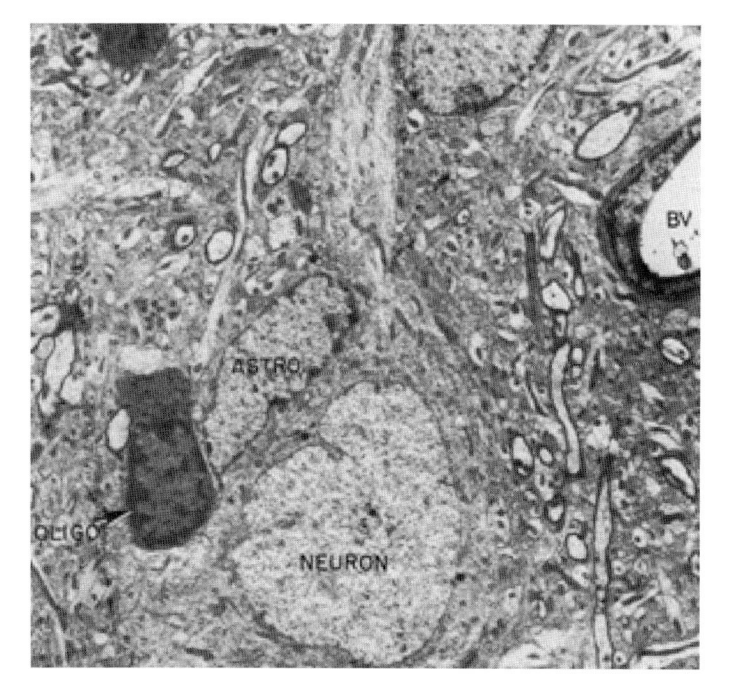

Fig. 2.4 Electron micrograph of the cerebral cortex showing the principal cell types in the central nervous system: neuron astrocytes (ASTRO), oligodendrocyte (OLIGO), and a blood vessel (BV) (From EM Marcus and S Jacobson, Integrated neuroscience, Kluwer, 2003)

G. Rough Endoplasmic Reticulum: Nissl Body (Fig. 2.4)

The rough endoplasmic reticulum or Nissl substance, which turns amino acids into proteins, is the chromodial substance found in light micrographs. It can be demonstrated by using a light microscope and basic dyes, such as methylene blue, cresyl violet, and toluidine blue. The appearance and amount vary from cell to cell. With electron microscopy, cisterns containing parallel rows of interconnecting rough endoplasmic reticulum are revealed (Fig. 2.4). Ribosomes (clusters of ribosomal RNA) are attached to the outer surfaces of the membranes and consist of a large and a small RNA–protein subunit. The Nissl substance is most concentrated in the soma and adjacent parts of the dendrite (Fig. 2.5). It is, however, also found throughout the dendrite (Fig. 2.7) and even in the axon hillock.

H. Mitochondria (Figs. 2.1, 2.4, and 2.7)

These organelles, found throughout the neuron, are the third largest organelles after the nucleus and endoplasmic reticulum and supply the energy for many activities in the cell. They are rod-shaped and vary from 0.35 to 10 µm in length and from 0.35 to 0.5 µm in diameter. The wall of a mitochondrion consists of two layers: an outer and inner membrane. The outer membrane contains pores that render the membrane soluble to proteins with molecular weights of up to 10,000. The inner membrane is less permeable and has folds called cristae that project into the center of the mitochondrial matrix. The interior of the mitochondrion is filled with a fluid denser than cytoplasm. On the inner membrane are found enzymes

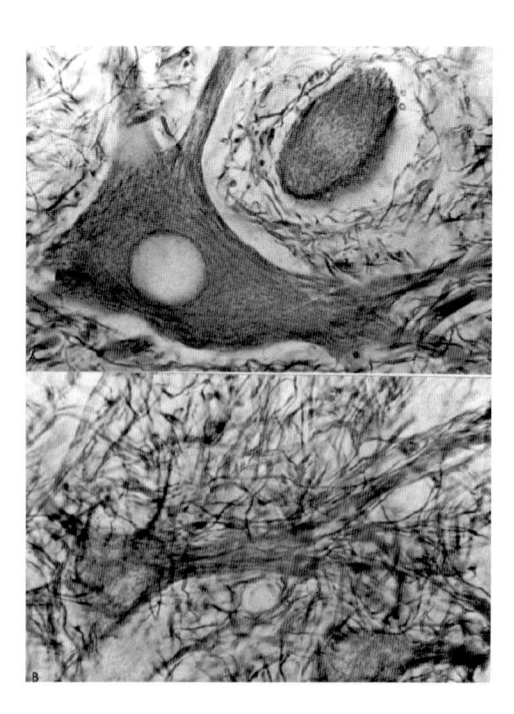

Fig. 2.5 Cytoskeleton: neurofibrillary stain of a ventral horn cell in the cat spinal cord, showing neurofibrillary network in soma and dendrites (**A**) and in the axons (**B**) (×400)

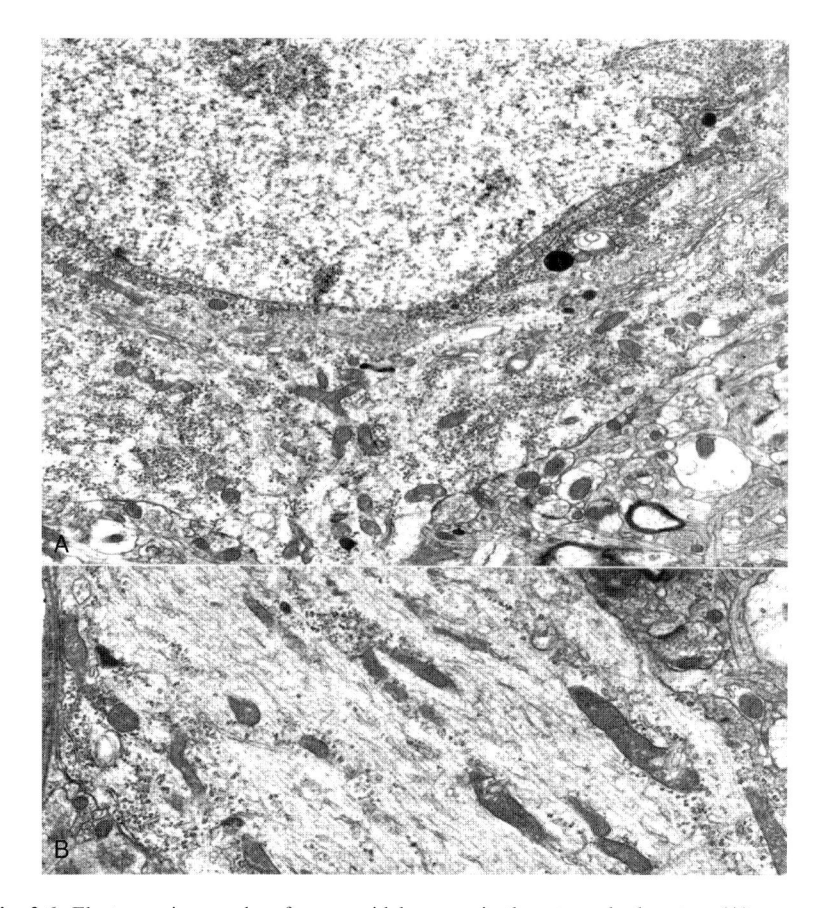

Fig. 2.6 Electron micrographs of a pyramidal neuron in the rat cerebral cortex: **(A)** soma and nucleus; **(B)** dendrite. Note the large amount of RER/Nissl substance in the soma and mitochondria. Dendrites have less Nissl substance and many microtubules. (×33,000). (From EM Marcus and S Jacobson, Integrated neuroscience, Kluwer, 2003)

that provide much of the energy required for the nerve cell. These respiratory enzymes (flavoproteins and cytochromes) catalyze the addition of a phosphate group to adenosine diphosphate (ADP), forming ATP. Cations and mitochondrial DNA have been demonstrated in the mitochondrial matrix. Mitochondrial DNA is derived from the mother. In the mitochondria is DNA from one's mother and an intriguing study links this mitochondrial DNA to a common human female ancestor, Lucy, who lived in Africa over 300,000 years ago. In the cytoplasm is ADP, providing the energy required for cellular metabolic functions. Also in the cytoplasm are found enzymes that break down glucose into pyruvic acid and acetoacetic acid. These substances are taken into the mitochondrial matrix and participate in the Krebs citric-acid cycle, which allows the mitochondria to metabolize amino acids and fatty acids.

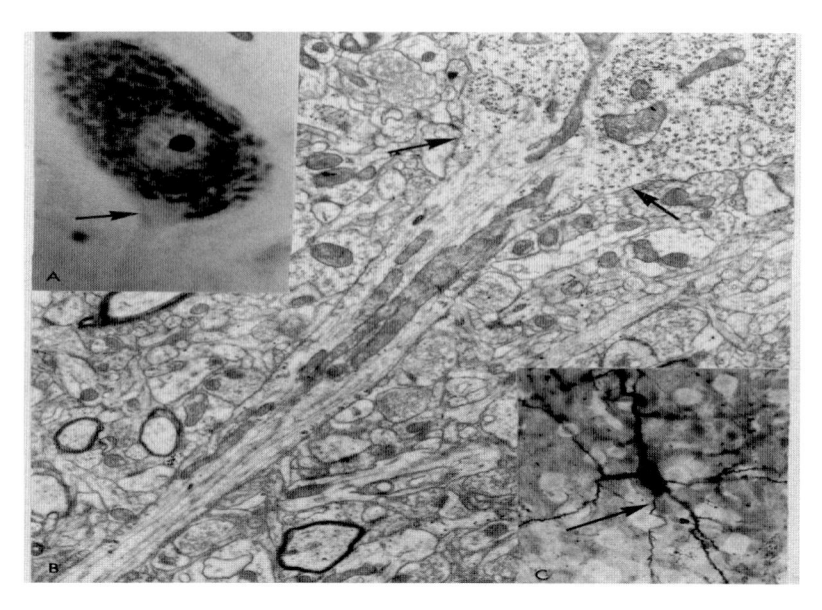

Fig. 2.7 Appearance of the axon hillock, axon origin: **(A)** in a Nissl stain (×400), **(B)** in an electron micrograph (×15,000), and **(C)** in a Golg–rapid stain (×350) (From EM Marcus and S Jacobson, Integrated neuroscience, Kluwer, 2003)

I. Neurosecretory Granules

Neurons in the supraoptic and paraventricular nuclei of the hypothalamus form neurosecretory material (Bodian, 1970, Palay, 1957 Scharrer, 1966). The axons of these cells form the hypothalamic–hypophyseal tract, which runs through the median eminence and down the infundibular stalk to the neurohypophysis (pars nervosa), where the axons end in close proximity to the endothelial cells. The secretory granules are 130 to 150 μm in diameter and are found in the tract (see Chapter 7, Fig. 7.4).

J. Neuronal Cytoskeleton

In silver-stained sections examined in a light microscope, a neurofibrillary network can be seen in the neurons (Fig. 2.5). Electron micrographs can distinguish microtubules, 3–30 μm in diameter, and neurofilaments, 1 nm in diameter (Fig. 2.6). It appears that fixation produces clumping of the tubules and filaments into the fibrillar network seen in light micrographs.

Neurons in common with other eukaryotic cells contain a cytoskeleton that maintains its shape. This cytoskeleton consists of at least three types of fiber:

I The Neuron

1. Microtubules 30 nm in diameter
2. Microfilaments 7 nm in diameter
3. Intermediate filaments 10 nm in diameter.

If the plasma membrane and organelle membrane are removed, the cytoskeleton is seen to consist of actin microfilaments, tubulin-containing microtubules, and criss-crossing intermediate filaments. Neurotubules (microtubules) predominate in dendrites and in the axon hillock, whereas microfilaments are sparse in dendrites and most numerous in axons (Figs. 2.6 and 2.7B). Microtubules and intermediate filaments are found throughout the axon.. Microfilaments form much of the cytoskeleton of the entire neuron.

K. Microtubules and Axoplasmic Flow

With the protein manufacturing apparatus present only in the soma, and to a lesser degree in the dendrites, a mechanism must exist to transport proteins and other molecules from the soma, down the axon, and into the presynaptic side. Weiss and Hisko (1948) placed a ligature on a peripheral nerve and this produced a swelling proximal to the tie, demonstrating that material flows from the soma, or trophic center, into the axon and, ultimately, to the axon terminal. The development of techniques that follow this axoplasmic flow has revolutionized the study of circuitry within the central nervous system. The ability to map this circuitry accurately has given all neuroscientists a better understanding of the integrative mechanisms in the brain. There are many compounds now available to follow circuitry in the brain and they include horseradish peroxidase, wheat germ aglutin, tetanus toxin, fluorescent molecules, and radiolabeled compounds.

Microtubules provide the structural basis for transport: axoplasmic flow. This mechanism of transport is not diffusion but, rather, retrograde axonal transport associated with the microtubule network that exists throughout the nerve cell. The rate of flow varies depends on the product being transported and ranges from more than 300 mm/day to less than 1 mm/day. The main direction of the flow is *anterograde* (from the cell body into the axon and synapse). There is also a very active *retrograde* flow from the synaptic region back to the cell body that is a source for recycling many of the substances found at the synaptic ending.

The particles that move the fastest (see Table 2.2) consist of small vesicles of the secretory and synaptic vesicles, and the slowest group is the cytoskeletal components. Mitochondria are transported down from the cell body at an intermediate rate. The retrograde flow from the synaptic telodendria back into the soma returns any excess of material for degradation or reprocessing. The retrograde flow permits any excess proteins or amino acids to recycle. It also permits products synthesized or released at the axonal cleft to be absorbed and then transported back

Table 2.2 Rate of axonal transport of cellular structures

Transport rate (mm/day)	Cellular structure
Fast: >300 and <400	Vesicles, smooth endoplasmic reticulum, and granules
Intermediate: <50 and >15	Mitochondria, filament proteins
Slow: >3 to <4	Actin, fodrin, enolase, component CPK, calmodulin B, and clathrin
Slow: >0.3 and <1	Neurofilament protein, component tubulin, and MAPs

Source: Data from Graftstein and Forman (1980), Mcquarrie (1988), and Wujek and Lasek (1983)

to the cell. The microtubules help to transport membrane-bound vesicles, protein, and other macromolecules. This orthograde transport, or anterograde axonal transport, is the means whereby these molecules formed in the soma/trophic center are transported down the axon into the axonal telodendria. The individual microtubules in the nervous system are 10–35 nm in length and together form the cytoskeleton. The intermediate filaments are associated with the microtubules. The wall of the microtubule consists of a helical array of repeating tubulin subunits containing the A and B tubulin molecules. The microtubule wall consists of globular subunits 4–5 nm in diameter; the subunits are arranged in 13 protofilaments that encircle and run parallel to the long axis of the tubule. Each microtubule also has a defined polarity. Associated with the microtubules are protein motors, kinesins and dyneins, which when combined with cAMP might well be the mechanism of transport in the central nervous system. The products transported down the microtubules move similar to a humans walking rather than like an inchworm or a locomotive. During mitosis, microtubules disassemble and reassemble; however, a permanent cytoskeleton lattice of microtubules and intermediate filaments in the neuron is somehow maintained. It is not yet known how long each microtubule exists, but there is evidence of a constant turnover.

L. *Neurofibrillar Tangles*

Neurofibrillar tangles are bundles of abnormal filaments within a neuron. They are helical filaments that are different from normal cytoskeletal proteins and they contain the tau protein—a microtubule-binding protein (MAP) that is a normal component in neurons. In Alzheimer's disease, there are accumulations of abnormally phosphorylated and aggregated forms the MAP tau. These large aggregates form the tangles that can be physical barriers to transport, might interfere with normal neuronal functions, and are probably toxic. Mutations in the human tau gene are found in autosomal domininant neuronal degenerative disorders isolated to chromosome. 17. These familial disorders are characterized by extensive neurofibrillar pathology and are often called "taupathies" (Hutton, 2000).

M. Axon and Axon Origin (Axon Hillock)

The axon originates at the hillock and is a slender process that usually arises from a cone-shaped region on the perikaryon (Fig. 2.7A). This region includes filaments, stacks of tubules, and polyribosomes (Fig. 2.7B). The initial segment of the axon, arising from the axon hillock, is covered by electron-dense material that functions as an insulating membrane. The axon contains some elongate mitochondria, many filaments oriented parallel to the long axis of the axon (Fig. 2.7B), and some tubules. In contrast, a dendrite contains a few filaments and many tubules, all arranged parallel to the long axis of the dendrite (Fig. 2.5). Polyribosomes are present, but the highly organized, rough endoplasmic reticulum is absent.

N. Myelin Sheath (Fig. 2.8)

In the nervous system, axons might be myelinated or unmyelinated. Myelin is formed by a supporting cell, wshich in the central nervous system is called the *oligodendrocyte* and in the peripheral nervous system is called the *Schwann cell*. The immature Schwann cell and oligodendrocyte have on their surface the myelin-associated glycoprotein that binds to the adjacent axon and might well be the trigger that leads to myelin formation. Thus, the myelin sheath is not a part of the neuron; it is only a covering for the axon. Myelin consists of segments approximately 0.5–3 mm in length. Between these segments are the nodes of Ranvier. The axon, however, is continuous at the nodes, and axon collaterals can leave at the nodes. The myelin membrane, like all membranes, contains phospholipid bilayers. In the central nervous system, myelin includes the following proteins:

Proteolipid protein (51%)
Myelin basic protein (44%)
Myelin-associated glycoprotein (1%)
3,3-Cyclic nucleotide (4%)

The oligodendrocytic process forms the myelin sheaths by wrapping around the axon. The space between the axonal plasma membrane and the forming myelin is reduced until most of the exoplasmic and cytoplasmic space is finally forced out. The result is a compact stack of membranes. The myelin sheath is from 3 to 100 membranes thick and acts as an insulator by preventing the transfer of ions from the axonal cytoplasm into the extracellular space. Myelin sheaths are in contact with the axon. In light microscopy, they appear as discontinuous tubes 0.5–3 mm in length, interrupted at the node.. The axon is devoid of myelin at the site of origin (the nodes) and at the axonal telodendria. At the site of origin, the axon is covered by an electron-dense membrane, and at the site of the synaptic telodendria, the various axonal endings are isolated from one another by astrocytic processes. In electron micrographs, each myelin lamella actually consists of two-unit membranes with the entire lamella being 130–180 Å thick (Fig. 2.8). Myelin is thus seen to

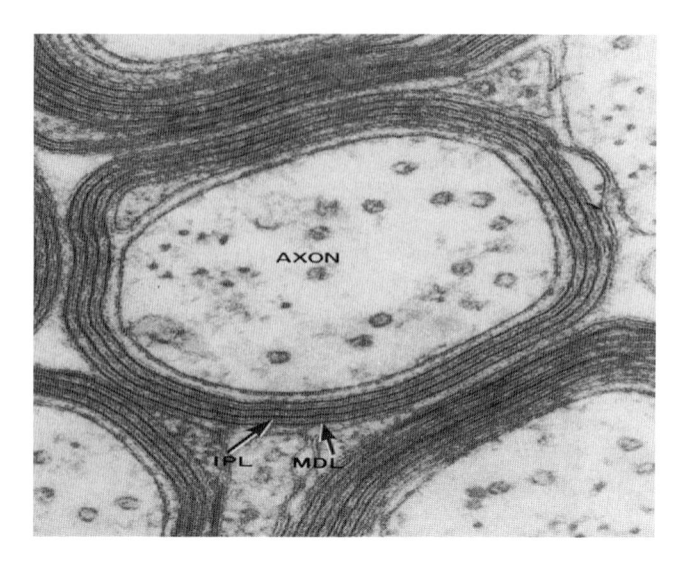

Fig. 2.8 Myelin sheath. Electron micrograph of myelin sheath from the optic nerve of the mouse demonstrating repeating units of the myelin sheath, consisting of a series of light and dark lines. The dark line, called the major dense line (MDL), represents the apposition of the inner surface of the unit membranes. The less dense line, called the interperiod line (IPL), represents the approximation of the outer surfaces of adjacent myelin membranes (×67,000) (Courtesy of Alan Peters, Department of Anatomy, Boston University School of Medicine)

consist of a series of light and dark lines. The dark line, called the major dense line, represents the apposition of the inner surface of the unit membranes. The less dense line, called the interperiod line, represents the approximation of the outer surfaces of adjacent myelin membranes. Only at the node of Ranvier is the axonal plasma membrane in communication with the extracellular space. The influx of Na^+ at each node causes the action potential to move rapidly down the axon by jumping from node to node.

O. Myelination

The process of covering a naked axon with myelin is myelination. An axon starts with just a covering formed by the plasma membrane of either the *Schwann cell* or the *oligodendrocyte*. More and more layers are added until myelination is complete. The myelin is laid down by the processes of either the Schwan or oligodendrocytel twisting around the axon (Geren, 1954; Robertson, 1955). In the peripheral nervous system, there is usually only one Schwann cell for each length or internode of myelin. In the central nervous system, each oligodendrocyte can form and maintain myelin sheaths on 30–60 axons. The unmyelinated axons in the peripheral nervous system are found in the cytoplasm of the Schwann cell; in the central nervous system, each oligodendrocyte enwraps many axons. There can be as many as 13 unmyelinated axons in one Schwann cell.

The *sequence of myelination* has been studied centrally in great detail (Jacobson 1963); it begins in the spinal cord, moves into the brain stem, and, finally, ends up with the diencephalon and cerebrum. A delay in myelination will produce developmental delays and might be a consequence of many factors, including genetic and nutritional ones (e.g., alcoholism associated with fetal alcohol syndrome), and is usually very harmful to the fetus. Breakdown of the myelin in a disease (e.g., multiple sclerosis) produces major functional deficits where they can affect the basic function of the cell, such as the signaling process. Fast axonal transport is associated with the microtubules. The slower components, including membrane-associated proteins (MAPs), are transported inside the microtubules, but the mitochondria actually descend in the axonal cytoplasm (Table 2.2).

P. Central Nervous System Pathways

The axons in the peripheral nervous systems are organized into nerves, whereas in the central nervous system, the axons run in groups called tracts, with each axon enwrapped in myelin, and groups of axons are bundled together by the processes of fibrous astrocytes. The axons vary in diameter (5–33 μm) and in length (0.5 mm to 1 m), but these axons cannot be separated into functional categories based on axonal diameter.

II. Synapse

Synapses can be seen at the light microscopic level (Fig. 2.10); however, to identify all of the components of a synapse, an electron microscope must be used. At the electron microscopic level, the synapse consists of the axonal ending, which forms the presynaptic side, and the dendritic zone, which forms the postsynaptic side (Fig. 2.10). Collectively, the presynaptic and postsynaptic sides and the intervening synaptic cleft are called the synapse.

At the synapse, the electrical impulse from one cell is transmitted to another. Synapses vary in size from the large endings on motor neurons (1–3 μm) to smaller synapses on the granule and stellate cells of the cortex and cerebellum (less than 0.5 μm). Synapses primarily occur between the axon of one cell and the dendrite of another cell. Synapses are usually located on the dendritic spins but are also seen on the soma and rarely between axons. At the synapse, the axon arborizes and forms several synaptic bulbs that are attached to the plasma membrane of the opposing neuron by intersynaptic filaments (Fig. 2.9).

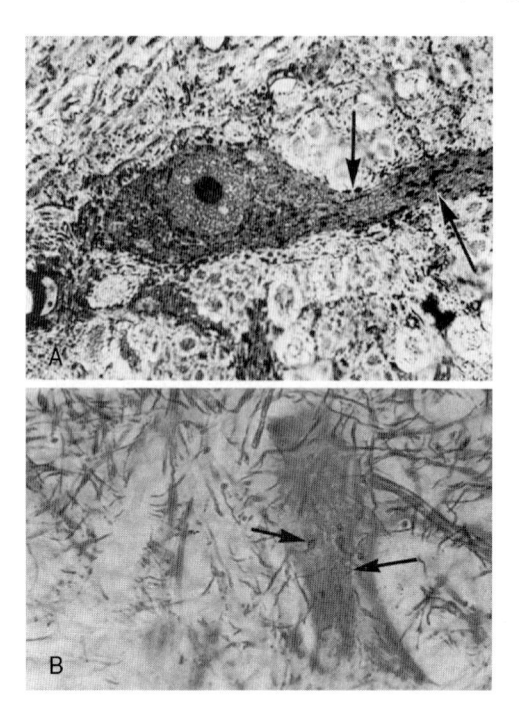

Fig. 2.9 Silver stain of a 1-μm plastic embedded section. (**A**) Synaptic boutons on neurons in the reticular formation; (**B**) boutons on ventral horn cells (×400) (From EM Marcus and J Jacobson, Integrated neuroscience, Kluwer, 2003)

A. Synaptic Structure

Synapses can be identified by light microscopy, but the electron microscope has revealed many details in synaptic structure (Bodian, 1970; Colonnier, 1969; Gray, 1959; Palay, 1967). In an electron micrograph, the presynaptic or axonal side of the synapse contains mitochondria and many synaptic vesicles (Fig. 2.10). Synaptic vesicles are concentrated near the presynaptic surface, with some vesicles actually seen fusing with a membrane (Fig. 2.10), illustrating that this site releases neurotransmitters. Neurofilaments are usually absent on the presynaptic side. Presynaptic and postsynaptic membranes are electron dense and are separated by a 30–40-nm space, the synaptic cleft, which is continuous with the extracellular space of the central nervous system.

B. Synaptic Types

There are two basic types of synapse: electrical and chemical; they differ in location and appearance. Most of the synapses in the mammalian central nervous system are

Fig. 2.10 Synapse in the sensory cortex of the rat demonstrating agranular synaptic vesicles (300–400 Å in the presynaptic axonal side). Note the electron-dense synaptic membranes and the intersynaptic filaments in the synaptic cleft. Electron micrograph (65,000). (From EM Marcus and J Jacobson, Integrated neuroscience, Kluwer, 2003)

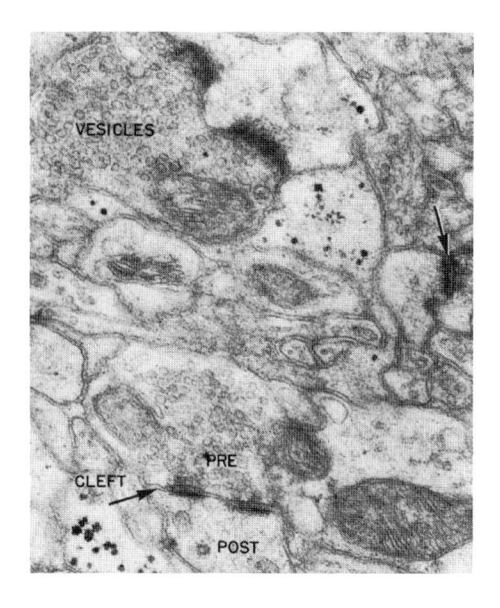

chemical. Electrical synapses are connected by membrane bridges (gap junction connections) that permit the electric impulse to pass directly from one cell to the other. Electric synapses have almost no delay and little chance of misfiring. These synapses are seen in many fish. Chemical synapses have a presynaptic side, containing vesicles and a gap, and the postsynaptic side, containing membrane receptors. The neurotransmitter released by the action potential is exocytosed and diffuses across the synaptic cleft and binds to the specific receptor on the postsynaptic membrane.

C. Synaptic Vesicles

The synaptic vesicles differ in size and shape and can be agranular, spherical, flattened, or round with a dense core. The method of fixation for electron micrographs affects the shape of a vesicle. Bodian (1970) has shown that osmium fixation produces only spheroidal vesicles. Aldehyde followed by osmium produces spheroidal and flattened vesicles. The shape of flattened vesicles can also be modified by washing the tissue in buffer or placing the tissue directly from the aldehyde into the osmium. The spheroidal vesicles retain their shape regardless of any manipulation.

Excitatory Synapses

Excitatory synapses depolarize the membrane potential and make it more positive and they appear asymmetrical, having a prominent postsynaptic bush with presynaptic vesicles (Fig. 2.10). This type of synapse is most commonly seen on dendrites. Glutamate has been identified in excitatory synapses. At the excitatory synapse, there is a change in permeability that leads to depolarization of the postsynaptic membrane and that can lead to the generation of an action potential.

Inhibitory Synapses

Inhibitory synapses hyperpolarize the membrane potential and make it more negative. They are symmetrical with thickened membranes on the presynaptic and postsynaptic side and vesicles only on the presynaptic side. GABA has been identified in the inhibitory synapses. At an inhibitory synapse, the neurotransmitter binds to the receptor membrane, which changes the permeability and tends to block the formation of the action potential. Synapses on the soma are symmetrical and they are considered inhibitory. Many spines are found on dendritic spines and these spines contain a spine apparatus that seems to function like a capacitor, charging and then discharging when its current load is exceeded (see Fig. 2.1).

D. Synaptic Transmission

Synaptic transmission in the mammalian central nervous system is primarily a chemical and not an electrically mediated phenomenon, based on the presence of the following:

1. A 30–40-nm cleft
2. Synaptic vesicles
3. Appreciable synaptic delay due to absorbance of the chemical onto the postsynaptic receptor site.

In contrast electrical synapses have cytoplasmic bridges that interconnect the presynaptic and postsynaptic membranes, resulting in a minimal synaptic delay, as transmission is ionic rather than by release of chemical from a vesicle.

Neurotransmitters

Excitatory neurotransmitters include glutamate and actylcholine. *Inhibitory* neurotransmitters include GABA, histamine, neurotensin, and angiotensin. Many other compounds have been identified as neurotransmitters. These substances are found in synaptic vesicles on the presynaptic side. Introduction of the compound into the synaptic cleft produces the same change in the resting membrane potential as stimulation of the presynaptic axon; the compound is rapidly degraded, and the membrane potential returns to the resting state.

The neurotransmitters are either amino acids or small neuropeptides. The classic neurotransmitters in the central nervous system include acetylcholine, epinephrine, norepinephrine, serotonin, glycine, glutamate, dopamine, and GABA.

Acetylcholine is found in the peripheral nervous system and forms the cholinergic synapses where it is liberated from preganglionic and postganglionic parasympathetic endings and its release produces many parasympathetic effects, including cardiac inhibition, vasodilation, and gastrointestinal peristalsis. Acetylcholine esterase has been found throughout the central and peripheral nervous systems and at postganglionic sympathetic endings.

Modulators of Neurotransmission

At certain synaptic sites the following compounds might also function as modulators (usually a slower transmitter form of neurotransmission): adenosine, histamine, octopamine, B-alanine, ATP, and taurine. Many of the neuropeptides, such as substance P, vasoactive peptide, peptide Y, and somatostatin, are also active in neurotransmission or neuromodulation. Catecholamines and 5-hydroxytryptamine are transmitters linked to synaptic transmission in the central nervous system. Noradrenaline is a transmitter at the preganglionic synapses. Many steroids and hormones have also been linked to synaptic transmission. It is still uncertain whether these compounds play a direct role in nervous transmission or if they are just related by their importance to the ongoing functions of the entire nervous system (Table 2.3).

Effectors and Receptors

Effectors. The motor nerves of the somatic nervous system end in skeletal muscles and form the motor end plates. Nerve endings in smooth and cardiac muscle and in glands resemble the synaptic endings in the central nervous system. Visceral motor

Table 2.3 Location and functions of neurotransmiters

Agent	Location	Function
L-Glutamine	Excitatory neurons	Excitation
GABA	Inhibitory neurons	Inhibition (fast/slow)
Acetylcholine	Motor neurons, basal forebrain, and midbrain and pontine tegmentum.	Excitation and modulation
Monoamines:	Brain stem, hypothalamus	Modulation
Norepinephrine	Locus ceruleus	
Serotionin	Locus ceruleus	
Histamine	Raphe nuclei	
Dopamine	Hypothalamus	
Neuropeptides	Limbic system, hypothalamus, autonomic and pain pathways	Modulation

endings are found on muscles in arterioles (vasomotor), muscles in hair follicles (pilomotor), and sweat glands (sudomotor).

Cutaneous Sensory Receptors. A stereogram of the skin is shown in Figure 1.3. Table 2.4 lists the mechanicoreceptors in the body. Sensory endings, found throughout the body, subserve pain touch, temperature, vibration, pressure, heat, and cold in the skin, muscles, and viscera as well as the specialized somatic and visceral sensations of taste, smell, vision, audition, and balance.

Visceral Sensory Receptors. These receptors are similar to somatic sensory receptors associated with the somatic nervous system, except that they are located in the viscera and their accessory organs.

III. Supporting Cells of the Central Nervous System

The central nervous system has billions of neurons, but the number of supporting cells exceeds them by a factor of 5 or 6. Supporting cells form a structural matrix and play a vital role in transporting gases, water, electrolytes, and metabolites from blood vessels to the neural parenchyma and in removing waste products from the neuron. In contrast to the neuron, the supporting cells in the adult central nervous system normally undergo mitotic division.

The supporting cells are divided into macroglia and microglia. The *macroglia* include astrocytes, oligodendrocytes, and ependyma and are the supporting cells or neuroglia (nerve glue) of the central nervous system (Fig. 2.4) *Microglia* include mesodermal microglia cells (of Hortega), the perivascular cells, and any white blood cells found within the parenchyma of the central nervous system. Schwann cells, satellite cells, and fibroblasts are supporting cells of the peripheral nervous system. Functions of the different supporting cells in the nervous system are summarized in Table 2.5.

A. Astrocytes (Fig. 2.4)

Astrocytes are of two types: fibrous (most common in white matter) or protoplasmic (most common in gray matter). All astrocytes are larger and less dense than the

Table 2.4 Mechanicoreceptor

Modality	Receptor
Sound	Cochlea in inner ear of temporal bone transforms mechanical.into neural impulses
Light touch and vibration	Encapsulated endings: Meissner's and Pacinian corpuscles
Proprioception: encapsulated sensory endings	Muscle spindles and Golgi tendon organs in joints, Meissner's and Pacinian corpuscles, Merkel's tactile discs
Pain and temperature	Free nerve endings, end bulbs of Krause and Golgi–Mason

oligodendrocytes. In light micrographs, astrocytes appear as pale cells with little or no detail in the cytoplasm. The nuclei are smaller than those of a neuron but larger and less dense than those of an oligodendrocyte. The astrocytes have many functions, as noted in Tables 2.5 and 2.6.

Electron micrographs demonstrate that *fibrous astrocytes* have many filaments, which in places appear to fill the cytoplasm. There are few microtubules, and the processes appear pale. The nuclei of these cells have some condensed chromatin adjacent to the nuclear membrane. Glycogen is also common in astrocytic processes. *Protoplasmic astrocytes* have nuclei that are a little darker than those of a neuron. They resemble fibrous astrocytes except that they have just a few filaments. Astrocytes and adjacent neurons form the *microenvironment* of the nervous system. There are approximately 100–1000 astrocytes per neuron. In areas with high glutamate concentration, such as the cerebral cortex, or areas with

Table 2.5 Functions of central nervous system supporting cells

Cell type	Functions
Astrocytes:	Major supporting cells in the brain, forming microenvironment for neurons
Fibrous type (white matter)	Act as phagocyte and enwrap axons contain many filaments;
Protoplasmic type (gray matter)	Isolate synapses, enwrap blood vessels, and form membranes on brain's inner and outer surfaces
Oligodendrocytes	Form and maintain myelin
Ependyma cells	Ciliated lining cells of the ventricles
Endothelial cells	Lining cells of blood vessels in the brain that form blood-brain barrier
Microglia (pericytes)	Supporting cells and multipotential cells found in the basement membrane of blood vessels and within brain parenchyma. Can become neurons??
Mononuclear cells	White cells from the circulation readily enter and stay in the brain (lymphocytes, monocytes, and macrophages) and function as sentinels for the immune system

Table 2.6 Role of astrocytes in the central nervous system

1. Form a complete membrane on the external surface of the brain called the external glial limiting membrane, which enwrap all entering blood vessels
2. Form the inner glial membrane, which fuses with the ependymal processes
3. Isolates neuronal processes
4. Form the skeleton of the central nervous system
5. Tend to segregate synapses and release or absorb transmitters
6. Help form the blood-brain barrier by enwrapping brain capillaries
7. If the brain is damaged by infarction, for example, astrocytes proliferate and form scars; the scar will interfere with any successful axonal regeneration

high dopamine content, such as the basal nuclei, the glia take on the chemical characteristics of the adjacent neuron. Glia are involved in neuronal functions because they absorb transmitters and modulators in their environment and often release them back into the synapse.

When central nervous system diseases affect only astrocytes, the reaction is called *primary astrogliosis*. More commonly, the disease process affects nerve cells primarily, which is called *secondary astrogliosis*. When an injury occurs to the nerve cell in the central nervous system without concomitant injury to blood vesicles and glia, the nerve cells are phagocytized and the astrocytes proliferate and replace the neurons, forming a glial scar (replacement gliosis). In the case of a more severe injury to the nervous system, such as an infarct that damages glia, nerve cells, and blood vessels, the astrocytes proliferate along the wall of the injury, and the dead neurons and glia are phagocytized, leaving only a cavity lined with meninges. In all other organs, there are enough fibroblasts to proliferate and form a scar, but in the central nervous system, there are only a few fibroblasts, so cavitation is a common sequela to extensive destruction.

B. Oligodendrocytes (Fig. 2.5)

In light micrographs, the oligodendrocyte has a small darkly stained nucleus surrounded by a thin ring of cytoplasm. In electron micrographs, oligodendrocytes are dense cells with many microtubules and few neurofilaments (Fig. 2.5). Dense clumps of rough endoplasmic reticulum and clusters of polyribosomes are seen in the cytoplasm, which is denser but scantier than that in neurons. The nucleus tends to be located toward one pole of the cell where the nuclear chromatin tends to be heavily clumped. In electron micrographs, oligodendrocytes can be distinguished from astrocytes because they have a darker cytoplasm and nucleus, few if any filaments, and more heavily condensed chromatin (Figs. 2.4 and 2.5). The role of the oligodendrocyte is to form and maintain myelin (although they can also be responsible for breaking down and then remyelinating in multiple sclerosis). Oligodendrocytes are usually seen in close proximity to astrocytes and neurons, and all three cell types are important in forming and maintaining myelin (Fig. 2.11).

C. Endothelial Cells (Fig. 2.4)

Endothelial cells form the lining of the capillaries in the central nervous system. They are of mesodermal origin and bound together by tight junctions. Their tight junctions and apinocytosis provide the basis of the blood-brain barrier (see below).

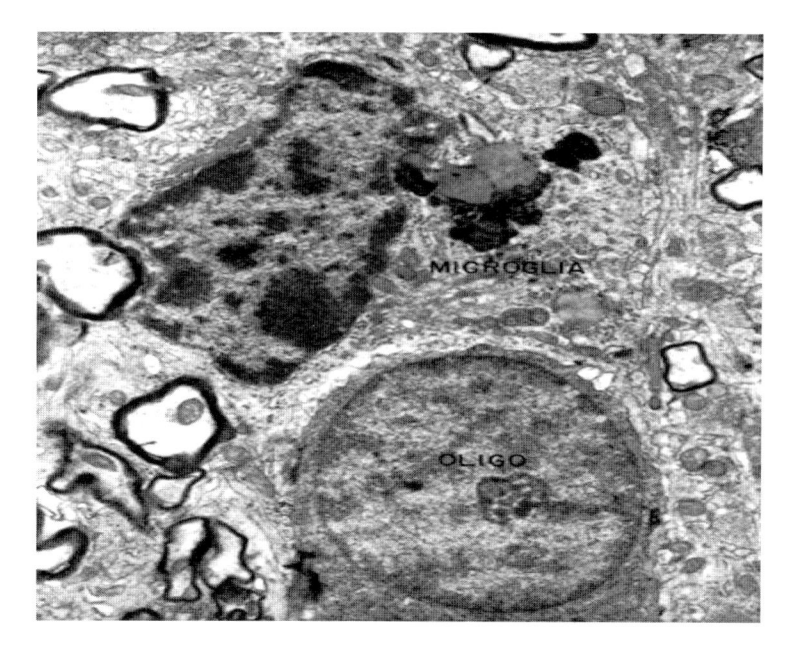

Fig. 2.11 Electron micrograph of a human cerebral cortex demonstrating differences in the density of the DNA in the nuclei of oligodendrocytes (OLIGO) and microglia (×30,000) (From EM Marcus and S Jacobson, Integrated neuroscience, Kluwer, 2003)

D. Mononuclear Cells

1. Mononuclear Cells

Mononuclear cells—lymphocytes, monocytes, and histiocytes—are found in the central nervous system, where they act as phagocytes, breaking down myelin and neurons. Myelin destruction always triggers intense macrophage reaction within 48 h, followed by infiltration of monocytes first and then lymphocytes. Note that astrocytes have also been shown to engulf degenerating myelin sheaths, axonal processes, and degenerating synapses.

The central nervous system was once considered an immunologically privileged site for the following reasons:

1. No specific lymph drainage from the central nervous system alerts the immune system of infection
2. Neurons and glia do not express the major histocompatibility complex
3. The major cell for stimulating the immune response (leukocyte dendritic cells) is not normally present in the disease-free nervous system.

However, recent studies have shown that there is a regular immune surveillance of the central nervous system, which is sufficient to control many viral infections (Sedgwick and Dorries, 1991). It is now known that immune cells regularly enter the brain through the capillaries and that macrophages infected with human immunodeficiency virus (HIV), for example, can infect the brain directly, the so-called Trojan-horse phenomenon (Haase, 1986; Price et al., 1988).

2. Microglial Cells

Microglia originate from monocytes that enter the brain (Table 2.7). Neurons, astrocytes, and oligodendrocytes are ectodermal in origin, but microglial cells are mesodermal in origin. The ovoid microglia cells are the smallest of the supporting cells and are divided into two categories: (a) perivascular pericytes cells and (b) the resting microglial cells in the brain parenchyma.

Pericytes are found in relation to capillaries but external to the endothelial cells and enwrapped in the basal lamina. In electron micrographs, they are not as electron dense as oligodendrocytes and lack the neurofilaments of the astrocyte and the tubules of the oligodendrocyte. The cytoplasm is denser than that of astrocytes and contains fat droplets and laminar-dense bodies. The granular endoplasmic reticulum consists of long stringy cisterns. Microglia are considered multipotential cells because, with the proper stimulus, they can become macrophages (Vaughn and Peters, 1968). The pericyte contains actin, and this cell might well be important in controlling the channels entering the endothelial cells (Herman and Jacobson, 1988).During early development, monocytes enter the brain and, after formation of the blood-brain barrier, become trapped (Davis et al., 1994; Ling and Wong, 1993). The monocytes pass through an intermediate phase of development, the amoeboid microglia, which evolve into a downregulated resting form, the ramified microglia. These resting microglia are found throughout the central nervous system and might well be the sentinels that alert the immune system to disease in the brain. With the appearance of any central nervous system disease (e.g., multiple sclerosis, stroke, trauma, or tumors), the resting microglia are upregulated and become activated microglial cells. The factor or gene that upregulates or downregulates these cells is

Table 2.7 Types of microglia cell

Cell type	Function
Monocytes	Enters brain during early development and is the stem cell of microglia
Pericytes	Found inside the brain in the (perivascular cell) basement membrane of the blood vessel; can act as a macrophage
Amoeboid microglia	Transitional form leads to resting microglia
Resting microglia	Downregulated from amoeboid (ramified) microglia; probably the sentinels in the brain that raise the alarm for invasive diseases
Activated microglia	Upregulated resting cell changes into partially activated macrophage with MHC class I
Reactive microglia	Fully activated macrophage with MHC class II and phagocytic properties, might become giant cells

currently unknown. Once the disease process has been resolved, the activated microglia can revert to resting microglial cells. The activated microglia cell is a partially activated macrophage containing the CR3 complex and class I major histocompatibility complex (MHC). These activated microglial cells evolve into giant multinucleated cells by the fusion of reactive cells. They are seen in viral infections and are considered the hallmark of AIDS dementia. These reactive multinucleated cells are associated with many viral brain infections. Reactive microglial cells are fully active macrophages containing class II MHC and phagocytic activity. These reactive cells also called gitter cells are very active during all major disease states in the brain. These giant multinucleated cells are often found in patients with Creutzfeldt–Jakob disease (Fig. 2.12), a disease caused by proteinaceous infectious particles, or prions.

E. Ependymal Cells (Fig. 2.13)

Ependymal cells line all parts of the ventricular system (lateral ventricles, third ventiricle, cerebral aqueduct, fourth ventricle, and spinal canal). They are cuboidal, ciliated, and contain filaments and other organelles. The processes of these cells extend in the central nervous system and fuse with astrocytic processes to form the

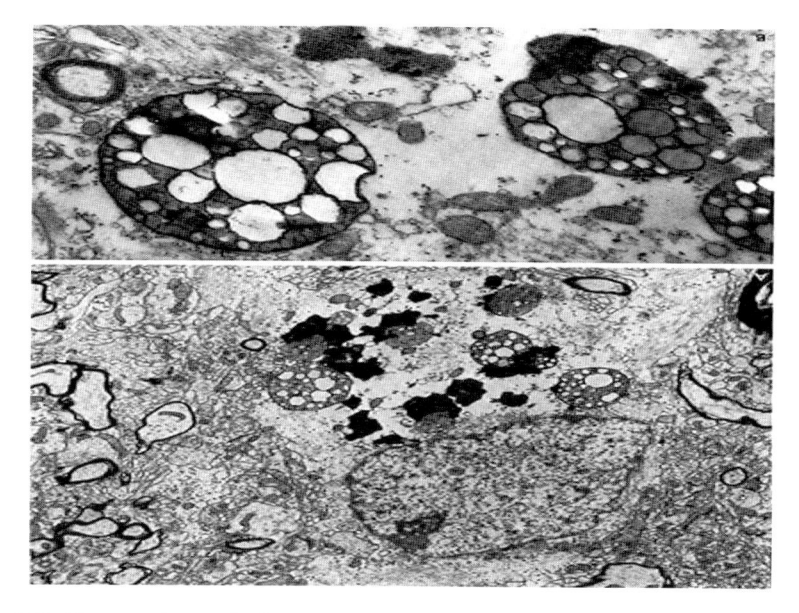

Fig. 2.12 Electron micrograph of a reactive astrocytes, gitter cell, in the cerebral cortex of a patient with Jakob–Creutfeld disease. Note the prominent digestion vacuoles in higher power A. (**B**, ×8000; **A**, ×35,000). (From EM Marcus and S Jacobson, Integrated neuroscience, Kluwer, 2003)

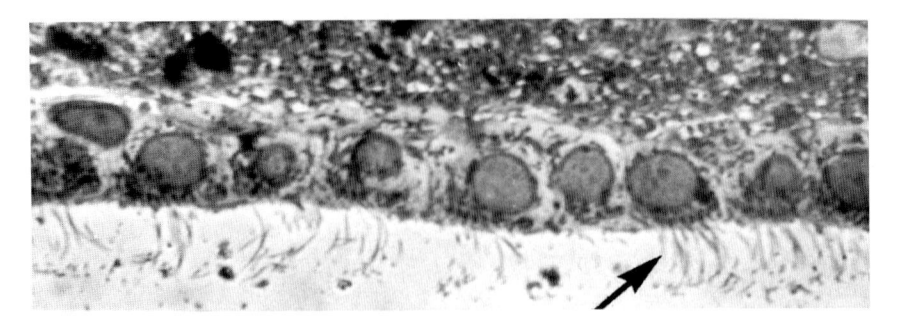

Fig. 2.13 Ependymal lining cell in the third ventricle of a rat. Note prominent cilia extending into the ventricle (arrow) in this 1-μm plastic section. (From EM Marcus and J Jacobson, Integrated neuroscience, Kluwer, 2003)

inner limiting glial membrane. Highly modified ependymal cells are found attached to the blood vessels in the roof of the body of the lateral ventricles, the inferior horn of the lateral ventricles, and the third and fourth ventricles. There they form the choroid plexus, which secretes much of the cerebrospinal fluid Ependymal cells originate from the germinal cells lining the embryonic ventricle, but they soon stop differentiating and stay at the lumen on the developing ventricles.

IV. Supporting Cells in the Peripheral Nervous System

A. Satellite Cells

Satellite cells, which are found only in the peripheral nervous system among sensory and sympathetic ganglia, originate from neural crest cells. Many satellite cells envelop a ganglion cell. Functionally, they are similar to the astrocytes, although they look more like oligodendrocytes.

B. Schwann Cells

Schwann cells are ectodermal in origin (neural crest) in the peripheral nervous system and function like oligodendrocytes, forming the myelin and neurilemmal sheath. In addition, the unmyelinated axons are embedded in their cytoplasm. Schwann cell cytoplasm stops before the nodes of Ranvier, leaving spaces between the node and Schwann cells. In an injured nerve, Schwann cells can form tubes that penetrate the scar and permit regeneration of the peripheral axons. Nerve growth factor is important to proliferation of the Schwann cells.

C. Neural Crest Cells

These cells originate embryologically as neuroectodermal cells on either side of the dorsal crest of the developing neural tube but soon drop dorsolaterally to the evolving spinal cord area. Neural crest cells migrate out to form the following: dorsal root ganglion cells, satellite cells, autonomic ganglion cells, Schwann cells of the peripheral nervous system, chromaffin cells of the adrenal medulla, and pigment cells of the integument.

V. Response of Nervous System to Injury

A. Degeneration

Neuronal death or atrophy could result from trauma, circulatory insufficiency (strokes), tumors, infections, metabolic insufficiency, developmental defects, and degenerative and heredodegenerative diseases.

1. Retrograde Changes in the Cell Body: Chromatolysis (Fig. 2.14)

Section of the axon or direct injury to the dendrites or cell body produces the following series of responses in the soma. The responses of neuronal soma to injury (chromatolysis) can be summarized in three steps:

a. **Swelling** of nucleus, nucleolus, and cytoplasm, with the nucleus becoming eccentric as a direct response to injury of the axon or dendrite. Nissl substance appears to dissolve. The nucleus, cell body, and nucleoli swell. The nucleus is displaced from the center of the cell body and might even lie adjacent to the plasma membrane of the neuron.

b. **Chromatolysis of RNA.** A slow dissolution of Nissl substance starts centrally and proceeds peripherally until only the most peripherally placed Nissl substance is left intact (which is probably essential to the protein metabolism of the rest of the cell). This dissolution of the Nissl substance (ribosomal RNA), called *chromatolysis*, allows the protein-manufacturing processes to be mobilized to help the neuron survive the injury. The mRNA then begins the manufacturing of membrane that is transported down the intact tubules into the growing axonal ending (growth cone). All other organelles in the cell body and dendrites also respond to the injury. The mitochondria swell, and the smooth endoplasmic reticulum proliferates to help in the formation of new plasma membrane and new myelin. These responses represent the increased energy requirements of the nerve cell and the need to form plasma membrane during the regenerative process.

c. **Recovery.** With successful recovery, the cell and its organelles return to normal size. If seriously injured, the cell becomes atrophic or might be phagocytized.

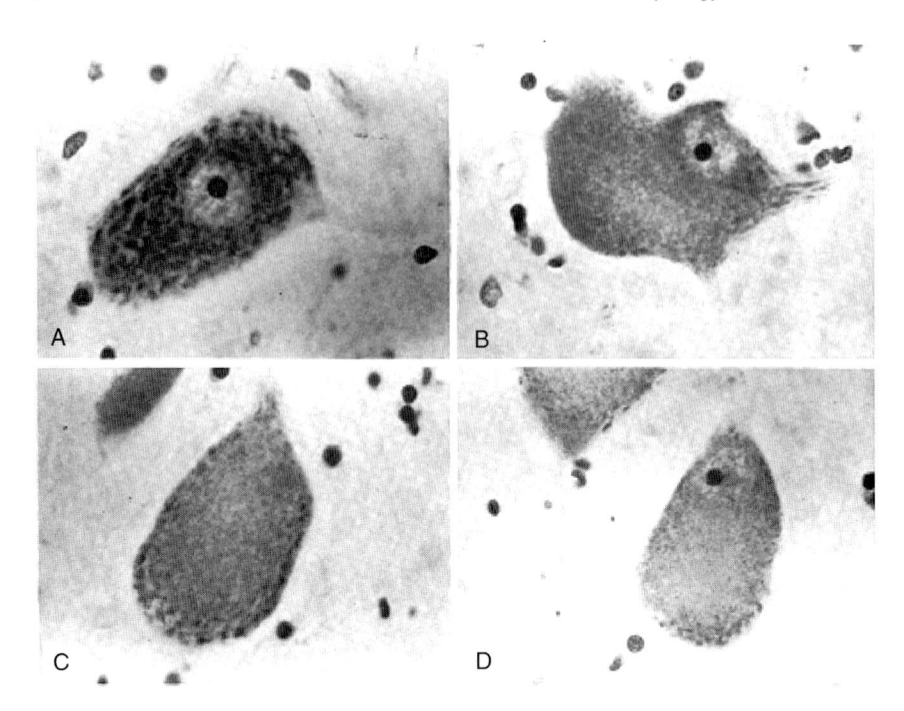

Fig. 2.14 Ventral horn cells in the human lumbar spinal cord: (**A**) normal and (**B** to **D**) Wallerian retrograde chromatolytic changes in ventral horn cells following injury to the peripheral nerve. (**B**) Chromatolytic neuron with eccentric nucleus and some dissolution of the Nissl substance; (**C**) chromatolytic neurons, showing a peripheral ring of Nissl substance (peripheral chromatolysis); (**D**) chromatolytic neuron, showing eccentric nucleus and only a peripheral ring of Nissl substance (Nissl stain, ×400) (From EM Marcus and J Jacobson, Integrated neuroscience, Kluwer, 2003)

2. Atrophic Change

In atrophic change, the nerve cell is too severely damaged to repair itself. Consequently, the cell body shrinks and becomes smaller. This response is similar to the response of a nerve cell to insufficient blood supply, which produces an ischemic neuron. and the Nissl substance begins to disperse, and after 7 days the nucleus becomes dark and the cytoplasm eosinophilic. Within a few days, these cells are phagocytized.

3. Wallerian Degeneration (Figs. 2.14 and 2.15)

When an axon is sectioned, the distal part that is separated from the trophic center (cell body) degenerates, a process called Wallerian, or anterograde, degeneration. At the same time, the cell body undergoes a process called axonal, or retrograde, degeneration. If the cell body remains intact, the proximal portion begins to regenerate. The distal stump is usually viable for a few days, but its degeneration begins

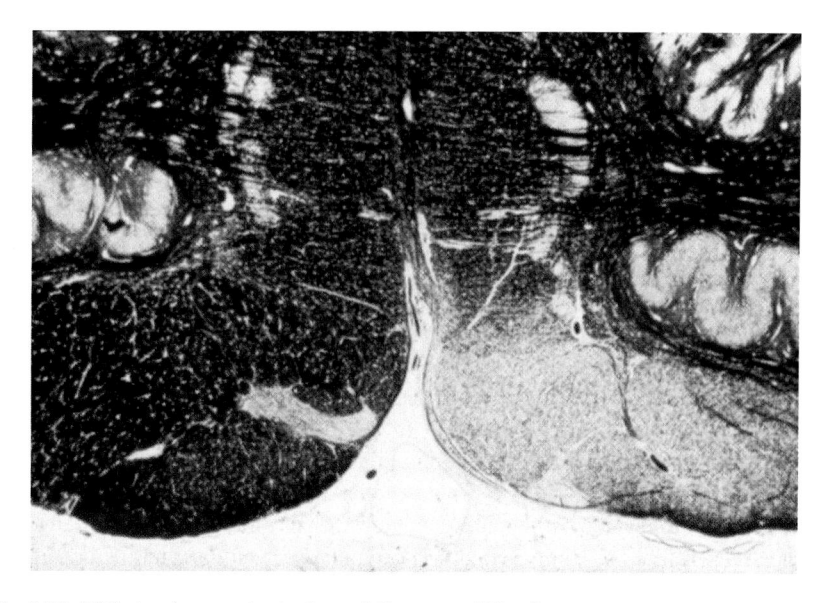

Fig. 2.15 Wallerian degeneration in the medullary pyramid in a human many months after an infarct in the contralateral motor–sensory strip. Left side is normal; note the absence of myelin on right side. (Weigert myelin stain; ×80). (From EM Marcus and J Jacobson, Integrated neuroscience, Kluwer, 2003)

within 13 h of injury. The axon starts to degenerate before the myelin sheath. In 4–7 days, the axon appears beaded and is beginning to be phagocytized by macrophages, which enter from the circulatory system. Fragments of degenerating axons and myelin are broken down in digestion chambers, and it might take several months before all of the fragments are ingested. In the proximal portion, degenerative changes are noted back to the first unaffected node. As the myelin degenerates, it is broken up into smaller pieces that can be ingested more easily.

B. Regeneration

1. Peripheral Nerve Regeneration

Within a few days after section, the proximal part (attached to a functional neuronal soma) of the nerve starts regrowing. Nerve growth factor is produced after injury to the axon, and it promotes the axonal sprouting. If the wound is clean (e.g., a stab wound), sewing the nerve ends together can dramatically increase the rate of recovery in the affected limb. The regenerating nerves might cross the scar within several weeks. The crossing is helped by the Schwann cells and fibroblasts, which proliferate from the proximal end of the nerve. The Schwann cells form a new basement membrane and provide tubes through which the regenerating axons can grow.

In certain peripheral nervous system diseases, only segmental degeneration occurs. One example is diphtheria: The myelin sheath degenerates but the axon remains intact. Phagocytes break down the myelin, and Schwann cells rapidly reform myelin. The rate of movement of the slow component of axoplasmic flow probably accounts for the rate of axonal regrowth, which is limited to about 1 mm/day. Slow components of the axoplasmic flow (–Scb) carry actin, fodrin, calmodulin, clathrina, and glycolytic enzymes that form the network of microtubules, intermediate filaments, and the axolemma, which limit the rate of daily axonal regeneration, although functional recovery might be a little faster (McQuarrie and Grafstein, 1988; Kandel, Wujek, and Lasek, 1983).

As the regenerating axon grows, the axonal end sprouts many little processes. If one axonal sprout penetrates the scar, the other sprouts degenerate, and the axon follows the path established by the penetrating sprout. If an axon reaches one of the tubes formed by the Schwann cells, it grows quickly, and after crossing the scar descends the distal stump at a rate of approximately 1 mm/day (Guth and Jacobson, 1966; Jacobson and Guth, 1965). When the motor end plate is reached, a delay occurs while the axon reinnervates the muscle and reestablishes function. At this stage, the average rate of functional regeneration is 1–3 mm/day. Only a small percentage of the nerves actually reach the effectors or receptors. The basal laminae helps direct the regenerating nerve to the motor end plate. If a sensory fiber innervates a motor end plate, it remains nonfunctional and probably degenerates, and the cell body atrophies. A sensory fiber that reaches a sensory receptor might become functional, even if the receptor is the wrong one. For example, after nerve regeneration, some patients complain that rubbing or pressing the skin produces pain. In these cases, it would appear that fibers sensitive to pain have reached a tactile or pressure-sensitive receptor.

Motor fiber might also reinnervate the wrong motor end plate, as when a flexor axon innervates an extensor. In such a case, the patient has to relearn how to use the muscle. Muscle that is denervated assists the regenerating axons by expressing molecules that influence the regenerating axons. Some of the molecules are concentrated in the synaptic basal lamina of the muscle. Other molecules are upregulated following denervation and help in attracting and reestablishing the synapse in the muscle. These upregulated molecules include growth factors (IGF-3 and FGF-5), acetylcholinesterase (AChE), agrin, laminin, s-laminin, fibronectin, collagen, and the adhesion molecules N-CAM and N-cadherin (Horner and Gage, 2000). Successful nerve regeneration also depends on an adequate blood supply; for example, in a large gunshot wound, nerves attempt to regenerate but might not succeed.

The following is a summary of the sequence of regeneration in the peripheral nervous system:

1. Peripheral nerve interrupted
2. The axon dies back to the first unaffected node of Ranvier, with the myelin and distal axon beginning to degenerate within 34 h
3. At the site of injury, the axon and myelin degenerates to form a scar. Phagocytosis begins within 48 h
4. Axons separated from the cell body degenerate. With an adequate blood supply, the portion of the axon still connected to the intact cell body begins regenerating by sprouting

5. Within 72 h, Schwann cells begin to proliferate and form basement membranes and hollow tubes. Nerve growth factor is also formed and released, which stimulates sprouting
6. From each of the severed axons, sprouts attempt to penetrate the scar. After one sprout successfully grows through the scar, the other sprouts die. Nerves take a month or more to grow through the scar
7. Once an axon penetrates the scar, it grows at 1 mm/day; about a third of the severed axons actually reinnervate muscle and skin.

2. Regeneration in the Central Nervous System

After an injury, axons in the central nervous system regenerate, but there seems to be no equivalent to the Schwann cell because oligodendrocytes and astrocytes do not form tubes to penetrate the scar. Instead, they form a scar that is nearly impenetrable. Even if the axons penetrate the scar, they have no means of reaching the neuron to which they were originally connected. Horner and Gage (2000) have reviewed the question of how to regenerate the damaged central nervous system in the brain that is inherently very plastic. They have noted that it is not the failure of neuronal regeneration, but it is rather a feature of the damaged environment; and it is now possible to reintroduce the factors present in the developing nervous system that produced this wonderful organ. The gene responsible for needed growth factors is probably missing or inactivated in adult tissue. Recently, a brain-derived neurotrophic factor has been identified, which might eventually help in finding a way to guide the axon (Goodman, 1994). Also, the identification of a potent inhibitor of neurite outgrowth associated with myelin, NoGo, has focused efforts to develop agents to counteract its effects (Chen, M.S., A.B. Huber, M.E. Vander Haar, M. Frank, L. Schnell, A.A. Spellman, F. Christarf, M.E. Schwab, 2000). Animal studies have shown that neurons have considerable plasticity; that is, if some axons in a region die off, bordering unaffected axons will sprout and form new synapses over many months, filling in where the synapses were and resulting in major functional reorganization. This reorganization might eventually produce some recovery of function.

3. Stem Cells: A Source of Replacement for Damaged Neurons?

In the adult brain, neuronal stem cells have been identified in the adult brain and spinal cord. These cells under the right conditions might well be activated and help to reverse the effects of lesions in the CNS (Kernakc and Rackic 1976). After the implantation of immature neurons (neuroblasts) in regions affected by certain diseases (e.g., the corpus striatum of patients with Parkinson's disease), there has been some recovery (Gage et al., 2002). In parkinsonian patients, the age of the individual receiving the transported cells seems to affect the outcome, with younger patients (less than 50 years of age) more likely to show some improvement.

4. Nerve Growth Factors

The first nerve growth factor was isolated by Levi-Montalcini and Angeletti in 1968, but only recently has biotechnology been able to produce these factors in large quantities. Attempts have been made to help regeneration in the central nervous system, for instance, by placing Teflon tubes on Schwann cells through the scarred portion of the spinal cord in the hope that the nerves would follow these channels. However, even though nerves do grow down these channels, no functional recovery occurs. Many factors that promote neuronal survival and axon outgrowth (e.g., brain derived neurotrophic factor; BDNF) have been identified and the focus is now on getting these cells to produce axons to grow into the injured areas and then to grow into the uninjured area. With the identification of neurotrophic factor (netrins 1 and 3) that produces axonal growth (Serifini et al., 1994) and with the studies of programmed cell death beginning to identify genes that might be responsible for premature neuron death (Oppenheim, 1991), we are now entering an era of brain research that offers great promise to help patients with neurodegenerative diseases such as Huntington's, Parkinson's, and Alzheimer's.

5. Glial Response to Injury

Neuronal death triggers an influx of phagocytic cells from the bloodstream, and microglia proliferate and break down the dying neurons.

In organs with numerous fibroblasts, necrotic areas are soon filled with proliferating fibroblasts, but in the central nervous system, there are few fibroblasts and the astrocytes do not proliferate in sufficient numbers. Within a few days of an ischemic attack with infarction, neutrophils are seen at the site of injury. Shortly thereafter, microglial cells and histocytes are seen in the region of the dying cells. Because the blood-brain barrier is usually compromised, monocytes can now migrate into the parenchyma of the central nervous system in greater numbers and assist in phagocytosis. The time it takes for the complete removal of injured cells depends on the size of the lesion. Large infarcts might take several years before phagocytosis is complete. If the lesion is huge, such as a large infarct in the precentral gyrus, a cavity lined by astrocytic scar will form. In small lesions, the neurons are phagocytized and glia proliferate, a process called replacement gliosis.

VI. Blood-Brain Barrier

A. *Blood-Brain Barrier*

Endothelial cells form this barrier in the central nervous system. The endothelial cells line the capillaries and the choroid plexus and are joined together by tight junctions (zonula occludens) and the capillaries are not perforated; the endothelial cells show very little pinocytosis or receptor-mediated endocytosis (Brightman, 1989). Studies have shown that the blood-brain barrier is impermeable to certain large molecules, including proteins, but substances such as small lipid-soluble compounds, including alcohol and anesthetics, gases, water, glucose, electrolytes (NA^+, K^+, and CL^-), and amino acids, can pass from the plasma into the intracellular space (inside neurons and glia) or into the extracellular space between neurons and glia.. In the peripheral nervous system, the endothelial cells are fenestrated and very active in *pinocytosis*. Fluid-phase endocytosis in the peripheral nervous system is relatively nonspecific; the endothelial cells engulf molecules and then internalize them by vesicular endocytosis. In *receptor-mediated endocytosis*, which is found in the central nervous system, a ligand first binds to a membrane receptor on one side of the cell. After binding to the ligand, the complex is internalized into a vesicle and transported across the cell; the ligand is usually released.

All vessels within the central nervous system are surrounded by a thin covering formed by astrocytic processes. However, the astrocytic processes do not fuse with the endothelial lining of the blood vessel or with the processes of other cells, so they have minimal effect on limiting the entry of solutes into the brain parenchyma. Thus, the extracellular space can be entered once the materials pass through the endothelium. The intravenous perfusion of various dye compounds (trypan blue, Evans blue, proflavin HCl, and horseradish peroxidase) demonstrate that the blood-brain barrier is leaky in certain midline regions of the third ventricle, in the *circumventricular* organs (pituitary, median eminence, organum vasculosum, subfornical organ, subcommissural organ, pineal gland, and the area postrema), and in the choroid plexus and locus ceruleus (Brightman, 1989) of the fourth ventricle. These open connections between the brain and the ventricular system permit neuropeptides from the hypothalamus, midbrain, and pituitary to enter the CSF and to be widely distributed in the brain and spinal cord, thus forming an alternate pathway in the neuroendocrine system.

Tumors within the central nervous system produce growth factors that cause blood vessels to sprout. These new blood vessels have immature tight junctions that are also quite leaky. Their have been many attempts with limited success to deliver chemotherapeutic agents specifically to the tumor. There have also been attempts to interfere with the formation of the blood vessel growth factors as a way to starve tumors. Stress has been shown to open the blood-brain barrier by activating the hypothalamic–hypophyseal–adrenal axis and releasing corticotropin releasing factor (CRH) (Esposito et al., 2001). Acute lesions of the central nervous system, including those caused by infections, usually increase the permeability of the barrier and alter the concentrations of water, electrolytes, and protein. In viral diseases, the infected leukocytes (macrophages) more easily penetrate into the brain by passing between the normally tight junctions in the endothelial cells, which is one way that HIV enters the brain directly from the blood.

B. Extracellular Space and the CSF

Between the cells in the central nervous system is the extracellular space, measuring between 30 and 40 nm and filled with cerebrospinal fluid (CSF) and other solutes. The CSF is formed primarily by the choriod plexus in the lateral ventricle, third ventricle, and fourth ventricle, and a small portion of the CSF appears to be formed by the diffusion of extracellular fluid. The amount of extracellular space in the brain is still a matter of controversy. Some solutes can readily pass from the blood plasma through the endothelial lining into the extracellular space, and the solutes present in this space (whether deleterious or not) affect the functions of the central nervous system. CSF might also be reabsorbed after temporary storage in the extracellular space. Fat-soluble compounds that readily pass through the blood-brain barrier can enter the extracellular space and might be useful in resolving infections in the central nervous system or in improving the function of certain brain cells.

Chapter 3
Spinal Cord

The spinal cord is tubular is shape, has 32 segments with the gray matter in the center and white matter on the outside. Each segment has nerve rootlets on the dorsal and ventral surface with ganglia attached to the dorsal nerve rootlets. These dorsal root ganglia contain the primary cell bodies of the general sensory systems: pain, temperature touch, and pressure.

The gray matter is shaped like a butterfly and the gray matter in all levels has three zones: a dorsal horn, a ventral horn, and an intermediate zone. A lateral horn is found in levels T1–L2.

The surrounding white matter is divided into three columns: dorsal, ventral, and lateral. The white matter is organized with the ascending systems on the outside and the descending systems closer to the gray matter.

I. Anatomy of the Spinal Cord

The spinal cord is cylindrical in shape and lies in the vertebral canal, with its cephalic end continuous with the medulla of the brain stem and the caudal end terminating on the inner border of the second lumbar vertebrae as the *conus medullaris*. From the conus, the filum terminale, which consists of connective tissue and pia, connects to the inner surface of the first coccyx vertebrae surrounded by the nerves exiting the lumbar and sacral levels of the spinal cord forming the cauda equinae. The spinal cord had two enlargements: the cervical (from vertebrae C3 to T2) and the lumbar (T9–T12). The cervical enlargement innervates the upper limb, whereas the lumbar enlargement innervates the lower limb.

The anterior median sulcus and the posterior median sulcus divide the cord into equal right and left halves (Fig. 3.1). The dorsal lateral sulcus is found at all levels and it marks the location of the dorsal/sensory rootlets. The actual gray and white matter of the spinal cord ends at L1.

Fig. 3.1 The posterior and anterior roots in relation to the gray and white matter of the spinal cord (From EM Marcus and S Jacobson, Integrated neuroscience, Kluwer, 2003)

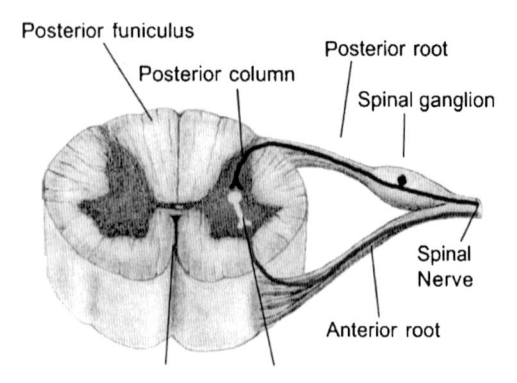

A. *Spinal Cord: Structure and Function*

1. Meninges/Coverings of the Spinal Cord

The spinal cord, just as the other regions in the central nervous system, is protected by three of layers of connective tissue coverings: the pia, archnoid, and dura (Fig. 3.2). The pia mater is adherent to the external surface of the spinal cord. The archnoid and pia form a membrane with cerebrospinal fluid occupying the subarachnoid space between the pia and archnoid. The arachnoid lies against the thick, tough dura. The spinal dura forms a protective tube beginning at the dura of the skull and tapering to a point in the region of the sacral vertebrae. Between the dura and the vertebrae lies the epidural space, usually filled with fat. The denticulate ligaments anchor the spinal cord to the dura. The spinal cord and, more importantly, the spinal roots are named in relation to the vertebral column. In humans there are 8 cervical, 12 thoracic, 5 lumbar, 5 sacral, and 2 coccygeal segments.

2. Nerve Roots

The anterior and posterior roots join to form a spinal nerve (Fig. 3.1) just before having the vertebral canal through a recess in the posterior process of each vertebra as shown in Figure 3.5. The canal, the intervertebral neural foramina, is only a little larger than the nerve, so that any swelling of the nerve or diminution of the diameter of the canal will pinch the nerve. It is not uncommon for the intervertebral disc to rupture posteriorly or laterally and press on a spinal nerve as it exits. The very intense pain this pressing causes is referred to the area of the skin where the nerve began. Frequent sites where the nerve roots are pinched are the intervertebral foramina between L2 and S1, which give rise to the sciatic nerve. The pinched nerve gives the patient the impression that a hot knife is being dragged along the posterior aspect of his calf. There are 32 segments of the spinal cord: 8 cervical, 12 thoracic, 5 lumbar, 5 sacral, and several coccygeal

I Anatomy of the Spinal Cord

Fig. 3.2 The relationships of the vertebral column, the meninges, and spinal cord. (From EM Marcus and S Jacobson, Integrated neuroscience, Kluwer, 2003)

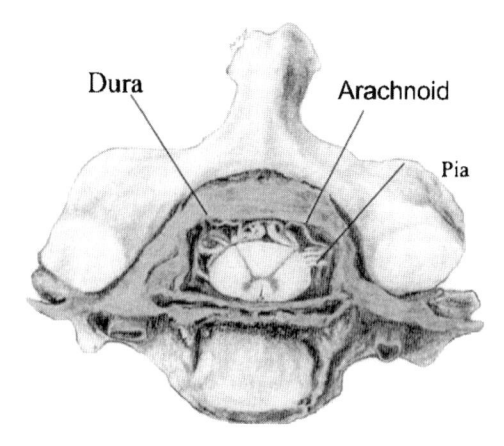

Fig. 3.3 The lateral aspect of the spinal cord exposed within the vertebral column. The spinous process and the laminae of the vertebrae have been removed and the dura mater has been opened longitudinally. (From C. Clemente (ed), Gray's anatomy, Lea & Febiger, 1988)

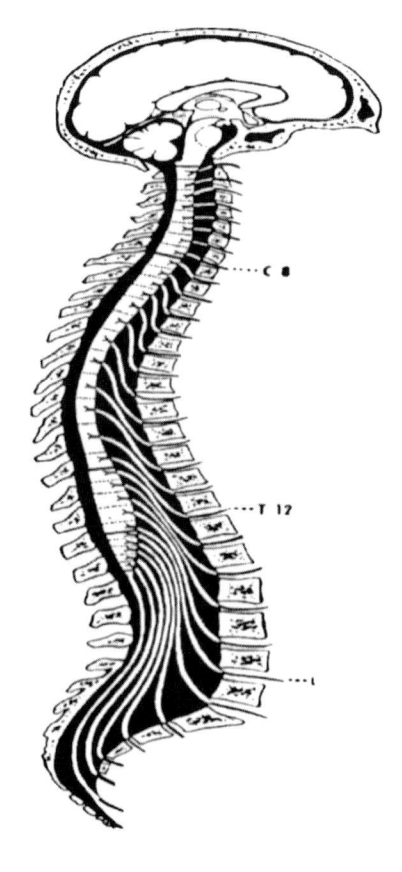

spinal cord segments. They are shown in relation to the vertebral column in Figure 3.2. There are seven cervical vertebrae; C8 in Figure 7.2 is the eighth cervical spinal nerve that arises from the anterior and posterior roots of the eighth cervical spinal cord segment.

The rootlet of the eighth cervical nerve emerges below the seventh cervical vertebrae and above the first thoracic vertebrae. The spinal nerves are always named for the spinal cord segment of origin. The intervertebral disc and foramen (Fig. 3.2) are normally named by the vertebrae above and below (i.e., L4, 5 disc; L5, Sl foramen). During development and growth, particularly intrauterine growth, the vertebral column grows faster than the spinal cord it contains so that, in the adult, the spinal cord does not extend the length of the vertebral column (Fig. 3.3). As a result, the spinal roots run interiorly before they leave the spinal canal. The spinal cord ends at the first or second lumbar vertebra, and below this level, the spinal canal contains only spinal roots. This mass of roots reminded early anatomists of a horse's tail, so it was named *cauda equina*. The nerve roots run through the subarachnoid space, which is filled with cerebrospinal fluid. When a needle is inserted through the space between the fourth and fifth lumbar vertebra (Fig. 3.5), the point penetrates the dura and pushes aside the spinal roots. If the needle were inserted above L2, there would be danger of damaging the spinal cord. Samples of cerebrospinal fluid can be taken through this needle; pressure can be measured or anesthetic agents or radio-opaque dyes injected (as in Fig. 3.5).

Dorsal Posterior Root: Sensory Nerves

Sensory nerve fibers enter the spinal cord gray matter (Fig. 3.1) and divide into a medial bundle of large, heavily myelinated fibers and a lateral bundle consisting of thinly myelinated and unmyelinated axons. The heavily myelinated medial bundles convey information from the large dorsal root ganglion cells subserving encapsulated somatic receptors (muscle spindles, Pacinian corpuscles, Meissner's corpuscles), carrying information on touch, position, and vibratory senses. Upon entering the spinal cord, the axons divide into ascending and descending processes, both giving off collaterals. Many of these collaterals end in the segments above or below the level of entry. Branches from the larger fibers enter the ipsilateral posterior funiculus and ascend to the medulla. The large myelinated axons also have numerous terminals in Clark's nucleus in lamina 7. Collaterals from posterior root fibers also enter laminae 7 and 9, ending on internuncial neurons and anterior horn cells for reflex activity. The smaller, lateral bundle of thin axons (A delta and C fibers) conveys information from free nerve endings, tactile receptors, and other unencapsulated receptors.

These fibers, conveying impulses of tissue damage, temperature, and light touch sensation, enter Lissauer's tract (fasciculus dorsal lateralis). Axons from Lissauer's tract enter laminae I, II, and III. Axons from this lamina ascend and descend in Lissauer's tract to reenter the same lamina.

Ventral Anterial Roots: Motor

These nerves leave the spinal cord and unite with a sensory root to form a peripheral nerve that will distribute into the skin, muscles and organs of the body.

I Anatomy of the Spinal Cord

3. White and Gray Matter

The gray matter is in the center of the spinal cord covered by the white matter that forms the connections to and from the spinal cord. .The relative amounts of white and gray matter vary with the spinal cord level (Fig. 3.4).

White matter decreases in bulk as the sections are further from the brain. Motor tracts from the brain descend into the white matter to enter the gray matter and synapse with motor neurons. Sensory fibers entering the cord, section by section, forming the ascending pathways within the white matter. A section of the cervical cord contains sensory fibers running up to the brain from the thoracic, lumbar, and sacral sections. It also contains motor fibers running down to innervate motor axons of the thoracic, lumbar, and sacral sections. The white matter of a section of the lumbar cord, however, only contains fibers to and from that level and lumbar segments inferior to it and the sacral segments. The white matter columns have the shape of pyramids with their bases at the foramen magnum and their tips at the last sacral section. The situation in the gray matter is quite different.

The gray matter is organized into a dorsal and ventral horn, and in C8–L2 there is an added intermediolateral horn that contains the preganglionic sympathetic neurons. Because the gray matter innervates a segment, the size of the gray matter (the number of cells it contains) is related to the complexity of the segment. The hand, for example, is innervated by cervical segments 6, 7, and 8 (C6, C7, and C8) and the first thoracic segment (T1). The hand has the highest

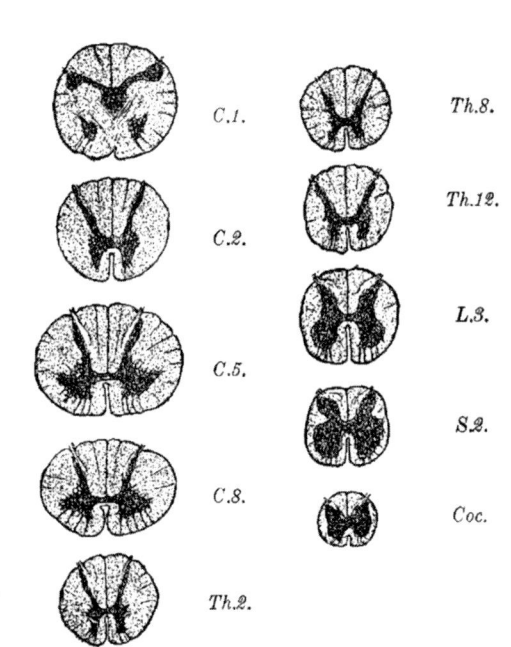

Fig. 3.4 Typical spinal cord cross sections (From Gross, C.M. (ed.): Gray's anatomy. Lea and Febiger. From EM Marcus and S Jacobson, Integrated neuroscience, Kluwer, 2003)

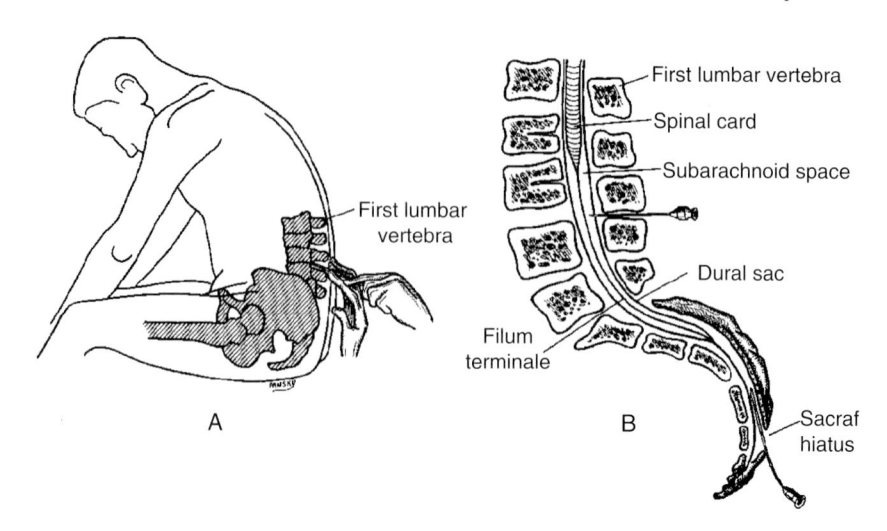

Fig. 3.5 The technique of lumbar puncture (Modified from House and Pansky: A functional approach to neuroanatomy, McGraw-Hill, 1960). (From EM Marcus and S Jacobson, Integrated neuroscience, Kluwer, 2003)

concentration of sensory receptors of any region of the body. All of these receptors send their axons into C6–T1, where they synapse, thus increasing the girth of the posterior gray matter. The muscles of the hand can carry out very fine and intricate movements. They are innervated by nerves having their cell bodies in the gray matter of C6–T1.

The innervation of the many muscles that constitute either the upper or lower, extremity requires an increase in the number of neurons in the dorsal and ventral horn. Consequently, the cord is enlarged at C6–T1 and T12–S1. The situation is much the same in the lumbar region. Sensory input from and motor output to the leg is complex and the gray matter is large. The thoracic and sacral segments, on the other hand, have very small gray matter areas because they innervate only a few muscles and receive relatively uncomplicated sensory messages. Segments can be recognized, then, by the amount of gray and white matter relative to the whole cross section. In the lower cervical segments, the section is large and oval, and white matter and gray matter are nearly equal. The thoracic segments have much more white matter than gray matter and the shape of the gray matter, a thin H, is very characteristic. The lumbar segments have more gray matter than white matter, and the sacral segments are very small and have much more gray matter than white matter. The segments can also be recognized by the shapes of the gray matter. A careful review of Figure 3.4 provides a basis for recognizing the segmental levels of spinal cord sections. The area of the body that sends sensory fibers into a given spinal cord segment is called a dermatome (Fig. 3.7). These have varying shapes and sizes. Figure 3.4 demonstrates the differences seen at these levels.

I Anatomy of the Spinal Cord

B. Laminar Organization of Central Gray

1. Laminar Organization

In order to clear up confusion about the terminology in the organization of the gray matter of the cord, Rexed (1952) proposed a laminar organization of the spinal cord in the cat that has since been extended to all mammals.

The gray matter is divided into 10 layers: 9 layers within the spinal gray and a thin region surrounding the central canal (Fig. 3.6). In this format, the dorsal horn includes laminae 1–6, the intermediate zone lamina 7, and the ventral horn is laminae 8 and 9.

Dorsal Horn: Laminae 1–6

Lamina 1 forms the cap of the dorsal horn and is penetrated by many fibers. It includes the nucleus posterior marginalis. Many intersegmental pathways arise from this layer, and Aα nociceptive fibers terminate in lamina 1.

Lamina 2 corresponds to the substantia gelatinosa. This nucleus extends the entire length of the cord and is most prominent in the cervical and lumbar levels. Cells in this lamina also form intersegmental connections. C pain fibers terminate in lamina 2, "The pain gate."

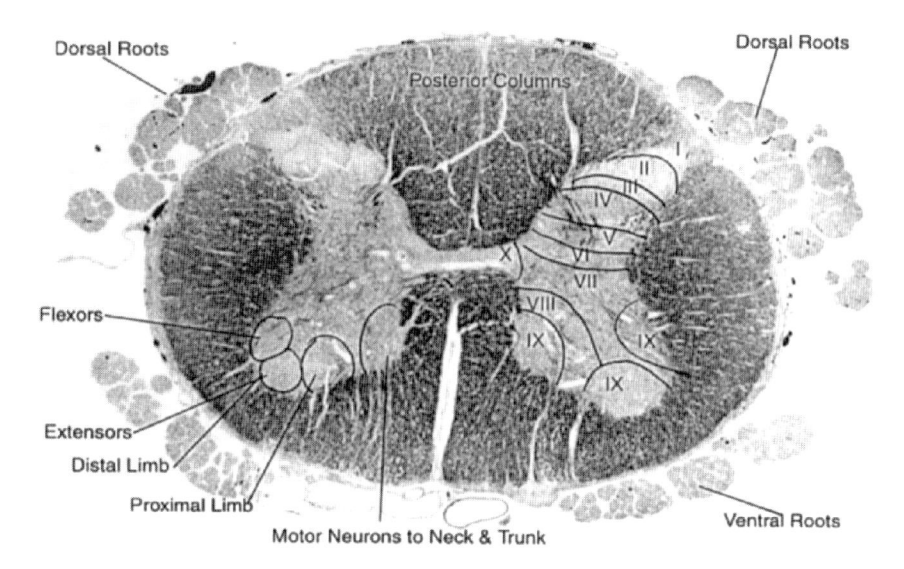

Fig. 3.6 Rexed's lamination pattern of the spinal gray 1954 matter on the right and the location of ventral horn cells on the left, lumbar section. myelin stain from,

Lamina 3 is the broad zone containing many myelinated axons and receives many synapses from the dorsal root fibers.

Lamina 4 is the largest zone and consists primarily of the nucleus proprius of the dorsal horn. This nucleus is conspicuous in all levels. Aα fibers from mechanoreceptors terminate in this layer.

Lamina 5 extends across the neck of the dorsal horn and in all but the thoracic region is divided into medial and lateral portions. The lateral portion consists of the reticular nucleus that is most conspicuous in the cervical levels Corticospinal and posterior root synapses have been identified in this lamina. The C pain fibers and the Aβ nociceptive fibers also terminate in this layer.

Lamina 6 is a wide zone most prominent in the cervical and lumbar enlargements. In these levels, it is divided into medial and lateral zones. Terminals from the posterior roots end in the medial region and descending fiber tracts project to the lateral zone.

Intermediate Region: Lamina 7

Lamina 7 includes most of the intermediate region of the gray matter in the spinal cord. In this lamina are found the intermediolateral and intermediomedial nuclei. The nucleus dorsalis or nucleus spinal cerebellar of Clark is obvious in C8 through L2 levels. The axons arising from this nucleus form the posterior spinocerebellar tract. In the cervical and lumbar enlargements, this lamina includes many of the internuncials and the cell bodies of gamma efferent neurons.

Axons from posterior roots, cerebral cortex, and other systems end in lamina 7. Cells in lamina 7 form tracts that project to higher levels, including the cerebellum and thalamus. In the thoracic and sacral regions, axons also leave this lamina to form preganglionic autonomic connections.

Ventral Horn: Lamina 8 and 9

Lamina 8 in the cervical and lumbar enlargement is confined to the medial part of the anterior horn. Many of the axons from these nerve cells form commissural fibers in the anterior white column. Axons from descending pathways originating in the brain stem terminate here.

Lamina 9 includes the largest cell bodies in the spinal cord, the ventral horn cells. The axons of these cells (alpha motor neurons) form much of the ventral rootlets that supply the extrafusal muscle fibers. The medial group innervates the muscles of the axial skeleton and the lateral group innervates the muscles of the appendicular skeleton. (Fig. 3.7).

Lamina 10 includes the commissural axons including and the crossing second fibers of the pain pathway (spinothalmic).

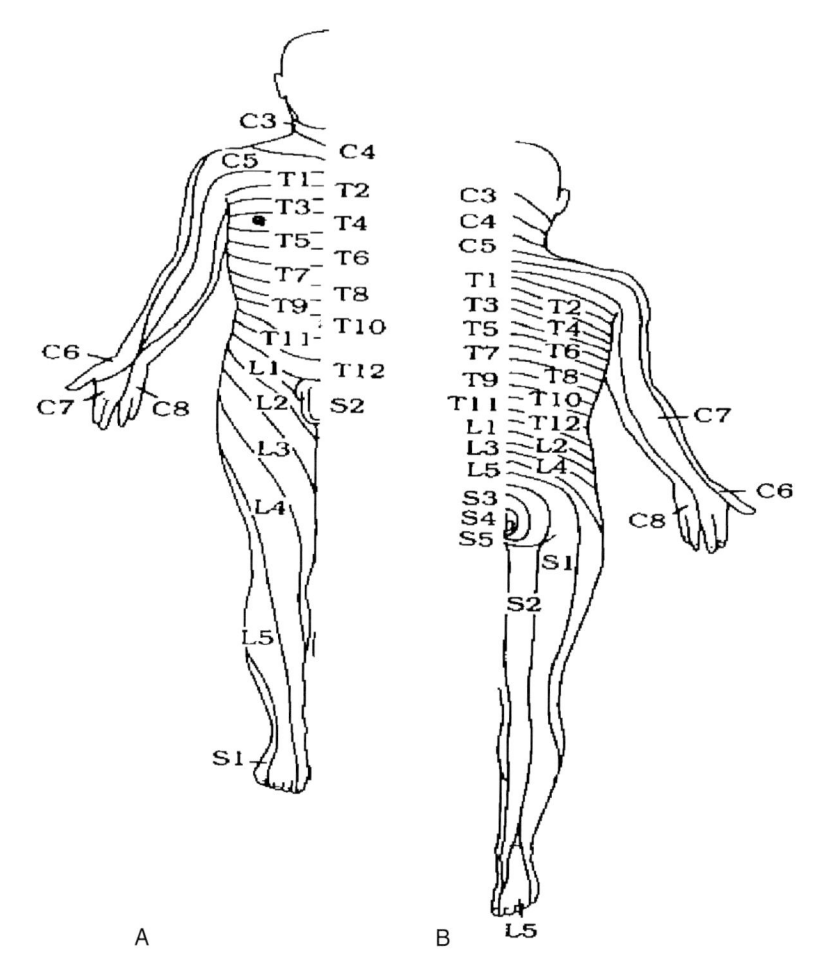

Fig. 3.7 Key dermatome boundaries in man: **A** Anatomical position. **B**-anterior surface, and **C** posterior surface (From J Zimmerman and S Jacobson, Gross anatomy, Little Brown, 1990)

2. Interneurons

Only a very small portions of the cells of the gray matter of the spinal cord are ventral motor horn cells or dorsal horn sensory cells; the vast majority are interneurons—small cells with short dendrites and axons. These cells interconnect incoming sensory axons and descending spinal cord tracts with each other and with anterior horn cells. An example of one such interneuron is the Renshaw cell, which is inhibitory; it depresses or totally inhibits firing. This inhibition is usually to opposing muscle groups. The Renshaw cell itself can be inhibited by a variety of pathways, such as squeezing the ipsilateral toes and stimulating the muscle nerve on the contralateral side. Note that, in each case, the response to the antidromic stimulation is much reduced. Because the Renshaw cell is inhibitory to anterior horn cells, inhibiting the Renshaw cell will remove inhibition from the anterior horn cell. The effect of inhibition of an inhibitory cell is called disinhibition.

C. Segmental Function

1. Motor/Ventral Horn Cells

The final effector cell of the spinal cord, the anterior horn cell, is probably the best place to begin a discussion of the segmental function of spinal cord segments. The extensive dendritic tree allows upward of 20,000–50,000 individual synaptic areas or knobs and the large cell body might have another 1000–2000 synaptic knobs. Incoming axons might have more than one synaptic connection and, hence, exert greater control. Each anterior horn cell gives rise to one large (8–12 µm) axon, called an *alpha motor neuron*, which innervates a motor group or unit made up of 10–200 individual extrafusal muscle fibers. There is only one motor end plate on each muscle for each alpha motor neuron. Therefore, for a motor group to contract, the anterior horn cell on the proximal end of the axon must fire an action potential. The muscle group is chained to its anterior horn cell. This anatomical relationship is often called the final common pathway. Activation of a specific anterior horn cell precedes activity of a motor group. There is also a smaller anterior horn neuron called the gamma motor neuron that innervates the intrafusal muscles fibers adjusting the sensitivity of the muscle spindles.

2. Organization of Neurons in Ventral Horn

There are many ways to activate this anterior horn cell, many thousand axons synapse upon it and its extensive dendritic tree. Neurons in the anterior horn can be divided into medial and lateral groups (Fig. 3.8).

The *medial nuclear division* is divided into posterior medial and anterior medial groups. The posterior medial nucleus is most prominent in the cervical and lumbar enlargement. The medial nucleus innervates the muscles of the axial skeleton.

The *lateral nuclear* division innervates the appendicular musculature. In the thoracic region, the intercostal and associated muscles are innervated by this region. In the cervical and lumbar enlargement, these nuclei become especially prominent and are divided into individual columns of nuclei (Fig. 3.4) and these nuclei columns include anterior, anterior lateral, accessory lateral, posterior lateral, and retroposterior lateral. These nuclei are represented functionally so that from medial to lateral, one passes from midline spine, to trunk, upper and lower limb girdle, upper leg and arm and lower leg and arm, to hand and foot. The most lateral nuclear groups innervate the muscles in the hand and foot.

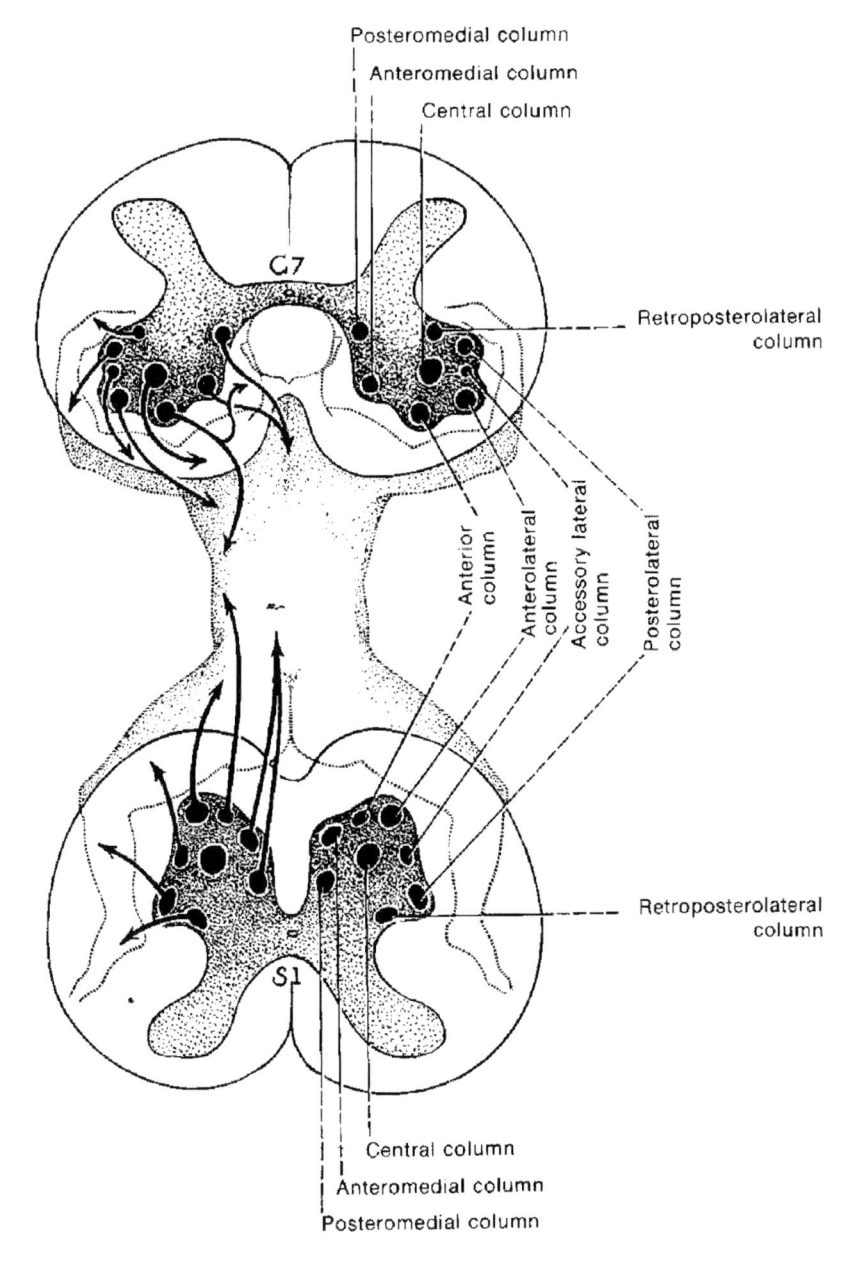

Fig. 3.8 Functional localization within the anterior horns (From Bossy, Atlas of neuroanatomy, WB Saunders 1974) (From EM Marcus and S Jacobson, Integrated neuroscience, Kluwer, 2003)

The *preganglionic autonomic nuclei* lie in the intermediolateral column, a prominent lateral triangle in the gray matter (Fig 3.4; T1–L2.).There are two distinct groupings.

1. Sympathetic. The intermediolateral nucleus from C8 to L2 is the origin of the preganglionic sympathetic neurons, which synapse again either in the sympathetic trunks or in remote ganglia.
2. Parasympathetic. The sacral nucleus in S2–S4 is the origin of preganglionic parasympathetic fibers that run to several ganglions in the walls of the pelvic organs.

3. Organization of Sensory Receptors

All of the sensory axons forming the posterior roots are from bipolar cells in the posterior root ganglion. The sensory information reaching the spinal cord subserves pain and temperature and tactile discrimination. Axons can be divided into many classes by the sensory modality they carry. As far as we know, each axon carries information about only one modality. The intensity of the pain or other sensation is coded as action potential frequency. For the present, two modalities will suffice (muscle stretch and pain). Pain is carried by very small axons, many of them unmyelinated, which have high thresholds and slow conduction velocities. The stretching of muscles, on the other hand, leads to a barrage of action potentials in large, heavily myelinated nerve fibers. These axons have low thresholds and rapid conduction velocities. Table 3.1 reviews the types and functions of the sensory receptors.

A. Reflex Response to Stretch

The classic stretch reflex is the knee jerk, produced by tapping the patellar tendon. This simple involuntary response is such a familiar part of the physical examination and has been used so frequently in medical humor that precise description seems

Table 3.1 Sensory Receptors

Type of receptor	Axon type	Function of receptor, records
Ia Primary spindle endings	12–20 μm myelinated	Muscle length and change of length
Ib Golgi tendon organ	12–20 μm myelinated	Muscle tension
II Secondary spindle endings	6–12 μm myelinated	Muscle length
III Free nerve endings 0.5-2 μm unmyelinated	2–6 μm myelinated	Pain, temperature, and chemical
Pacinian corpuscle Merkel's discs, Meissner's corpuscles, organ of Krause, corpuscle of Ruffini, plexus around hair follicles.	12t20 μm myelinated	Tactile discrimination

I Anatomy of the Spinal Cord

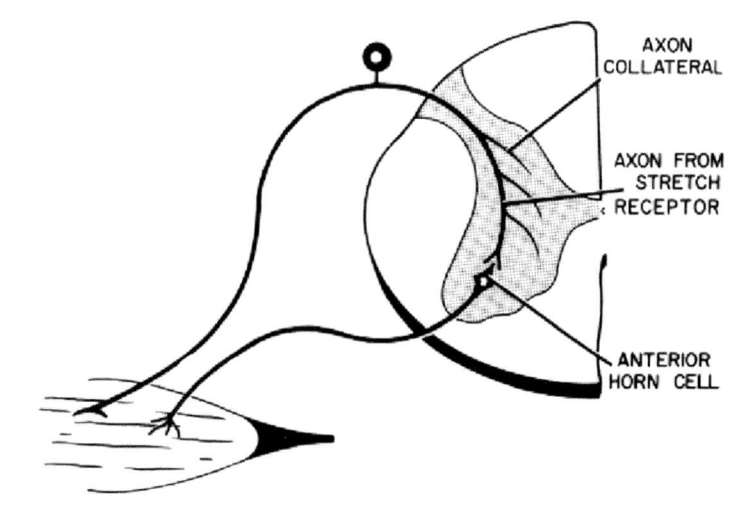

Fig. 3.9 The pathway for the monosynaptic stretch reflex (From EM Marcus and S Jacobson, Integrated neuroscience, Kluwer, 2003)

unnecessary. This simple test can give the astute examiner many clues to the function and dysfunction of the levels of the central nervous system (Fig. 3.7).

If the nerve to the muscle is cut, the response is abolished. When the spinal cord is severed from the rest of the nervous system, the response remains. The pathway of this response is very simple. The axon from the stretch receptor runs into the posterior horn, and while it branches many times in the gray matter, it eventually synapses directly on the cell body of the anterior horn cells of the muscle stretched. The stretch reflex is often referred to as a monosynaptic reflex. The only muscle that contract is the one stretched. The reflex continues for as long as the muscle is stretched (Fig. 3.9). There are stretch reflexes in all muscles, but they are much stronger in the antigravity muscles, particularly the leg extensors. In addition to the familiar knee jerk from the quadriceps group, brisk stretch reflexes can be elicited from the ankle (the Achilles tendon), the jaw, and the biceps and triceps muscles in the arm. Although the reflex loop is completed within a few segments of spinal cord, the magnitude of the response can be drastically altered by input from other levels. For example, grasping the hands together and pulling will greatly enhance the knee-jerk response (reinforcement).

B. Stretch Receptors: The Muscle Spindle

The major dynamic stretch receptor is the muscle spindle. It is an encapsulated sensory receptors bundle of modified muscle fibers that lie parallel to the rest of the muscle fibers and it signals changes in the length of the muscle. Its structure is shown in Figure 3.10; it is composed basically of three to five small muscle fibers,

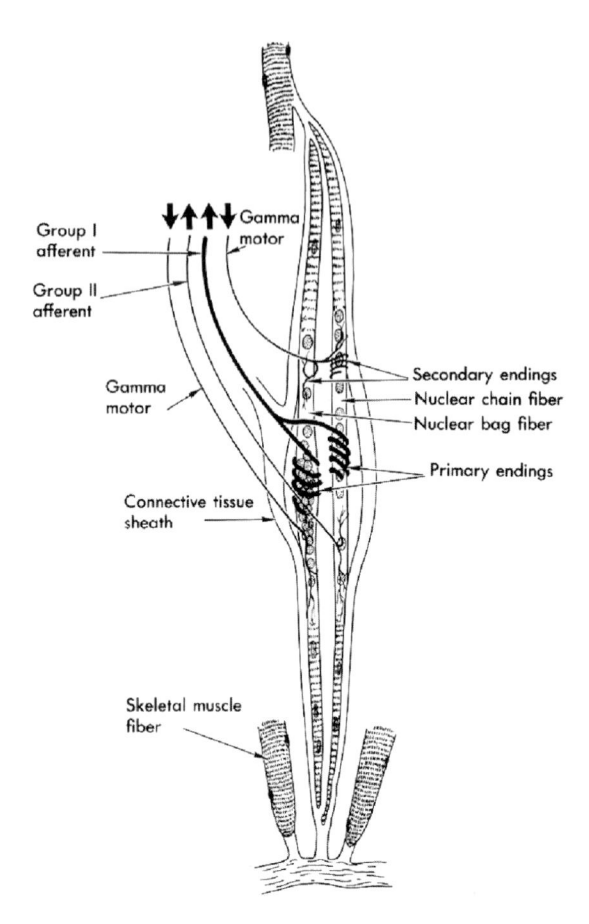

Fig. 3.10 A muscle spindle (From H Gardner, Fundamentals of neurology. Saunders, 1968 (From EM Marcus and S Jacobson, Integrated neuroscience, Kluwer, 2003) In contrast to this type of movement, which requires constant attention if the desired position is to be achieved, many of our movements, such as walking, require little attention beyond the decision to walk along the sidewalk. Most of us can even chew gum at the same time! These movements are probably carried out through mediation of the *gamma system*, which allows the brain to set a desired position and then forget about it. Compare then a corticospinal system, which produces a force, and a second system, the gamma system, which produces a new position. The gamma anterior horn cells are innervated by descending motor tracts other than the corticospinal. This system is controlled by neurons in the brain stem: the reticular formation, the vestibular nucleus, and the red nucleus

each containing a specialized, nonstriate region in its center. The striated ends of the muscle fibers can contract and are innervated by small, gamma motor neurons. These gamma motor neurons have cell bodies in the anterior horn, just as do the large alpha fibers that innervate the bulk of the muscle.

The small muscle fibers in the spindle are called *intrafusal fibers*, and the large muscle fibers that make up the bulk of the muscle are referred to as *extrafusal fibers*. The extrafusal fibers are primarily innervated by large (8–12 μμ) axons.

I Anatomy of the Spinal Cord

The pattern is not exclusive and there is some dual innervation: alpha motor fibers to both intrafusal and extrafusal fibers.

There are two sensory nerves that take origin from the unstriated center region of the muscle spindle. The largest (l2μμ), classified IA, comes from the center of the sensory region. The unmyelinated ends of the nerve wrap around each of the muscle fibers and are called primary or annulospiral endings. From these endings arise the action potentials that stimulate the stretch reflexes. This axon gives off many types of collateral within the gray matter of the cord that then travel up and down the cord for several segments.

The second ending gives off a smaller nerve, classified IIa, from specialized, secondary or flower-spray endings on either side of the annulospiral ending. These axons rise to the ipsilateral cerebellum, carrying information on "unconscious position sense." They apparently play no part in the stretch reflex. The basic function of the stretch receptor is to fire when the muscle is stretched. The response of the annulospiral ending is of short latency. The frequency of firing is at first high; it then slows down, but never adapts completely.

4. The Gamma System

The function of the intrafusal fibers is complicated. The first function is to keep the muscle spindle tight as the extrafusal fibers contract. At the rest length, the spindle is tight. Any further stretch will set up a volley of action potentials in the I sensory nerve that will lead to a stretch reflex: contraction of the extrafusal fibers. When the extrafusal fibers contract because of firing in the alpha motor neuron, the spindle goes slack and is no longer responsive to stretch. Indeed, the muscle would have to be pulled out slightly further than the rest position for the spindle to react. To rectify this situation, the gamma motor neuron fires, thus contracting the intrafusal fiber, and the spindle is tight again. Through the mechanism of the gamma motor neurons the sensitivity of the stretch reflex is maintained throughout the entire range of the limb movements.

Let us digress a moment and consider the simplest type of muscle movement:– opposing the thumb to the palm. Cells in the anterior horn are activated by fibers from the large descending corticospinal tract, a direct pathway, of which more will be said later. The intensity of stimulation and the number of motor groups activated are determined by the motor cortex. The motor cortex basically sets the tension that is to be developed. The distance moved is a function of the resistance to movement and the duration of the stimulus. The extent of movement is visually controlled; when the thumb reaches the desired position, the motor cortex stops stimulating the anterior horn cells.

5. Golgi Tendon Organs, Ib

Golgi tendon organs are a second type of stretch receptor found in the tendinous insertion of muscles. It contains no muscle fiber system. This receptor is in series with the muscle fibers and signals tension via Ib sensory fibers. It is much more sensitive to tension generated by the muscle than to stretch of the muscle. Activation of the sensory fibers from the tendon organ causes an inhibition of the contraction of the muscle. Among the functions of the tendon organ is to protect the insertion of the muscle from too great a stress that might tear the insertion from the bone. Tension information from the tendon organ is transmitted to higher centers via both dorsal and ventral spinocerebellar tracts.

II. Nociception and Pain

A. Nociceptive Stimulus

Any stimulus, such as heat, trauma, or pressure, that produces tissue damage or irritation is a *nociceptive* stimulus. Nociceptive stimuli are the afferent arm of many reflexes; locally, a red wheal develops around a cut, withdrawal of a limb from a hot pipe is a spinal cord reflex, whereas tachycardia from an electric shock to the finger is a brain stem autonomic reflex. Only when this nociceptive information reaches the thalamus and cerebral cortex can we talk about pain. Pain perception by these higher centers triggers affective responses and suffering behaviors. The pain experience varies enormously from person to person and the circumstances might alter the response. Contrast the pain of your thumb being hit with a hammer when you are doing a "chore" around the house to that of hitting your thumb while fixing your car or boat. As a gross oversimplification, we perceive pain in two ways: (1) as pricking, itching, or sharp and easily localizable, such as a razor blade cut or a mosquito bite; this type of pain is usually short-lasting, up to a day or so, and is usually tolerated. (2) In contrast, pain might be described as dull, aching, or burning and is poorly localized; this type of pain is longer-lasting (rheumatoid arthritis, dental pain) or repetitive (menstrual cramps) and is often poorly tolerated, frequently coloring the person's entire view of life.

B. Pain Receptors

Tissue damage releases a variety of typical intracellular substances, including K^+, H^+, bradykinin, as well as specialized compounds such as serotonin from blood platelets, substance P from nerve terminals, and histamine from mast cells. All of these substances directly stimulate the free endings of small myelinated A delta fibers (1–5 μm in diameter) and smaller, nonmyelinated C fibers (0.25–1.5 μm in diameter). A second group of released substances, including prostaglandin precursors and leukotrienes, act as sensitizers. Many of these small fibers have collaterals ending in regions containing neurotransmitter vesicles that apparently amplify nociceptive stimuli by releasing substance P when the free nerve ending is stimulated. In addition to these relatively nonspecific receptors, the upper end of the range of stimuli to temperature and pressure receptors generates nociceptive responses. For example, heating a patch of skin to 40 °C with a heat lamp gives a comfortable, warm sensation. Heating to 47 °C generates a painful, but tolerable experience; higher temperatures are perceived as unbearable pain. Cold is a diverse sensation and, a cold channel has been identified. Cold and warmth are sensed by thermoreceptor proteins on the free nerve endings of the somatosensory neurons; one channel, vanilloid receptor subtype 1 (VR1), is activated by temperatures above 43 °C and by vanilloid compounds including capsaicin, the component of hot chili peppers. The vanilloid receptor type 1 (VRL-1) is sensitive to temperatures above 50 °C. The cation channel that is methanol activated (CMR1: cold methanol type 1) is activated by temperatures of 8 °C–30 °C and is a member of the transient receptor potential family of ion channels, with VR1 and VRL-1 also members of this family. These nerve fibers, carrying nociceptive information, join peripheral nerves and segregate into spinal nerves and once inside the spinal dura, they join posterior roots. Their cell bodies are in the posterior root ganglia while the axon continues and enters the spinal cord in the lateral bundle.

C. Projection Fibers

Nociceptive information entering on one posterior root projects for two to three segments up and down the cord in laminae I–III. After extensive processing, large-diameter, myelinated axons with cell bodies in laminae I and V cross the neuroaxis near the central canal (Fig. 3.7) to join the anterior–lateral white matter. These fibers form the spinothalamic tract that projects to the brain stem, thalamus, and, ultimately, the cerebral cortex.

D. Modulation of Pain Transmission

Transmission of pain information through the chain of neurons in the posterior root is influenced by many factors, some of them originating within the segment and some from higher centers. Most of us, after cutting or bruising ourselves, rub the surrounding area and obtain relief of the pain for as long as we rub. An animal that has been hurt will lick the wound, presumably to obtain relief of pain. This subjective phenomenon is frequently spoken of as counterirritation; stimulation of the large, myelinated touch fibers reduces the magnitude of transmission of pain sensation through the posterior horn. This modulation takes place in the substantia gelatinosa, lamina II, within the "gate" proposed by Wall and Melzak (1989).

Several clinical observations point to the importance of large fiber inhibition of nociceptive transmission, even in the absence of apparent nociceptive stimulation. Perhaps the most painful of these conditions is avulsion (traumatic tearing out) of posterior roots. Instead of giving pain relief, surgical severing of posterior roots often results in intolerable pain. In both cases, large fiber input is lost. In contrast, destruction by a laser beam of the lateral dorsal root entry zone (for small fibers) is often highly successful in blocking the flow of nociceptive information up the neuro-axis. The large fiber input is largely spared because they enter in the medial zone.

Severed peripheral nerves often generate itching or burning sensations (causalgia), which can be alleviated by stimulating the large-diameter axons central to the cut. The large- diameter axons have low thresholds so they can be stimulated without stimulating the smaller, high-threshold nociceptive fibers. Activity in the large fibers suppresses background nociceptive inflow. TENS (Transepithelial nerve stimulation) is often effective in preventing or reducing peripherally generated nociception from becoming painful; the stimulation is effective in closing the gate. As mentioned earlier, branches of many incoming large axons enter the posterior column.

Stimulation of the posterior column sends antidromically conducted action potentials into the posterior horn and often reduces the pain experience. Damaged small axons are prone to developing $\alpha 2$ adrenergic receptors, so the axon becomes sensitive to norepinephrine liberated by the peripheral sympathetic nerves.

Pharmacological blocking or surgically severing sympathetic outflow often reduces pain. The major excitatory neurotransmitter in the pain system is substance P, which is found in abundance in the posterior horn, particularly laminae I and II. The mechanism of postsynaptic action is a reduction in potassium permeability leading to depolarization. The major inhibitory transmitters in the pain pathway are the enkephalin/endorphin series of peptides released from cells entirely within the posterior horn, mostly lamina II. Their action is closely mimicked by the opioid peptides.

Many studies suggest that these compounds also compete with the excitatory transmitters of the region (substance P and glycine) for postsynaptic sites. Other studies suggest reduction of Ca^{2+} influx into the presynaptic terminal in response to incoming depolarization and consequent reduction of the amount of excitatory transmitter released. Other studies implicate presynaptic inhibition. Injection of opioids into the lumbar cerebrospinal fluid often reduces nociceptive transmission and reduces the pain experience.

The effect is blocked by the universal narcotic blocker naloxone. Successful acupuncture increases endorphins in the lumbar cerebrospinal fluid. Descending axons in both the anterolateral and posterior white columns influence transmission across this chain of synapses. Supraspinal control of these encephalin/endorphin neurons is from the brain stem, particularly from the locus ceruleus.

The Norepinephrine and Raphe nuclei (5HT) are both in the medulla. These areas, in turn, are modulated by the periaqueductal gray of the midbrain that also contains an inhibitory endorphin/enkephalin system. Stimulation of these structures results in profound anesthesia (electroanesthesia). Although the descending pathways are not clear, we can suppress the flow of nociceptive information through the posterior horn by positive thinking; for example, all "pain" studies are bedeviled by a 40% placebo effect. Yes, the third postsurgical morphine injection can be replaced with saline and still "work" half the time.

E. White Matter Tracts

The white matter of the cord contains axons that run up and down the spinal cord, connecting segment to segment (intrasegmental) and the segments to the brain. The white matter is divided into three columns that are delimited by the presence of the dorsal roots that separate the posterior from the lateral funiculi and the ventral roots that separates the lateral from the anterior funiculi (Fig. 3.11). The posterior funiculus consists primarily of the posterior columns. The lateral funiculus is organized so that the ascending fibers are on the outside and the descending fibers are closer to the gray.

The only way that the location of individual tracts has been determined in humans is to observe degeneration caused by discrete lesions (Fig. 3.12). In this subsection we will provide an overview of the pathways in the spinal cord. Detailed discussions of the individual sensory and motor pathways are included in Chapter 12 on the somatosensory system. The discussion in this subsection will focus on topics relevant to only the spinal cord.

1. Descending Tracts in the Spinal Cord

Basic Principal of the Descending Motor Control. Commands for voluntary movement travel from the motor cortex (the upper motor neurons) and this pathway descends through the brain stem into the lower motor neurons in the ventral horn of the spinal cord.

Voluntary Control of Lower Motor Neurons in the Spinal Cord via the Corticospinal Tracts

This tract has its origin in layer V of the cerebral cortex, most prominently from the motor and premotor cortex of the frontal lobe (see Chapter 9). At the junction

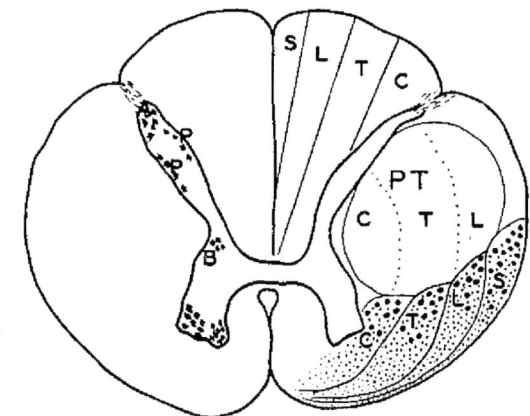

Fig. 3.11 The lamination pattern of the major tracts of the spinal cord (From EA Walker, Arch Neurol. Psychiat (Chicago), 43:284, 1940) (From EM Marcus and S Jacobson, Integrated neuroscience, Kluwer, 2003)

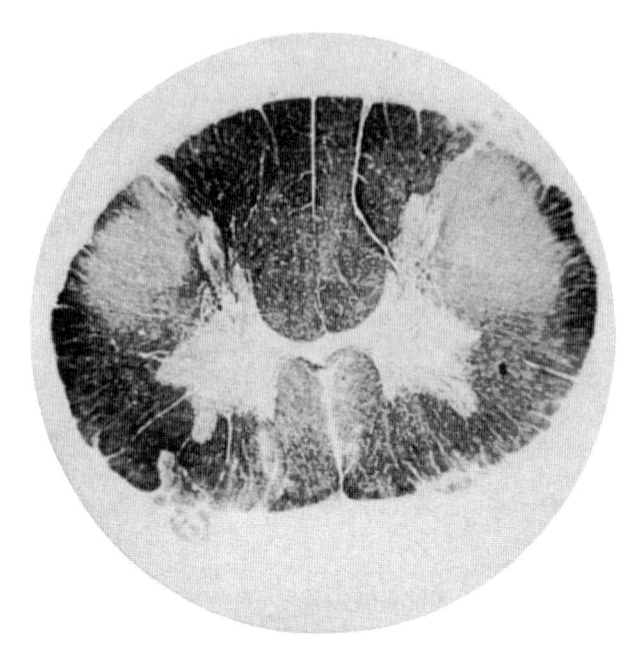

Fig. 3.12 The location of the corticospinal tracts as shown by degeneration caused by a lesion in the internal capsule (From IS Wechsler,: Clinical neurology, Saunders, 1963) (From EM Marcus and S Jacobson, Integrated neuroscience, Kluwer, 2003)

between the medulla and the spinal cord (the level of the foramen magnum), most of the fibers *cross* the neuroaxis, decussate, and move laterally and posteriorly to form the lateral corticospinal tract, which terminates on ventral form cells in the spinal cord (the lower motor neurons). The location of this pathway in the human is determined by analyzing cord sections obtained at autopsy from cases where cortical destruction has occurred in the motor areas of the precentral gyrus or in the internal capsule and axons degenerated following death of the cell body (Fig. 3.12).

2. Ascending Tracts in the Spinal Cord

Basic Principle. All ascending sensory systems have three orders of neurons:

1. *Primary sensory neurons* in the dorsal root ganglion attached to each segment of the spinal cord—origin of the sensory system.
2. *Second-order neurons* in the dorsal horn of spinal cord or gracile and cuneate nuclei of the medulla. The axons of the second-order sensory neuron cross the neuro-axis and form ascending pathways within the spinal cord and brain stem that terminate on third-order neurons in the thalamus of the diencephalon.
3. *Third-order neurons* in the thalamus, which project to the ipsilateral sensory area of the cerebral cortex.

F. Motor and Sensory Pathways

The motor pathways are discussed in Chapter 9. The pathways conveying tactile information are discussed in Chapter 12. The visual pathway is detailed in Chapter 13.

1. The Anterolateral Pathway and Pain (Fig. 3.15)

Fibers carrying information on pain and temperature sense from the body form contralateral spinothalamic tract. Axons carrying in light touch are also found in this region, and together they form the anteriolateral pathway. Compression, intrinsic disease, or deliberate section all result in anesthesia of the contralateral body beginning three segments below the level of disruption.

2. Pain and Temperature

Cutaneous receptors for pain and temperature (free nerve endings) send axons to small- and medium-sized dorsal root ganglion cells. These axons enter the spinal cord via the lateral aspects of the dorsal root entry zone. Most of these fibers enter Lissauer's tract and branch extensively (over two or three segments on either side of the segment of entry) before entering the posterior horn—laminae I, II, and III (Fig. 3.6).

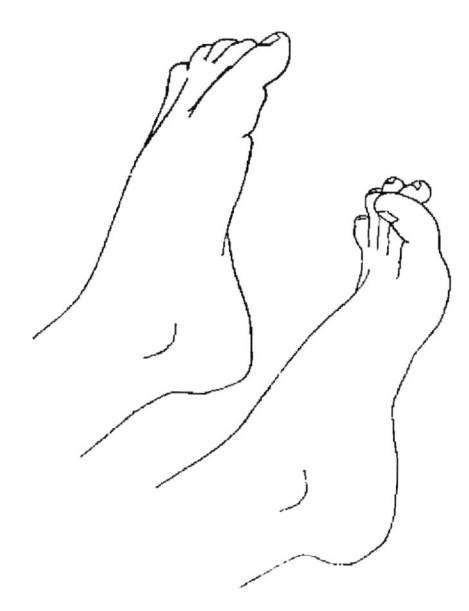

Fig. 3.13 The Babinski response. Upper: The normal adult response to stimulation of the lateral plantar surface of the foot; lower: the normal infant and upper motor neuron lesion effects in an adult (From EM Marcus and S Jacobson, Integrated neuroscience, Kluwer, 2003)

The secondary neurons arise primarily from cell bodies in laminae I and V that give origin to large axons which cross in the anterior white commissure and ascend in the contralateral anterior–lateral funiculus. As discussed earlier, many factors modulate information transfer across the posterior horn: between incoming nociceptive fibers and outgoing spinothalamic fibers. There is a somatotopic arrangement in the spinothalamic tract (Fig. 3.11), with the most lateral and external fibers representing sacral levels and the most intermediate and anterior fibers representing cervical levels. Pain fibers are located anteriorly to the more posteriorly placed temperature fibers. Perhaps as many as half the fibers in the tract, often called the spinoreticular tract, end in the brain stem and many of the rest send collaterals into the brain stem on their way to the thalamus (Chapter 15). There are endings in two major nuclei of the thalamus: the ventral posterior medial nucleus and the intralaminar nuclei. (See Chapter 5.) Recordings from individual spinothalamic neurons reveal four major functional categories, all contralateral:

1. Low-threshold units, activated only by gentle stimuli (e.g., by stroking hairs on the arm)
2. Wide dynamic range units, activated by many types of stimulus of both high and low intensity with response graded by intensity
3. High- threshold units, activated only by nociceptive stimuli
4. Thermosensitive units, with high action potential frequency signaling nociception.

All of these units have small receptive fields peripherally and larger fields toward the midline. The posterior horn seems to tease apart what stimulus from what part

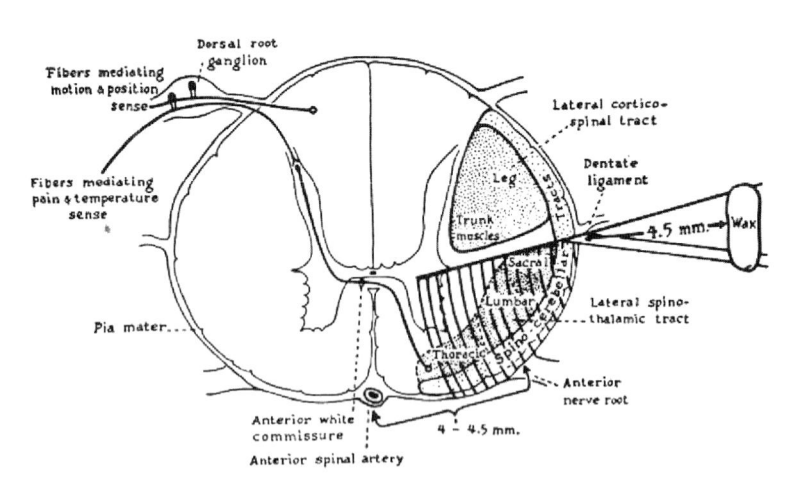

Fig. 3.14 Diagram illustrating a chordotomy. The cross section of the spinal cord shows the lamination of the spinothalamic tract, the position of the pyramidal tract in relation to it, and the presence of other tracts in the lower quadrant. A piece of bone wax is mounted 4.5 mm. from the tip of the knife as a depth gauge. Heavy curved lines in the ventral quadrant indicate the sweep of the knife. Note that a desire to spare the lateral corticospinal tract would result in sparing the sacral dermatomes. (From A Kahn and S Rand, J. Neurosurg 9:611–619, 1952) (From EM Marcus and S Jacobson, Integrated neuroscience, Kluwer, 2003)

of the body. Recording from the thalamic endings of these fibers also shows this separation of function as well as adding a wake-up function: ouch! A unilateral section of this tract (Fig. 3.14) for relief of pain produces a complete absence of pain and temperature from the opposite side of the body lasting r 6–9 months, but pain sensation slowly returns.

Nociceptive information probably rises in Lissauer's tract until it is above the cut and then crosses. There are several "pain" responses to nociceptive stimuli. A direct spinothalamic pathway to the contralateral ventral posterior medial nucleus of the thalamus with third-order projection to the postcentral gyrus probably mediates the tract running from the spinal cord to the midbrain. Stimulation of the postcentral gyrus, however, rarely generates the sensation of pain. Pain is rarely reported by patients during epileptic (cortical) seizures. Apparently, the thalamus tells us what (pain) and the postcentral gyrus tells us where. We test this pathway with the light prick of a pin and expect the patient can tell us, or point to, the location of the pin prick. The dull throbbing quality of the pain probably ascends by a multisynaptic pathway via brain stem synapses to the midbrain and then to intralaminar thalamic nuclei with a much wider cortical projection, including the limbic system.

III. Upper and Lower Motor Neuron Lesions

Causes of muscle weakness or paralysis can be grouped functionally into two categories. (1) If the difficulty is located in the corticospinal or other descending motor tract, the problem is called an *upper motor neuron lesion*. (2) If the problem is in the ventral horn cell, its axon, or its motor group, the problem is called *a lower motor neuron lesion*.

A. Upper Motor Neuron Lesion

Descending motor tracts normally exert an inhibitory effect on spinal cord reflexes. Hyperreflexia is a sure sign of decreased corticospinal function. A descending motor tract normally exerts an inhibitory effect on spinal cord reflexes to move for a short interval; then resistance to movement increases rapidly. Upon further pressure, the resistance suddenly gives way. This latter phenomenon is often referred to as a clasp-knife reflex. In the leg, spasticity is greatest in the extensors, whereas in the arm, the flexors are more affected.

The lateral corticospinal tract is the major tract for voluntary control of skeletal muscle. Destruction of this tract leads to upper motor neuron signs: spasticity of skeletal muscle and the loss of voluntary movement. This paralysis is usually total distally in the hand and somewhat less severe in the trunk musculature. *Hemiplegia*, is a paralysis of the arm and leg on the same side. *Monoplegia* is the paralysis of a single limb.

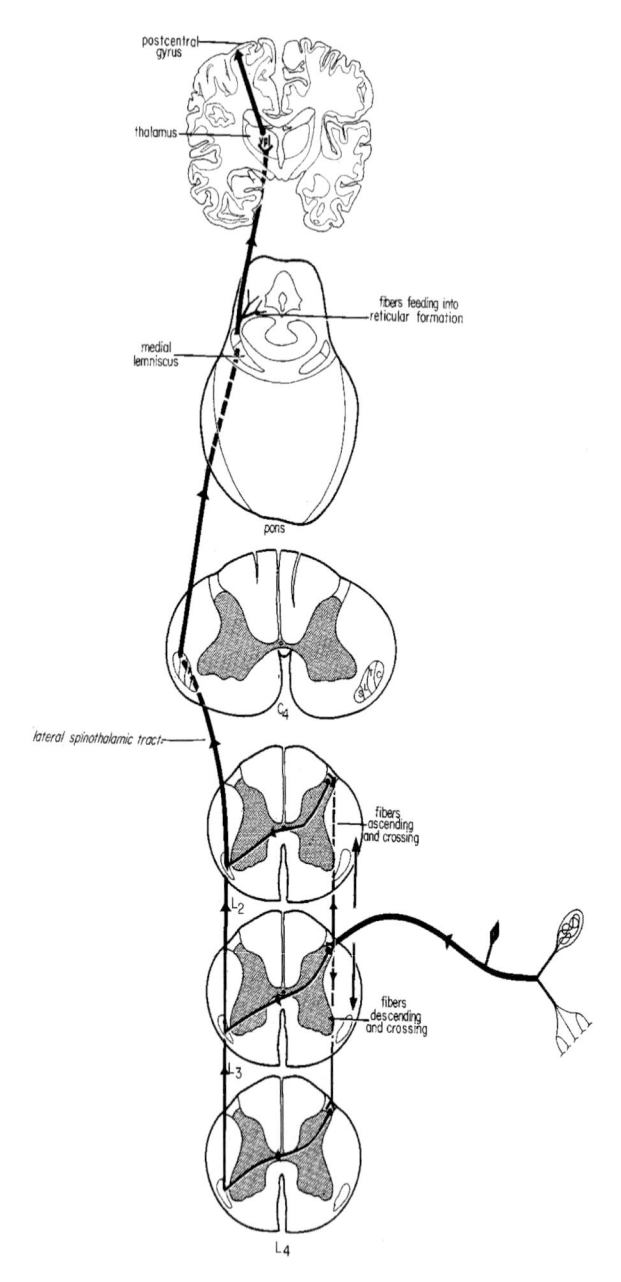

Fig. 3.15 Anterolateral pathway. The lateral and anterior spinothalamic tract Incoming fibers that originate in the dorsal root ganglion are activated by tissue-damaging stimuli. The axonal endings rise ipsilaterally from up to three spinal cord segments (as the fiber entering on the left; the fiber entering on the right synapses at the level of entry, one axon crosses the neuro-axis and rises and the other axon enters the anterior horn to participate in local reflexes, such as withdrawal from a hot surface after synapsing in the dorsal horn on the second-order neuron). The axons from the second-order sensory neuron cross and enter the contralateral spinothalamic tract.

Decreased function of the corticospinal tract also produces a curious reflex of the foot (Babinski's sign). If a moderately sharp object such as a key is drawn over the lateral boundary of the sole of the foot, the toes of an adult flex (curl). In very young children and in adults with corticospinal tract destruction, the toes extend and fan out as shown in Figure 3.13. Passively moving or rotating the ankle, knee, hip, shoulder, elbow, or wrist easily demonstrates the increased tone. When spasticity is present, the join is easy to move for a short interval then resistance to movement increase rapidly and upon further movement the resistance suddenly gives way- the clasp-knife reflex. In the leg spasticity is greatest in the extensor while in the arm the flexors are most affected. If a moderately sharp object such as a large key is drawn over the lateral surface of the sole of the foot causing the toes of an adult flex (curl). In young children an in adults with corticospinal tract destruction the toes extend and fan.

B. Lower Motor Neuron Lesion

The patient would present with weaknesses, loss of reflexes, extreme muscle wasting, and a flaccid tone (hyporeflexia) to the muscles, which characterize the lower motor neuron syndrome. The reason for the lower motor neuron lesion relates to the loss of the anterior horn cell or the motor axon. The loss of reflexes and the flaccid tone of the muscles relate to the loss of the motor side of the stretch reflex pathway. Often, it is possible to observe spontaneous twitching of the muscle (fasciculation). The following two case histories compare and contrast the effects of compression of a nerve rootlet to that of actual compression of the spinal cord. Case History 3-1 demonstrates the effect of a ruptured cervical disc, producing compression of a cervical nerve root. Figure 3.16 demonstrates the effects of Amyotrophic lateral sclerosis on the corticospinal tract.

Case History 3.1

A 45-year-old right-handed, married, white female employed as an educational coordinator was referred for neurological consultation with regard to pain in the neck radiating into the left arm with extension into the index and middle fingers of the left hand. The symptoms had been present for 2 weeks. The patient estimated

Fig. 3.15 (continued) The major projections of the spinothalamic tract are the following: midline medulla, periaqueductal gray of the midbrain, and thalamus. In the thalamus, these end in either the ventral posterior lateral (VPL) nucleus or the intralaminar or dorsomedial thalamic nucleus. The information from VPL projects onto the postcentral gyrus and as a result, one knows where in the body the pain is originating. The information that reaches the prefrontal cortex via the DM nucleus permits one to decide how to respond to a painful stimulus. (From EM Marcus and S Jacobson, Integrated neuroscience, Kluwer, 2003)

that approximately 30% of her pain was in the neck, 30% in the scapular area, and 40% in the arm. The pain would shoot into the arm if she coughed or sneezed or strained to move her bowels (Valsalva maneuver). The patient had also experienced tingling in the index and middle fingers of the left hand but was not aware of any weakness in the hand. Four months previously the patient had experienced pain in the neck extending to the left shoulder area, but that symptom had cleared. Cervical spine X-rays at the time demonstrated narrowing of the C 6–C7 interspace with encroachment on the neural foramen at that level. She had no leg or bladder symptoms. In the last several days prior to consultation the patient had developed twitching of biceps and triceps muscles.

Neurological examination: Mental status, cranial nerves, and motor system: All were intact.

Reflexes: The left triceps deep tendon stretch reflex was absent or reversed. All other deep tendon reflexes and plantar responses were intact.

Sensory system: Pain sensation decreased over the left index and middle fingers.

Neck: Rotations were limited to 45° and pain was present on hyperextend sion. There was tenderness over the spinous processes of C 7 and T1 and over the left supraclavicular area.

Clinical diagnosis: Cervical 7 radiculopathy secondary to lateral rupture of disc.

Subsequent course: The use of cervical traction, cervical collar, nonsteroidal anti-inflammatory agents, and various pains and anti-muscle-spasm agents and various measures in physical therapy failed to produce any relief. Epidural injection produced no relief. Finally, 4 weeks after her initial evaluation, the patient now indicated she was willing to consider surgical therapy. This had been strongly recommended since per pain was now predominantly in the arm and her examination now demonstrated significant weakness at the left triceps muscle in addition to a persistence of her earlier findings demonstrated a lateral disk rupture at the C 6 -7 interspace. The patient underwent a left sided laminectomy at that level with removal of ruptured disk material. When seen in follow up 3 months after surgery she no longer had pain in the neck and arm. Strength had returned to normal she continued to have a minor residual tingling and decrease in pain sensation over the index finger. There had been a minor return of the left triceps deep tendon stretch reflex.

Comment: Radiation of pain and numbness in his case suggested involvement on the C7 nerve root. The triceps muscle is innervated by the cervical nerve roots 6, 7, and 8.The C7 nerve root provides the major supply. The sensory examination confirmed the involvement of the C7 root. The laboratory data, particularly the MRI, confirmed the level of root involvement and indicated the specific pathology. At the time of surgery, 80% of pain was present in the arm, and the patient had significant benefit from the surgical procedure.

The following case history provides an example of epidural metastatic tumor compressing the spinal cord.

Case History 3.2

A 55-year-old white housewife first noted pain in the thoracic, right scapular area, 5 months prior to admission. Three months prior to admission, a progressive weakness in both lower extremities developed. One month prior to admission, the patient had been able to walk slowly into her local hospital. Over the next 2 weeks, she became completely bedridden, unable to move her legs or even to wiggle her toes. During the last 2 weeks prior to her neurological admission, the patient had noted a progressive pins-and-needles sensation involving both lower extremities. At the same time, she developed difficulty with the control of bowel movements and urination.

Past history: At age 40, 15 years prior to this admission, a left radical mastectomy had been performed for an infiltrating carcinoma of the breast with regional lymph node involvement. A hysterectomy had also been performed at that time.

Neurological examination: The following abnormal findings were present: Motor system: A marked relatively flaccid weakness was present in both lower extremities with retention of only a flicker of flexion at the left hip.

Reflexes: Patellar deep tendon stretch reflexes were increased bilaterally with a 3+ response. Achilles reflexes were 1+ bilaterally. Plantar responses were extensor bilaterally (bilateral sign of Babinski). Abdominal reflexes were absent bilaterally. Sensory system: Position sense was absent at toes, ankles, and knees and impaired at the hip bilaterally. Vibratory sensation was absent below the iliac crests bilaterally. Pain sensation was absent from the toes through the T6–T7 dermatome level bilaterally, with no evidence of sacral sparing.

Spine: Tenderness to percussion was present over the midthoracic vertebrae (T4 and T5 spinous processes)

Clinical diagnosies: Spinal cord compression at the T4–T5 vertebral level, most likely secondary to metastatic epidural tumor.

Subsequent course: X-rays of the chest demonstrated multiple nodular densities in both hilar regions of the lung, presumably metastatic in nature. In addition, the T4 and T5 vertebrae were involved by destructive (lytic) metastatic lesions. Lytic metastatic lesions were also present in the head and neck of the left femur. An emergency myelogram demonstrated a complete block to the flow of contrast agent at the T5 level. An extradural lesion was displacing the spinal cord to the left. The neurosurgeon, Dr. Peter Carney, performed an emergency thoracic T3–T4 laminectomy for the purpose of decompression. A gelatinous and vascular tumor was present in the epidural space displacing the spinal cord. Adenocarcinoma presumably metastatic from breast was removed from the vertebral processes, laminae, and epidural space. A slow moderate improvement occurred in the postoperative period. Examination 3 weeks following surgery indicated that movement in the lower extremities had returned to 30% of normal and pain sensation had returned to the lower extremities. Radiation therapy was subsequently begun.

Comment: This patient presents a typical pattern of the evolution of an epidural metastatic tumor compressing the spinal cord. Midthoracic–scapular pain had been present for 2 months when a slowly progressive weakness of both lower extremities

Table 3.2 Major pathways in the spinal cord

Pathway	Origin	Termination	Function
Corticospinal, lateral column, contralateral in cord	Area 4 3-1-2 Layer V	Laminae 7 and 9	Voluntary movements
Cuneocerebellar; ipsilateral Information for upper extremity and neck	Accessory cuneate nucleus medulla	Anterior lobe and pyramis and uvula of cerebellum	Unconscious tactile and proprioception
Descending autonomics	Hypothalamus and brain stem	Sympathetics: lamina 7 in C8–L2. Parasympathetics: sacral nucleus of S2–S4.	Modulation of movement
Dorsal columns Ipsilateral in cord	Dorsal root ganglion	Gracile and cuneate nuclei in medulla	Tactile discrimination
MLF anterior column	Cranial nerves III–XII	Cervical cord	Coordination of eye and head movements
Reticulospinal Lateral and anterior columns	Nuclei: reticularis gigantocelluaris, reticularis pontis caudalis, and oralis.	Laminae 7 and 8 of all levels	Influence gamma motor system. Facilitory to extensor motor neurons. Inhibitory to arm flexors
Rubrospinal Lateral column	Magnocellular red nucleus in midbrain	Laminae 7 and 9	Facilitates flexor, Inhibits extensor motor in arm flexors.
Tectospinal Anterior column	Contralateral deep layers superior colliculus	Deep layers of 7 and 9, cervical cord	Support corticospinal pathway
Spino-spinal	All segments and dorsal roots	All laminae	Intersegmental- and 2-neuron reflex arcs.
Post. spinocerebellar from Golgi tendon organs, stretch receptors, and muscle spindles.	Clarks nucleus from C2-L 2 (Largest myelinated afferent fibers)	Ipsilateral vermis of cerebellum	Unconscious tactile and proprioceptive information for movements
Ant. spinocerebellar; contralateral	Laminae 5, 6, 7 most levels	Contralateral vermis of cerebellum	Unconscious tactile
Spinothalamic/Anterolateral	Laminae I–III	Thalamus DM, VPL, VPM	Nociceptive
Vestibulospinal Uncrossed Anterior column	Lateral vestibular nucleus of CN VIII	All levels of cord laminae 7 and 9.	Coordination of eye and head movements.

evolved over the subsequent 2½ months. At that point, the patient had a marked flaccid weakness of both lower extremities, with only minimal preservation of the hip flexors. The flaccid nature of the paraparesis might well have reflected the continued effects of spinal shock related to the final progressive events. In many cases, the evolution of motor and sensory deficit is much more rapid. When finally transferred to a neurological service, a dense sensory deficit for pain and touch was present almost up to the actual level of compression. Such findings plus the flaccid paralysis carry a very poor prognosis for any recovery of function following surgery. At the present time, in most neurological centers, radiation therapy combined with high-dosage corticosteroids is considered to be the treatment of choice when the nature of the epidural compressive lesion is clear-cut. Also, an MRI study of the spine would be included, allowing studies not only of the specific area of compression but also of the spinal cord above and below the area of compression. Thus, other epidural lesions might be identified.

IV. Other Spinal Pathways

Table 3.2 lists all of the major sensory and motor tracts in the spinal cord.

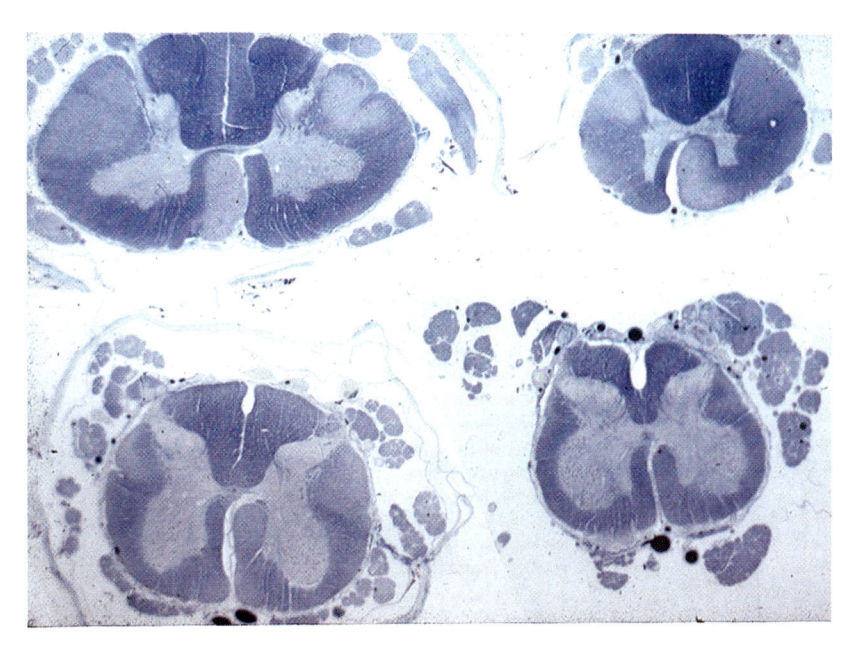

Fig. 3.16 Effects of ALS on the corticospsinal treat and nentral horn cells. Courtesy of Klüver- Berrera Stain

Chapter 4
Brain Stem

The brain stem is located in the posterior cranial fossa and consists of medulla, pons, and midbrain. The spinal cord shows the most straightforward organization for the student, with the gray matter in the center surrounded by the white matter and there is a segmental arrangement or organization as demonstrated by the nerve rootlets from the 32 segments of the cord. The brain stem lies between the spinal cord and the diencephalon and the gray and white matters are intermixed. There are three major regions in the brain stem the medulla, pons, and midbrain, but there is no segmental organization. Each of the regions has cranial nerve rootlets Throughout the brain stem there are centers related to cranial nerves that undertake functions as diverse as breathing, swallowing, chewing, eye movements, and setting the level of consciousness. There is a major tract that originates in the motor cortex (the corticobulbar tract) that controls the motor cranial nerves.

I. Gross Anatomical Divisions

The medulla, pons, and midbrain are subdivided by their relationship to the IV ventricle, with each level consisting of the tectum, tegmentum, and basis. The tectum forms the roof and walls or the posterior surface; the tegmentum forms the floor or anterior surface. The basis forms the anterior surface of the tegmentum.

The tectum in the medullary, pontine, and midbrain levels consists of regions that have highly specialized functions related to the special senses and movement: the cerebellum inferior and superior colliculus. The tegmentum (Fig. 4.2) contains cranial nerve nuclei, the reticular formation, and tracts that interconnect higher and lower centers as well as tracts that interconnect the brain stem with other portions of the central nervous system. The basis consists of the descending fibers from the cerebral cortex to the brain stem, cerebellum, and spinal cord.

The tegmentum of the brain stem (medulla) is continuous with the gray and white matters of the spinal cord. (Fig. 4.1). Based on clinical observations, we have divided the tegmentum into five zones: ventricular, lateral, medial, central, and

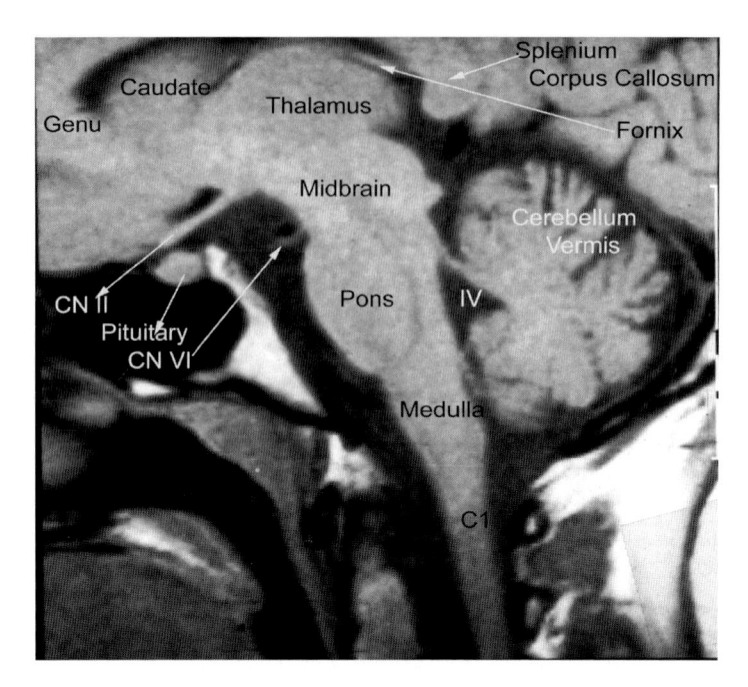

Fig. 4.1 Gross view of the brain stem in sagittal MRI-T1. (From EM Marcus and S Jacobson, Integrated neuroscience, Kluwer, 2003)

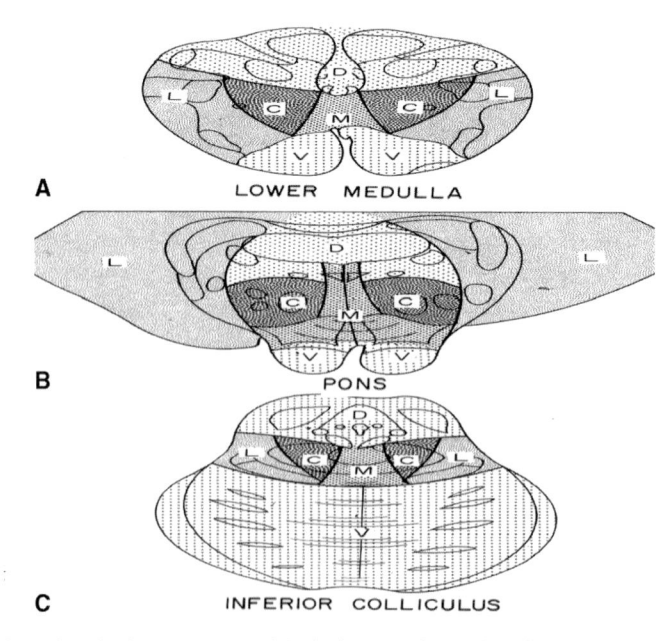

Fig. 4.2 A Regions in the tegmentum of the brain stem. Coronal sections: Zones: d = ventricular, l = lateral zone, c = central zone, m = medial, v = basilar zone (From EM Marcus and S Jacobson, Integrated neuroscience, Kluwer, 2003)

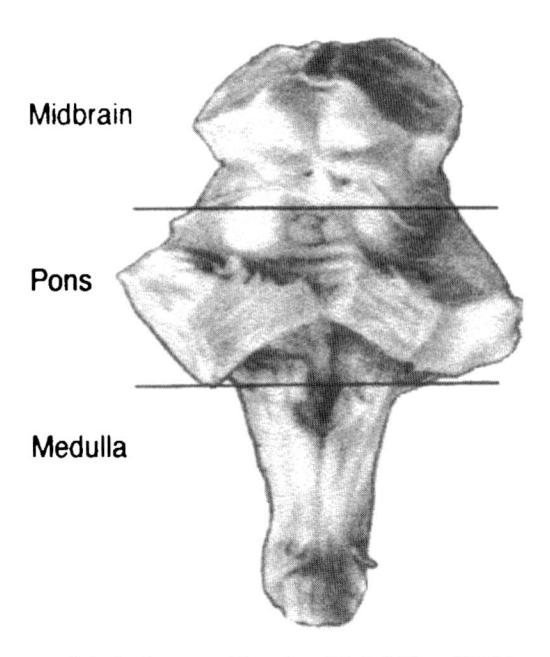

Fig. 4.2 B Tegmentum of the brain stem with regions labeled (From EM Marcus and S Jacobson, Integrated neuroscience, Kluwer, 2003)

basilar. Each half of the brain stem should be considered organized like an apple, with the core being the reticular formation and the pulp formed by the four surrounding zones (see Tables 4.1 and 4.2).

Although the discussion in this chapter of each level is confined to the coronal section under examination, it should be noted that most nuclei and all tracts extend through more than just one level. Therefore, when reading this chapter, the student should keep in mind the origin and destination of structures identified in each of these sections. It is especially important that the student understand the function and clinical importance of the anatomical structures they are studying. The cranial nerve nuclei and many pathways (e.g., corticospinal, corticonuclear, and spinothalamic) in the brain stem are clinically important because any functional abnormality in these systems help to identify where the disease process is occurring in the central nervous system.

II. Functional Localization in Coronal Sections of the Brain Stem

In this section, we will discuss the functional anatomy of the brain stem by examining coronal sections at representative levels through the medulla, pons, and midbrain. Each level will contain the following illustrations:

Table 4.1 Regions in the brain stem

Region	Contents
Tegmentum: anterior floor of the IV ventricle and cerebral aqueduct	Cranial nerve nuclei and all ascending and descending tracts except for tract in basis
Basis: anterior surface of the tegmentum	Corticospinal, corticonuclear, and corticopontine tracts
Tectum: posterior surface of the IV ventricle and aqueduct	Cerebellum attached to medulla and pons; superior and inferior colliculi attached to tegmentum of midbrain

Table 4.2 The five zones in the tegmentum

Zone	Gray matter	White matter
Dorsal/ ventricular	Medulla: hypoglossal all levels – & vagal trigones, autonomics solitary nucleus medulla – vestibular & cochlear solitary tract nerves & nuclei Pons: median eminence with facial colliculus Midbrain: CN II and III, periaqueductal gray	All levels descending autonomics, medulla solitary tract
Lateral	Medulla: inferior olive, Medulla and pons descending nucleus of V.	All levels- spinothalamic, rubrospinal. Medulla: spinocerebellars trigeminal afferents inferior cerebellar peduncle Pons: lateral lemnsicus. Midbrain: lateral lemniscus and medial lemnsicus
Medial	All levels reticular formation and Rraphe nuclei.	All levels tectospinal. medulla – medial lemniscus
Central (core)	Reticular formation, central and lateral descending reticular nuclei. Medulla: tracts, ambiguous nucleus of IX, X, and XI Pons: motor nucleus of V, parapontine nucleus of VI, facial nucleus of VII, Midbrain:- red nucleus and substantia nigra	Ascending and descending reticular tracts and ascending and descending autonomics.
Basilar	Medulla: Arcuate nucleus. Pons: pontine gray Midbrain: substantia nigra	Medulla: pyramidal tract, Pons: middle cerebellar peduncle. Midbrain: cerebral peduncle with its tracts.

In many of the following brain stem sections part A is a myelinated section with the nuclei and tracts labeled on an accompanying cartoon, while B is an MRI at a similar level (Please remember that the MRI is from a living patient, whereas the brain sections were obtained from a postmortemsection).

Table 4.3 Gray and white matter equivalents in spinal and brain stem structures

Spinal cord region	Equivalent brain stem region
Dorsal (sensory) horn	Ventricular zone, lateral region with sensory cranial nerves VII, IX, X; lateral zone cranial nerve V
Ventral (motor) horn	Ventricular zone, medial region with motor cranial nerves nuclei III–VI, X–XII
Intermediate zone	Reticular formation with motor nuclei of cranial nerves V, VII, IX, and X and associated functional centers.
Intermediolateral cell column.(Sympathetic) T1–T12, L1, and L2; Parasympathetic S2–S4	Parasympathetic Nuclei with Cranial Nerves III, VII, IX & X.
Lateral column of white matter corticospinal/crossed, rubrospinal, &– Spinothalamic	Basilar zone = corticospinal/not crossed
	Lateral zone =spinothalamic and rubrospinal
Posterior column of white matter = uncrossed gracile and cuneate fasciculi	Medial lemniscus: crossed tactile information
Anterior column of white matter = MLF, Vestibulospinal Reticulospinal Spinothalamic	Medial zone = MLF, Tectospinal, reticular formation reticulospinal, Lateral zone: vestibulospinal

III. Differences between the Spinal Cord and Brain Stem

The separation of the gray matter and white matter seen in the spinal cord gradually changes to a mixture of gray and white matter in the brain stem. Table 4.3 identifies the nuclei and tracts in the spinal cord and lists their functional equivalents in the brain stem. The student should be aware that the development of the cerebellum, midbrain, and cerebrum has greatly affected the neural contents of the brain stem and has lead to the formation of many nuclei and tracts (e.g., the following nuclei: inferior olive of the medulla, pontine gray of the pons, red nucleus and substantia nigra of the midbrain, and the corticonuclear and corticopontine tracts).

A. Medulla

1. Blood Supply: Vertebral Arteries and Its Branches

The first level we will discuss is at the sensory and motor decussation (Fig. 4.3). This section has mostly spinal cord elements: the ventral horns and the three white matter columns; however, one can also note brain stem elements: the gracile and cuneate nuclei This region is a transitional zone and is the size and shape of the spinal cord. As we move up the brain stem, we come to the widened medulla with the opening of the fourth ventricle. In the brain stem tegmentum, the gray and white matter appear intermingled; however, there is a basic organization of the tracts in the brain stem zones, as shown in Table 4.4.

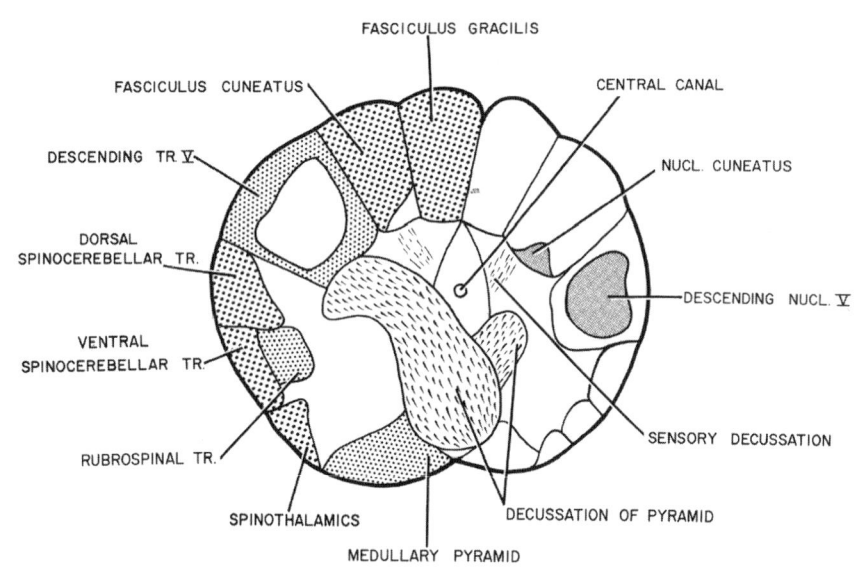

Fig. 4.3 Brain Stem Level 1 at transition level between cervical spinal cord and low medulla, with motor and sensory decussation. Coronal. (From EM Marcus and S Jacobson, Integrated neuroscience, Kluwer, 2003)

Table 4.4 Major tracts in the tegmental zone

Ventricular zone: Solitary tract and autonomic tracts
Medial zone: MLF, tectospinal, medial lemniscus
Lateral zone: Spinothalamics, rubrospinal, cerebellar peduncles, and lateral lemniscus. Medial lemniscus enters this zone in the pons and midbrain
Central core: Descending autonomics and (reticular formation) central tegmental tract
Basilar zone: Corticospinal and corticonuclear. Hallmarks of this level are the medullary pyramids and the inferior olivary prominence

III Differences Between the Spinal Card and Brain Stem 91

2. Gross Landmarks in the Medulla

The tegmentum of the medulla has two distinct levels: a narrow lower portion that is similar to the spinal cord and an broader upper portion that includes the widened IV ventricle.

Brain Stem Level 1: Spinomedullary Junction at Motor Decussation (Fig. 4.3)

Gross Features. This level resembles the spinal cord, as the dorsal columns are conspicuous; the trigeminal funiculus is more laterally placed.

On the anterior surface of the spinal cord, the medullary pyramids (corticospinal tracts) are evident (Fig. 4.3A). Note that this descending pathway on the right side has crossed first (decussated) with the fibers, shifting from the anterior surface of the medulla, and then entered the lateral funiculus of the spinal cord. The narrow spinal canal is present in the gray matter above the pyramidal decussation.

Motor Cranial Nerve Nuclei (Fig. 4.3A). Only the cranial portion of nerve XI is present at the lateral margin of the reticular formation. It innervates the sternocleidomastoid and trapezius muscles that rotate the head and elevate the shoulders, respectively.

Sensory Cranial Nerve Nuclei (Fig. 4.3A). Primary axons convey pain and temperature from the head and neck into the medulla and are located in the spinal tract of the fifth cranial nerve; these fibers have descended from the pons. The second order axons originate from the underlying descending nucleus and their axons cross and then ascend contralaterally adjacent to the medial lemniscus and terminate in the ventral posterior medial nucleus of the thalamus. The nucleus lateral to the pyramidal tract is the inferior reticular nucleus, a portion of the reticular formation.

White Matter (Fig. 4.3A). At this level the tracts are still in the same positions as in the spinal cord.

Posterior columns. Tactile discrimination (fine touch, pressure, vibration sensation, and two-point discrimination) and proprioception from the extremities, thorax, abdomen, pelvis, and neck are carried via the dorsal columns. The nuclei can now be seen that are the second order/2° neurons in this pathway: the gracile and cuneate nuclei.The gracile and cuneate tracts (fasciculi) have reached their maximum bulk with the addition of the last of the fibers from the uppermost cervical levels. The somatotopic arrangement of the tactile fibers in the posterior column is as follows: The most medial fibers are sacral and then the lumbar, thoracic, and, most laterally, cervical fibers.

Spinothalamics/anterolateral column. Pain and temperature from the extremities, abdomen, thorax, pelvis, and neck are carried by the lateral spinothalamic tract located at the lateral surface of the medulla. Light touch from the extremities, thorax, abdomen, pelvis, and neck is carried via the anterior spinothalamic tract, which is seen on the surface of the medulla just posterior to the corticospinal tract. In the spinothalamic pathways, sacral fibers are on the outside and cervical fibers are on the inside. The dorsal and ventral spinocerebellar tracts are found in the lateral funiculus. The rubrospinal and tectospinal tracts, which are important in supporting voluntary motor movements, are found in the lateral funiculus and near the midline in the medullary and pontine levels).

Brain Stem Level 2: Lower/Narrow Medulla at Sensory Decussation (Fig. 4.4)

Gross Features. At this level, the fourth ventricle is narrow. The funiculus gracilis and the funiculus cuneatus are conspicuous on the posterior surface of the spinal canal and the medullary pyramids are prominent on the anterior surface (Figs. 4.4A and 4.4B).

Ventricular Zone: Motor Cranial Nerve Nuclei (Fig. 4.4A). This section contains the inferior extent of the hypoglossal nerve (XII) and the dorsal motor nucleus of the vagus nerve (X). The hypoglossal nucleus innervates the intrinsic and extrinsic musculature of the tongue, whereas the dorsal motor nucleus of the vagus nerve provides parasympathetic preganglionic innervation of the viscera. This level marks the superior extent of the cranial portion of the 11th cranial nerve in the ambiguus nucleus.

Lateral Zone: Sensory Cranial Nerve Nuclei (Fig, 4.4A). Pain and temperature from the head are conveyed by the descending nucleus and tract of nerve V that is prominent anterior to the cuneate nucleus. Note that myelinated ipsilateral primary axons are on the outside of the second-order neurons while the second-order neurons are leaving the inner surface of the nucleus, crossing and entering the medial lemniscus, and forming the trigeminothalamic (quintothalamic) tract.

White Matter (Fig. 4.4A). Pain and temperature from the extremities, thorax, abdomen, pelvis, and neck are carried via the lateral spinothalamic tract. The spinothalamics are located in the lateral funiculus throughout the spinal cord and brain stem! The anterior spinothalamic tract is adjacent to the lateral spinothalamic tract. The tract carrying unconscious proprioception from the upper extremity originates from the external cuneate nucleus and its fibers enter the cerebellum through the posterior spinocerebellar tract. The function of this nucleus is similar to that of Clark's column in the spinal cord. The vestibulospinal tract is found internal to the spinothalamic tracts. The dorsal and ventral spinocerebellar tracts are seen on the surface of the medulla covering the spinal tract of nerve V and the spinothalamic tracts. The rubrospinal tract is an important afferent relay to the alpha and gamma neurons in the spinal cord. It originates from the red nucleus in the tegmentum of the midbrain, crosses the midline, and descends. It is found in the lateral funiculus of the medulla internal to the spinocerebellar tract and anterior to spinal nerve V throughout the pons and medulla. It is important in postural reflexes. The tectospinal tract is phylogenetically an old tract, being the equivalent of the corticospinal

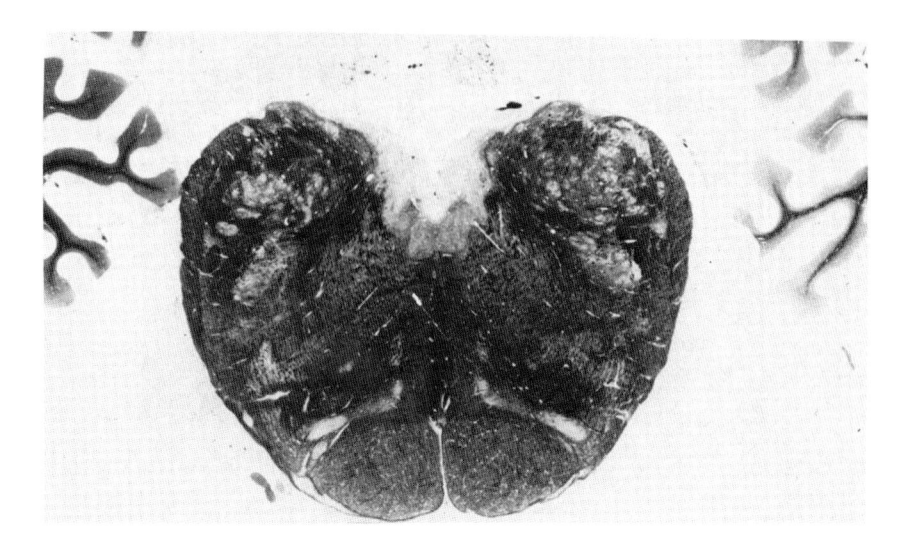

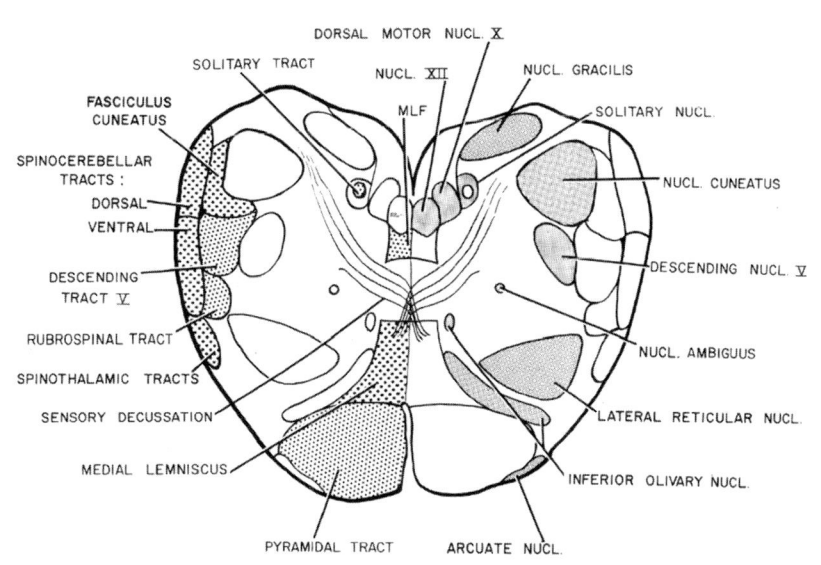

III Differences Between the Spinal Card and Brain Stem

Fig. 4.4 A Brain Stem Level 2 at lower medullary level. Coronal (From EM Marcus and S Jacobson, Integrated neuroscience, Kluwer, 2003)

tract in nonmammalian vertebrates. It originates from the deep layers of the superior colliculus and, to some extent, from the inferior colliculus. It is seen anterior to the medial longitudinal fasciculus throughout the medulla, pons, and midbrain and is important in coordinating eye movements and body position.

Medial Zone. The medial longitudinal fasciculus is located in the midline of the tegmentum just below the hypoglossal nucleus and above the tectospinal and

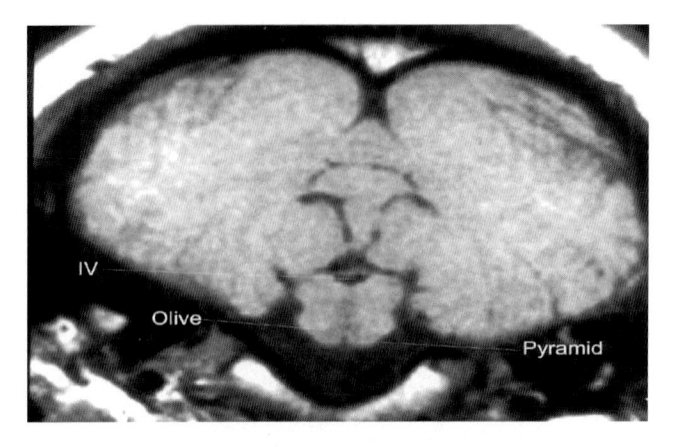

Fig. 4.4 B MRI, T_2 Brain stem at lower medullary level MRI-T1. (From EM Marcus and S Jacobson, Integrated neuroscience, Kluwer, 2003)

medial lemniscal pathways. The medial lemniscus is the largest pathway in the medial zone. However, as we progress superiorly in the brain stem, these fibers migrate laterally to finally enter the lateral zone in the midbrain. This migration is necessary for the medial lemniscus to be correctly positioned as it enters the thalamus. Tactile discrimination and proprioception are conveyed from the extremities, thorax, abdomen, pelvis, and neck by the axons in the fasciculus gracilis and cuneatus. The nucleus gracilis and the nucleus cuneatus form the second-order neurons in this pathway and are conspicuous in this section. Internal arcuate fibers are seen leaving the inner surface of the gracile and cuneate nuclei, curving around the ventricular gray, crossing the midline (sensory decussation), and accumulating behind the pyramid and beginning the formation of one of the major ascending highways—the medial lemniscus. The serotonin-containing raphe nucleus of the reticular formation is found in this zone throughout the brain stem. In all levels of the brain stem, it will be in this subventricular position.

Central Zone. The central core of the medulla, pons, and midbrain consists of the reticular formation, which is important for many vital reflex activities and for the level of attentiveness.

Basilar Zone. This level contains the corticospinal fibers just before they cross in the medullospinal junction. Note that the pyramidal system is named for the passage of the corticospinal fibers through this pyramidal-shaped region respectively (Figs. 4.4A and 4.4B). One of the important concepts from this section is that axons that form the volitional motor pathway, the pyramidal system, originate in the cerebral cortex. These tracts are located in the basis of the brain stem and include the corticospinal and corticonuclear fibers that run in the medullary pyramid. Motor system fibers that do not run in the medullary pyramid are considered extrapyramidal (e.g., cerebellar peduncles, rubrospinal tectospinal, etc.).

Brain Stem Level 3: Wide Medulla at Level of Inferior Olive *(Fig. 4.5)*

Gross Features. At this level the medullary tegmentum expands laterally as the fourth ventricle enlarges. The prominent hallmark of this level is the inferior olive located above the pyramids.

Ventricular Zone. The floor of the fourth ventricle contains the median eminence and the vestibular and cochlear tubercles. The median eminence consists of aqueductal gray and cranial nerve nuclei, whereas the vestibular tubercle is formed by the medial vestibular nucleus and the descending root and nucleus of cranial nerve VIII. The dorsal cochlear nucleus forms the cochlear tubercle. The ventricle at this level is at its widest extent. (Figs. 4.5A and 4.5B). The sulcus limitans in the ventricular floor separates the medially placed motor cranial nuclei from the laterally placed sensory cranial nuclei.

Motor Cranial Nerve Nuclei (Fig. 4.5A). This level marks the superior extent of nerve XII and the dorsal motor nucleus of nerve X. The ambiguous nucleus in the lateral margin of the reticular formation cotains cell bodies innervating the pharynx and larynx (nerve X).

Lateral Zone: Sensory Cranial Nerve Nuclei

Trigeminal. Pain and temperature are conveyed from the face by nerve V. The descending nucleus and tract of cranial nerve V are nearly obliterated by the olivocerebellar fibers.

Solitary Tract and Nucleus. Taste and visceral sensations are found in the solitary nucleus and tract in the tegmental gray below the medial vestibular nucleus. Fibers from cranial nerves VII, IX, and X ascend and descend in this tract, carrying general sensations from the viscera (nerve X) and gustatory sensations from the taste buds in the tongue (nerves VII and IX) and epiglottis (nerve X).

Vestibular. The medial and descending vestibular nuclei are present with first-order axons (Fig. 4.5B) found in the descending root of cranial nerve VIII, interstitial to the descending nucleus. Fibers can be seen running from the vestibular nuclei into the medial longitudinal fasciculus.

Auditory. The dorsal and ventral cochlear nuclei are second-order nuclei in the auditory pathway and receive many terminals from the cochlear portion of nerve VIII seen within its borders.

Cerebellum. The conspicuous inferior olivary nucleus in the anterior portion of the medullary tegmentum consists of the large main nucleus and the medial and dorsal accessory nuclei. The entire complex is important in supplying information to the cerebellum. Removal of the olive in animals produces a contralateral increase in tone and rigidity in the extremities, with concomitant uncoordinated movements. The olive connects to the contralateral cerebellar

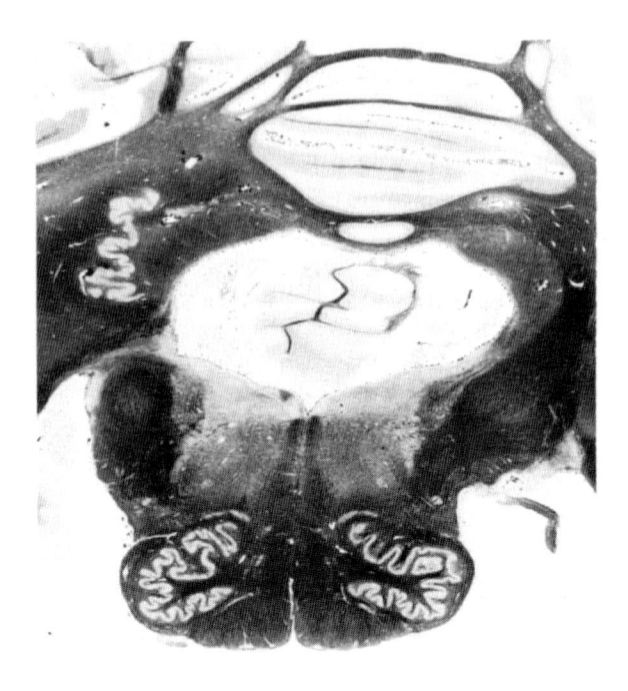

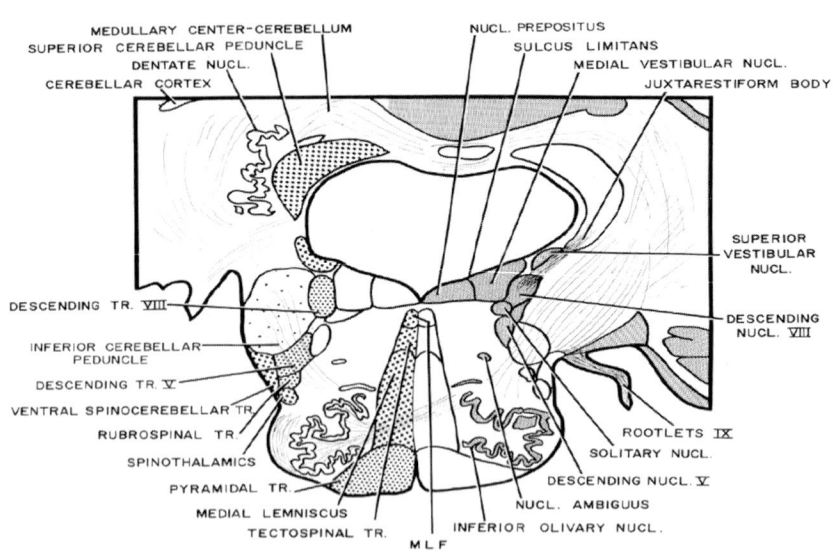

Fig. 4.5 A Brain stem level 3 at midmedullary level. (From EM Marcus and S Jacobson, Integrated neuroscience, Kluwer, 2003)

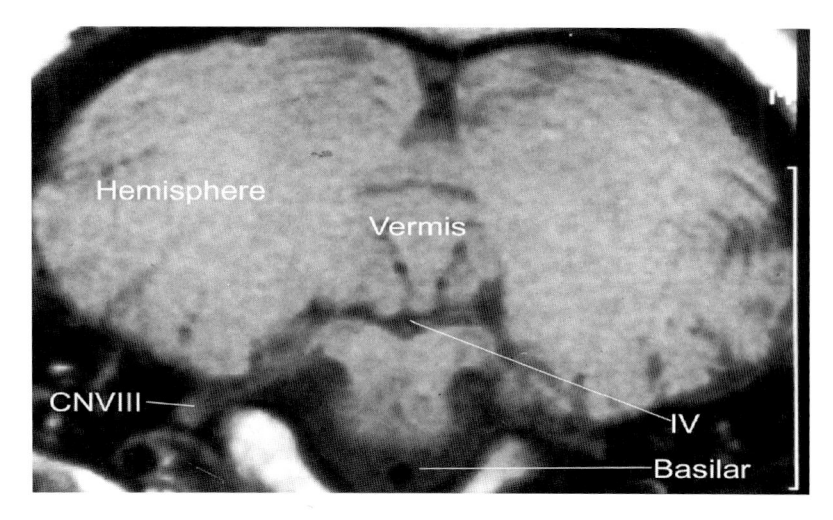

Fig. 4.5 B Brain stem at medullary level, inferior olive. MRI, T1

hemispheres via the inferior cerebellar peduncle. The climbing fibers found on the dendrites of Purkinje's cells in the cerebellum originate in the inferior olive. The olive also has strong connections with the red nucleus and receives input from the spinal cord, cerebellum, red nucleus, intralaminar nuclei, and basal ganglia. The central core of this section, as in other levels, consists of neurons of the reticular formation.

Medial Zone (Fig. 4.5A). The serotinergic raphe nuclei of the reticular formation are found in this zone. The position of the medial longitudinal fasciculus (MLF), tectospinal, and medial lemniscus corresponds to that in the previous level.

Basilar Zone (Figs. 4.5A and 4.5B). The medullary pyramids are prominent at this level. The rootlets of cranial nerve XII exit lateral to the pyramid.

B. PONS

1. Blood Supply: Basilar Artery and Its Branches

2. Gross Landmarks in the Pons

The hallmark of pontine levels are the prominent basis ponti, middle cerebellar peduncle, and pontine gray, which is present on the anterior surface. The cerebellum forms the tectal/posterior surface. The tegmentum takes up much less of these levels due to the massive enlargement of the pontine basis. With the cerebellum removed, the posterior surface of the tegmentum demonstrates the three cerebellar peduncles (inferior cerebellar peduncle, middle cerebellar peduncle, and superior cerebellar peduncle– posterior surface) and the fourth ventricle is now opened.

Brain Stem Level 4: Lower Pons at Level of Facial Nerve and Facial Colliculus (Fig. 4.6) Ventricle Equivalent Level MRI

Gross Features. The bulk of this section consists of the middle cerebellar peduncle (brachium pontis) and the cerebellum. In Figure 4.6, the bulk of the cerebellum has been removed. The fourth ventricle narrows as it nears the cerebral aqueduct. The medullary pyramids are present at the anterior surface. Note that the inferior olive is no longer present.

Ventricular Zone

Motor Cranial Nerve Nuclei (Fig. 4.6B). The nucleus of nerve VII is conspicuous at the lateral margin of the tegmentum. This nucleus innervates the muscles of facial expression. In the pons, in close proximity to the ventricular floor, the rootlet of this nucleus swings around the medial side of the nucleus of nerve VI, forming the internal genu of nerve VII. These fibers then pass lateral to the nucleus of nerve VII and exit from the substance of the pons on the anterior surface in the cerebellopontine angle.

The nucleus of nerve VI innervates the lateral rectus muscle of the eye. The rootlet leaves the anterior surface of the nucleus and has the longest intracerebral path of any nerve root. The fibers finally exit on the anterior surface of the brain stem near the midline at the pontomedullary junction. The close proximity of the nucleus of nerve VI to the rootlets of nerve VII demonstrates why any involvement of the nucleus of nerve VI usually produces a concomitant alteration in function of nerve VII. The nucleus of nerve VI and the internal genu of the rootlets of nerve VII form the prominent facial colliculus on the floor of the fourth ventricle.

Cerebellar Pathways (Fig. 4.6B). At this level, the middle and inferior cerebellar peduncles form the lateral walls of the ventricle as well as the bulk of the cerebellar medullary center. The ventral spinocerebellar tract is seen lateral to the superior cerebellar peduncle that it enters and then follows back into the cerebellum. The superior cerebellar peduncle consists primarily of axons carrying impulses from the dentate nucleus of the cerebellum to the red nucleus and the ventral lateral nucleus in the thalamus. This tract is also called the dentatorubrothalamic tract. The tractus uncinatus connects the deep cerebellar nuclei with the vestibular nuclei and reticular formation bilaterally.

Sensory Cranial Nerve Nuclei (Fig. 4.6). Pain and temperature are conveyed from the head. At pontine levels, the descending nucleus of nerve V is small, whereas the tract is large. The superior vestibular nucleus is seen at the lateral margin of the ventricle with primary vestibular fibers present in its substance.

Auditory sensation at this level is related to the superior olive that is seen inferior to the motor nucleus of nerve VII. The superior olive is one of the secondary nuclei in the auditory pathway. The auditory fibers are seen accumulating inferior to the

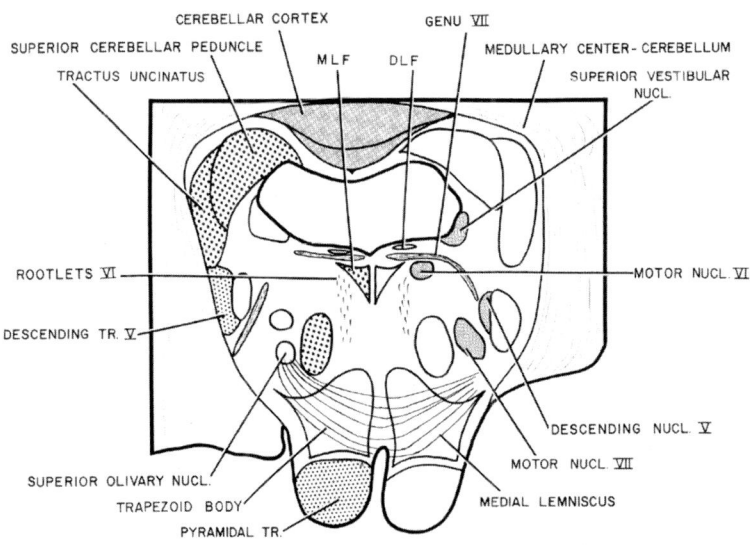

Fig. 4.6 Brain stem level 4 at pontine level of cranial nerve VI and VII. Coronal lateral zone (From EM Marcus and S Jacobson, Integrated neuroscience, Kluwer, 2003)

superior olive and cutting through the medial lemniscus to form the lateral lemniscus.

White Matter (Fig. 4.6B). At this pontine level, some of the ascending tracts are starting to move more laterally. The medial lemniscus will shift laterally and approach the rubrospinal and spinothalamic tracts. Tactile discrimination and proprioception from the limbs, thorax, pelvis, abdomen, and neck are conveyed by fibers in the medial lemniscus, which is seen posterior to the corticospinal tract.

Pain and temperature from the head are conveyed by the secondary trigeminothalamic tracts, which are found in the medial lemniscus and ascend to the nucleus ventralis posteromedialis in the thalamus. Pain and temperature from the extremities, thorax, abdomen, pelvis, and neck are carried in the lateral spinothalamic tract, which is near the anterior surface of the pons, close to the medial lemniscus.

The anterior spinothalamic tract that is mixed in with the lateral spinothalamic tract carries light touch.

The gustatory fibers from the tongue are found in the solitary tract. The secondary fibers ascend bilaterally in the medial lemniscus. Fibers carrying visceral sensations also synapse in the solitary nucleus.

Medial Zone. The raphe or the medial reticular nuclei are present in this level. The medial lemniscus at this level contains secondary fibers from the dorsal columns and the spinothalamic, trigeminothalamic, and solitary tracts. The medial longitudinal fasciculus and tectospinal tracts are still present near the midline in the floor of the fourth ventricle with the medial longitudinal fasciculus conspicuous under the floor and the tectospinal tract below it.

Central Zone (Figs. 4.6A and 4.6B). In the core of the pontine tegmentum, the main efferent ascending tract of the reticular system (the central tegmental tract) occupies the bulk of the reticular formation. Internal to the nucleus of VI is the parapontine reticular nucleus that coordinates eye movements. Fibers descend the opposite frontal eye fields, synapse here, and then connect via the MLF to the nuclei of IV and III.

Basilar Zone (Fig. 4.6B). The descending corticospinal and corticonuclear pathways are nearly obscured by the fibers of the middle cerebellar peduncle and the pontine gray.

Brain Stem Level 5: Upper Pons at the Motor and Main Sensory Nuclei of Nerve V (Fig. 4.7)

Gross Landmarks. The bulk of this section consists of the middle cerebellar peduncle and cerebellar cortex. The fourth ventricle begins to narrow as it nears the cerebral aqueduct.

Ventricular Zone. In this level, there are no motor cranial nerve nuclei on the floor of the ventricle; the nuclei in the ventricular floor being for visceral functions.

Lateral Zone

Motor Cranial Nerve Nuclei (Fig. 4.7A). The rootlet of CN V and motor nucleus of nerve V is conspicuous bilaterally in the reticular formation, medial to the

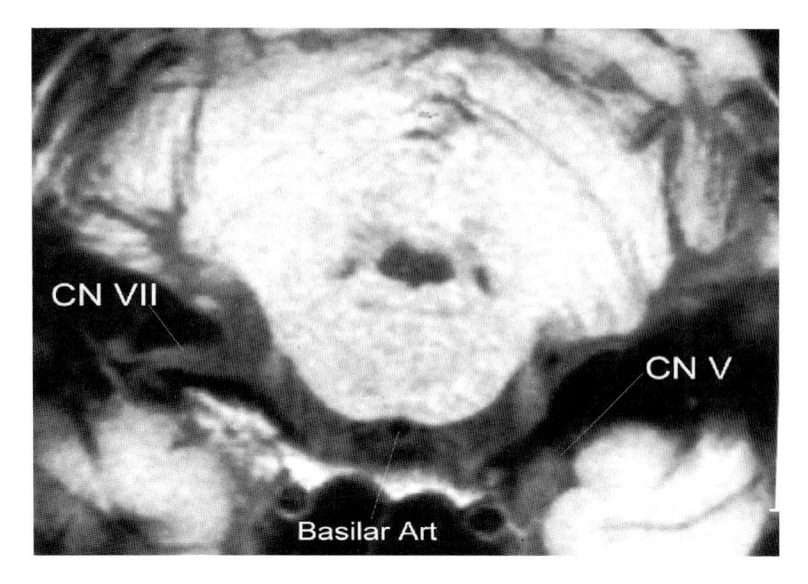

Fig. 4.7 A Brain stem at pontine level of cranial nerves V and VII. MRI, T1

sensory nucleus and the entrance of the root of nerve V. Each of these nuclei provides innervation to the ipsilateral muscles of mastication.

Sensory Cranial Nerve Nuclei (Fig. 4.7A). Tactile discrimination from the face is carried by the trigeminal root into the main sensory nucleus of nerve V, seen lateral to the intrapontine root of nerve V. The secondary fibers originate from this nucleus and ascend crossed and uncrossed. The crossed fibers run adjacent to the medial lemniscus (ventral trigeminothalamic), whereas the uncrossed fibers run in the dorsal margin of the reticular formation (dorsal trigeminothalamic). Both fiber pathways terminate in the ventral posterior medial nucleus of the thalamus. Proprioception from the majority of the muscles in the head is carried in by the mesencephalic root and ends in the mesencephalic nucleus of nerve V, located lateral to the walls of the fourth ventricle in the upper pontine and midbrain levels. These neurons are the only primary cell bodies (dorsal root ganglion equivalent) in the central nervous system. These axons synapse on the chief nucleus of V. The jaw jerk is a monosynaptic stretch reflex. The receptor is the mesencephalic nucleus of nerve V and the effector is the motor nucleus of nerve V.

The superior olive is one of the nuclei in the auditory system. Auditory fibers are seen running through the superior olive or inferior to it, cutting through the medial lemniscus and crossing the midline and forming the trapezoid body. The fibers are then found lateral to the medial lemniscus and are known as the lateral lemniscus. The pontine nuclei are conspicuous in the basilar portion of the pons.

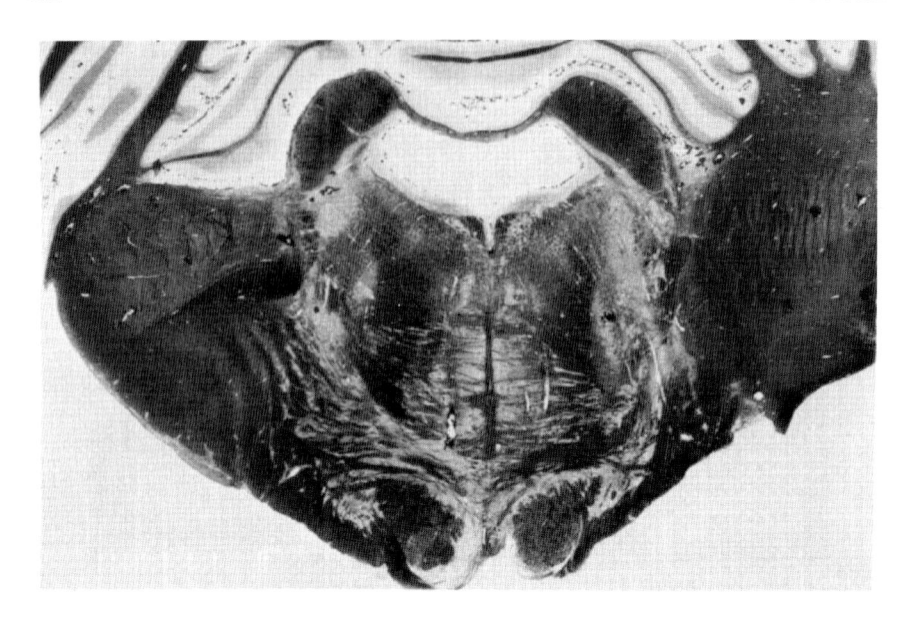

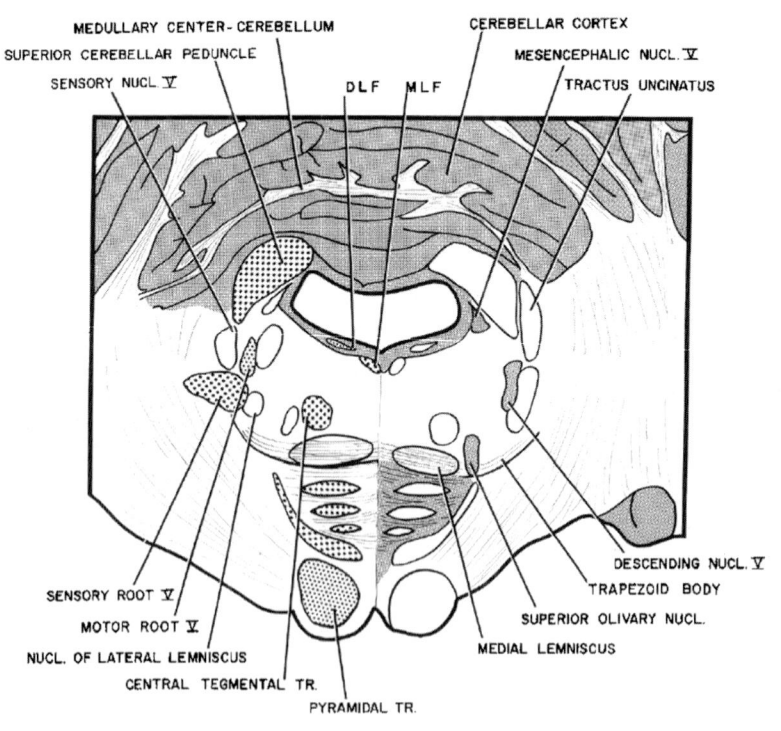

Fig. 4.7 B Brain stem at pontine level of cranial nerve V. Coronal (From EM Marcus and S Jacobson, Integrated neuroscience, Kluwer, 2003)

White Matter (Fig. 4.7A). At this level, the medial lemniscus has moved from the midline to a more lateral position, and many of the ascending sensory systems are now either in the medial lemniscus or adjacent to it.

Medial Zone. The raphe nuclei of the reticular formation are evident in this level. The medial longitudinal fasciculus is larger in this section between the motor nuclei of cranial nerves III and VI because it contains many vestibular and cerebellar fibers necessary for accurate eye movements. The tectospinal fibers always lie anterior to the MLF.

Central Zone. In the reticular formation, the central tegmental tract is conspicuous.

Basilar Zone. The corticospinal and corticobulbar tracts are found in the pontine gray and white matter.

C. Midbrain

1. Blood Supply: Basilar Artery and Posterior Cerebral Arteries

2. Gross Landmarks in the Midbrain

On the anterior surface of the midbrain we find the principal landmark of this level: the cerebral peduncles with the third cranial nerve exiting from their medial surface in the interpeduncular fossa. On the posterior surface of the midbrain we find the corpora quadragemini: the superior and inferior colliculi. The fourth cranial nerve exits from the posterior surface of the midbrain and the junction with the pons. In a coronal section through the midbrain, one finds the red nucleus in the tegmentum of the midbrain (Fig. 4.8B). Also in the inferior levels in the tegmentum just before the red nucleus, one finds the crossing of the superior cerebellar peduncle.

Brain Stem Level 6: Inferior Colliculus and Pontine Basis (Fig. 4.8)

In this section the midbrain forms the roof and floor of the narrow cerebral aqueduct, whereas the pons makes up the basilar portion. The roof is the inferior colliculus, an important nucleus in the auditory pathway.

Tectum

Inferior Colliculus (Figs. 4.8A and 4.8B). The gray matter of the inferior colliculus forms the tectum at this level. The inferior colliculus is divided into three nuclei: a large central nucleus, a thin dorsal nucleus)the paracentral or cortical nucleus), and an external nucleus. The central and cortical nucleus functions as a relay center for the cochlea. The external nucleus functions for the acousticomotor reflexes

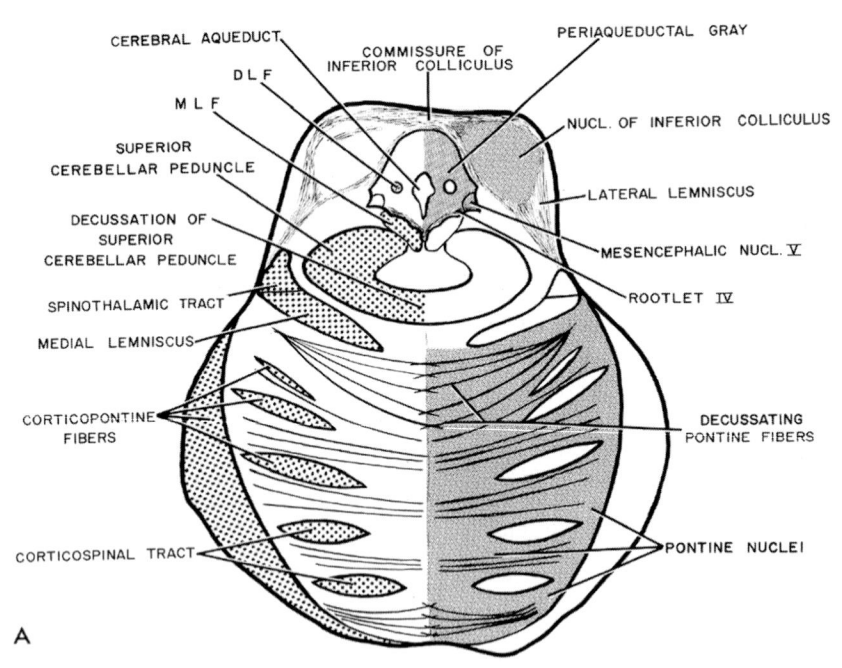

Fig. 4.8 A Brain stem level 6 at midbrain inferior collicular level, Coronal (From EM Marcus and S Jacobson, Integrated neuroscience, Kluwer, 2003)

through the tectospinal pathway. The brachium of the inferior colliculus carries auditory information onto the medial geniculate nucleus of the diencephalon. The tectospinal pathway descends from this region and terminates on the motor nuclei of the lower cranial nerves and upper cervical ventral horn cells.

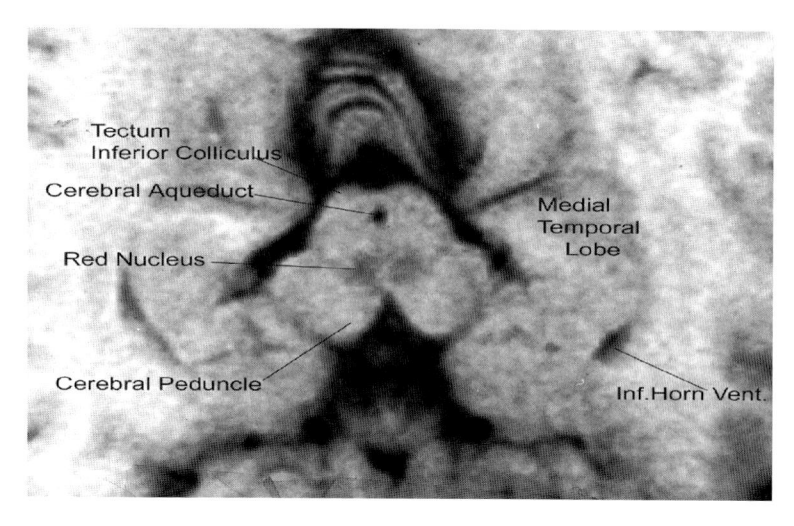

Fig. 4.8 B Brain stem at inferior collicular level. Coronal MRI, T1

Tegmentum

Ventricular Zone (Figs. 4.8A and 4.8B). The periaqueductal gray is conspicuous in midbrain levels. Descending and ascending tracts associated with the visceral brain are found here. In the midbrain, the lower half of the periaqueductal gray and some other nuclei in the tegmentum are part of the midbrain limbic area. This zone is important in our level of attentiveness. Bilateral lesions to the periaqueductal gray and midbrain tegmentum usually produce comatose patients.

Motor Cranial Nerve Nuclei (Fig. 4.8A). The nucleus of cranial nerve IV is seen indenting the medial longitudinal fasciculus in lower regions of the midbrain. The axons of this nucleus pass posteriorly (unique for a cranial nerve) and exit the brain at the midbrain pontine junction, where the fibers then proceed anteriorly to reach the superior oblique muscle through the superior orbital fissure.

Sensory Cranial Nerve Nuclei (Fig. 4.8A). Proprioception is conveyed from the muscles in the head and neck. The mesencephalic nucleus of nerve V is located at the lateral margin of the periaqueductal gray. Remember that these are the only primary sensory neurons in the central nervous system; their axons join the medial lemniscus and ascend to the ventral posterior medial nucleus in the thalamus.

Lateral Zone (Fig. 4.8A). At this level, the medial lemniscus and the other major ascending tracts are stretched over the lateral surface of the tegmentum of the midbrain. The spinothalamic and trigeminothalamic fibers are found lateral to the medial lemniscus and the auditory fibers are located above them. Thus, the following sensory modalities for the entire body are located here: tactile discrimination, proprioception, pain and temperature, gustatory, and visceral sensations.

Auditory Fibers lateral lemniscus (Fig. 4.8A). The lateral lemniscus is seen superior to the medial lemniscus. These secondary auditory fibers are now entering the inferior colliculus. Many of these fibers synapse in the inferior colliculus and then continue up into the medial geniculate nucleus MGN,. as the brachium of the inferior colliculus.

Medial Zone (Fig. 4.8A). The raphe nuclei of the reticular formation are very conspicuous in these levels. The medial longitudinal fasciculus is prominent in the floor of the ventricle as it nears cranial nerve III; the fibers located within it are important in coordinating ocular movements.

Central Zone (Fig. 4.8A). This zone in the lower tegmentum of the midbrain is nearly completely filled with the crossing fibers of the superior cerebellar peduncle. Ultimately, all of these fibers decussate and continue up to the red nucleus, where some fibers synapse, but the majority of these axons bypass the red nucleus and terminate on the ventral lateral thalamic nuclei.

Basilar Zone (Fig. 4.8A). The bulk of this region is taken up by the pontine gray and descending fibers from the cerebrum to the pontine gray (the corticopontine system). The corticospinal and corticobulbar fibers are also present here.

Brain Stem Level 7: Midbrain Superior Collicular Level and Pontine Basis (Fig. 4.9)

Tectum

Superior Colliculus (Fig. 4.9). The superior colliculi form the rostral portion of the midbrain tectum. The parenchyma of the superior colliculus is organized in layers. There are four layers containing gray and white matter from inside out: stratum zonale, stratum cinereum, stratum opticum, and stratum lemnisci. The superficial layers of the superior colliculus receive their input from the retina and visual cortex, with the contralateral upper quadrant medially and the contralateral lower quadrants laterally. In contrast, the deeper layers receive their input from polysensory sources, including the cerebellum, inferior colliculus, spinal cord, reticular formation gracile, cuneate, trigeminal nuclei, and visual regions. A portion of the tectospinal pathway originates at this level and descends in the lower brain stem to terminate on the lower cranial nerves and ventral horn cells of the spinal cord. (Remember that the tectospinal and tectobulbar fibers terminate on the same motor nuclei as the corticonuclear and corticospinal pathways). The zone of the tectum above the superior colliculus (the pretectal zone) is important in light reflexes.

Tegmentum

Gross Features. In this section, the roof and floor are formed by the midbrain, whereas the basilar portion is made up of peduncles and the pons. The roof is the superior colliculus, an important station in the visual pathway. The cerebral aqueduct forms the narrow ventricular lumen (Figs. 4.9A and 4.9B). The superior col-

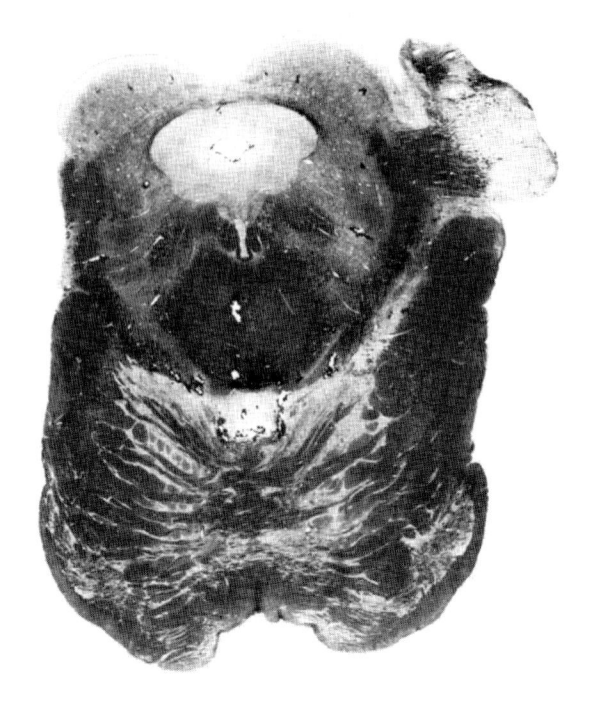

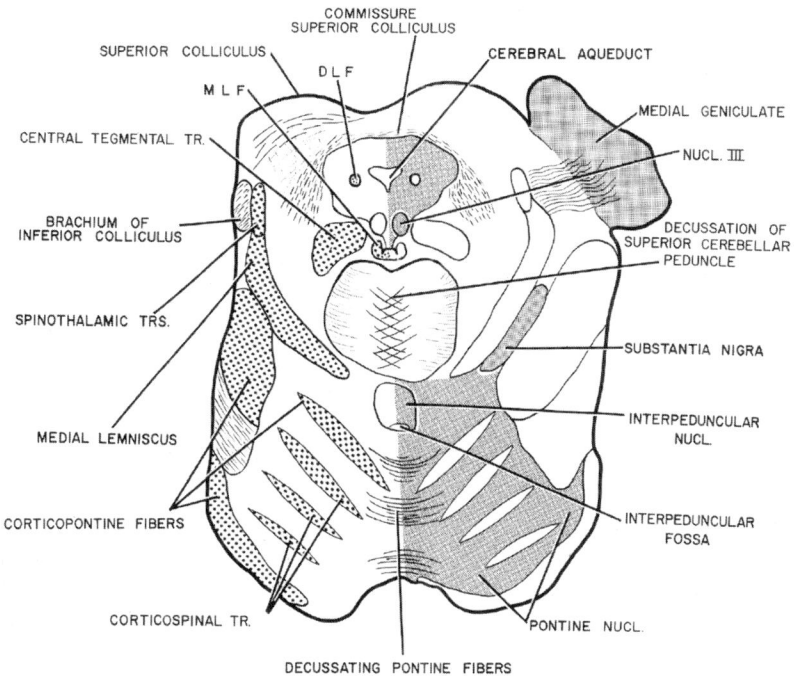

III Differences Between the Spinal Cord and Brain Stem

Fig. 4.9 A Brain stem level 7; midbrain: superior collicular level

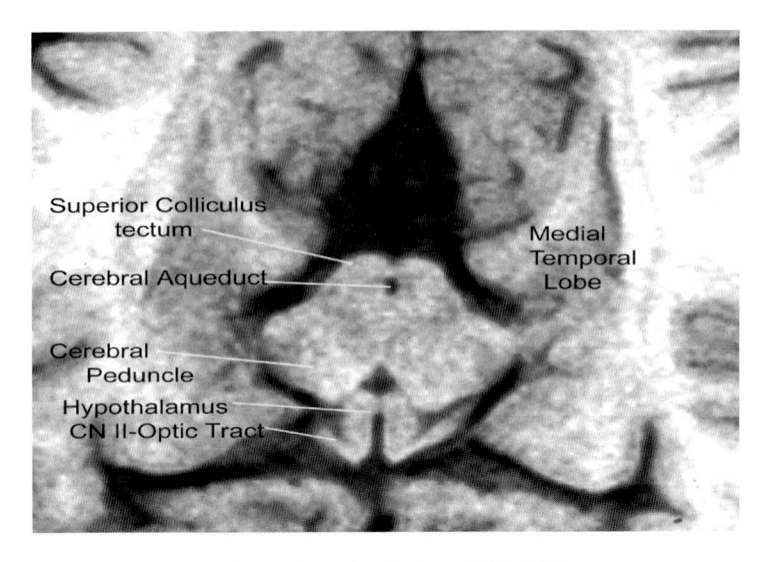

Fig. 4.9 B Brain stem at superior collicular level. Coronal, MRI, T1

liculus is a laminated structure important in relating eye movements and body position. The tectospinal tract originates from its deepest layer and provides connections onto certain cranial and spinal neurons.

Ventricular Zone

Sensory Cranial Nerve Nuclei (Fig. 4.9A). Proprioception is conveyed from the muscles in the head and neck by the mesencephalic nucleus of nerve V, located at the lateral margin of the periaqueductal gray in the superior collicular levels as well as in the upper pontine and inferior collicular levels.

Motor Cranial Nerve Nuclei (Fig. 4.9A). Cranial nerve III is visible in the floor of the cerebral aqueduct, adjacent to the medial longitudinal fasciculus. This nerve supplies the medial, inferior, and superior rectus muscles and the inferior oblique and superior levator muscles of the eyelid. The Edinger–Westphal nucleus of nerve III is also present; it provides preganglionic parasympathetic innervation to the constrictor muscle of the pupil via the ciliary ganglion.

Lateral Zone

White Matter (Fig. 4.9A). Medial lemniscus (At this level, in close proximity to the medial lemniscus, the white matter includes most of the ascending sensory fibers).

This tract is stretched out over the inferior and lateral surface of the tegmentum, with the fibers that mediate pain and temperature from the extremities, thorax, abdomen, and pelvis located at its superior extent. The trigeminal fibers form its middle portion, and the fibers that mediate tactile, proprioceptive, and visceral sensations are placed medially.

Anterolateral Column: Lateral Spinothalamic. Pain and temperature from the limbs, thorax, and pelvis are found in this pathway.

Anterior spinothalamic. Light touch fibers from the limbs, abdomen, and neck are seen in close proximity to the lateral spinothalamic pathway.

Auditory Pathway. The fibers from the inferior colliculus (called, at this level, the brachium of the inferior colliculus) are seen at the inferior surface of the superior colliculus. On the left side, they enter the medial geniculate nucleus of the metathalamus, at which point they reach their final subcortical center.

Cerebellar Fibers. The superior cerebellar peduncle is ascending and continues to cross in the tegmentum of the midbrain. Just above this level, many of these fibers synapse in the red nucleus, whereas others will continue to the VL nucleus in the thalamus.

Central Zone (Fig. 4.9A). The central tegmental tract is conspicuous in the midbrain tegmentum in the reticular formation. The red nucleus is found in the medial edge of the reticular formation. In the human, it consists of a small magnocellular region and a large parvocellular region. Fibers from all cerebellar nuclei, but especially the dentate, synapse here.

Corticorubral fibers run bilaterally from the motor cortex synapse also terminate here in a somatotopic relationship. Fibers from the red nucleus terminate directly or indirectly in the cerebellum and on many of the motor nuclei in the brain stem and spinal cord, which also receive input from the corticospinal and corticonuclear pathways. The red nucleus has minimal, if any, projections onto the thalamus in the human, although it most likely influences the cerebellar input onto the thalamus.

The substantia nigra is also located in this level, just behind the cerebral peduncle. Its functions are discussed in the relationship between the basal ganglia and movement (see Chapter 11).

Medial Zone (Fig. 4.9A). The serotinergic raphe nuclei of the reticular formation are present in this level. The medial longitudinal fasciculus in the floor of the cerebral aqueduct is conspicuous and connected across the midline. The oculomotor complex indents the MLF and receives ascending fibers from cranial nerves VI and VIII through the MLF. This might be significant if herniation of the temporal lobe produces compression of the cerebral peduncle and cranial nerve III, with the accompanying cranial nerve and upper motor neuron signs. The sympathetic nucleus of Edinger–Westphal, associated with cranial nerve III to the pupillary constrictor, is also present. The interstitial nucleus of Cajal and the nucleus of Darschewtiz, which are important pupillary reflexes, are also found here.

The tectospinal fibers at this level are primarily from the superior colliculus.

Basilar Zone (Figs. 4.9A and 4.9B): Cerebral Peduncles. At this level, the frontopontine fibers, which occupy the most medial part of the cerebral peduncles, enter the pons. The corticospinal fibers and many of the corticobulbar fibers are still in the peduncle. In Figure 4.9B, a coronal MRI section through the midbrain, the

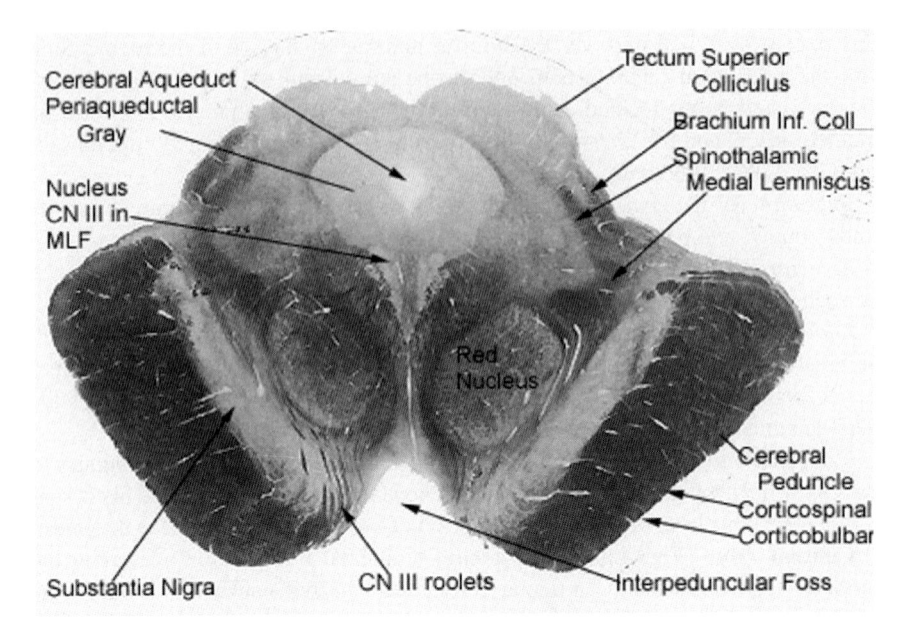

Fig. 4.10 Brain stem at level of cerebral peduncles and III cranial nerve. Weil myelin stain. Coronal section. (From EM Marcus and S Jacobson, Integrated neuroscience, Kluwer, 2003)

relationship of the cerebral peduncle to the medial temporal lobe demonstrates that the cerebral peduncles are adjacent to the medial temporal lobe and that herniation of the temporal lobe will affect the cerebral peduncle signs in the III cranial nerves and descending pathways.

Figure 4.10 is a myelin-stained section through the cerebral peduncles demonstrating the relationship between the peduncles and the III cranial nerve.

IV. Functional Centers in the Brain Stem

Now that the anatomical features of the brain stem have been discussed, it is appropriate to identify some important functional centers located in the brain stem and diencephalon. During this discussion, it will become evident that the cranial nerves are the pivotal point for many of these activities.

A. *Reticular Formation*

The central core of the medulla, pons, and midbrain tegmentum consists of the reticular formation. Upon microscopic examination, this region is seen to consist of groupings of neurons separated by a meshwork of medullated fibers. With a Golgi neuronal stain, for the Golgi type I cells with long axons, neurons are shown with their dendrites extending transversely and the axons bifurcating into ascending and descending branches, which run throughout the system. Each neuron receives input from at least 1000 neurons, and each neuron connects to as many as 10,000 neurons in the reticular formation. The reticular formation of the brain stem blends inferiorly into lamina VII of the cord, and superiorly, it is continuous with the hypothalamus and dorsal thalamus of the diencephalon.

1. Nuclei

Many nuclei have been identified in the reticular formation. Functionally, the nuclei can be divided into cerebellar and noncerebellar nuclei (Table 4.5). The cerebellar portion of the reticular formation includes primarily the lateral and paramedian nuclei of the medulla and the tegmental nucleus of the pons.

The noncerebellar portion nuclei are divided anatomically into three columns: the raphe, central, and lateral groupings. Most of the nuclei in the reticular formation consist of large cells with ascending and descending axons. Functionally, the reticular formation in the medulla is especially important because the descending reticulospinal fibers and much of the ascending reticular system originate there.

2. Descending Reticulospinal System (Table 4.6)

The *nucleus reticularis gigantocellularis* is found at the rostral medullary levels, dorsal and medial to the inferior olive. This nucleus gives origin to much of the lateral reticulospinal tract, which is primarily an ipsilateral tract, running in the lateral funiculus of the spinal cord in all levels and terminating on internuncial neurons.

The axons from the *nucleus reticularis pontis oralis and caudalis* form much of the medial reticulospinal tract, which runs in the anterior funiculus in all levels and terminates on internuncial neurons.

The *lateral reticular nucleus* is located in the lateral margin of the reticular formation, dorsal to the inferior olive, whereas the ventral reticular nucleus is found in the caudal end of the medulla, dorsal to the inferior olive. These two nuclei, in conjunction with the nuclei in the pons and midbrain, form much of the ascending reticular fibers in the *central tegmental tract*, distributing to neurons in the thalamus (intralaminar and reticular), hypothalamus, and corpus striatum. The

Table 4.5 Functional groupings of nuclei in the reticular formation

Noncerebellar	Cerebellar related
Raphe, central, and lateral in the tegmentum of brain stem	Lateral and paramedian nuclei of medulla and tegmental nucleus of the pons

Table 4.6 Major pathways of the reticular formation

Tract	Origin	Termination of pathway
Lateral reticulospinal	Nucleus gigantocellularis	Spinal cord and autonomic interneurons
Medial reticulospinal	Pontis oralis and caudalis	Spinal cord and autonomic interneurons
Central reticulospinal	Lateral and ventral	Thalamic–intralaminar and reticular, hypothalamus, corpus striatum, substantia nigra

paramedian reticular nucleus is found near the midline at midolivary levels, dorsal to the inferior olive, and provides direct input into the anterior lobe of the cerebellar vermis.

3. Input to Reticular Formation

The reticular formation receives information via the ascending spinal tracts (spinotectal, spinoreticular, spinothalamics, and spinocerebellar) from the brain stem itself (olivoreticular, cerebelloreticular, and vestibulospinal), from the cerebral hemispheres (corticoreticular), and from the basal ganglia and hypothalamus. The cranial nerves are another important source, especially nerves I, II, V, VII, VIII, and X, for sensory information that appears to project most heavily onto the central nuclei. The hypothalamus and striatum also project to the reticular system via the dorsal longitudinal fasciculus on the floor of the cerebral aqueduct and fourth ventricle and the more diffuse descending fiber systems in the core of the reticular formation.

4. Output of Reticular System

The medial lemniscus is a specific point-to-point relay system with few synapses (a closed system), while the fiber tracts of the reticular system are a multisynaptic nonspecific system (an open system)…

a. Central Tegmental Tract

The central tegmental tract is the principal fiber tract of the reticular formation. Its descending portions are located in the medial tegmentum, and its ascending portions are located in the lateral tegmentum. The ascending system projects to the thalamic intralaminar and reticular nuclei, hypothalamus, basal ganglia, substantia nigra, and red nucleus. The reticular system is functionally important in controlling our "posture" by its reflex relationship to the position of our body in space and in controlling our internal milieu by maintaining the stability of our viscera. The

IV Function Centers in the Brain Stem

descending system synapses via the reticulospinal tract onto interneurons that mediate their effects through alpha and gamma motor neurons in the spinal cord and via autonomic pathways and cranial nerves onto visceral neurons.

b. Role of Ascending Reticular System

This fiber system is the structural and functional substrate for maintaining consciousness. It also receives proprioceptive, tactile, thermal, visual, auditory, and nociceptive information via the spinal and cranial nerves. Many sensations (cutaneous, nociceptive, and erotic) activate the system. The sensory information ascends via the central tegmental tract into the limbic midbrain area from which information can be more directly passed into the thalamus and hypothalamus. This midbrain to diencephalons to telencephalon circuit seems especially important in setting our level of consciousness.

c. Role of Descending Systems

The reticular formation has many roles and is especially important in controlling posture and orientation in space. Stimulation of the nucleus reticularis gigartocellularis of the medulla inhibits the knee jerk. Stimulation of reticularis oralis/caudalis of the pons facilitates the knee jerk. The lateral part of the reticular formation is the receptor area, whereas the medial portion is the effector zone and the origin of the central tegmental and reticulospinal tracts. Many of the functions vital to the maintenance of the organism are found in the medulla.

d. Neurochemically Defined Nuclei in the Reticular Formation Affecting Consciousness

Cholinergic Nuclei. The cholinergic nuclei are located in the dorsal tegmentum of the pons and midbrain, in the mesopontine nuclei, and in the basal forebrain region and they project diffusely to the cerebral cortex through the thalamus and have a modulating influence on cerebral cortical activity and wakefulness.

Monoamine Nuclei. In the reticular formation, we find cells containing norepinephrine and serotonin. The *norepepinephrine*-containing cells are found in the locus ceruleus (blue staining) in the upper pons and midbrain and they project widely upon nuclei in the spinal cord, brain stem, thalamus, hypothalamus, and corpus striatum, which are important for maintaining attention and wakefulness.

Serotinergic Nuclei are found in the raphe of the medial tegmental zone in the medulla, pons, and midbrain. The nuclei in the pons and medulla project onto the spinal cord and brain stem, whereas the nuclei in the upper pons and midbrain project onto the thalamus, hypothalamus, corpus striatum, and cerebral cortex. The serotinergic system facilitates sleep.

An area outside of the reticular formation, the *histamine*-containing area of the posterior hypothalamus, is also important in maintaining wakefulness.

B. Respiration Centers

Respiration is under the control of neurons in the respiratory center in the upper medulla and pons. Sensory fibers from the lungs ascend via cranial nerve X and enter the solitary tract and proceed onto the neurons in the reticular formation in the medulla and pons. Specific regions in the medulla control either inspiration or expiration. The medullary respiratory center itself is responsive to the carbon dioxide content in the blood. Increased carbon dioxide produces increased respiration and decreased carbon dioxide produces decreased respiration. In the pons, the pneumotaxic centers are related to the frequency of the respiratory response. These centers play onto the medullary respiratory center to determine the respiratory output. The axons from cells in the medullary reticular center descend to the appropriate spinal cord levels to innervate the diaphragm (phrenic nerve) and the intercostal and associated muscles of respiration. The vagus nerve itself provides preganglionic innervation of the trachea, bronchi, and lungs. The respiratory response is also modified by cortical and hypothalamic control; that is, emotionally stimulated individuals breathe more rapidly because of the hypothalamic influence on the brain stem and spinal cord. The actual respiration occurs with the intercostals muscles, the accessory muscles, and the lungs expanding and contracting in concert with associated vascular changes (autonomic nervous system).

C. Cardiovascular Centers

The *carotid body* (innervated by sensory fibers of cranial nerve IX) is found distal to the bifurcation of the common carotid artery, and the aortic body, innervated by the sensory fibers of nerve X, is found near the origin of the subclavian arteries. These carry information on blood pressure into the solitary tract and then into the medullary respiratory centers. These sites are sensitive to oxygen–carbon dioxide pressure: When oxygen is reduced, ventilation increases; when oxygen increases, ventilation decreases. Neural control over the tone of the arterioles is exercised through vasoconstrictor fibers (and sometimes vasodilator fibers). A vasomotor center in the medulla extends from the midpontine to the upper medullary levels. In the lateral reticular formation of the medulla, stimulation elicits an increase in vasoconstriction and an increase in heart rate—pressor center. In the lower medulla, stimulation of the depressor center produces a decrease in vasoconstriction and a decrease in heart rate. The carotid body (cranial nerve IX) and the aortic body (cranial nerve X) contain specialized pressure receptors stimulated by an increase in the size of these blood vessels. This information runs via nerves IX and X to the medulla and depresses activity. Most sensory nerves, cranial and spinal, contain some nerve fibers, which, when stimulated, cause a rise in arterial pressure by exciting the pressor region and inhibiting the depressor centers. Higher centers also influence respiratory and vascular centers in the medulla and alter blood flow, depending on the psychic state; that is, mental activity decreases peripheral blood (constricts vessels), whereas emotional states can produce increased blood flow, blocking vasodilation.

D. Deglutition

The act of swallowing starts volitionally but is completed by reflex activity.

1. The food is first masticated (motor nucleus of nerve V) and reduced to smaller particles, called the bolus, which is lubricated by saliva from the salivary glands (nerves VII and IX).
2. The bolus is propelled through the oral pharyngeal opening when the tongue is elevated against the soft palate (nerve XII), and the facial pillars are relaxed.
3. The oral pharyngeal pillars close and the superior and middle pharyngeal constrictor muscles (ambiguous nucleus of nerve X) force the bolus along into the laryngeal pharynx, where the pharyngeal walls contract, closing the superior opening.
4. At the same time, the tracheal opening is closed by the epiglottis and glottis, as the larynx, trachea, and pharynx move up. This movement is noted externally as the bobbing of the thyroid eminence (Adam's apple).
5. Finally, the inferior constrictor muscle contracts, pushing the bolus into the esophagus, where peristaltic waves and gravity carry it through the esophagus. Cranial nerves V, IX, X, and XII perform the motor part of the activity, whereas nerves V, VII, IX, and X form the sensory function.

E. Vomiting

Vomiting is produced by many stimuli and is usually a reflex activity. The vomitus is composed of the gastric contents.

1. Vomiting is commonly caused by irritation of the oropharynx, gastrointestinal mucosa, and genitourinary and semicircular canals. Stimulation of the vestibular system, including the semicircular canals and the nerve itself, could also produce vomiting.
2. The afferent nerves travel via the vagus and glossopharyngeal tracts into the solitary tract. After synapsing on the dorsal motor nucleus of cranial nerve X, the information is conveyed via the vagus nerve to the stomach. Impulses are also passed down to the cervical and thoracic levels onto the ventral horn cells

that control the simultaneous contraction of the intercostal, diaphragmatic, and abdominal musculature.
3. Nausea and excessive salivation precede the deep inspiration associated with retching. The glottis is closed and the nasal passages are sealed off. The descent of the diaphragm and the contraction of the abdominal muscles exert the pressure that causes the stomach to contract in a direction contrary to normal peristalsis and forces vomitus through the relaxed cardia of the stomach and into the esophagus.

F. Emetic Center

The area postrema is found in the caudal end of the fourth ventricle above the obex of the medulla. This area is very vascular and contains many venous sinuses. The blood-brain barrier is lacking in this region. The vagal, glossopharyngeal, and hypoglossal nuclei are strongly interconnected with the area postrema. This region is very sensitive to changes in pressure or to drugs whose passage into this area is not inhibited by any blood-brain barrier.

G. Coughing

Irritation of the lining of the larynx or trachea produces coughing. The stimulus is picked up by free nerve endings associated with the internal laryngeal branch of cranial nerve X, which carries it up into the solitary tract, following the same pathway as in vomiting, except that the muscles contract alternately rather than simultaneously, with the intercostals muscle contracting suddenly.

H. Taste

The taste cells, just as the receptive neurons in the olfactory system, are in a very hostile environment and they are constantly being replaced by new cells, and yet their functions continue unabated. Taste plays a critical role in determining what an infant eats. In most animals, gustatory sensations from each foodstuff greatly influence the animal's intake, but in humans, this is not necessarily the case. We are affected by environmental pressure that produces bizarre eating habits that could lead to extremes such as obesity or anorexia.

1. Lingual Taste Buds

The tongue is a very mobile muscular organ that is located in the oral a cavity. It is divided into a root, a body, an apex, a dorsal surface, and an inferior surface. The body and apex of the tongue are very mobile. The tongue has at least three main functions: (1) forming words, (2) moving food from the oral cavity to pharynx by passing the bolus along the palate, and (3) taste.

1. Lingual Papillae

a. Circumvallate papillae are large and flat, lie in the terminal sulcus, and consist of a wall with a deep central depression. The walls contain- taste buds.
b. Foliate papillae are small lateral folds of the lingual mucosa, which contains the taste buds.
c. Filiform papillae are long and numerous and contain sensory nerve endings from cranial nerve V3 (anterior two-thirds of the dorsum and apex of tongue) and CN IX (posterior third of tongue); taste buds are not present in these processes.
d. Fungiform papillae (mushroom-shaped) are pink in color and are located among the filiform papillae contain taste buds; they are most numerous at the apex and sides of the body of the tongue. There are also a few taste buds in the oral surface of the soft palate, the posterior wall of oral pharynx, and the epiglottis.

2. Subjective Taste

Classically, there are four groups of subjective taste qualities, each localized to a zone of the tongue: sweet and salt at the tip of tongue, sSour at the sides, bitter at the pharyngeal end. In reality, all parts of the tongue respond to the four kinds of taste. Bitter compounds are especially protective because they make the individual aware of toxic substances, and they use both ligand-gated channels and G-protein-coupled receptors. That permits them to be activated by many compounds.

Chorda tympani of cranial nerve VII is more sensitive to sweet or salty, whereas glossopharyngeal fibers are more sensitive to sour or bitter. The gustatory nerve endings usually end in many taste buds, with there being one predominate response in each nerve that matches the function of that region (e.g., sweet from the apex and bitter from the pharyngeal border)

3. Second-Order Neurons

The second-order neurons originate in cranial nerves VII, IX, and X and participate in reflexes, including swallowing, or coughing. The axons from the second-order ascend uncrossed in the central tegmental tract to the medial part of the ventral posterior medial nucleus (VPM) of the thalamus that then projects onto the gustatory cortex in the insula and parietal operculum.

Gustatory information projects from these areas onto the orbital cortex, where there is an overlap of olfactory, gustatory, and hypothalamic sensations that produce a strong response to many gustatory stimuli. Gustatory information also reaches the -parabrachial nucleus in the pontine reticular formation and is then distributed with nociceptive and visceral information onto the hypothalamus, amygdale, and septum.

4. Cortical Neurons

As one ascends in the gustatory system, the neurons are more sensitive to particular chemical sensations. Cells in the brain stem and thalamus respond to more than one taste, whereas in the cortical regions, cells show a selectivity that is not seen in the subcortical regions.

V. Guidelines for Localizing Disease to and within the Brain Stem

1. Mental status is not directly involved (except that lesions involving the reticular formation in the upper brain stem might alter the level of consciousness).
2. No muscle atrophy is presented, except that relevant to local involvement of cranial nerves. Limb weakness, if present, involves central control and spasticity.
3. Ipsilateral lower motor neuron findings relevant to cranial nerve motor function are often associated with contralateral upper motor neuron findings.
4. Deep tendon reflexes are increased in a bilateral or unilateral manner. A unilateral or bilateral sign of Babinski might be present.
5. Extraocular movements are affected. Involvement of horizontal eye movements is indicative of lower pontine involvement. Involvement of vertical eye movements is indicative of an upper midbrain level of involvement.
6. Vertigo nausea and vomiting in association with other neurological symptoms is almost always indicative of brain stem pathology at the lower pontine or upper medullary level. Vertigo alone or in association with tinnitus and ipsilateral hearing deficit is indicative of disease outside of the brain stem at the level of the cochlea or cranial nerve VIII.
7. The combination of selective ipsilateral deficits in pain and temperature over the face in association with selective deficits in pain and temperature over the contralateral body is always indicative of a lateral tegmental upper medullary or lower pontine level, indicating involvement of the spinal tract and nucleus of V.

8. Horner's syndrome in association with ipsilateral cranial nerve findings and the selective pain deficits indicated in item 7 similarly localizes the disease to the lateral tegmental areas of the lower pons or upper medulla.
9. The combination of cranial nerves V, VII, and VIII findings with ipsilateral cerebellar findings localizes the disease to the cerebellar–pontine angle and strongly suggests a vestibular Schwannoma or, less likely, a meningioma in this location.
10. Example of consequences of vascular lesions in the brain stem. Other examples are included on the CD that accompanies this text.

Case History 4.1: Lateral (Dorsolateral) Medullary Syndrome (Wallenberg's Syndrome)

A 53-year-old male on the morning of admission while seated had sudden onset of vertigo accompanied by nausea and vomiting, incoordination of his left arm and leg, and numbness of the left side of his face. On standing, he leaned and staggered to the left. Several hours later, he noted increasing difficulty in swallowing and his family described a minor dysarthria. He had a two-pack–a-day cigarette smoking history for more than 10 years.

Physical examination: Elevated blood pressure of 166/90.

Neurologic examination: Mental status: Intact.

Cranial nerves: The right pupil was 3 mm in diameter; the left pupil was 2 mm in diameter. Both were reactive to light. There was partial ptosis of the left eyelid. Both of these findings were consistent with a Horner's syndrome (ptosis, miosis, anhydrosis). Horizontal nystagmus was present on left lateral gaze, but extraocular movements were full. Pain and temperature sensation was decreased over the left side of the face, but touch was intact. Speech was mildly dysarthric for lingual and pharyngeal consonants. Secretions pooled in the pharynx. Although the palate and uvula elevated well, the gag reflex was absent.

Motor system: A dysmetria (abnormal movement) was present on left finger to nose test and heel to shin tests. He tended to drift to the left on sitting and was unsteady on attempting to stand, falling to the left.

Reflexes: Deep tendon reflexes were physiologic and symmetrical except for ankle jerks, which were absent. Plantar responses were flexor.

Sensory system: Pain and temperature sensations were decreased on the right over the right arm and leg, but all other modalities were normal.

Clinical diagnosis: Dorsolateral medullary infarct.

Laboratory data: The MRI indicated a significant acute or subacute infarct involving the PICA territory of the left dorsolateral medulla and left posterior inferior cerebellum. The MRA suggested decreased flow in the left vertebral artery. Note that prior to the era of modern imaging, the additional involvement of the PICA would not have been detected based only on the clinical findings.

Comments

This syndrome reflects the infarction of the territory of the posterior inferior cerebellar artery (PICA) supplying the rostral lateral medulla. The actual point of occlusion is more often located at the intracranial segment of the vertebral artery than at the PICA.

In the pure or modified form, it is one of the more common of the classical brain stem syndromes. Included within this territory are the following structures: the inferior cerebellar peduncle (ipsilateral finger to nose and heel to shin dysmetria and intention tremor), the vestibular nuclei (vertigo and nausea and vomiting plus nystagmus), the descending spinal tract and nucleus of the trigeminal nerve (ipsilateral selective loss of pain over the face), the spinothalamic tract (contralateral loss of pain and temperature over the body and limbs), the descending sympathetic pathway (ipsilateral Horner's syndrome), and the nucleus ambiguus (dysarthria, hoarseness, ipsilateral vocal cord paralysis, difficult in swallowing with ipsilateral failure in palatal elevation). In Chapter 17 on the CD, other illustrative cases with lesions in the brain stem are presented and discussed in greater detail.

Chapter 5
The Cranial Nerves

There are 12 cranial nerves and they originate from the brain and upper spinal cord and they innervate the special sense organs in the head (eye, ear, nose, and taste buds), the skin over the face and neck, and muscles that permit us to speak, eat, and produce facial expressions. These cranial nerves also provide parasympathetic innervation to the eye and glands in the head and neck and via the X cranial nerve, the vagus, to the organs of the cardiovascular system, pulmonary system, gastrointestinal system, and urogenital system.

HINT: When you study the cranial nerves, first learn the functions of the following "easy" cranial nerves: special sensory, I (smell), II (vision), and VIII (hearing and balance); eye movements, III, IV, and VI (LR6SO4); pure motor functions of XI (shrug shoulders and turn neck) and XII (tongue movements), parasympathetic functions: III, VII, IX, and X; Figure 5.11. Then learn the more difficult mixed ones: V, VII, IX, and X.

I. How the Cranial Nerves Got Their Numbers

The cranial nerves are numbered in sequence from superior to inferior based on the location of their nerve rootlets within the three cranial fossa of the skull: anterior, middle, or posterior cranial fossa.

A. Anterior Cranial Fossa (CN I and II)

Cranial nerve I is the most rostral nerve and hence its listing as number I. It originates in the nasal mucosa and it is found on the base of the frontal lobe and enters the anterior cranial fossa through the foramen in the ethmoid bone, terminating in the olfactory bulb at the base of the frontal lobe.

Cranial nerve II is actually a pathway in the central nervous system and originates in the retina of the eye, passes into the brain through the optic canal in the lesser wing of the sphenoid, and terminates in the diencephalon.

B. Middle Cranial Fossa (CN III, IV, V, and VI)

Cranial nerves III and IV leave the midbrain, whereas CN VI leaves the pons, passes through the medial most portion of the middle cranial fossa, close to the internal carotid artery, enters the cavernous sinus, then exits the cranial cavity via the superior orbital fissure with the ophthalamic branch of the trigeminal, V1, and then terminates on muscles in the orbit.

C. Posterior Cranial Fossa (CN VIII–XII)

Cranial nerves VIII–XII exit the medulla, with CN VII and CN VII exiting through the internal acoustic meatus of the petrous portion of the temporal bone high in the posterior fossa. Just below the internal acoustic meatus, one finds the jugular foramen with CN IX, CN X, and CN XI found in the medial part of the jugular foramen, and, finally, CN XII is the most inferior root exiting through the hypoglossal foramen in the occipital bone just above the foramen magnum.

II. Functional Organization of Cranial Nerves (Table 5.1)

In the spinal cord, the neurons are arranged in continuous columns of cells, each of which is connected to a structure with a specific function; this grouping of neurons with similar anatomic and physiologic functions is called a *nerve component* (Table 5.1). In the spinal cord, the neurons innervate general structures such as skin, skeletal muscles, blood vessels, and viscera. In the brain stem, the cranial nerves innervate not only general structures but they also innervate the special sense organs (the eye, ear, nose, or taste buds).

The cranial nerves that innervate these structures are organized in columns in the tegmentum of the brain stem, with their cell bodies in either the floor of the IV ventricle or in the reticular formation that forms the core of the brain stem.

The following are other useful facts on the cranial nerves (see Figs. 5.1 and 5.2):

Cranial nerve I originates from the olfactory mucosa and terminates in the olfactory bulbs on the base of the frontal lobe.

Cranial nerve II is actually a pathway in the central nervous system and it originates in the retina and exits the eye and runs on the base of the diencephalon on its way to the lateral geniculater nucleus (LGN) of the diencephalon.

Cranial nerves III, IV, and VI are centrally wired together to provide conjugate movement of the eyes.

Cranial nerves V, VI, and VII are within the pons, while cranial nerves VIII–XII are within the medulla

Table 5.1 Components of the cranial nerves

Function	Cranial nerve
1. Motor to skeletal muscles of somite origin	III in midbrain, IV in midbrain, VI in pons, and XII in medulla
	Tegmental zone = in ventricular floor
2. Motor to the parasympathetic Organs (Fig. 5.1)	Edinger–Westphal nucleus of III to pupil of eye: superior salavatory nucleus of VII in pons; iInferior salavatory nucleus of IX in medulla; dorsal motor nucleus of X in medulla
	Tegmental zone = in ventricular floor
3. Motor to skeletal muscles of pharyngeal origin	Motor nucleus of V and VII in pons
	Ambiguous nucleus of X and XI in medulla
	Tegmental zone =in core
4. Sensory: visceral and gustatory sensations	Solitary nucleus in medulla with primary cell bides in ganglia of VII, IX, and X
	General sensation from the viscera and taste also found here
	Tegmental zone = in ventricular floor
5. Sensory: cutaneous and proprioceptive from skin and muscles in the head and neck	Trigeminal CN V nuclei: mesencephalic in midbrain; chief sensory in midpons, descending nucleus in pons: upper cervical cord
	General sensations carried in by III, IV, VI, VII, XI, and XII synapse in the descending nucleus of V
	Tegmental zone =in. In lateral
6. Special somatic sensory	Vision: II; -hearing and balance: VIII
7. Special visceral sensory	Olfaction: I; taste: VII, IX. and X

Cranial nerve V, the trigeminal nerve, innervates the skin on the face and in the oral and nasal cavities, the teeth, sinuses in the cranial bones, muscles of mastication, and the dura,

Cranial nerve VII, the facial nerve, controls the muscles of facial expression, taste buds on the anterior two-thirds of the tongue, and many glands in the head and neck. General sensation from the tongue for the anterior two-thirds is from the lingual nerve of the trigeminal, whereas sensation to the posterior third of the tongue is from CNIX.

Cranial nerve IX, the glossopharyngeal nerve, innervates the taste buds in the posterior third of the tongue and initiates swallowing by controlling the stylopharyngeus muscle.

Cranial nerve X, the vagus, is the longest cranial nerve and controls all of the types of muscle in the body: skeletal muscle in the larynx and pharynx, smooth muscle in the viscera of the thorax and abdomen, and cardiac muscle. It provides sensory innervation from these same regions, including the larynx and pharynx and viscera and it also innervates taste buds in the pharynx and larynx.

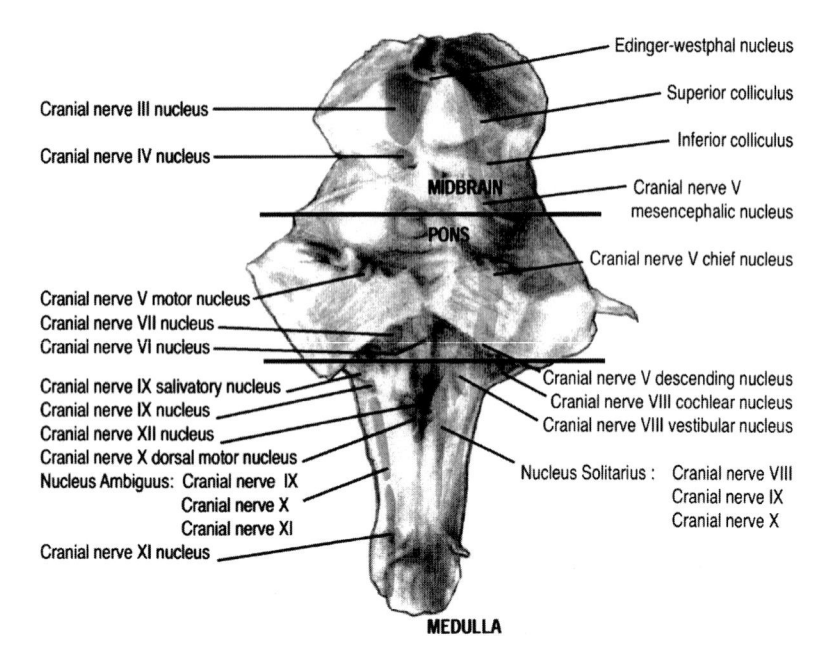

Fig. 5.1 Locations of the cranial nerve nuclei in the gross brain stem. Motor nuclei of CN III–VII and CN IX–XII on the left; sensory nuclei of CN V and CN VII–X on the right. See Figure 5.3 for a demonstration of the functions of each of the 12 cranial nerves. (From EM Marcus and S Jacobson, Integrated neuroscience, Kluwer, 2003)

III. Embryological Considerations

The distribution of each cranial nerve is related to embryological structures derived from the pharyngeal arches, occipital somites, and paraxial mesoderm or placodes. The following lists the origins of the cranial nerves and associated muscles.

A. Cranial nerve to each pharyngeal arch. There are six branchial arches with a core of mesoderm lined on the outside with ectoderm and lined on the inside by endoderm that begin on day 22 and form in a craniocaudal sequence:

Arch 1: Maxillary and mandibular swelling by CN V.
Arch 2: Facial stapes, styloid process, lesser horn of hyoid by CN VII with muscles of facial expression
Arch 3: Hyoid, greater horn hyoid by CN IX
Arches 4 and 6 fuse, with arch 5 regressing
Arch 4: Upper Laryngeal cartilage innervated by superior laryngeal nerve of CN X also innervates recument layugeal branch of CN X also innervates constrictors of the pharynx, cricothyroid and levator veli palatini
Arch 6: Lower portion of laryngeal cartilage innervated by recurrent laryngeal branch of CN X also innervates intrinsic muscles of larynx that originated from occipital somites I and 2.

III Embryological Considerations

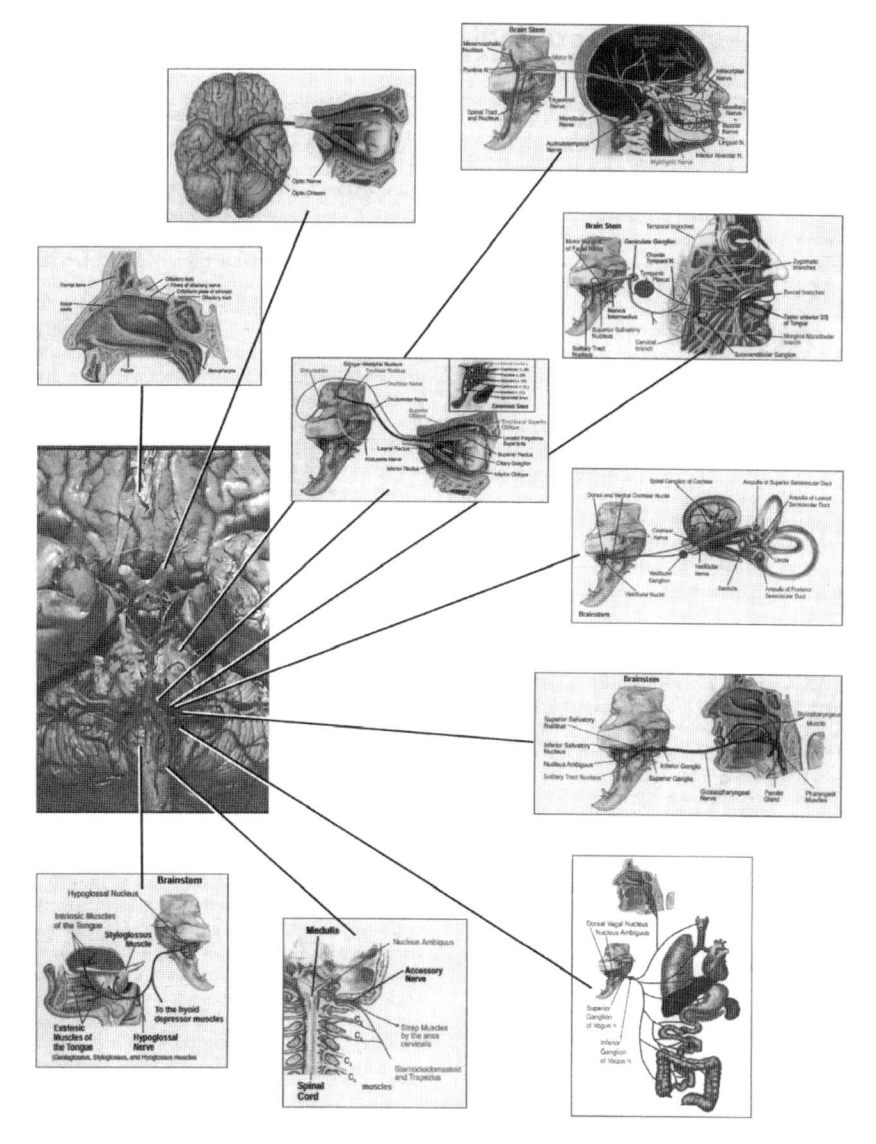

Fig. 5.2 The 12 cranial nerves (From EM Marcus and S Jacobson, Integrated neuroscience, Kluwer, 2003)

    B. Occipital somites. These somites form muscles of tongue- XII is to intrinsic & extrinsic muscles of tongue & X to palatoglossal muscle.

    C. Pariaxial mesenchyme condenses near optic cup for recti muscles innervated by muscles of CN III, CN IV, and CN VI

    D. Placodes, origin of special sensory nerves. Ectodermal condensations form olfactory (I), optic (II) and vestibulo-cochlear (VIII) nerves.

IV. The Individual Cranial Nerves

Table 5.2 summarizes the components of each of the 12 cranial nerves.

Table 5.2 Cranial nerve components (see Fig. 5.3 for a visual summary)

Cranial Nerve	Location of Cell Bodies	Function
I Olfactory	Neuroepithelial cells in nasal cavity	Olfaction
II Optic-central pathways discussed in Chapter 21	Ganglion cells in retina	Vision
III Oculomotor	Oculomotor nucleus in tegmentum of upper midbrain	Eye movements, levator palpekal supenored, superior medial and inferior rectus, inferior oblique
	Edinger-Westphal nucleus in tegmentum of upper midbrain – preganglionic to ciliary ganglion	Pupillary constriction and accommodation of lens for near vision through ciliary ganglion
IV Trochlear	Trochlear nucleus in tegmentum of lower midbrain	Eye movements (contralateral), superior oblique muscle
V Trigeminal	Primary trigeminal ganglion, Secondary Mesencephalic (midbrain), chief sensory (Pons), and descending spiral nuclei (pons, Medulla, upper cervical levels)	Cutaneous and proprioceptive sensations from skin and muscles in face, orbit, nose, mouth, forehead, teeth, paranasal sinuses, meninges, and anterior two-thirds of tongue.
	Motor nucleus in pons	Muscles of mastication
VI Abducens	Abducens nucleus in tegmentum of pons	Eye movements, ipsilateral, involving lateral rectus muscle
VII Facial	Primary geniculate gangllon: Secondary Nucleus solitarius	Gustatory sensations from taste buds in anterior two-thirds of tongue
	Motor nucleus in lateral margin of pons	Muscles of facial expression and platysma; extrinsic and intrinsic ear muscles, stapedius muscle
	Superior salivatory nucleus in pons preganglionic parasympathetic to ganglia associated with these glands	Glands of nose and palate and the lacrimal, submaxillary and sublingual glands
VIII Vestibulo-acoustic. (Central pathways discussed in chapter 15)	Primary vestibular & cochlear ganglion, temporal bone;	Audition
	Secondary cochlear nuclei in medulla	Equilibrium, coordination, orientation in space
	Secondary vestibular nuclei in medulla and pons	

Table 5.2 (continued)

Cranial Nerve	Location of Cell Bodies	Function

(continued)

IX Glossopharyngeal	Primary inferior ganglion; Secondary Nucleus solitarius Inferior salivatory nucleus in medulla Nucleus ambiguous Preganglionic to otic ganglion	Interoceptive palate and posterior one-third of tongue, carotid body; Gustatory sensations from taste buds in posterior one-third of tongue Secretions from parotid gland through otic ganglion Swallowing, stylopharyngeus mm Visceral Sensations
X Vagus	Primary Superior inferior ganglion; Dorsal motor nucleus – preganglionic (Parasympathetic innervation) Nucleus ambiguous in medulla	Visceral sensations from pharynx, larynx aortic body and thorax and abdomen Gustatory Sensation from taste buds in epiglottis and pharynx Skeletal muscles in pharynx & larynx Smooth muscles in heart, blood vessels, trachea, bronchi, esophagus stomach, intestine to lower colon
XI Spinal Accessory	Cranial Portion Ambiguous nucleus in medulla Spinal portion: C2 C3 & 4	Innervation Stemomastoid Trapezius and mm of pharynx, larynx, except cricothyroids
XII Hypoglossal	Hypoglossal nucleus in medulla	Innervation of intrinsic and extrinsic muscles of the tongue except palatoglossal MM

A. *Cranial Nerve I: Olfactory (Fig. 5.3)*

Special Visceral Sensory: Olfaction. Originates in olfactory mucosa; enters anterior cranial fossa via 20 foramina in cribiform plate of ethmoid; found at base of frontal lobe in anterior cranial fossa.

Clinical Disorders. Defects in smell = anosmia. The most frequent causes of unilateral or bilateral anosmia are as follows:

1. Disease of nasal mucosa producing swelling and preventing olfactory stimuli from reaching the olfactory receptors
2. Head trauma, resulting in shear effects tearing the filaments of olfactory receptor cells passing through the cribiform plate
3. Olfactory groove meningioma.

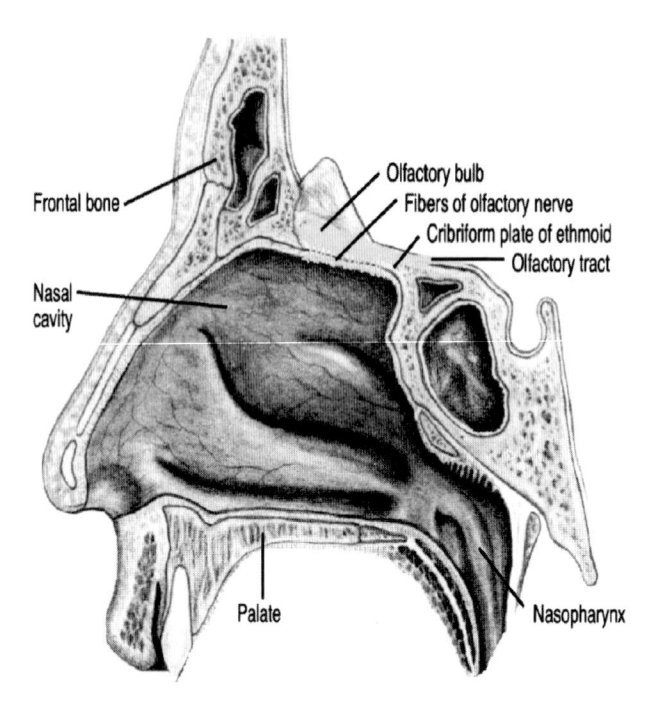

Fig. 5.3 Cranial nerve I, the olfactory nerve (From EM Marcus and S Jacobson, Integrated neuroscience, Kluwer, 2003)

B. Cranial Nerve II, Optic (Fig. 5.4)

Special Somatic Sensory. Originates from ganglion cell layer of retina of eye; enters cranial vault via optic foramen found at base of diencephalon next to pituitary and enters LGN of diencephalon.

Clinical Disorders. A complete or partial unilateral loss of vision is produced by retinal or optic nerve disease. Partial lesions produce a scotoma "a hole in the visual field."

1. Retinal disease.

    a. Vascular: Transient: carotid occlusive disease and central retinal artery thrombosis: sudden and persistent: -central venous thrombosis
    b. Degeneration:retinitis pigmentosa

2. Optic nerve disease. Optic neuritis: demyelination as manifestation of multiple sclerosis, ischemic optic neuropathy, toxic effects (e.g., methanol usually bilateral, thiamine and B_{12} deficiency states usually bilateral, familial degeneration (Leber's optic neuropathy), compression: by masses; Inner third sphenoid wing or olfactory groove: Meningioma, glioma of optic nerve; long-standing papilledema; pituitary masses (usually involve optic chiasm resulting in a bitemporal hemianopia).

IV The Individual Cranial Nerves
129

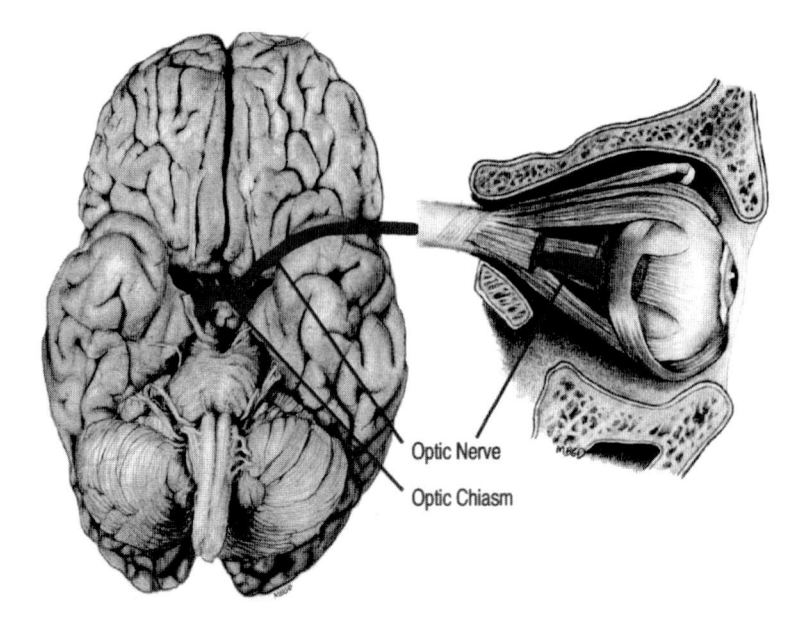

Fig. 5.4 Cranial nerve II, the optic nerve (From EM Marcus and S Jacobson, Integrated neuroscience, Kluwer, 2003)

C. Cranial Nerve III, Oculomotor (Fig. 5.5)

Pure Motor [somatic and parasympathetic (only III)] Eye movements controlled by CN III, CN IV, and CN VI work together to move the eyes-controlled from frontal eye fields.

1. Originates from nucleus of III in tegmentum within medial longitudinal fasciculus (MLF) below superior colliculus of midbrain, Fibers run in middle cranial fossa and enter orbit via superior orbital fissure.

   a. General somatic efferent to levator palpebrae superioris, medial rectus, inferior rectus, superior rectus, and inferior oblique
   b. Parasympathetic: general visceral efferent, preganglionic to ciliary ganglia in orbit, postganglionic parasympathetic to ciliary muscles and constrictor of pupil (Fig. 5.8).

Clinical Disorders (Table 5.3)

1. Complete paralysis of nerve III produces ptosis of the lid, paralysis of the medial and upward gaze, weakness in the downward gaze, and dilation of the pupil. The eyeball deviates laterally and slightly downward. The dilated pupil does not react to light or accommodation. See also Chapter 13.

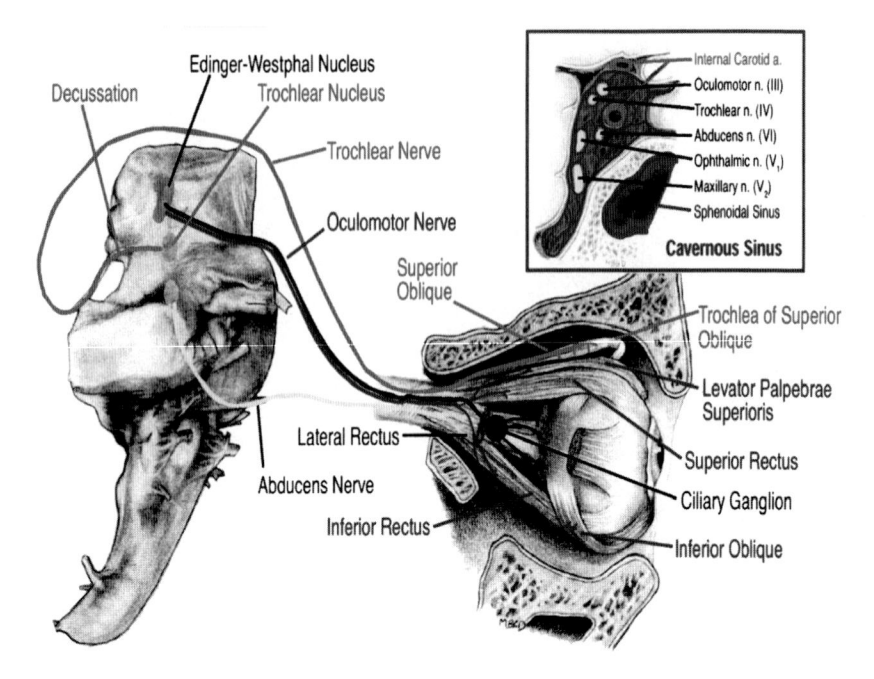

Fig. 5.5 Cranial nerves III, IV, and VI (From EM Marcus and S Jacobson, Integrated neuroscience, Kluwer, 2003).

Table 5.3 Dysfunction in the eye due to lesions in the brain stem

Dysfunction in eye movements	Location of clinical deficits
Role of MLF	
Ascending fibers coordinate III and VI	Pons or midbrain
Descending fibers coordinate eye movements with head and neck movements	Pons or medulla
Coordinate movements of III and VI	Lateral gaze center in pons
Disturbance of conjugate gaze	Pons or midbrain
Convergence	Damage in midbrain to pretectal area or the Edinger–Westphal nucleus
Pupil dilated, paralysis levator, med, sup. Inf recti muscles and inf oblique muscles Paralysis ips. Lat. Rectus, and no coodinated lateral gaze with Med rectus	CN III in midbrain by herniation or vascular -CN VI in pons due to vascular or tumor
Pupillary dysfunction	
Dilation-Constriction	Dilation of CN III in midbrain or peduncle results in fixed dilated pupil (blown pupil); only sympathetic control. Constriction lesion of sympathetic fibers and now only parasympathetic control through III.

Paralysis of the intrinsic muscles (the ciliary sphincter, or pupil) is called *internal ophthalmoplegia*, whereas paralysis of the extraocular muscles (recti and obliques) is called *external ophthalmoplegia*.

2. Causes of cranial nerve III dysfunction:

   a. A compression syndrome involves both pupillary and extraocular muscles. Initially, only the pupil may be involved: aneurysm of posterior communication internal carotid artery junction

      (i) Lesions within the cavernous sinus also effect CN V1. -.(ii) Herniation of temporal lobe

   b. Tumors: meningioma of tuberculum sellae or sphenoid wing; noncompression syndromes

   c. Diabetic cranial neuropathy (vascular). Pupil may be spared, at time of extraocular involvement.

   d. Ophthalmaloplegic migraine

D. Cranial Nerve IV: Trochlear (Fig. 5.5), Pure Motor

Originates from nucleus of CN IV in tegmentum below inferior colliculus of midbrain in MLF. Axons exit on posterior surface of tegmentum and are crossed (both unique conditions for motor cranial nerves). Fibers run in middle cranial fossa and enter orbit via superior orbital fissure, controls superior oblique mm.

Clinical Disorders. Trauma is the primary cause of defects in CN IV. A nuclear lesion of nerve IV paralyzes the contralateral superior oblique muscle, wereas a lesion in the nerve roots after decussation involves the ipsilateral muscle. A thrombosis or tumor in the cavernous sinus might also affect this nerve and cranial nerves III, V, and VI.

E. Cranial Nerve VI: Abducens (Fig. 5.5), Pure Motor

Originates from nucleus of CN VI in pons with fibers of CN VII hooking around the nucleus of CN VI. The fibers from CN VI pass ventral and run in the middle of the cranial fossa via superior orbital fissure to lateral rectus in orbit.

Clinical Disorders. This nerve has a long intracranial course and is thus commonly affected by increases in intracranial pressure (either unilateral or bilateral) and by nasopharyngeal tumors invading the base of the skull in adults. Pontine gliomas produce similar effects in children. The major non-neoplastic cause is a diabetic neuropathy (vascular).

F. Cranial Nerve V: Trigeminal (Fig. 5.6), Mixed Nerve (Sensory and Motor But No Parasympathetic)

1. Motor. Branchial Motor. Nerve to first pharyngeal arch, the mandibular arch to muscles of mastication
2. Sensory fibers originate in middle cranial fossa from large trigeminal ganglion in Meckel's cave.
    a. General somatic afferent fibers terminate in second sensory nuclei in the brain stem (midbrain, pons, and medulla and upper cervical cord)
    b. Peripheral Branches:
        (i) V1 Ophthalmic: sensory, into orbit via superior orbital fissure to meninges (dura), falx cerebri, and tentorium cerebelli
        Cutaneous to forehead, upper eyelid
        (ii) V2 Maxillary: sensory, leave middle cranial fossa via foramen rotundum, passes through pterygopalatine fossa and finally enters orbit through inferior orbital fissure where it become the infraorbital nerve.
        Cutaneous to side of the nose, nasal septum, lower eyelid, upper lip, and upper teeth, meninges mucous membranes in nasopharynx, maxillary sinus, soft palate, tonsial, and roof of the mouth.
        (iii) V3 Mandibular (largest branch): only branch of CN V with motor functions; mixed sensory and motor, enters infratemporal fossa through foramen ovale

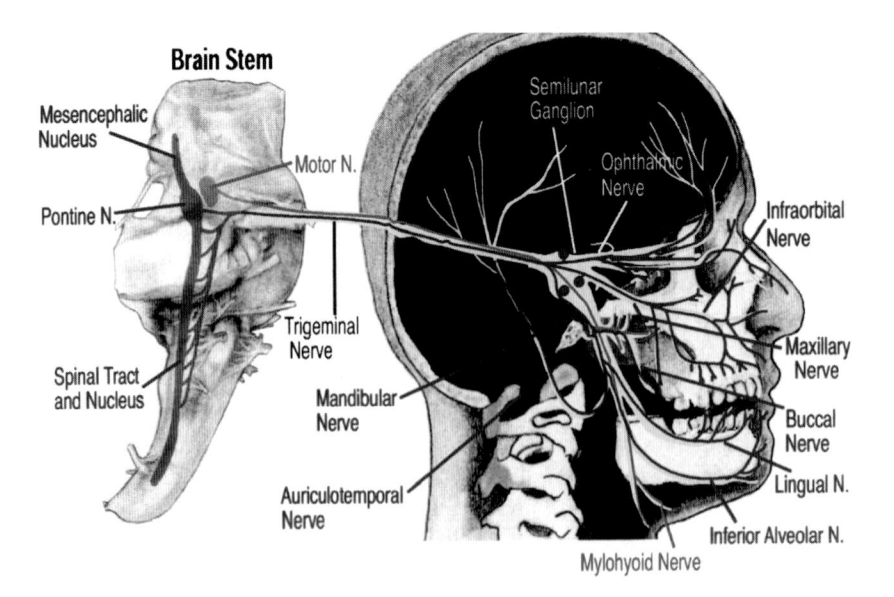

Fig. 5.6 Cranial nerve V, trigeminal nerve (From EM Marcus and S Jacobson, Integrated neuroscience, Kluwer, 2003)

IV The Individual Cranial Nerves 133

   (a) Sensory branch: Cutaneous from lower lip, jaw and lower teeth, external ear and skin over cheek, and meninges; general sensation to anterior two-thirds of tongue via lingual nerve

   (b) Motor branch: Originates from motor nucleus in pons; (branchiomeric/special visceral efferent) to muscles of mastication = temporal, masseter, lateral and medial pterygoid, mylohyoid, anterior belly of digastric, tensor tympani, and tensor veli palatini.

Clinical Disorders. Injury to the trigeminal nerve paralyzes the muscles of mastication, with the jaw deviating toward the side of the lesion. Injury might also block sensation of light touch, pain, and temperature in the face and results in the absence of the corneal and sneeze reflexes.

Localization of Sensory Dysfunction of the Trigeminal Nerve:

1. Trigeminal neuralgia (tic, douloureux) is a disorder characterized by recurrent paroxysms of stabbing pains along the distribution of the involved branches.

   a. Clinical Signs of Trigeminal Neuralgia:

      (i) Etiology: "Idiopathic": with mean age of onset in the fifties
      (ii) Trigger points produce the pain on tactile stimulation. The pain is of short duration; lancinating pain, usually occurring in clusters or paroxysms. Pain might resolve spontaneously only to recur months later.
      (iii) Location of pain: The second (maxillary division) is most frequently involved. The third (mandibular division) is next in frequency. The first (ophthalmic division) is least common. The pain is unilateral and rarely bilateral.
      (iv) Clinical findings:. There are no sensory or motor findings on examination, although after frequent repeated episodes the patient might complain of minor numbness and a more persistent aching type pain.
      (v) Treatment: Many of these cases respond to medication– such as carbamazepine; some do not and require surgical therapy. Patients with onset after age 40 often have compression of the trigeminal nerve by an anomalous loop of an artery compressing the nerve or root entry zone. Treatment in such cases that fail to respond to medications involves relocating the loop of the superior cerebellar artery or insulating the nerve from the pulsation of the vessel by inserting a soft implant between nerve and vessel. Other surgical procedures are designed to alter transmission through the Gasserian ganglion.

2. Other causes of trigeminal symptoms.

   a. Compression or invasion of nerves by nasopharyngeal tumors that have infiltrated the base of the skull through the foramen rotundum (maxillary) or foramen, ovale (mandibular), or other points of erosion.
   b. Dental pathology or maxillary sinusitis might produce pain within the appropriate divisions of the trigeminal nerve. Tingling and numbness in the

mandibular or, less often, maxillary division might occur spontaneously or follow dental procedures.

c. Herpes zoster involvement of the Gasserian ganglion. Acute pain almost always in the ophthalmic division followed in 4–5 days by cutaneous lesions. Treatment with acyclovir will decrease the period of pain. Note: The *H. zoster/varicella* virus is often dormant in the Gasserian ganglion.

G. Cranial Nerve VII, Facial (Fig. 5.7), Mixed Nerve (Sensory, Motor, Parasympathetic)

Nerve to second pharyngeal arch, the hyoid arch. Large motor nucleus in pons. Rootlets in posterior cranial fossa (exits via internal acoustic meatus). Peripheral nerve fibers enter internal acoustic meatus, run in the facial canal in petrous temp, to geniculate ganglia here sensory fibers exit via stylomasotid foramen, then gives off branches to post belly of digastric, stylohoyoid and post auricular muscles, passes through parotid gland (does not innervate) where it divides into terminal branches: Temp, Zygomatic, Buccal, Marginal, Mandibular, and Cervical (platysma).

1. Motor

    a. Motor visceral efferent from nucleus in pons to muscles of facial expression and platysma, extrinsic and intrinsic ear muscles, post belly digastric, stapedius, and stylohyoid.

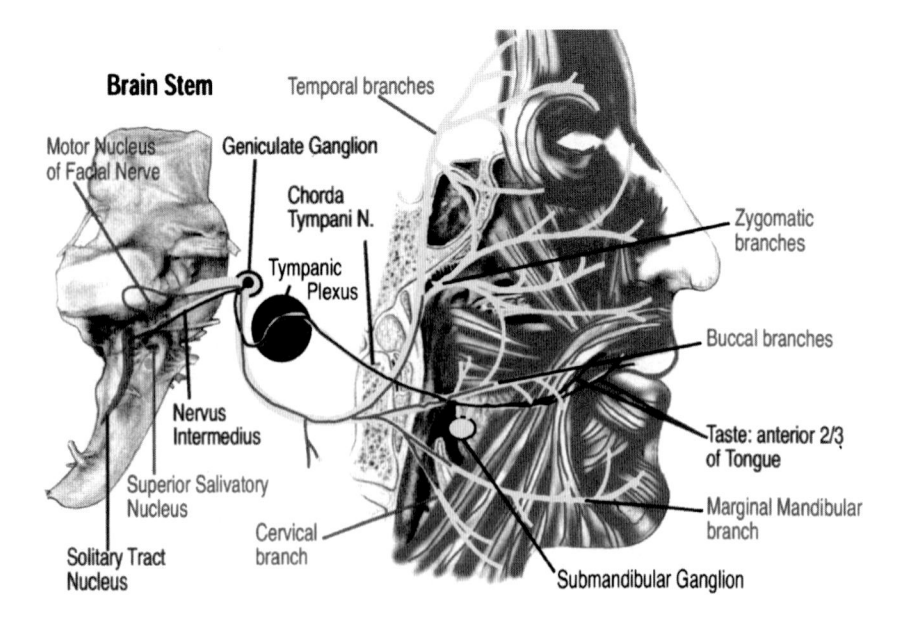

Fig. 5.7 Cranial nerve VII, facial nerve (From EM Marcus and S Jacobson, Integrated neuroscience, Kluwer, 2003)

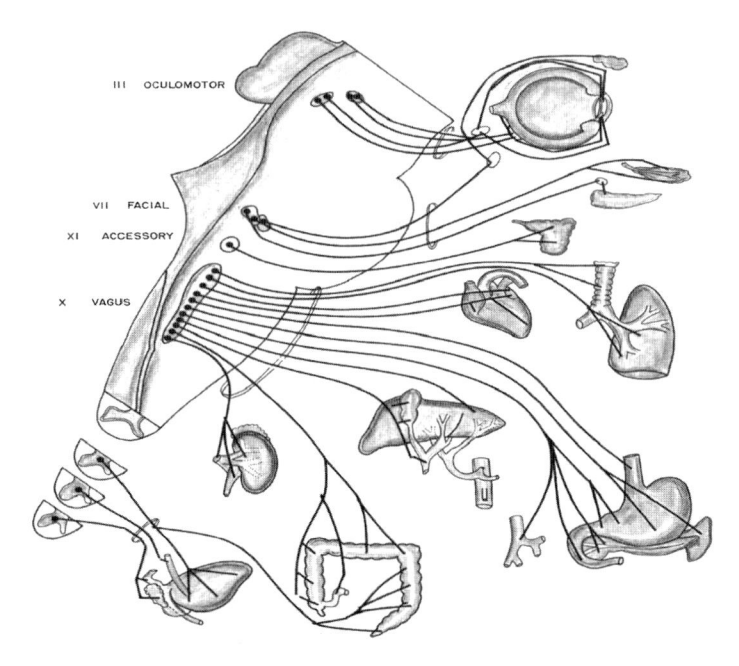

Fig. 5.8 Parasympathetic innervations: Cranial from cranial nerves III, VII, IX, and X and sacral from spinal cord levels S2–S4 (From Curtis, Jacobson, & Marcus, 1974)

  b. General visceral efferent preganglionic/parasympathetic to pterygopalatine ganglion from superior salavatory nucleus in pons to glands in nose, palate, cranial sinuses, and to ganglia for submandibular and submaxillary gland and sublingual glands (Fig. 5.8).
2. Sensory
  a. Intermediate nerve (between CN VII and CN VII). Special visceral afferent taste buds (chorda tympani) from ant 2/3 of tongue via petrotympanic fissure.

Clinical Disorders:

1. Central injury in the brain stem. A central injury to the facial nucleus in the pons produces ipsilateral paralysis of the facial muscles. Injury to just the nucleus or tractus solitarius, although rare, produces a loss of taste on half of the tongue because the taste fibers from nerves VII and IX are found together in the tractus solitarius.
2. Peripheral injury. Injury to the facial nerve near its origin or in the facial canal produces:
  a. Paralysis of the motor, secretory, and facial muscles.
  b. A loss of taste in the anterior two-thirds of the tongue.
  c. Abnormal secretion in the lacrimal and salivary glands. The patient has sagging muscles in the lower half of the face and in the fold around the lips and nose and widening of the palpebral fissure.

    d. Voluntary control of facial and platysmal musculature is absent. The patient is unable to close the eyelid. When the patient smiles, the lower portion of the face is pulled to the unaffected side. Saliva and food tend to collect on the affected side. When the injury is distal to the ganglion, an excessive accumulation of tears occurs because the eyelids do not move, but the lacrimal glands continue to secrete.
    e. Injury to the facial nerve might also interrupt the reflex arc for the corneal blink reflex, which includes the motor component to the orbicularis from nerve VII and the sensory corneal nerves of V1. The muscles eventually atrophy.
    f. Injury to nerve VII at the stylomastoid foramen produces ipsilateral paralysis of the facial muscles without affecting taste.
    g. Injury to the chorda tympani results in absence of taste in the anterior two-thirds of the tongue.
3. Other peripheral sites of injury to CN VII: Tumors and infections caused by mumps virus (rubulavirus) of the parotid gland often involve the facial nerve
4. Central/cerebral innervation of CN VII. The corticobulbar pathway to the lower half of the facial nucleus is a crossed unilateral innervation, whereas the upper half of the facial nucleus receives a bilateral innervation. A unilateral lesion in the appropriate part of the precentral gyrus will thus paralyze the lower half on the opposite side, whereas a bilateral cortical involvement paralyzes the upper and lower facial musculature. The cerebral cortical control of the muscles of facial expression through the motor strip is especially important in communication.

If these upper motor neuron fibers are destroyed, other cortical regions, including the limbic system, will now cause inappropriate facial expressions (pseudobulbar palsy). Pseudobulbar palsy represents excessive uninhibited brain stem activity. External stimuli will then trigger excessive response (e.g., excessive crying to a sad story or excessive laughter).

H. Cranial Nerve VIII, Vestibulo-cochlear (Fig. 5.9), Special Somatic Sensory, Receptive Organs in Petrous Temporal Bone, Nerve Exits via Internal Acoustic. Nerve in Posterior Cranial Fossa (Exits Internal Acoustic Meatus in Petrous Temporal)

Terminates in nuclei in pons/medulla: hearing and balance

A. Clinical Disorders of Auditory System

1. Auditory: Symptoms and Signs
    a. Deafness: Classification

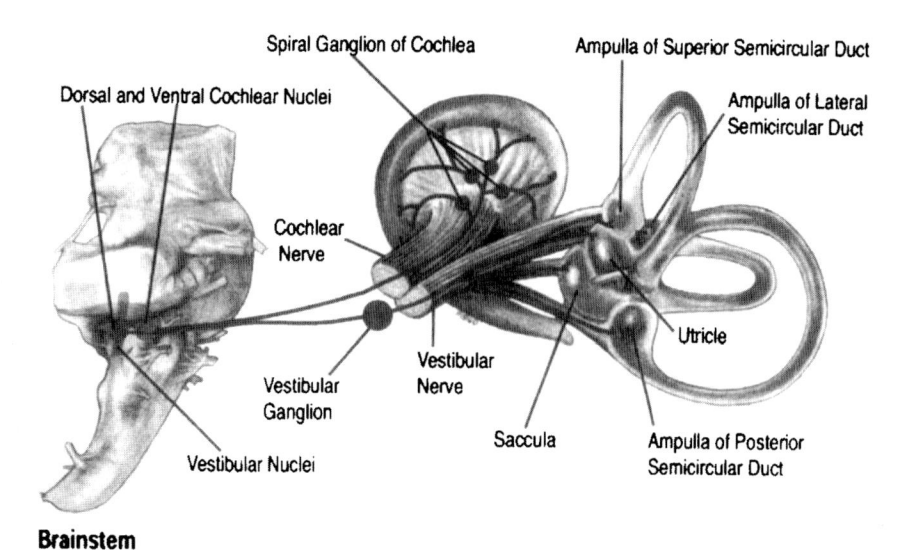

Fig. 5.9 Cranial nerve VIII, vestibulo-acoustic nerve, (From EM Marcus and S Jacobson, Integrated neuroscience, Kluwer, 2003)

   (i) Conductive: External and middle ear disease (loss of low frequency sounds)
   (ii) Sensory neural (nerve deafness, loss of high-frequency sounds)
  b. Disease of cochlea, often bilateral, hereditary, or related to loud sound exposure.
  Antibiotic us: aminoglycosides, dihdrostreptomycin and vancomycin.
  If sudden and unilateral, consider vascular etiology: occlusion of the internal auditory artery.
  c. Disease of eighth nerve: Small tumors (acoustic neuroma) of eighth nerve within the bony canal might produce limited CN VIII symptoms. Large tumors, such as acoustic neuromas, meningiomas, and so forth, at the cerebellar pontine angle produce involvement of a cluster of cranial nerves (VIII, VII, ±V, ±IX, X).

Note that these "acoustic neuromas" arise from the Schwann cells of the vestibular divisions of the nerve. A more correct term is vestibular schwannoma.

1. Tinnitus: various forms of ringing in the ear. Etiology: middle or inner ear or eighthh nerve usually associated with hearing loss.
2. *Central lesions of cochlear pathway*. Note that beyond the cochlear nuclei, central connections of the auditory system are both crossed and uncrossed. Therefore, a unilateral lesion is unlikely to produce a unilateral hearing defect. Massive intrinsic lesions are required to produce "brain stem deafness."

B. Clinical Disorders: Vestibular

Symptom:

1. Vertigo: a sensation of movement (usually rotation) of the environment or of the body and head
   a. When due to involvement of the horizontal (lateral) canals, the sense of movement is usually rotation in the horizontal plane.
   b. When due to involvement of the posterior or vertical canals, the sense of movement might be in the vertical or diagonal planes with a sense of the floor rising up or receding.
   c. Vertigo must be distinguished from the more general symptom of dizziness (or light-headedness), which has many causes. Simple lightheadedness as a symptom of presyncope (the sensation one is about to faint) might accompany any process that produces a general reduction in cerebral blood flow (cardiac arrhythmia, decreased cardiac output, peripheral dilatation of vessels, decreased blood pressure, or hyperventilation). Hypoglycemia and various drugs also produce dizziness. Some patients cannot make the distinction and a series of operational tests might be necessary to reproduce symptoms.

   Rotation, hyperventilation, or orthostatic changes in blood pressure are other causes. Additionally, caloric testing, cardiac auscultation, EKG, and Holter monitor might be necessary in selected cases.

2. Nausea and vomiting: Vertigo of vestibular origin is accompanied by these symptoms because of the connections of the vestibular system to the emetic center of the medulla.
3. Nystagmus (a jerk of the eyes): A slow component alternating with a fast component is associated with stimulation or disease of the vestibular system. Nystagmus is named from the fast component and the direction of movement is specified: rotatory, horizontal, or vertical.
4. Unsteadiness: Severe or mild. Etiology: Multiple labyrinthine disease is the most common

Meniere's disease: Paroxysmal attacks of vertigo, ± vomiting and ataxia, are associated with tinnitus and deafness. The etiology relates to endolymphatic hydrops—an increase in volume and distention of the endolymphatic system of the semicircular canals.

Vestibular Nerve: Vestibular neuropathy acute, self-limited syndrome, which follows a viral infection. Symptoms are vestibular without cochlear involvement. There is unilateral vestibular paresis.

Tumors: Arising from or compressing the vestibular nerve, "acoustic neuroma." Note that this tumor actually arises from the Schwann cells of the vestibular nerve.

Brain stem: Infarcts involving the vestibular nuclei will produce vertigo and/or nystagmus. Almost always other brain stem signs are present.

Cerebral cortex: Stimulation of temporal lobe in patients with seizures of temporal lobe origin produces dizziness.

I. Cranial Nerve IX, glossopharyngeal, (Fig. 5.10), Mixed (Sensory, Motor, Parasympathetic). Nerve to third pharyngeal arch. Nerve in posterior cranial fossa (exits via jugular foramen)

1. Motor
   a. Parasympathetic/general visceral efferent from inferior salavatory nucleus in pons to parotid gland via otic ganglia (Fig. 5.8)
   b. Special visceral efferent, from ambiguous nucleus in medulla to the styopharyngeus mm that initiates swallowing

2. Sensory:
   a. General visceral afferent visceral sensations from posterior third of the tongue, and carotid body and sinus terminate in solitary nucleus in medulla.
   b. Special visceral afferent to taste buds posterior third of the tongue. Terminates in solitary nucleus in medulla.
   c. Reflexes
      1. Gag reflex. Forms sensory portion; CN X the motor portion.
      2. Carotid sinus nerve X Carotid sinus monitors BP feeds back to vagus nerve while carotid body monitors O_2 an CO_2 levels.

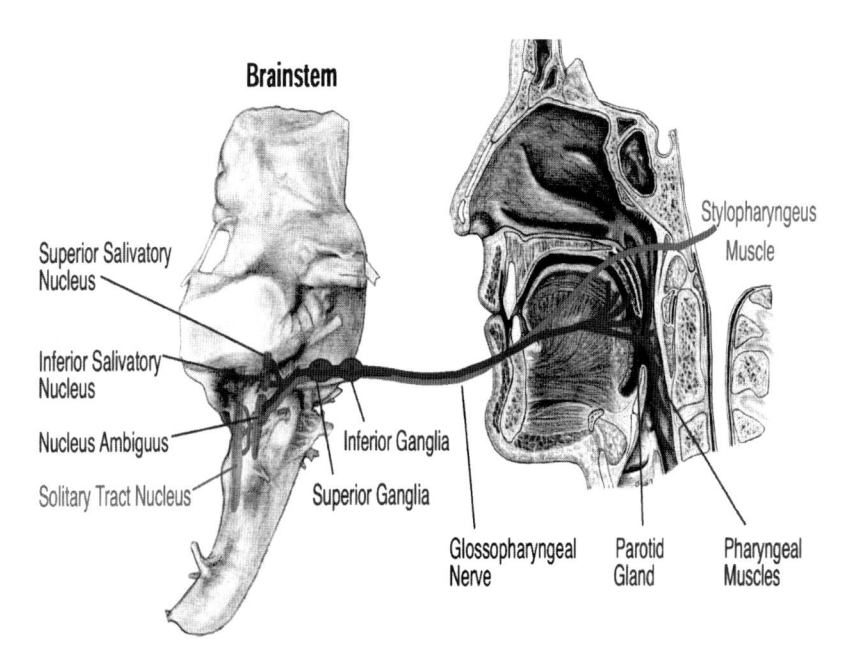

Fig. 5.10 Cranial nerve IX, glossopharyngeal (From EM Marcus and S Jacobson, Integrated neuroscience, Kluwer, 2003)

J. Cranial Nerve X. vagus (Fig. 5.11), Mixed (Sensory, Motor, Parasympathetic), and Longest Cranial Nerve. Nerve to Fourth and Sixth Pharyngeal Arch.

Vagus nerve exits posterior cranial fossa exits via jugular foramen with CN IX and CN XI. The vagus. A mixed nerve. Motor (skeletal and smooth); Sensory special visceral afferent to taste buds and general afferents from viscera and aexternal acoustic meatus and back of ear. Efferent to muscles in pharynx and larynx for phonation and glutination from ambiguous nucleus in medulla.

1. Motor:

    a. Parasympathetic. General visceral efferent to smooth muscles in heart, blood vessels, trachea, bronchi, esophagus, stomach, small intestine to lower half of large intestine from dorsal motor nucleus of X in medulla (Fig. 5.8)
    b. Branchial motor. To muscles in larynx and pharynx

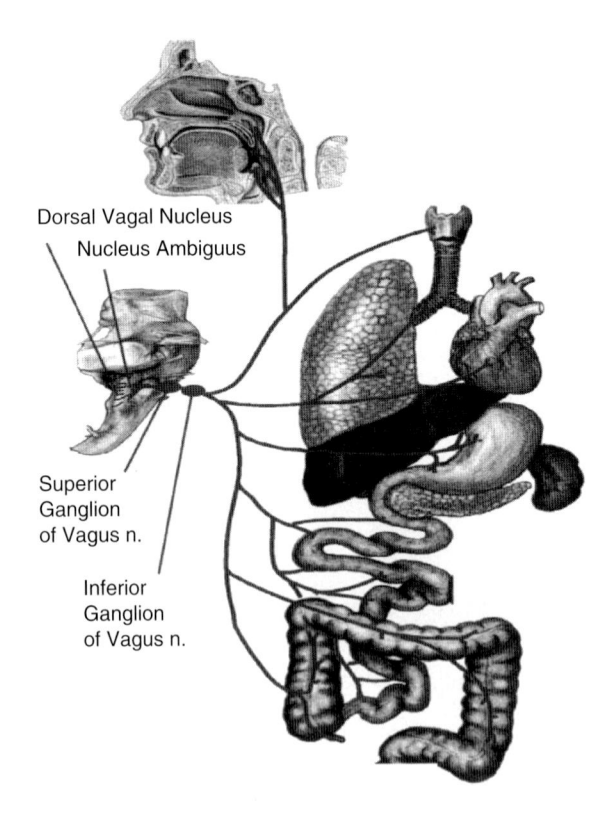

Fig. 5.11 Cranial nerve X, vagus. Motor nuclei distribute to branchial motor muscles in larynx and pharynx (ambiguous of CN X) and parasympathetic smooth muscles in gastrointestinal system (dorsal motor of CN X. (From EM Marcus and S Jacobson, Integrated neuroscience, Kluwer, 2003)

2. Sensory:
   a. General visceral afferent carries visceral sensation from pharynx, larynx, aortic body, thorax, and abdomen to solitary nucleus, medulla.
   b. Special visceral afferent carries gustatory sensation from taste buds on epiglottis and pharynx to solitary nucleus of medulla.

K. Cranial Nerve XI, Spinal Accessory (Fig. 5.12), Pure Motor

Somatic Nerve in posterior cranial fossa and exits via jugular foramen.

1. Spinal portion: Somatic motor, to sternomastoid from nucleus in ventral horn of C2, and to trapezius from C3 and C4
2. Cranial portion: General visceral efferent, accessory to vagus and distributes with pharyngeal branch of vagus to muscles in soft palate (muscula uvulae, levator veli palatini), and constrictorpharyngeus and through superior and inferior laryngeal of CN X to muscles of larynx and esophagus (due to cranial portion just makes it as a cranial nerve)

L. Cranial Nerve XII, Hypoglossal (Fig. 5.13), Pure Motor

Originates in hypoglossal nucleus in medulla; nerves exit in posterior cranial fossa (exits via hypoglossal foramen). General somatic from medulla to intrinsic and extrinsic muscles of the tongue except palatoglossal by CN X.

V. Cranial Nerve Dysfunction

A. Motor Cranial Nerve Lesion

The effects of a lesion involving the motor nuclei of a cranial nerve or of the motor fibers of a cranial nerve will be ipsilateral to the lesion. The effects will be those of a lower motor lesion: atrophy, weakness, and a loss of segmental reflex activity. With regard to the facial nerve, the effects are often referred to as peripheral. In a local lesion of the brain stem, it is the segmental ipsilateral motor findings, in association with the contralateral corticospinal, and corticobulbar findings below the level of the lesion, that allow for localization (termed *hemiplegia alternans* or *alternating hemiplegia*).

The nuclei of the somatic motor cranial nerves (III, IV, VI, and XII) are located close to the midline. These nuclei tend to be involved, then, by paramedian lesions rather than by lateral lesions. Intrinsic lesions in a paramedian location tend to produce, early in their course, a bilateral involvement of these nuclei. These nuclei are

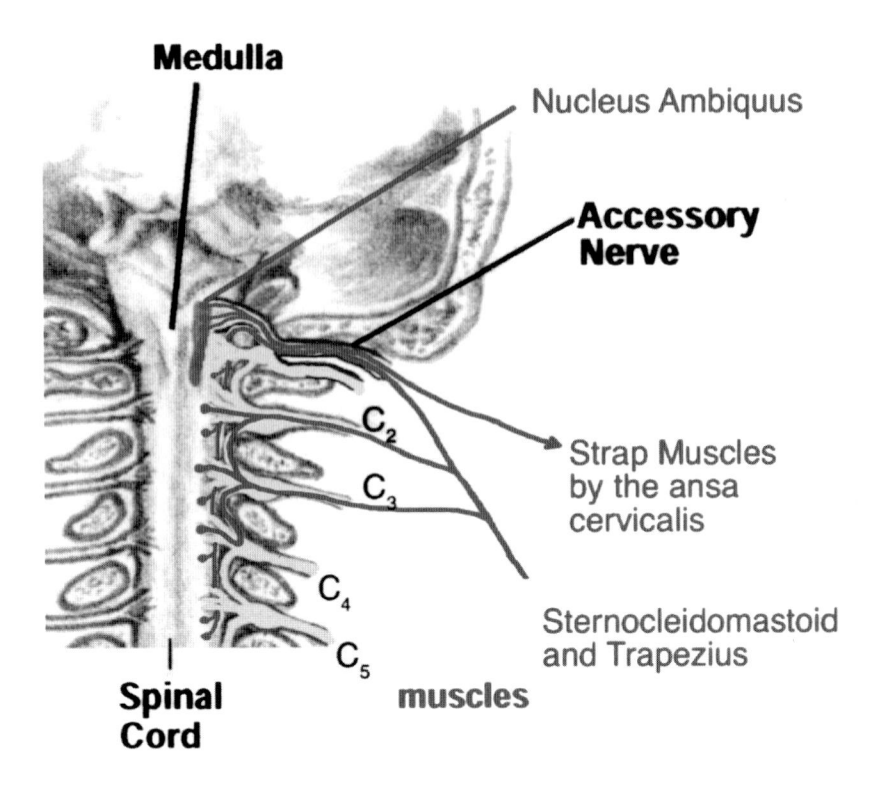

Fig. 5.12 Cranial nerve XI, accessory (From EM Marcus and S Jacobson, Integrated neuroscience, Kluwer, 2003)

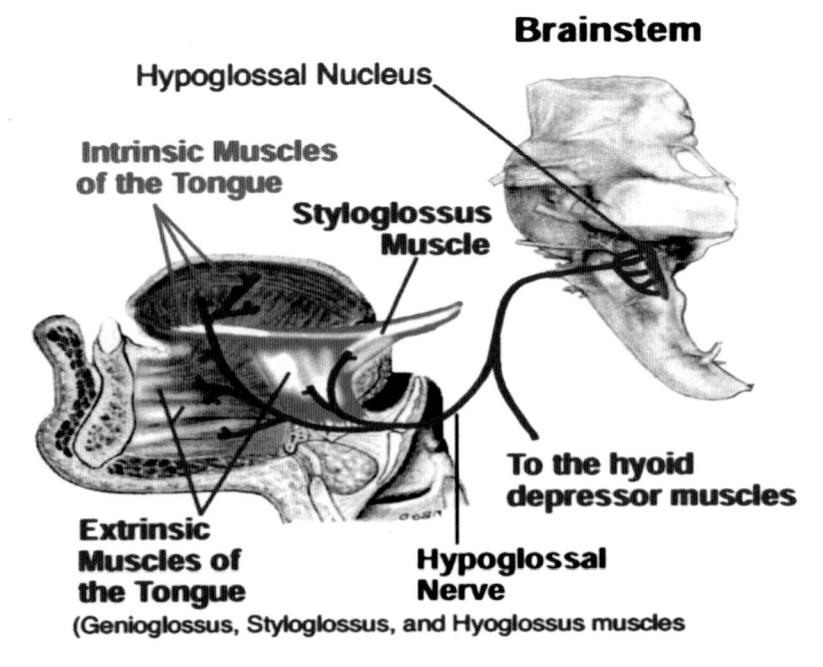

Fig. 5.13 Cranial nerve XII, hypoglossal (From EM Marcus and S Jacobson, Integrated neuroscience, Kluwer, 2003)

also all located in a relatively dorsal position. These nuclei would be involved early by a lesion in a dorsal or tegmental location. Except for CN IV, the fibers of these cranial nerves all exit in a paramedian ventral location.

B. Sensory Cranial Nerve Lesion

Cranial Nerve V Lesions:

a. In the main sensory root of the trigeminal nerve, in its course within or external to the midpons, cranial nerve V lesions will produce ipsilateral deficits in pain, temperature, and touch over the face.
b. In the descending spinal tract of the trigeminal nerve or of its associated nucleus, cranial nerve V lesions will produce an ipsilateral deficit in pain and temperature sensation over the face, with sparing of facial touch sensation. These structures are situated close to the lateral spinothalamic tract; the combination of contralateral pain and temperature deficit over the body and extremities with ipsilateral pain and temperature deficit over the face is frequently found as a consequence of lesions that involve the dorsal lateral tegmental portion of the medulla (e.g., infarction of the territory of the posterior inferior cerebellar artery, which supplies this sector). A similar combination might be found in the infarction of the dorsolateral tegmentum of the caudal pons—the territory of the anterior inferior cerebellar artery.
c. The descending spinal tract (analogous to and, in a sense, the direct continuation of Lissauer's tract) and associated nucleus (analogous to the substantia gelatinosa) descend into the upper cervical spinal cord. Although there is considerable overlap, the primary trigeminal fibers synapse on the nucleus of the descending spinal nucleus at the following levels:

   Mandibular division fibers at a lower pontine–upper medullary level
   Maxillary division at a medullary level
   Ophthalmic division fibers at a lower medullary and upper cervical cord level

   There is overlap at the upper cervical cord level between the upper cervical segment pain fibers and the ophthalmic division pain fibers. It is not surprising, then, that pain originating in C2 and C3 roots or segments of the spinal cord and, for example, the pain of C2 and C3 (occipital neuralgia) is often referred to the orbit.
d. The secondary pain fibers decussate and join the ventral secondary trigeminal tract (also labeled trigeminothalamic or quintothalamic tract), which, at a medullary level, is located just lateral to the medial lemniscus. At a pontine and midbrain location, the secondary trigeminal tract is just lateral to and essentially continuous with the medial lemniscus and adjacent to the lateral spinothalamic path. A lesion that involves this tract would then usually produce contralateral deficits in pain and temperature over a variable portion of the face in association with a contralateral deficit in pain and temperature and/or position and vibration over the affected arm, leg, and trunk.

VI. Cranial Nerve Case Histories

Cranial Nerve Case due to an Aneuyrsm

A cranial nerve case resulting from a subarachnoid hemorrhage secondary to an aneurysm at the junction of the posterior communicating and internal carotid arteries is presented in Fig. 17.8, Case 17.4.

Cranial Nerve Case History 5.1

This 28-year-old single, black, right-handed female reported approximately 21 days of sudden lancinating jabs of pain plus a duller more steady pain involving the right maxillary and mandibular distribution. She was very aware that she could trigger her pain by touching her nose or the skin over the maxillary or mandibular area. The pain was so severe that talking; chewing, or eating would l trigger episodes. As a result, she has not been eating a great deal. There was no clear-cut extension of pain into the eye, although some of the background pain did spread into the supraorbital area and back toward the interauricular line.

Two years prior to admission, she had a less severe episode in the same distribution that lasted 3 weeks. She denied any tearing of the eye or nasal stuffiness. She had no tingling of the face, change in hearing, facial paralysis, or diplopia.

Neurological examination: Mental status, motor system (including gait and cerebellar system) and sensory system were intact.

Cranial nerves were not remarkable except for findings relevant to the right fifth cranial nerve. Although touch and pain sensation over the face was normal, there were clear-cut exquisite trigger points producing sharp pains on light tactile stimulation of the nose or other maxillary areas as well the mandibular skin areas and the lower molars.

Subsequent course: Evaluation by a dentist indicated no relevant dental disease.

Clinical diagnosis: Trigeminal neuralgia: maxillary and mandibular divisions.

Cranial Nerve Case History 5.2

A 44-year-old, right-handed physician developed, over 24 hours, increasing pain in the right supraorbital and orbital areas. Edema of the eyelid developed within 36 hours. Erythematous, edematous skin lesions developed over the right supraorbital area and the anterior half of the right side of the scalp. Edema of the lid was sufficient to produce closure of the lid. Significant cervical lymphadenopathy, neck pain, and low-grade fever developed. The acute pain gradually subsided over 2 weeks, but the skin lesion persisted. The skin lesion crusted and gradually cleared over 4 weeks. Occasional episodes of right supraorbital pain continued to occur for 1 year. Faded, depressed scars were still present 24 years after the acute episode.

Clinical diagnosis: Ophthalmic *herpes zoster*.

Cranial Nerve Case History 5.3

A 37-year-old, right-handed, white housewife was referred for emergency room consultation regarding left facial paralysis. Three days prior to evaluation, the patient had developed a dull pain above and behind the left ear. She then noted occasional muscle twitching about the left lower lip and some vague stiffness on the left side of the face. On the morning of the evaluation, the patient noted she had difficulty with eye closure on the left. She noted that she was drooling from the left side of her mouth. She denied any alteration in sensation over the face and any actual deficit in hearing and denied any tinnitus. She had no crusting in the ear and no actual pain within the ear.

Neurological examination: Cranial nerves: The patient had an incomplete paralysis of the entire left side of the face. She had sufficient eye closures sitting or recumbent to cover the cornea. She was able to have minor elevation of the eyebrow in forehead wrinkling. She had minimal elevation of the lower lip in smiling. Taste for sugar was slightly decreased but not absent on the left side of the tongue (anterior two-thirds). All other cranial nerves were intact. Examination of the external canal and of the tympanic membranes demonstrated no abnormality.

Motor system, reflexes, sensory system, and mental state: All intact.

Clinical diagnosis: Bell's palsy related to involvement of the peripheral portion of the facial nerve. This usually transient impairment of a facial nerve is the result of swelling of the nerve within the relatively narrow bony facial canal. The etiology of this swelling is often uncertain, but infections with the herpes simplex virus or the spirochete agents producing Lyme's disease have been implicated. In many cases, treatment with steroids will relieve these symptoms. More precise localization depends on whether motor function alone is involved, as opposed to additional involvement of taste from the anterior two-thirds of the tongue, or of parasympathetic supply to the facial and salivary glands.

Chapter 6
Diencephalon

The diencephalon appears at the upper end of the brain stem between the brain stem and cerebrum. All of the ascending fiber pathways from the spinal cord and brain stem terminate upon nuclei in the diencephalon and this is the basis for crude awareness of many of these senses at this level. The thalamus is the primary provider of information from the spinal cord and brain stem onto the cerebral cortex, permitting the cerebral cortex to do its many functions

The diencephalon consists of four distinct regions: the thalamus, hypothalamus, subthalamus, and epithalamus (Figs. 6.1 and 6.2). The epithalamus and hypothalamus are discussed in Chapter 7 and the subthalamus is included in the discussion on the basal ganglia in Chapter 10.

This chapter will concentrate on the thalamus. As can be seen from the terminology associated with the diencephalon, all structures are described by their spatial relationship to the thalamus, being above, below, or behind it. Glutamate is the principal transmitter in the thalamus.

The thalamus has several major roles:

1. To receive input from the brain stem, cerebellum, and corpus striatum
2. To project this information, after some processing, onto the cerebrum, and
3. To receive reciprocal projections from the same cerebral area.

Due to the convergences of all major ascending somatic sensory, motor, and reticular systems, consciousness for general sensations are first realized at this level. In addition, the ascending motor information from the cerebellum mixes with the striatal and cortical motor fibers in the dorsal thalamus, producing a motor thalamic region. Finally, there is a region in the diencephalon that receives input from the hypothalamus and ascending reticular systems that projects emotional tone (from autonomic nervous system) onto the cerebral cortex.

All nuclear structures superior to the hypothalamic sulcus are included in the dorsal thalamus, and all nuclei below it are included in the hypothalamus (Fig. 6.3).

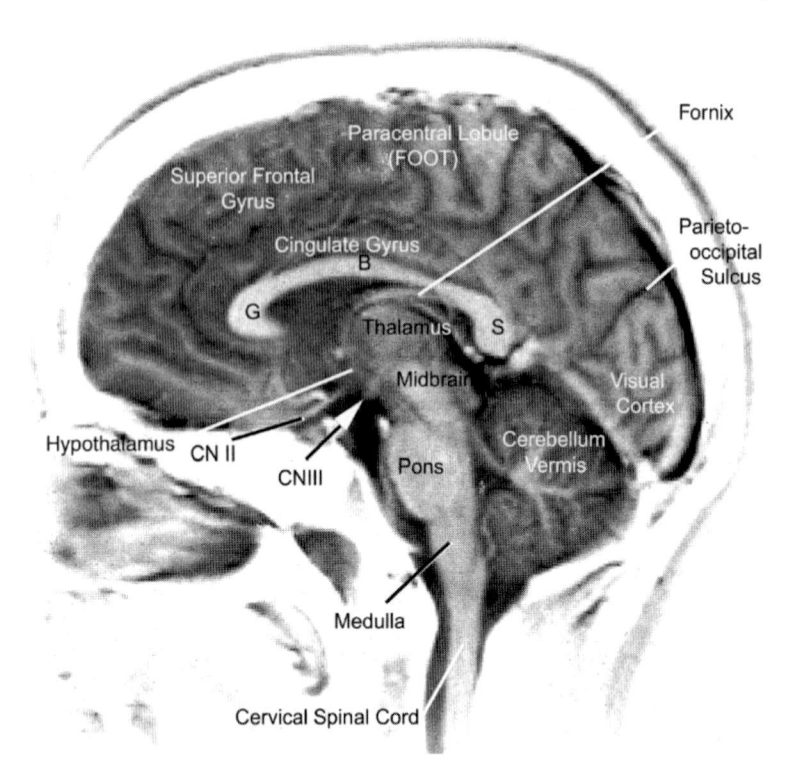

Fig. 6.1 Brain in sagittal plane demonstrating the relationship between brain stem, diencephalons, and cerebrum (MRI T1.) Location of the thalamus. MRI. Diencephalon about a third of the way up from the floor of the third ventricle. (From EM Marcus and S Jacobson, Integrated neuroscience, Kluwer, 2003)

I. Nuclei of the Thalamus

The external medullary lamina forms the lateral boundary of the thalamus and separates it from the reticular nucleus of the thalamus. There is also an intrinsic bundle of white matter, the *internal medullary lamina*, that divides the thalamus into three major nuclear masses: the *anterior, medial,* and *lateral*. The anterior mass is the smallest and includes the anterior nuclei; the medial mass includes the dorsomedial nuclei and the midline nuclei; the lateral mass is the largest and includes the lateral dorsal, ventrobasal, and pulvinar nuclei. The thalamus is the largest portion of the diencephalon. It is ovoid in shape. Its boundaries are the posterior limb of the internal capsule, medially by the third ventricle, laterally inferiorly by the subthalamus and hypothalamus, and superiorly by the body of the lateral ventricle and corpus callosum. The boundary between the thalamus and hypothalamus (Fig. 6.3) is the hypothalamic sulcus on the medial wall of Figure 6.3A, a schematic coronal section showing nuclei of thalamus, hypothalamus, subthalamus, basal ganglia, and the white matter

II Functional Organization of Thalamic Nuclei   149

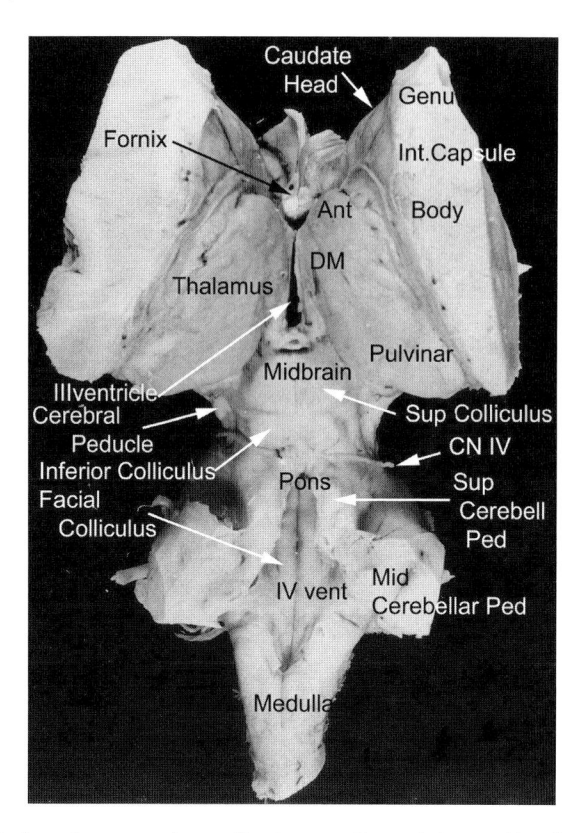

Fig. 6.2 Dorsal view of gross specimen of brain stem, diencephalon, and basal ganglia. Medulla, pons, midbrain, thalamic regions, and internal capsule are labeled. (From EM Marcus and S Jacobson, Integrated neuroscience, Kluwer, 2003)

internal capsule. Various stains are used to further subdivide the thalamus. With these stains, the following nuclei are identifiable: anterior, medial, midline, intralaminar, lateral, posterior, reticular, and metathalamus. Each of these nuclear groupings can be further subdivided, as presented in the next section.

II. Functional Organization of Thalamic Nuclei (Table 6.1)

A. Sensory and Motor Relay Nuclei: The Ventrobasal Complex and Lateral Nucleus

These nuclei (Fig. 6.6) are part of the somatic brain as they receive direct input from the ascending sensory and motor systems (from the cerebellum) and relay that information onto the cortex with only minimal modifications. The lateral nuclear mass is

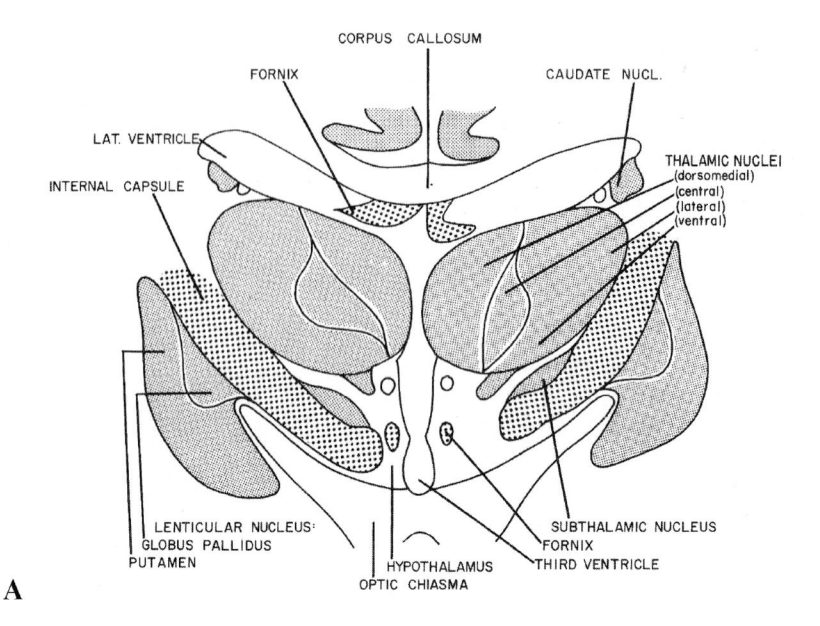

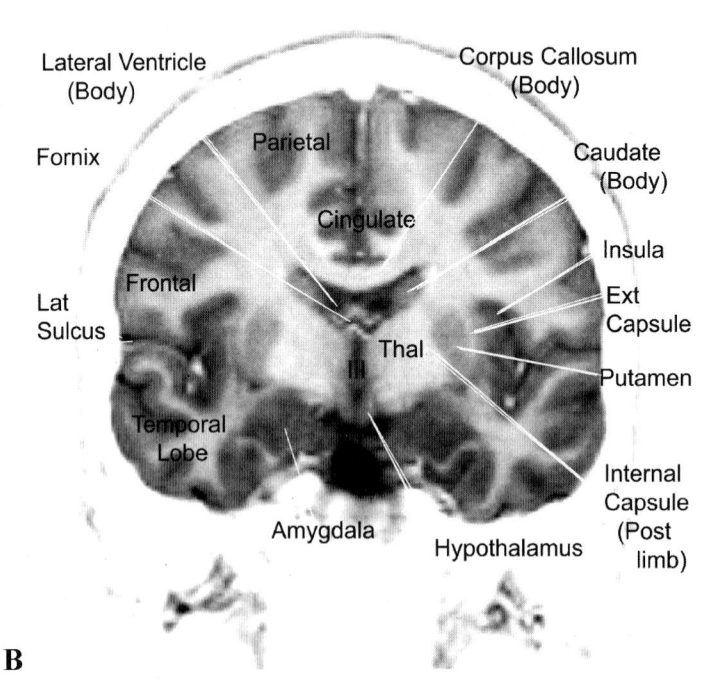

Fig. 6.3 Midthalamic level. **A** Schematic coronal section showing nuclei of the diencephalon: thalamus, epithalamus, hypothalamus, and suthalamus, and internal capsule and basal ganglia; **B** MRI, T2 weighted (From EM Marcus and S Jacobson, Integrated neuroscience, Kluwer, 2003)

Table 6.1 Nuclei in the thalamus

A. Anterior Nuclei (Limbic)
   1. Anterior dorsal
   2. Anterior medial
   3. Anterior ventral (principalis)

B. Medial Nuclei (Limbic and Specific Associational)
   1. Nucleus dorsal medialis
      a. Pars parvocellular → the large parvicellular division is interconnected with the temporal, orbitofrontal and prefrontal cortices.
      b. Pars magnocellular → receives from hypothalamus, amygdala, and midline nuclei and connects to the orbital cortex

C. Midline Nuclei (Nonspecific Associational)
   1. Nucleus parataenials
   2. Nucleus paraventricular
   3. Nucleus reuniens
   4. Nucleus rhomboid

D. Intralaminar Nuclei (Nonspecific Associational)
   1. Nucleus centrum medium
   2. Nucleus parafascicular
   3. Nucleus paracentral
   4. Nucleus centralis lateral
   5. Nucleus centralis medial

E. Lateral Nuclei
   1. Pars dorsal (specific associational)
      a. Nucleus lateral posteriors
      b. Nucleus lateral dorsalis
      c. Pulvinar (polysensory)
         (1) Pars inferior extrastriate occipital and temporal cortex
         (2) Pars lateral → supramarginal, angular, posterior temporal
         (3) Pars medial → prefrontal and anterior temporal
   2. Pars ventralis/ventrobasal complex (Motor and sensory relay)
      a. Ventral anterior → from basal ganglia
      b. Ventral lateral → from cerebellum and red nucleus
      c. Ventral ventral → from basal ganglia and cerebellum
      d. Ventral posterior:
         (1) Posterolateral → from posterior columns → upper postcentral gyrus
         (2) Posteromedial → from trigeminal system → lower postcentral gyrus

F. Metathalamic Nuclei (Special Sensory Relay)
   1. Lateral geniculate (vision: opticothalamic) → visual cortex
   2. Medial geniculate (audition-auditothalamic → auditory cortex

G. Reticular Nuclei (Nonspecific Associational)

H. Posterior Nuclei (Nonspecific Associational)
   1. Limitans
   2. Suprageniculate
   3. Posterior (nociceptive system)

divided into three parts: lateral dorsal, lateral posterior, and pulvinar. It occupies the upper half of the lateral nuclear grouping throughout the thalamus. It is bounded medially by the internal medullary lamina of the thalamus and laterally by the external medullary lamina of the thalamus. In a stained preparation its ventral border with the ventral nuclear mass is distinct. The lateral nuclear mass starts near the anterior end of the thalamus and runs the length of the thalamus with its posterior portion, the pulvinar, overhanging the geniculate bodies and the midbrain. The lateral posterior part of the lateral nuclear mass projects to the superior parietal lobule and receives input from specific thalamic nuclei. The lateral dorsal part projects onto the posterior cingulated gyrus and is included in the discussion of the limbic nuclei.

1. Ventral Basal Nuclear Complex

The **ventral basal nuclear complex** consists of three distinct regions: ventral anterior (VA), ventral lateral (VL), and ventral posterior (VP).

Ventral Anterior

The VA is the smallest and most rostral nucleus of this group. It can be identified by the presence of numerous myelinated bundles. The magnocellular portion of this nucleus receives fibers from the globus pallidus, via the lenticular and thalamic fasciculi, and from the intralaminar nuclei and has some projections to areas 4 and 6. Stimulation of this nucleus produces the same effects as stimulation of the intralaminar nuclei.

Ventral Lateral

The VL (somatic motor thalamus) occupies the middle portion of the ventral nuclear mass. It is a specific relay nucleus in that it receives fibers from the contralateral deep cerebellar nuclei (via the superior cerebellar peduncle/dentatorubrothalamic tract) and the ipsilateral red nucleus. This nucleus also receives fibers from the globus pallidus and from the intralaminar nuclei. This nucleus projects to areas 4 and 6 and has a topographic projection to the motor cortex. The L and VP nuclei contain cell bodies that project onto the motor and sensory cortex. In these nuclei, there is a somatotopic organization, with the head found medial and posterior and the leg anterior and lateral. This nucleus is important in integrative somatic motor functions because of the interplay among the cerebellum, basal ganglia, and cerebrum in this nucleus.

2. Ventral Posterior Nuclear Complex

The VP nucleus occupies the posterior half of the VN mass and consists of two parts: the VP medial nucleus and the VP lateral nucleus. Both of these nuclei receive input from the specific ascending sensory systems.

II Functional Organization of Thalamic Nuclei 153

Ventral Posterior Medial Nucleus

The ventral posterior medial nucleus (VPM; the arcuate, or semilunar, nucleus) is located lateral to the intralaminar centromedian nucleus and medial to the ventral posterior lateral nucleus. The secondary fibers of the trigeminal and gustatory pathways terminate in this nucleus; thus this nucleus is concerned primarily with taste and with general sensation from the head and face.

The basal region of the ventroposterior medial nucleus receives the taste fibers and probably some of the vagal input. General sensation from the face, including pain, temperature, touch, pressure, and proprioception, terminate in the other portions of this nucleus. This nucleus projects information from the face, ears, head, and tongue regions onto the inferior part of the postcentral gyrus (areas 3, 1, and 2)

Ventral Posterior Lateral

The ventral posterior lateral (VPL) nucleus receives fibers from the posterior columns and the direct spinothalamic tracts, with the bulk of the input originating in the upper extremity. The VPL nucleus projects to the body and neck regions on the postcentral gyrus (areas 3, 1, and 2).

Lateral Posterior

The lateral posterior (LP) nucleus projects onto the superior parietal lobule and provides sensory input used in recognition of the parts of the body.

B. Limbic Nuclei: Anterior, Medial, Lateral Dorsal, Midline, and Intralaminar Nuclei (Fig. 6.4)

The *limbic nuclei* can also be called the nuclei of the emotional brain; they were named by their location on the limb or medial surface of the hemisphere.

The *anterior nuclei* (Fig. 6.4) are at the most rostral part of the thalamus and form the prominent anterior tubercle in the floor of the lateral ventricle. The tubercle includes a large main nucleus (anteroventral) and small accessory nuclei (anterodorsal and anteromedial). The internal medullary lamina of the thalamus surrounds the anterior nuclei. The mammillothalamic tract forms the bulk of the fibers in the internal medullary lamina, providing input to the anterior nuclei from the subiculum, presubiculum, and mammillary bodies.

The anterior nuclear complex is part of the Papez circuit (Figs. 14.6 and 14.7): from hippocampal formation → fornix → mammillary body → mammillothalamic tract → anterior thalamic nucleus → anterior thalamic radiation → cingulate cortex → cingulum → hippocampus. This nucleus projects to the anterior cingulate gyrus (areas 23, 24, ans 32) and receives input from the hypothalamus, habenula, and cingulate cortex.

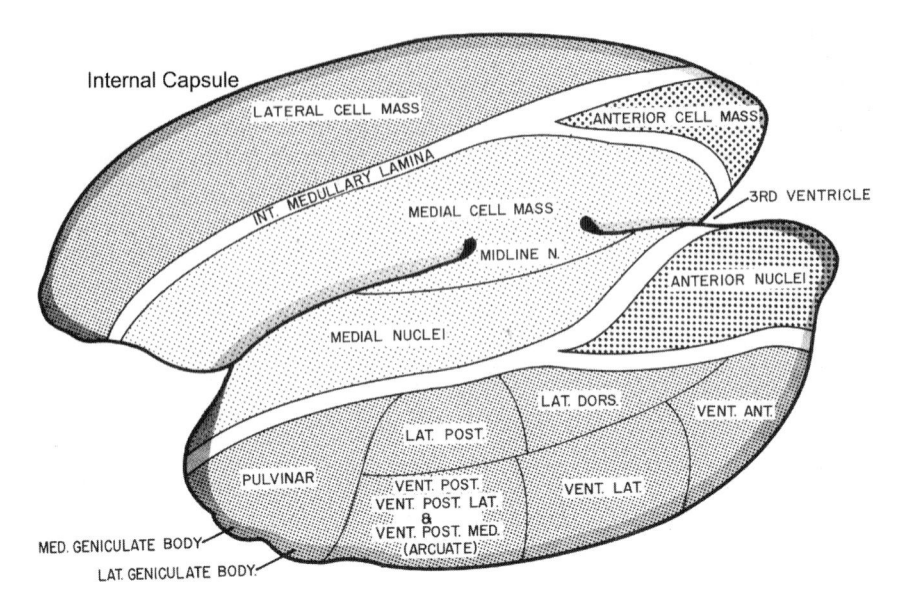

Fig. 6.4 Three-dimensional reconstruction of the thalamic nuclei with the upper portion showing the three major nuclear masses and the lower portion showing individual nuclei (Modified from M.B. Carpenter, Core text, Williams & Wilkins, 1992) (From EM Marcus and S Jacobson, Integrated neuroscience, Kluwer, 2003)

Stimulation or lesions of this nuclear complex alters blood pressure and might have an effect on memory. The *lateral dorsal nucleus*, like the anterior nucleus, is surrounded by the rostral portion of the internal medullary lamina of the thalamus and should be considered a caudal continuation of the anterior nuclear group. This nucleus projects to the rostral cingulate cortex and onto the parahippocampus.

The *medial nuclear complex* or the *dorsomedial nucleus* (DN) consists of both a large-celled and a small-celled division. This medial region is found between the internal medullary laminae and the midline nuclei lining the third ventricle (Fig. 6.4). The magnocellular portion of the dorsomedial nucleus is interconnected with the hypothalamus, amygdala, and midline nuclei and connects to the orbital cortex; the large parvicellular division is interconnected with the temporal, orbitofrontal, and prefrontal cortices (areas 9, 10, 11, and 12; see Chapter 10)

The medial nuclear complex is an important relay station among the hypothalamus, amygdale, and the prefrontal cortex and is concerned with the multimodal associations between the limbic (visceral) and somatic impulses that contribute to the emotional makeup of the individual. In humans, ablation of the medial nucleus or a prefrontal lobotomy has been used as a therapeutic procedure to relieve emotional distress.

C. Specific Associational Polymodal/Somatic Nuclei: The Pulvinar Nuclei (Fig. 6.4)

The pulvinar (Fig. 6.4) forms the largest nucleus in the thalamus and is continuous with the posterior part of the lateral division. All of the major ascending sensory and motor pathways have some terminations in the medial and lateral nuclei. The pulvinar therefore is the major multimodal association nucleus of the thalamus, integrating sensory, motor, visual, and limbic information.

The pulvinar is divided into three major nuclei: the medial pulvinar nucleus, the lateral pulvinar nucleus, and the inferior pulvinar nucleus. The *medial pulvinar nucleus* projects onto the prefrontal and anterior temporal limbic cortex. The *lateral pulvinar nucleus* projects onto supramarginal, angular, and posterior temporal lobes. The lateral pulvinar nucleus has an extensive projection onto the cortex surrounding the posterior end of the lateral sulcus; consequently, lesions here can produce disturbances in language. The *inferior pulvinar nucleus* is related to the visual system and contains major extrastriate visual projections onto the visual association cortex (areas 18 and 19).

D. Special Somatic Sensory Nuclei: Vision and Audition. Lateral Geniculate and Medial Geniculate

The lateral geniculate and medial geniculate are nuclei of the metathalamus (Fig. 6.4). The special somatic sensory cranial nerves are cranial nerves II and VIII.

1. Audition

The medial geniculate nucleus (MGN) is located on the most caudal portion of the thalamus (metathalamus) and receives the ascending auditory fibers originating from the following nuclei: cochlear, trapezoid, superior olive, lateral lemniscal, and inferior colliculus. The auditory fibers end on the medial geniculate pars parvocellulares. This nucleus projects to area 41 in the temporal lobe—the transverse temporal gyrus of Heschl. The ventral nucleus of the medial geniculate receives input from the inferior colliculus. Vestibular fibers end in the magnocellular portion of the medial geniculate and are projected onto postcentral gyrus

2. Vision (Also See Chapter 13)

The lateral geniculate nucleus (LGN) is a horseshoe-shaped six-layered nucleus with its hilus on its ventromedial surface. The LGN receives the optic tract and projects to areas 17 and 18. Figure 6.5 demonstrates the visual radiation onto the calcarine cortex of the occipital lobe. This nucleus contains the six neuronal layers separated by bands of myelinated axons. The four outer layers contain small- to

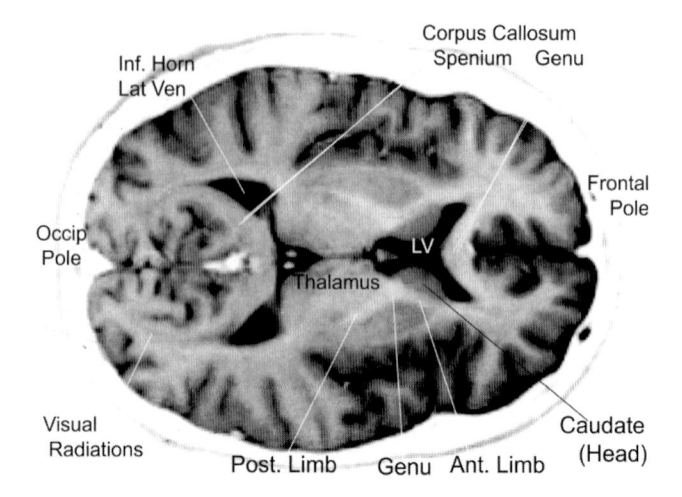

Fig. 6.5 Horizontal section demonstrating the anterior limb, genu, and posterior limb of the internal capsule. Note that the putamen and globus pallidus are external to the internal capsule. MRI, T2 weighted (From EM Marcus and S Jacobson, Integrated neuroscience, Kluwer, 2003)

medium-sized cells that project onto layer IV of the striate cortex, with the two innermost layers contain large cells (magnocellular) projecting onto layer I and the deep portion of layer IV. *Crossed fibers* of the optic tract terminate in laminae 1, 4, and 6, whereas *uncrossed fibers* end in laminae 2, 3, and 5. This nucleus also connects with the inferior pulvinar, ventral, and lateral thalamic nuclei. Some optic fibers proceed directly to the pretectal area and the superior colliculus with or without synapsing in the lateral geniculate nucleus. There they partake in the light reflex and accommodation reflex (see Chapter 13).

E. Nonspecific Associational

1. Reticular Nucleus of Thalamus

The thalamic reticular nucleus forms a cap around the thalamus, medial to the internal capsule. All fibers systems leaving the thalamus and projecting onto the cerebral cortex and the fibers coming back from these same cortical areas pass through this nuclear complex with many specific systems terminating in this nucleus. This nucleus contains GABA-ergic neurons and has a modulating effect on thalamic neurons.

2. Midline Nuclei

The midline thalamic nuclei (Fig. 6.4) are located in the periventricular gray above the hypothalamic sulcus. They are small and difficult to delimit in human beings, but these nuclei are intimately connected to the hypothalamus and the intralaminar nuclei, and their function must be interrelated.

3. Intralaminar Nuclei

The intralaminar nuclei (Figs. 6.3 and 6.4) are found in the internal medullary laminae of the thalamus. The most prominent intralaminar nucleus is the posterior placed *centromedian*, which is located in the midthalamic region between the medial and ventral posterior nuclei and receives many nociceptive fibers. The centromedian nucleus has a strong projection to ventral anterior and ventral lateral nuclei (Fig. 6.5). The *parafascicular* nucleus is also easily delimited because it is at the dorsomedial edge of the habenulopeduncular tract. The intralaminar nuclei receive input from many sensory and motor regions throughout the central nervous system, including ascending axons from the reticular nuclei, indirect spinothalamic pathways, midbrain limbic nuclei, subthalamus, hypothalamus, as well as other thalamic nuclei. The intralaminar nuclei have some direct projections onto the frontal lobe; however, their major influence is on the cerebral cortex by their connections to the specific thalamic nuclei. Electrical stimulation of the intralaminar nuclei activates neurons throughout the ipsilateral cerebral hemisphere. As the stimulation continues, more and more cortical neurons fire, which is called the recruiting response, and there is a waxing and waning; the response finally peaks and then decreases but might increase again.

In Table 6.2, the principal thalamic nuclei are grouped by the type of modality that reaches the individual nuclei. This table demonstrates that the thalamus receives input from nuclei in the spinal cord, brain stem, cerebellum, and cerebrum. The type of input is identified functionally as sensory, motor, limbic, or associational (polysensory).

Table 6.2 Functional groups of thalamic nuclei

Modality	Nuclei
Sensory relay	Nuclei: ventral posterior medial (touch and taste), ventral posterior lateral (touch and vestibular), medial geniculate (hearing), and lateral geniculate (vision). Input from ascending specific sensory modalities and project to cortical area subserving that modality.
Motor relay	Nuclei: ventral lateral and ventral anterior. Input from the contralateral cerebellar hemispheres, superior cerebellar peduncle, medial globus pallidus, and the nigrostriatal fibers from the area compacta of the substantia nigra.
Limbic	Nuclei: dorsal medial, anterior, and lateral dorsal. Receive a well-defined input from the mammillothalamic tract of the Papez circuit and project onto the prefrontal and cingulate gyrus.
Specific associational	Nuclei: dorsal medial, lateral, and pulvinar nuclei. Form the bulk of the thalamus and receives a multimodal input and project onto cortical areas that also receive a multimodal intracortical associational input. VPM
Nociceptive	Dorsomedial, VPL, intralaminar, and posterior: Receive input from ascending anterolateral system, spinothalamic, spinoreticular spinomesencephalic, and trigeminothalamic.
Nonspecific Associational	Intralaminar, midline, and thalamic reticular. Receive input from diverse subcortical regions, including the reticular formation, visceral brain, and corpus striatum; project to specific thalamic nuclei or associational areas in the cortex.

III. White Matter of the Diencephalon (Table 6.3)

A. Internal Capsule (Fig. 6.6)

At the lateral margin of the diencephalon is the internal capsule, a major white matter bundle that contains fibers that connects the diencephalon and the cerebrum. The internal capsule is a large myelinated region that marks the border between the diencephalon and telencephalon. The internal capsule, consisting of the anterior limb, genu, and posterior limb, is best visualized in a horizontal section, where it forms an obtuse angle. Note that the putamen and globus pallidus of the corpus striatum are external to the internal capsule (MRI, weighted T2).

All fibers projecting from the thalamus onto the cerebral hemispheres must pass through the internal capsule, whereas all fibers leaving the cerebral cortex and going either to the diencephalon, basal nuclei, brain stem, or spinal cord must also pass through its parts. In Fig. 6.6, the three major parts of the internal capsule are identified and their constituent fiber systems discussed: the anterior limb, genu, and posterior limb.

B. Anterior Limb of the Internal Capsule (Figs. 6.5 & 6.6)

The anterior limb lies between the caudate nucleus and putamen, and this portion of the capsule contains the anterior thalamic radiation and the frontopontine fibers.

C. Genu of the Internal Capsule (Figs. 6.5 & 6.6)

This region is found at the rostral end of the diencephalon on the lateral wall of the lateral ventricle and is the point at which the fibers are displaced laterally by the increasing bulk of the diencephalon. The corticobulbar fibers are found in the genu.

Table 6.3 Intrinsic fiber bundles of the diencephalon

Pathway	Connections
External medullary lamina	Includes fibers from medial lemniscus, superior cerebellar peduncle, spinothalamic, and pathways between cortex and thalamus
Fornix Papez circuit	Subiculum and hippocampus to septum → hypothalamus
Habenulopeduncular	Habenula to interpeduncular nucleus and tegmentum of midbrain
Internal capsule	Between thalamus and cerebral cortex, and cerebral cortex to brain stem and spinal cord
	Divisions: anterior limb, genu, posterior limb
Internal medullary lamina	Contains mammillothalamic, ascending reticular, and spinothalamic fibers
Mammillothalamic	Mammillary bodies with anterior thalamic nuclei (part of Papez circuit)
Mammillary peduncle	Mammillary nuclei with tegmentum of midbrain

III White Matter of the Diencephalon

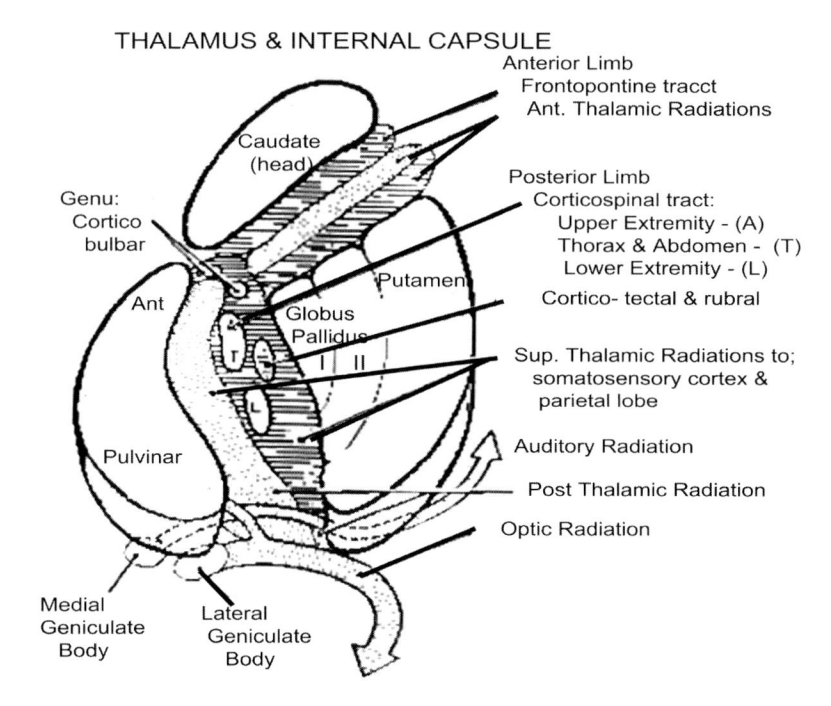

Fig. 6.6 Internal capsule (right). Horizontal section with the major tracts and radiations labeled. Note the relationship between the lenticular nuclei (globus pallidus, putamen) and caudate to the capsule. (Modified M.B. after Carpenter, Core text, Williams & Wilkins, 1992) (From EM Marcus and S Jacobson, Integrated neuroscience, Kluwer, 2003)

D. *Posterior Limb of the Internal Capsule (Figs. 6.5 & 6.6)*

The posterior limb of the internal capsule consists of three subdivisions: thalamolentiform, sublenticular, and retrolenticular

a. The *thalamolenticular portion* (between the thalamus and lenticular nuclei; putamen and globus pallidus) contains the corticospinal, corticorubral, corticothalamic, thalamoparietal, and superior thalamic radiations.
b. The *sublentiform portion* (passing inferiorly and posterior) includes the posterior thalamic radiations that include optic radiation (Fig. 6.6), acoustic radiation, corticotectal, and temporal pontine fibers.
c. The *retrolenticular portion* (passing posterior into the temporal and occipital lobes) contains the posterior thalamic radiation to the occipital and temporal lobes and the parietooccipital fasciculus. Many other myelinated fiber bundles can be seen in the intrinsic thalamic nuclei.

IV. Relationship between the Thalamus and the Cerebral Cortex (Fig. 6.8)

The relationship between many thalamic nuclei and cortical areas is so intimate that when a cortical region is destroyed, the neurons in a particular thalamic nucleus atrophy (Walker, 1938). This degeneration is a consequence of the thalamic projection and its reciprocal from the cortex, being restricted to only a single specific cortical region, and is seen in portions of the following nuclei: anterior, ventral lateral, ventral posterior, lateral geniculate, medial geniculate, and magnocellular portion of the medial dorsal and pulvinar. The intralaminar, midline, and posterior portions of the pulvinar and medial nuclei are unaffected by cortical lesions.

A. *Thalamic Input onto the Cortical Layers*

The most important sensory input the cerebral cortex receives is from the thalamus. The densest input from the specific relay projectional nuclei (e.g., VPL, VPM) is to layer 4, but as indicated earlier, pyramidal cells in layers 3 and 5 might also receive direct or indirect inputs. There is a modality specific columnar arrangement. With many nuclei having an extensive input onto the cortex, there is always a reciprocal cortical input. The nonspecific nuclei (midline and intralaminar) have a more diffuse input and project onto layer 1.

B. *Thalamic Radiations and the Internal Capsule*

The fibers that reciprocally interconnect the cortex and the thalamus form the thalamic radiations (Fig. 6.6). The thalamic radiations are grouped in the internal capsule into four peduncles: anterior, superior, posterior, and inferior.

1. *Anterior radiations* connect the frontal lobe and cingulate cortex with median and anterior thalamic nuclei. This is an extensive projection.
2. *Superior radiations* connect the motor sensory strip and adjacent frontal and parietal lobes with fibers from the ventrobasal nuclei. This is also a large projection.
3. *Posterior radiation* connects the occipital and posterior parietal areas with the pulvinar, and lateral geniculate nucleus (geniculocalcarine radiations - optic radiations). This is an extensive radiation due to the optic radiations.
4. *Inferior radiations* connects from medial geniculate (auditory radiations) and pulvinar with the temporal lobe. This is the smallest group of fibers between the thalamus and cortex.

   In Figure 6.7 the afferents and efferents of the individual thalamic nuclei are presented. In Figure 6.8 A and B we show the cortical projections of these same

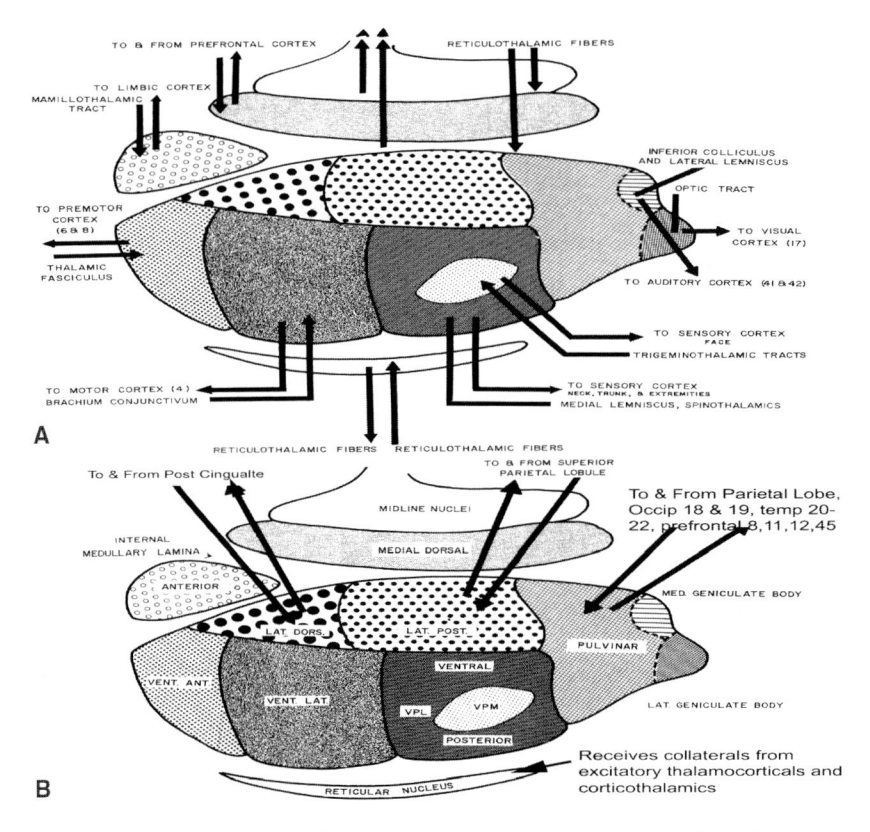

Fig. 6.7 Representation of the thalamic nuclei: **A** Connections of ventrobasal, midline, anterior, and lateral nuclei and geniculate nuclei; **B** connections of lateral dorsal, lateral posterior, and pulvinar nuclei. (Modified from RC Truex and MB Carpenter, Human neuroanatomy, Williams and Wilkins, 1970)

thalamic nuclei. In Figure 6.9 we show the major afferent pathways to the thalamus and metathalamus. Table 6.2 is a summary of the cortical projections of the major thalamic nuclei and also include the functions of these regions.

C. Other Possible Inputs to the Thalamus

In addition to the major pathways discussed earlier there are other subcortical sources of input onto the cerebral cortex that pass through the diencephalon, and they might well have modulating effects on the thalamus. These includethe following:

1. Noradrenergic (norepinephrine) pathway from the locus ceruleus of the midbrain projecting in primates predominantly to layer 6 of the motor and somatosensory cortex and related frontal and parietal association cortex.

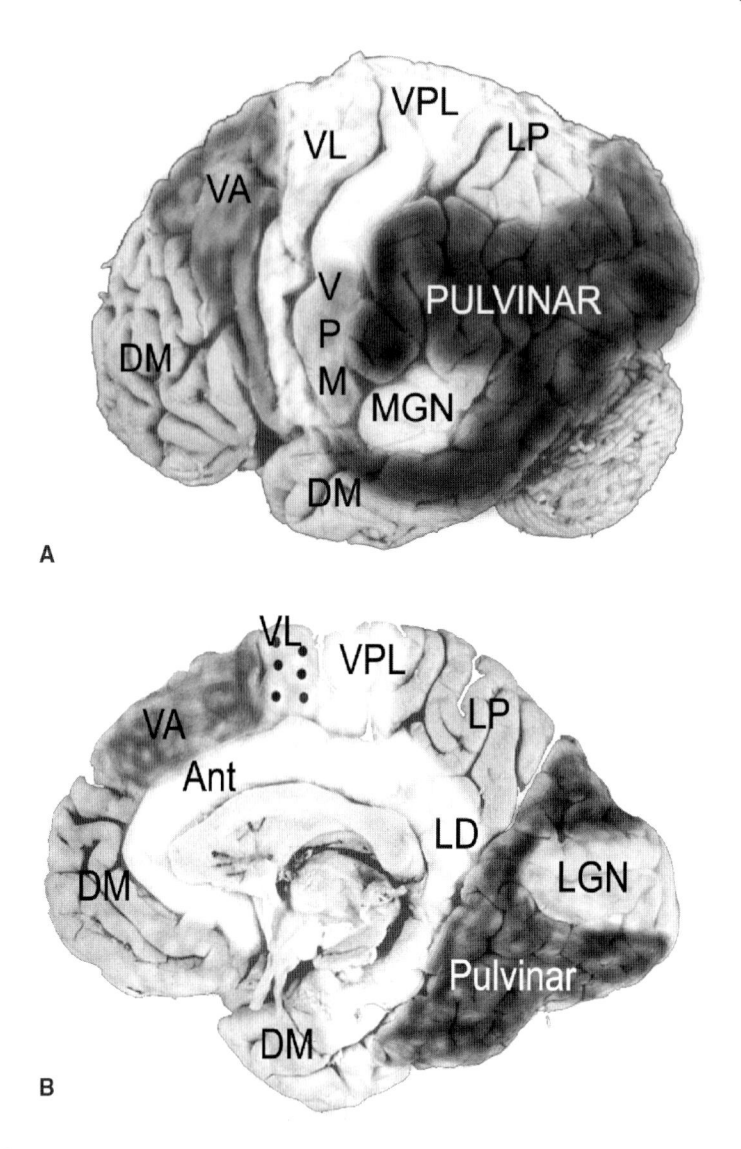

Fig. 6.8 Major thalamic projections onto the cerebral cortex. **A** Projections onto the lateral surface of the hemisphere; **B** thalamic projections onto medial surface (From EM Marcus and S Jacobson, Integrated neuroscience, Kluwer, 2003)

2. Serotoninergic pathway from the raphe nuclei in the pons and medulla and the pontine reticular formation.
3. Dopaminergic pathways from ventral tegmental–rostral mesencephalic nuclear groups. The strongest projection is to the prefrontal cortex and limbic system.

4. Cholinergic pathways from the basal forebrain nucleus of Meynert project widely to the cerebral cortex.
5. GABA-ergic pathways from basal forebrain, ventral tegmental area, and zona incerta to sensory and motor cortex.

The thalamus is the final processing station for systems that project to the cerebral cortex; thus, it serves an important integrative function. Many sensations are first crudely appreciated at thalamic levels, including pain, touch, taste, and vibration. The discriminative processes associated with these sensations, as well as tactile discrimination, vision, audition, and taste, are elevated to consciousness in the cerebral hemisphere.

In the human, a large lesion in the thalamus results in the thalamic syndrome (of Dejerine), causing diminished sensation on the contralateral half of the head and body. (Complete anesthesia results from injury to the ventral posterior lateral nucleus.) Some pain and temperature sensation from the contralateral side of the body might be retained. The internal capsule is usually involved, producing an upper motor neuron lesion of the contralateral limb. The thresholds for pain, temperature, touch, and pressure are elevated contralaterally.

A mild sensory stimulus now produces exaggerated sensory responses on the affected side and might even cause intractable pain. A change in emotional response might also be noted.

Lesions restricted to the medial thalamic nucleus produce memory and personality disturbances; lesions in the lateral nuclei produce sensory deficits.

V. Subthalamus (Fig. 6.3A)

The subthalamus region is included in the diencephalon, but functionally it is part of the basal ganglia. The subthalamic nucleus, zona incerta, and prerubral field together form the subthalamus. This region is medial to the posterior limb of the internal capsule, lateral to the hypothalamus, and below the thalamus. The ansa lenticularis, subthalamic fasciculus, lenticular fasciculus, and thalamic fasciculus provide input from the globus pallidus to the subthalamus.

The zona incerta lies between the thalamus and lenticular fasciculus. The adjacent substantia nigra also projects onto the subthalamus. The medial portion of the globus pallidus provides the principal input to the subthalamus via the subthalamic fasciculus, which runs through the internal capsule. The motor and premotor cortex also project to this region. The subthalamus projects back to both segments of the globus pallidus and to substantia nigra via the subthalamic fasciculus, and it projects to the thalamus and contralateral subthalamus. The glutaminergic cells in the subthalamus form the main excitatory projection from the basal ganglia.

In the human, a lesion in the subthalamus, usually vascular in origin from a penetrating branch of the Posterior cerebral artery (PCA). produces hemiballismus—

purposeless, involuntary, violent, flinging movements of the contralateral extremity. These movements persist during wakefulness, but disappear during sleep. This lesion represents underactivity in the indirect pathway (see Chapter 10). The neurons of the lateral globus pallidus give rise to inhibitory fibers again utilizing GABA, which connect to the subthalamic nucleus. Decreasing the excitatory drive from the subthalamic nucleus acting on the medial globus pallidus and the substantia nigra reticularis would decrease the inhibitory drive from those structures acting in the thalamus. Therefore, the thalamic/premotor supplementary motor region A (SMA) circuit would be more active and more movement would result.

The thalamic terminations of the major ascending pathways are listed in Table 6.4.

Table 6.4 Thalamic terminations of major ascending pathways of the central nervous system

Pathway and thalamic terminations	Function
Anterolateral/spinothalamics → VPL and DM Trigeminothalamics → VPM and DM	Pain and temperature from body, head, and neck
Solitary tract → VPM	Taste and general sensation from the viscera
Cerebellar peduncles: Superior (cerebellum → VA and VL → cerebrum) Middle (cerebrum → pons to contralateral cerebellum) Inferior (cerebellum → brain stem and spinal cord and cranial nerve VIII)	Unconscious sensory and motor
Optic pathway (special sense): from retina via lateral geniculate to calcarine cortex (Chapter 13).	Vision and light reflexes
Auditory pathway (special sense): cochlear nuclei via lateral lemniscus auditory to inferior colliculus, brachium of inferior colliculus to medial geniculate and to auditory cortex	Audition
Vestibular pathway (special sense: part of proprioceptive pathway): proprioceptive lateral vestibular nucleus to VPL and MGN, then to postcentral gyrus.	Balance (from inner ear) and proprioception
Gustatory pathway, special visceral from taste buds (cranial nerves VII, IX, and X) in tongue, pharynx, and larynx to VPM in thalamus → cortical areas in parietal operculum	Taste
Reticular ascending portion → intralaminar, (dorsomedial) DM	Emotions, consciousness
Monamine-containing: → hypothalamus, intralaminar, DM, midline. Norepinephrine, dopaminergic, serotonergic	Sets tone in motor system and affect limbic system
Ascending sympathetic and parasympathetic to limbic thalamus → anterior, DM	Autonomic functions

Chapter 7
Hypothalamus, Neuroendocrine System, and Autonomic Nervous System

The hypothalamus is a very small nuclear mass found at the base of the third ventricle in the diencephalon. Its functional potency is related to its control of the hypophysis and the descending autonomic fibers, which affect preganglionic parasympathetic and sympathetic nuclei in the brain stem and the thoracic, lumbar, and sacral spinal cord. In addition, this diencephalic region, with its important tracts, forms part of the Papez circuit and must be included in any discussion of the limbic (or emotional) brain (Chapter 14).

I. Hypothalamus

The hypothalamus forms the medial inferior part of the floor of the diencephalons and is located on the wall of the third ventricle, extending from the anterior margin of the optic chiasm to the posterior margin of the mammillary bodies. It is bounded dorsally by the dorsal thalamus and laterally by the subthalamus and internal capsule. It is connected to the hypophysis by the hypothalamic–hypophyseal tract and the hypothalamic–hypophyseal portal system. The hypothalamus consists of more than a dozen nuclei with their associated fiber systems.

Although the hypothalamus weighs only 4 g, this small area controls our internal homeostasis and influences emotional and behavior patterns.

A. *Hypothalamic Nuclei*

Anatomically, the hypothalamus is divided into a medial and a lateral zone (Figs. 7.1 and 7.2). The medial zone consists of three regions:

1. The anterior zone, including the preoptic, supraoptic, pariventricular, anterior, and suprachiasmatic nuclei (Fig. 7.2).
2. The middle zone, including the dorsomedial ventromedial, lateral, and tuberal nuclei (Fig. 7.3).

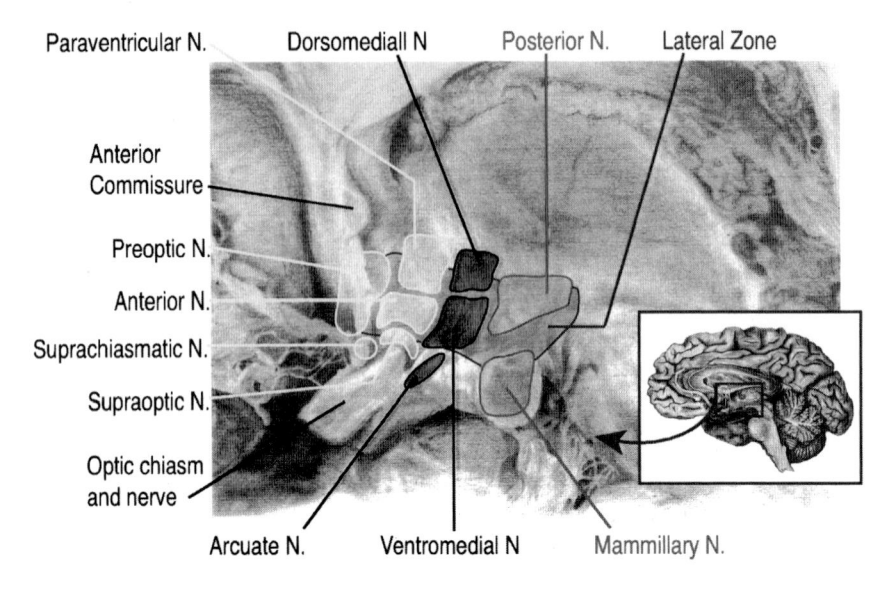

Fig. 7.1 The zones and nuclei of the hypothalamus (From EM Marcus and S Jacobson, Integrated neuroscience, Kluwer, 2003)

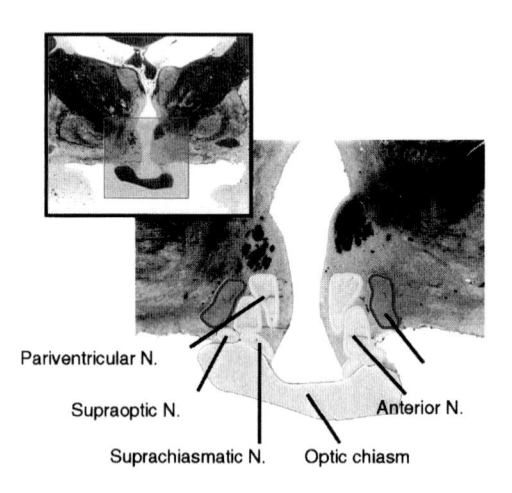

Fig. 7.2 Myelin stain of diencephalon at the level of the anterior hypothalamus and optic chiasm showing corpus callosum, precommissural anterior commissure, and corpus striatum (From EM Marcus and S Jacobson, Integrated neuroscience, Kluwer, 2003)

3. The posterior zone (Fig. 7.5), including the posterior hypothalamic nuclei and mammillary nuclei (both large and small celled). The lateral zone includes the lateral nucleus, the lateral tuberal nucleus, and the medial forebrain bundle. In addition, periventricular nuclei are adjacent to the third ventricle at all levels of the medial region.

I Hypothalamus

1. Anterior Group (Fig. 7.2)

The preoptic, suprachiasmatic, supraoptic, and paraventricular nuclei are found in this region. The supraoptic and paraventricular nuclei contain secretory granules (containing oxytocin and vasopressin) that leave these nuclei and pass down the hypothalamic–hypophyseal pathway, which terminates at the capillaries in the posterior lobe of the pituitary, or neurohypophysis.

The supraoptic nucleus functions as an osmoreceptor; it releases vasopressin in response to an increase in osmotic pressure in the capillaries. Vasopressin or ADH (antidiuretic hormone) increases water resorption in the kidneys. The suprachiasmatic nucleus, which receives direct input from optic fibers in the chiasm, is important in the endocrine response to light levels and might well be the location of the biological clock.

2. Neurosecretory Granules (Fig. 7.3)

Neurons in the supraoptic and paraventricular nuclei of the hypothalamus form neurosecretory material. The axons of these cells form the hypothalamic–hypophyseal tract, which runs through the median eminence, down the infundibular stalk to the neurohypophysis (pars nervosa), where the axons end in close proximity to the endothelial cells. The secretory granules are 130–150 millimicrons in diameter and are found in the tract. They increase in size as one approaches the endothelial end of the axons. The protein in the secretory granules is made in the Nissl substance; the granules are formed in the Golgi apparatus and transported by the axons of the hypothalamic–hypophyseal tract to the infundibulum, where they are stored in the neural lobe. The sites of storage are called Herring bodies. Interruption of the hypothalamic–hypophyseal tract produces diabetes insipidus.

3. Middle Group (Fig. 7.4)

This region includes the dorsomedial, ventromedial arcuate (infundibular), lateral, and lateral tuberal nuclei. The ventromedial nucleus has extensive projections onto the diencephalon, brain stem and spinal cord. Lesions here affect feeding behavior (e.g., obesity and anorexia). Arising from the tuberal nuclei, the tuberoinfundibular pathway terminates on the hypophyseal portal veins, permitting the releasing factors produced in the hypothalamus to reach the anterior lobe of the pituitary.

4. Posterior Group (Fig. 7.5)

The consists of the mammillary body (medial and lateral nuclei) and posterior hypothalamic nuclei. A major part of the fornix projects onto this region, with the mammillary nuclei then projecting via the mammillothalamic tract onto the anterior thalamic nuclei. The mammillary nuclei also project onto the tegmentum

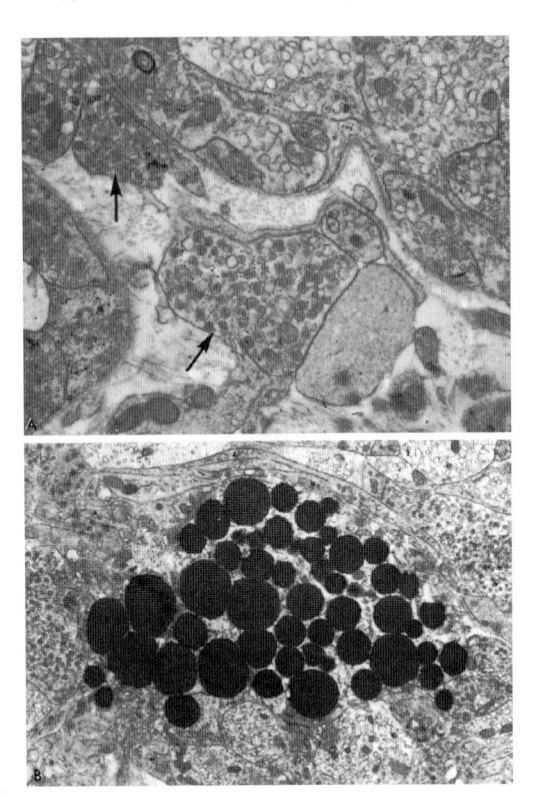

Fig. 7.3 A Electron micrograph of the neurohypophysis of an albino rat showing neurosecretory granules in the axoplasm of fibers in the hypothalamic–hypophyseal tract. (×30,000); **B** a Herring body, a storage site of neurosecretory material (×8000) (From EM Marcus and S Jacobson, Integrated neuroscience, Kluwer, 2003)

through the mammillary peduncle. Lesions in these nuclei are seen in Wernicke's encephalopathy.

B. Afferent Pathways (Fig. 7.6)

The hypothalamus is so situated that it functions as an important integrating center between the brain stem reticular system and the limbic forebrain structures. The major subcortical (afferent) inputs into the hypothalamus include the following:

1. Reticular input from the ascending multisynaptic pathways in the mesencephalic reticular nuclei via the medial forebrain bundle, dorsal longitudinal fasciculus, and mammillary peduncle.
2. The globus pallidus through the lenticular fasciculus and ansa lenticularis (see Chapter 10).
3. Thalamic input from the dorsal medial nucleus and intralaminar nuclei via direct connections onto the hypothalamus.

I Hypothalamus

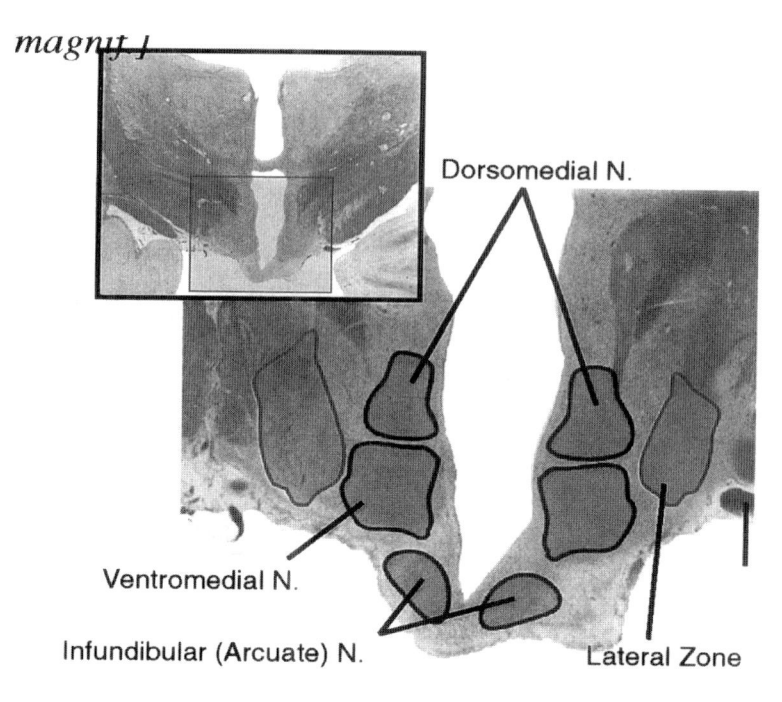

Fig. 7.4 Myelin and cell stain (Kluver–Berrera) of diencephalon at midhypothalamus, demonstrating tuberal nuclei and mammillary bodies (From EM Marcus and S Jacobson, Integrated neuroscience, Kluwer, 2003)

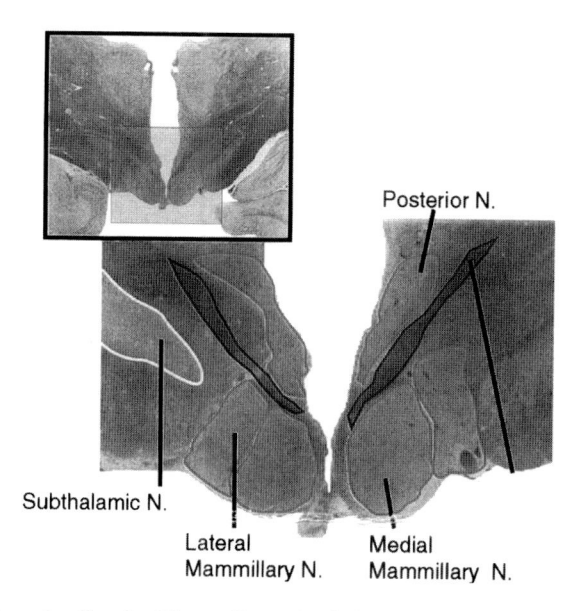

Fig. 7.5 Myelin and cell stain (Kluver–Berrera) of diencephalon at posterior hypothalamus (corpus callosum removed) (From EM Marcus and S Jacobson, Integrated neuroscience, Kluwer, 2003)

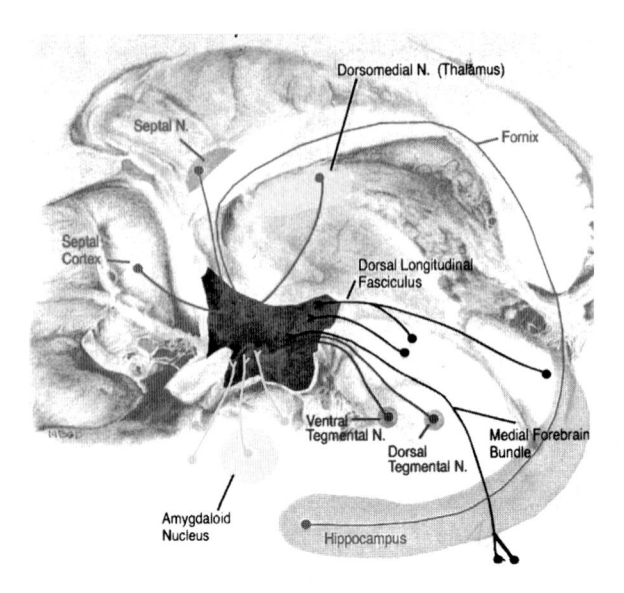

Fig. 7.6 Afferent pathways into the hypothalamus (From EM Marcus and S Jacobson, Integrated neuroscience, Kluwer, 2003)

4. Fibers from the optic nerve (retinohypothalamic fibers), which end in the suprachiasmatic nucleus.
5. The major cortical input into the hypothalamus comes from the temporal lobe: hippocampal formation (fornix), amygdala (stria terminalis), from cingulate gyrus (cingulohypothalamic), and frontal associational cortex (corticohypothalamic fibers).

Thus we see that the hypothalamus is so situated that it serves as an important integrating center between the brain stem reticular system and the limbic forebrain structures.

Fiber tracts and the neurosecretory system mediate the output of the hypothalamus. The efferent pathways (Fig. 7.7) from the hypothalamus are as follows:

1. *Mammillothalamic tract*, which interconnects he medial mammillary nucleus with the anterior thalamic nucleus and, ultimately, with the cingulate cortex
2. *Periventricular system*, which arises in the supraoptic, tuberal, and posterior nuclei in the hypothalamus to supply the medial thalamic nucleus and joins the medial forebrain bundle and dorsal longitudinal fasciculus
3. *Mammillotegmental tract*, which connects the medial mammillary nuclei with the midbrain reticular nuclei and autonomic nuclei in the brain stem
4. *Medial forebrain bundle and dorsal longitudinal fasciculus*, which receive fibers from the lateral hypothalamic nucleus and paraventricular nuclei and connect to the midbrain tegmentum and autonomic neurons in the brain stem associated with cranial nerves VII, IX, and X and the the spinal cord.

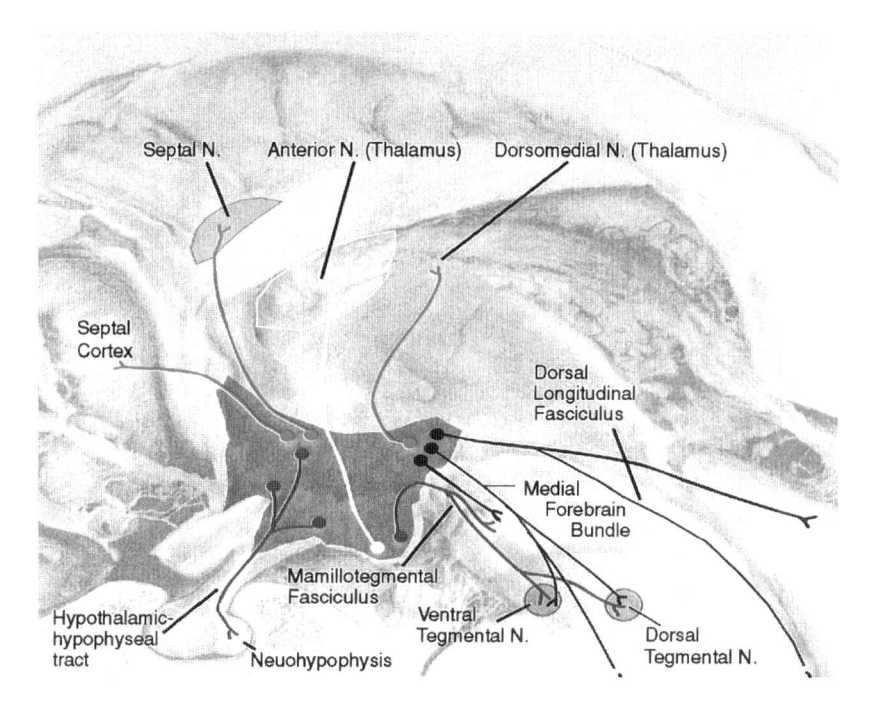

Fig. 7.7 Hypothalamic efferent pathways (From EM Marcus and S Jacobson, Integrated neuroscience, Kluwer, 2003)

5. *Hypothalamic–hypophyseal tracts: neurosecretory system* (Fig. 7.8), which connects the supraoptic, paraventricular, and magnocellular nuclei via the hypophyseal stalk with the neural hypophysis.
6. *Hypothalamic–hypophyseal portal system*, which permits releasing factors made in the hypothalamus to reach the adenohypophysis

The hypothalamus, through its relation with the autonomic nervous system and hypophysis, maintains a more-or-less stable, controlled, internal environment regardless of the external environment. It affects body temperature, water balance, neurosecretion, food intake, sleep, and the parasympathetic and sympathetic nervous systems.

II. Neuroendocrine System: The Hypothalamus and Its Relation to Hypophysis (Table 7.1)

In addition to the direct control of the autonomic nervous system by central pathways, the hypothalamus has a powerful influence on other homeostatic mechanisms of the human by its direct control of the hypophysis. The discussion that follows is an overview of this information flow that is the basis of the neuroendocrine system.

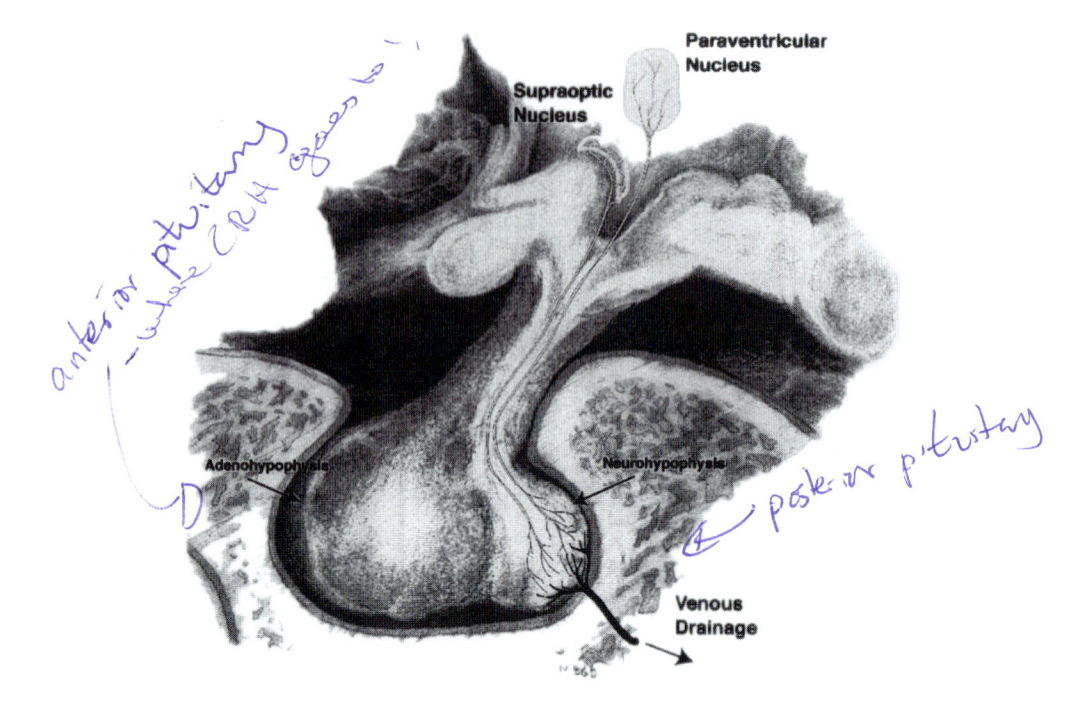

Fig. 7.8 The hypothalamic–hypophyseal tract (neurosecretory system). This tract provides a direct connection between the hypothalamus and the neurohypophysis. The neurosecretory granules pass down the axons of this tract and are stored in the neural lobe until they are released into the bloodstream. (From EM Marcus and S Jacobson, Integrated neuroscience, Kluwer, 2003)

Table 7.1 Functional localization in hypothalamic nuclei

Hypothalamic nuclei	Function
Supraoptic and paraventricular	Osmoregulation:release of vasopressin or lesion here produces diabetes insipidus oxytocin.
Ventromedial and lateral hypothalamus: feeding centers	Lesion in ventromedial nucleus produces overeating; lesion in lateral hypothalamus produces anorexia.
Anterolateral region; excites parasympathetic and inhibits sympathetic	Stimulation of anterolateral region excites the parasympathetic nervous system and inhibits the sympathetic nervous system (posterior region) and produces slow-wave sleep.
Posterior hypothalamus: excites sympathetic and inhibits parasympathetic	Stimulation here excites sympathetic and inhibits parasympathetic; important in arousal and waking from sleep. Lesion in anterolateral produce sympathetic response; lesion in posterior produces parasympathetic response.

A. Hypophysis Cerebri

The hypophyseal gland or pituitary gland is found in the sellae turcica of the sphenoid bone. It consists of two major divisions: the anterior lobe or the adenohypophysis and the posterior lobe or the neurohypophysis. There is also a smaller pars intermedia associated with the anterior lobe. The *adenohypophysis* is the larger component and contains cords of different cell types closely associated with capillaries of the hypothalamic–hypophyseal portal system The *neurohypophysis* consists of the infundibular stalk and the bulbous neurohypophysis and includes pituicytes (modified neuroglia), blood vessels, and some finely myelinated axons. The hypophysis is directly connected to the hypothalamus by the hypophyseal stalk (Fig 7.8) and indirectly connects to the hypothalamus by the hypothalamic hypophyseal venous portal system

The hypophysiotrophic area of the hypothalamus is necessary for hormone production in the adenohypophysis. The hypophysiotropic area includes parts of the ventromedial and other hypothalamic nuclei. These peptidergic neurons are termed *parvicellular* (composed of small cells). The cells forming the neurohypophyseal tract are magnocellular neurons.

B. Hypothalamic–Hypophyseal Portal System

The *hypothalamus* and *adenohypophysis* are connected by the hypothalamic– hypophyseal venous portal system (Figs. 7.8 and 7.9). Releasing factors travel from the hypothalamus to the adenohypophysis by this means.

The *hypophysis* receives its blood supply from two pairs of arteries: the superior hypophyseal arteries and the inferior hypophyseal arteries:

1. *Superior hypophyseal arteries* leave the internal carotid and posterior communicating arteries and supply the median eminence, infundibular stalk, and adenohypophysis. In the infundibular stalk and median eminence, the arteries terminate in a capillary network, which empties into veins that run down the infundibular stalk. In the anterior lobe, these veins form another capillary network, or sinusoid work, which empties into veins that run down the infundibular stalk. In the anterior lobe, these veins form another capillary network, or sinusoid. The *capillary loop*s in the infundibular stalk and neural hypophysis drain in close proximity to the neurosecretory axons of the hypothalamic–hypophyseal tract.

2. *Inferior hypophyseal arteries* leave the internal carotid and distribute to the posterior lobe, where they form a capillary network that flows into the sinus in the adenohypophysis. The hypophyseal veins finally drain into the cavernous sinus. The releasing factors enter the venous drainage of the hypothalamus, which drains into the median eminence and hypophyseal stalk. These factors are then carried via the hypophyseal portal system to the adenohypophysis.

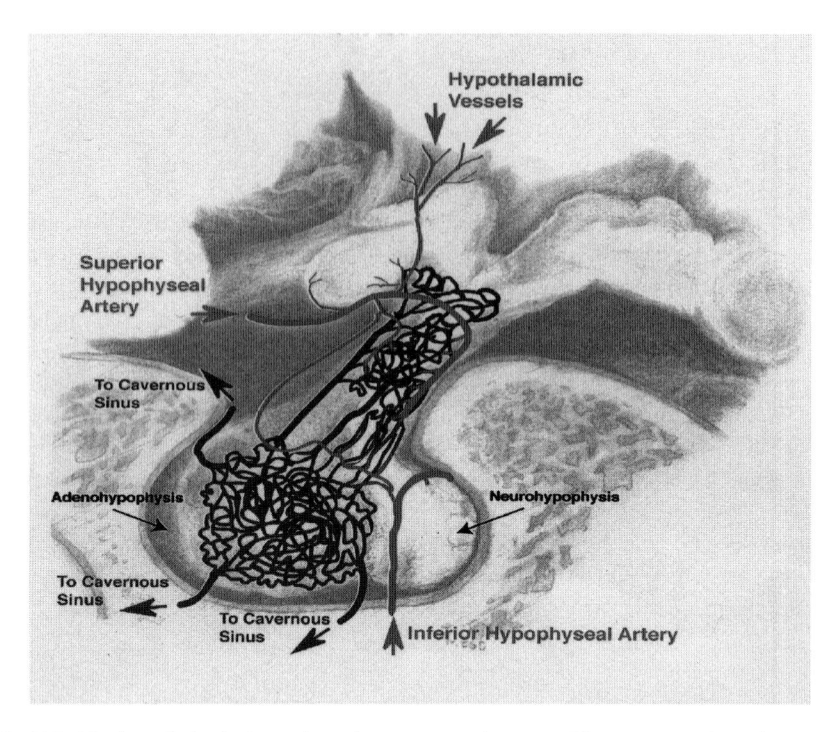

Fig. 7.9 The hypothalamic–hypophyseal venous portal system. These venous channels provide vascular continuity between the hypothalamus and the adenohypophysis, with releasing factors from the hypothalamus draining into the veins of the hypothalamus, which connect to the capillary bed in the infundibular stalk and median eminence. The factors are then carried to the adenohypophysis. (From EM Marcus and S Jacobson, Integrated neuroscience, Kluwer, 2003)

C. Hypophysiotrophic Area

The hypophysiotrophic area is a paramedian zone that extends rostrally at the level of the optic chiasm dorsally from the ventral surface to the paraventricular nuclei. Caudally, this area narrows down to include only the ventral surface of the hypothalamus at the level of the mammillary recess of the third ventricle. The hypophysiotrophic center includes neurons in the following hypothalamic nuclei: paraventricular, anterior, arcuate, ventromedial, and tuberal. Releasing and inhibitory factors, listed in Table 7.2, are produced in the parvicellular hypophysiotrophic area and stimulate the adenohypophysis to produce and release hormones into the circulation.

D. Hormones Produced by Hypothalamus

Two hormones, vasopressin and oxytocin, are manufactured in the hypothalamus, carried down the hypothalamic–hypophyseal tract, and stored in the neurohypophysis before being released into the bloodstream. Vasopressin is an antidiuretic

II Neuroendocrine System: The Hypothalamus and Its Relation to Hypophysis

Table 7.2 Releasing factors and hormones produced in the hypothalamus

1. Releasing factors produced in parvicellular region of hypothalamus:
   Corticotrophin
   Growth hormone (somatotropin hormone)
   Growth hormone-inhibiting factor
   Luteinizing hormone
   Melanocyte-stimulating hormone
   Melanocyte-stimulating hormone-inhibiting factor
   Prolactin-inhibitory factor
   Thyrotropin
2. Hormones produced or stored in neurohypophysis:
   Vasopressin from paraventricular nucleus
   Oxytocin from supraoptic nucleus

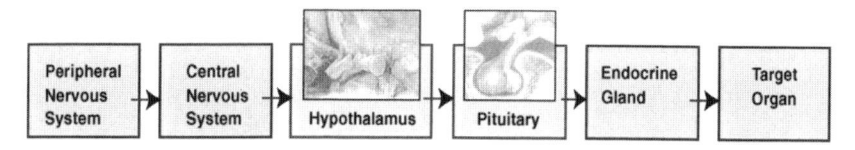

Fig. 7.10 Information flow between the brain and endocrine system

hormone (ADH) formed primarily in the supraoptic nucleus and, to a lesser degree, in the paraventricular nucleus, carried down the hypothalamic–hypophyseal tract. This substance acts on the distal convoluted tubules and collecting ducts in the kidney, causing resorption of water.

Oxytocin is formed primarily in the paraventricular nucleus and, to a lesser degree, in the supraoptic nucleus and is carried down the hypothalamic–hypophyseal tract. This substance stimulates the mammary gland, causing it to contract and expel milk concomitant with suction on the nipple. During delivery, oxytocin also contracts the smooth muscle in the uterus and dilates the uterus, cervix, and vagina.

E. Hormones Produced in Adenohypophysis (Figs. 7.10 and 7.12)

Releasing factors from the hypothalamus traverse the hypothalamic–hypophyseal portal system and trigger the formation and eventual release of the glandotropic or tropic hormone (listed in Table 7.3) Figure 7.11 outlines the information flow between the brain and endocrine system. All of the tropic (switch on) hormones produced in the anterior lobe are, at the same time, trophic (growth promoting or maintaining) hormones, in whose absence their target organs atrophy.

Table 7.3 Hormones produced in adenohypophysis

1. Releasing Hormones:
Adrenocorticotropic (ACTH)
Lactogenic (prolactin);
Gonadotrophic:
Luteinizing (LH) in females and interstitial cell-stimulating in males (ICSH)
Follicle-stimulating (FSH) in females
Melanocyte-stimulating (MSH)
Somatotropin-somatotropin (STH)
Thyrotropic hormone (TSH)

2. Inhibitory Hormones:
Lactogenic(dopamine)
Melanocyte
Somatotopic

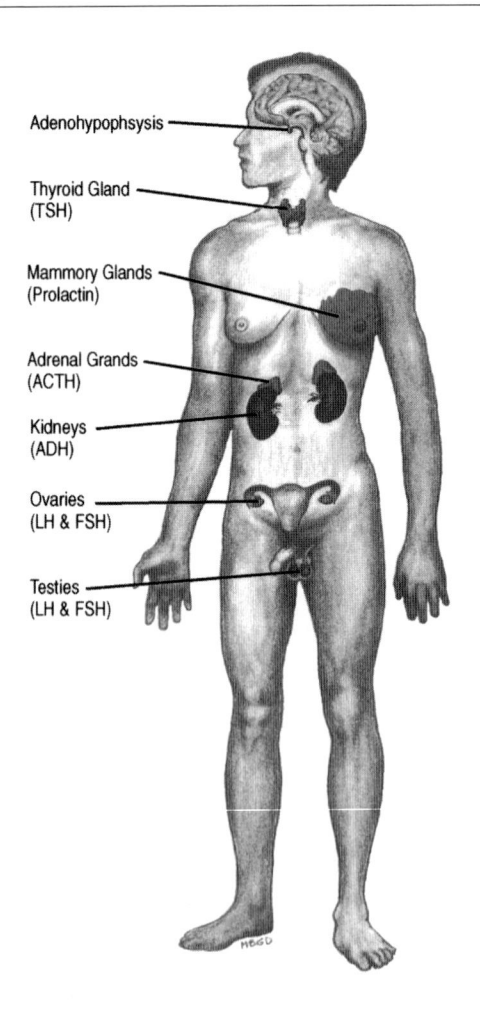

Fig. 7.11 Target glands for adneohypophyseal homones in an androgynous human (From EM Marcus and S Jacobson, Integrated neuroscience, Kluwer, 2003)

1. Adrenocorticotrophic Hormone (ACTH)

Adrenocorticotrophic hormone (ACTH) is produced by basophilic cells and stimulates the formation of adrenal glucocorticoids (hydrocortisone) and, to a lesser degree, mineral corticoids (aldosterone). In conditions of stress (cold, heat, pain, fright, flight, inflammation, or infection), ACTH is produced. Cushing's disease is caused by excessive secretion of ACTH and Addison's disease results from ACTH insufficiency.

2. Lactogenic Hormone

Lactogenic hormone (Prolactin) in combination with the growth hormone promotes development of the mammary glands and lactation.

3. Gonadotropin

Gonadotropin is luteinizing hormone (LH) in females and interstitial cell-stimulating hormone (ICSH) in males. In females, gonadotropin is necessary for ovulation. It causes luteinization of follicles after they are ripened by FSH (follicle-stimulating hormone). In males, gonadotropin activates the interstitial cells of the testis to produce testosterone that is essential for the development and maturation of the sperm as well as male characteristics.

4. Follicle-Stimulating Hormone

Follicle-stimulating hormone (FSH stimulates growth of the ovarian follicle in the female and stimulates spermatogenesis in males.

5. Growth Hormone

Growth hormone (somatotropin, or STH). is important for growth after birth. It specifically affects the growth of the epiphyseal cartilage but is also important in the metabolism of fat, protein, and carbohydrates. Cessation of growth follows hypophysectomy. The inhibitory form of this hormone controls the release of the hormone. Gigantism results from an excessive somatotropin before the epiphyseal plates close. Acromegaly results after epiphyseal plate closure.with enlargement of facial bones and hands.

Case History 7.1 (Fig. 7.12)

A 25-year-old white male auto body mechanic developed numbness in the median nerve distribution of the left hand with pain at the left wrist, shooting

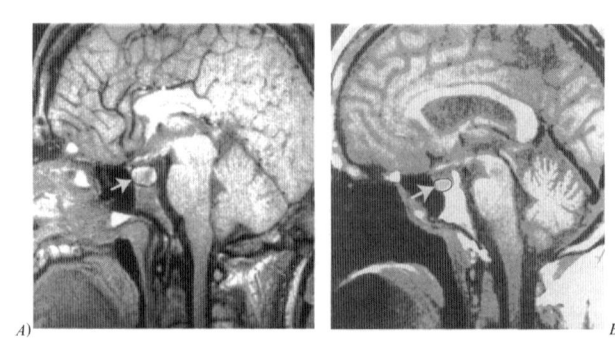

Fig. 7.12 Case 7.1. Adenomas: **A** MRI, T1-weighted midline sagittal sections; **B** control MRI, T1, weighted (From EM Marcus and S Jacobson, Integrated neuroscience, Kluwer, 2003)

into the index, middle, and ring fingers and aching pain in finger joints. To a lesser degree, similar symptoms had developed in the right hand. The clinical impression of bilateral carpal tunnel syndrome was confirmed by nerve conduction studies. In addition, the patient presented many of the facial features of acromegaly: coarse lips, large jaw, and nose. None of his family had similar facial features, and according to his mother, these features had emerged about age 12.

Growth hormone levels were consistently elevated to 17–20 ng/ml (normal is less that 5 ng/ml fasting and recumbent). Tthe prolactin level was moderately increased to 37 ng/ml (normal in men if less that 1–20 ng/ml). MRI demonstrated an enlarged pituitary gland with a hypodense lesion, consistent with an adenoma (Fig. 16.1). A partial agenesis of the corpus callosum (splenium and posterior one-third of the body) was also present and might be related to longstanding learning disabilities. A transsphenoid excision of the pituitary adenoma was performed. Except for transient diabetes insipidus, he did well and all symptoms disappeared.

Comments. The patient presented with a chief complaint relevant to compression of the median nerve at the carpal tunnel. Usually, carpal tunnel syndrome occurs in one hand or the other and might reflect local swelling in the tendons and soft tissues. This is, at times, a result of occupation-related repetitive motion at the wrist. In occasional cases, an underlying systematic conditions is present; these include rheumatoid arthritis, pregnancy, myxedema of the hypothyroid state, and acromegaly associated with excessive secretion of growth hormone by pituitary adenomas. In this case, the neurologist noted the facial feature characteristic of acromegaly. In retrospect, the patient's mother was able to date these changes to his teen years. Excessive production of growth hormone, prior to closure of the epiphyseal plates in long bones, result in excessive growth or gigantism results in changes of acromegaly with enlargement of the facial bones And hand. after surgery on the pituitary diabetes may occur. Excessive production of growth hormone after epiphyseal closure insipidus often is present, reflecting temporary damage to the posterior pituitary or pituitary stalk.

6. Thyrotrophic Hormone

Thyrotrophic hormone (TSH) stimulates the thyroid gland to produce triiodothyronine (T3) and thyroxine (T4), which regulate the rate of breakdown of fats and carbohydrates. When circulating levels of T3 and T4 rise, thyrotropic-releasing hormone (TRH) release from the adenohypophysis is inhibited by a drop in TRH production in the hypothalamic neurons until the preferred levels returns

F. Hypothalamus and the Autonomic Nervous System (Fig. 7.14)

An individual performing daily functions usually has a balance between the parasympathetic and sympathetic nervous system.

1. Autonomic Effects of Stimulation of the Anterior Hypothalamus

Stimulation of the anterior hypothalamus (anterolateral region) excites the parasympathetic nervous system and inhibits the sympathetic nervous system. The heart beat and blood pressure decrease (the vagal response), the visceral vessels dilate, peristalsis and secretion of digestive juices increase, the pupils constrict, and salivation increases (Table 7.4).

2. Autonomic Effects of Stimulation of the Posterior Hypothalamus

Stimulation of the posterior hypothalamus (posteromedial region) excites the sympathetic nervous system and inhibits the parasympathetic nervous system. The heart beat and blood pressure increase, the visceral vessels constrict peristalsis and secretion of gastric juices decrease, pupils dilate, and sweating and piloerection occur (Table 7.4).

Table 7.4 Effects of hypothalamic stimulation on the autonomic responses

Stimulation of anterior hypothalamus = parasympathetic response	Stimulation of posterior hypothalamus = sympathetic response or anxiety response
Decrease in heart rate, respiration, blood pressure	Increase in heart rate, respiration, and blood pressure
Increase in peristalsis and in gastric and duodenal secretions	Decrease in peristalsis and in gastric and duodenal secretions,
Increased salivation and, depending on the internal milieu, even evacuation of the bowels and bladder	Decreased salivation with sweating, and piloerection
Constriction of Pupils	Dilation of Pupils
Concomitant with the parasympathetic excitation, there is sympathetic inhibition.	Concomitant with sympathetic excitation, there is parasympathetic inhibition.

A lesion in the anterior or parasympathetic zone produces a sympathetic response; destruction of the posterior or sympathetic zone might produce a parasympathetic response. The effects on blood vessels are produced by stimulation of the smooth muscles in the tunica media, and the effects on the sweat glands result from the stimulation of the glands through the cranial or peripheral nerves.

G. Functional Localization in Hypothalamus

The hypothalamus, through its relation with the autonomic nervous system and hypophysis, maintains a more-or-less stable, controlled, internal environment regardless of the external environment. It affects body temperature, water balance, neurosecretion, food intake, sleep, and the parasympathetic and sympathetic nervous system.

1. Food Intake

Bilateral lesions restricted to the ventromedial hypothalamic nuclei produce hyperphagia, resulting in obesity. Stimulation of this center inhibits eating. Experiments have shown that animals with ventromedial lesions are not motivated to eat more; they just do not know when to stop eating. For this reason, it is thought that the satiety center is located here. Bilateral lesions in the lateral hypothalamic region produce anorexia; the animal refuses to eat and eventually dies. However, if this region is stimulated but not destroyed, the animal overeats. Appetite, the stimulus for eating, is dependent on many factors, including glucose concentration in the blood, stomach distension, smell, sight, taste, and mood. Lesions in other portions of the limbic system might also affect appetite. Although obesity in humans usually does not have a neurological cause, in rare cases some overweight individuals have had a ventromedial nuclei lesion.

2. Sleep Cycle

The hypothalamus helps set our level of alertness with the assistance of input from the reticular system and cerebral cortex. The medial preoptic region can initiate slow wave sleep. The lateral hypothalamus receives input from the ascending reticular system that produces arousal. The posterior portion, including the mammillary body, is important in controlling the normal sleep cycle. Lesions in this region might produce an inversion in the sleep cycle, or hypersomnia. Lesions in the anterior hypothalamus might produce insomnia.

The amount of light is monitored in the hypothalamus by input from the optic nerve directly to the suprachiasmatic nucleus. The sympathetic innervation to the pineal monitors the dark levels and melatonin release from the pineal is part of the sleep mechanism.

3. Body Temperature

Maintenance of a constant body temperature is vital to a warm-blooded animal. The hypothalamus acts as a thermostat, establishing a balance among vasodilation, vasoconstriction, sweating, and shivering. It does this by means of two opposing thermoregulatory centers: One regulates heat loss and the other regulates heat production. Lesions in either of these centers interfere with the regulation of body temperature.

4. Control Center for Heat Loss

A destructive lesion in the anterior hypothalamus at the level of the optic chiasm or in the tuberal region renders an animal incapable of controlling its body temperature in a warm environment; a tumor in the third ventricle might produce a similar result. The human can no longer control the heat-loss mechanism, which consists of cutaneous vasodilation and sweating. Without this mechanism, the body temperature rises (hyperthermia) and the patient becomes comatose and could even die.

5. Control Center for Heat Production and Conservation

Lesions in the posterior hypothalamus above and lateral to the mammillary bodies disrupt an animal's ability to maintain normal body temperature in either a warm or cold environment. Normally, exposure to cold produces vasoconstriction, shivering, piloerection, and secretion of epinephrine in mammals, all of which tend to increase body temperature. Following a lesion in the posterior hypothalamus, the body temperature tends to match that of the environment, whether warmer or colder, relative to $37\,°C$ (poikilothermy). Poikilothermy probably reflects damage to the fibers descending from the heat-loss center as well as direct injury to the heat production and conservation center.

The thermoreceptors are located in the skin (end bulbs of Krause and end bulbs of Golgi–Mazzoni) or in the hypothalamus itself, which monitors the temperature of the cranial blood. Some receptors respond to cold and others to warmth. Cold receptors excite the caudal heat production and conservation center and inhibit the rostral heat-loss center. Warmth receptors reverse this pattern.

6. Water Balance and Neurosecretion

Water balance is regulated by the hypothalamic–hypophyseal tract, which originates in the supraoptic and paraventricular nuclei and runs through the median eminence and hypophyseal stalk to terminate in close proximity to blood vessels in the neural hypophysis. A lesion in the region of the supraoptic nucleus or in the hypothalamic–hypophyseal system causes diabetes insipidus with polyuria (excessive formation of urine). The patient drinks copious amounts of water (polydipsia) and might excrete 20 liters of urine a day. The symptoms of diabetes insipidus might be relieved by the administration of vasopressin (Pitressin), an antidiuretic hormone (ADH), or extract from the hypophysis.

The hypothalamic–hypophyseal tract controls the formation and release of ADH, which causes water to be reabsorbed from the distal convoluted tubules and collecting duct of the kidneys into the bloodstream, thereby limiting the amount of water lost in the urine. Ingestion of large amounts of water causes a drop in the ADH level and reduces tubular reabsorption, with resultant diuresis. Water deprivation increases the ADH level and causes increased reabsorption and concentrated urine.

The cells in the supraoptic nuclei form much of the ADH or its precursors. ADH is formed in the cell body, travels down the axons of the hypothalamic–hypophyseal tract, and is stored in the posterior lobe of the hypophysis as neurosecretory granules. The osmoreceptors in the aortic and carotid body and in the hypothalamus itself detect the concentration of water in the blood. This information is passed on to the hypothalamic nuclei, which then control the release of ADH into the circulation.

Water intake is also controlled by hypothalamic neurons, with the area dorsal to the paraventricular nuclei being one of the water intake centers.

7. Hypothalamus and Emotions

The hypothalamus sets the foundation for emotional responses and controls visceral functions. It exerts these powerful influences through the multisynaptic descending autonomic pathways in the lateral margins of the reticular formation and by controlling the release of the catecholamine—pinephrine (adrenaline) and norepinephrine (noradrenalin)—from the adrenal medulla.

Most postganglionic sympathetic fibers release a substance that is 80% epinephrine and 20% norepinephrine, which acts on α- and β-adrenergic receptors in the organs.

Norepinephrine causes constriction of the peripheral vessels, with resultant increased blood flow in the large vessels. The blood pressure and pulse rate rise and blood flow to the coronary muscles increases.

Epinephrine increases the heart rate and the force, amplitude, and frequency of heart contractions. It also dilates the pupils and the urinary and rectal sphincters, causing voiding and defecation. It inhibits the motility of the gut and causes the bronchial musculature to relax, thereby dilating the bronchial passages, and it increases the rate of synthesis of lactic acid to glycogen, thereby prolonging the contraction of the skeletal muscle.

8. Hypothalamus and Light Levels

Light produces dramatic effects that are mediated by the hypothalamus via the retinohypothalamic tract to the suprachiasmatic nucleus. In many animals, abnormal variations in light levels alter the sexual cycle and prevent estrus. Light levels also affect the formation of melanin.

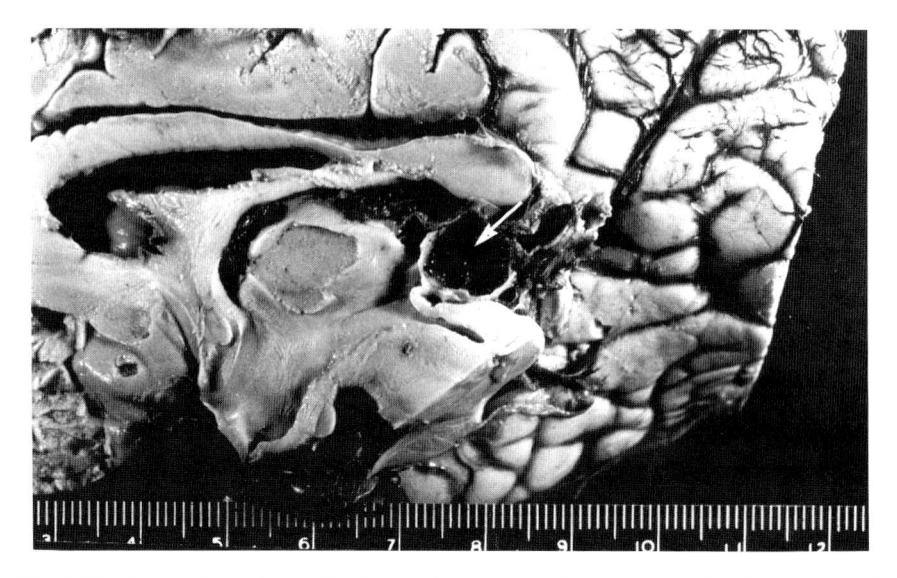

Fig. 7.13 Tumor of the pineal. This is an enlarged nonmalignant cyst arising from the pineal gland (arrow). (Courtesy of Dr. John Hills and Dr. Jose Segarra) (From EM Marcus and S Jacobson, Integrated neuroscience, Kluwer, 2003)

The pineal, or epiphysis cerebri, is a small pinecone-shaped gland found in the roof of the third ventricle (Fig. 7.13). This richly vascularized gland contains glial cells and pinealocytes and is innervated by the sympathetic nervous system from the superior cervical sympathetic ganglion with synapses on the pinealocytes. *Pinealocytes* contain serotonin, melatonin, and norepinephrine. The pineal makes melatonin from serotonin; the enzyme levels necessary for this transformation are controlled by the suprachiasmatic nucleus. This nucleus sends fibers via the descending sympathetic pathway into the medial forebrain bundle and via the descending spinal fibers to the intermediolateral cell column of the upper thoracic cord and then on to the superior cervical ganglion. The pinealocytes are related to the photoreceptor cells seen in this region in fish and amphibians. Pinocyte secretion of serotonin and melatonin is a cyclic response to stimulation from the optic system. These 24-hour circadian rhythmns might well be the biological clock.

III. Autonomic Nervous System (Fig. 7.14)

Langley's description of the autonomic nervous system includes three parts: the sympathetic, parasympathetic, and enteric nervous systems. The enteric nervous system is recognized as a separate portion of the autonomic nervous system because peristalsis and other spontaneous movements persist after it is isolation from all nervous input (Gershon 1981, 2000). Much of the effectiveness of the hypothalamus is due to its innervation of the nuclei in the central nervous system

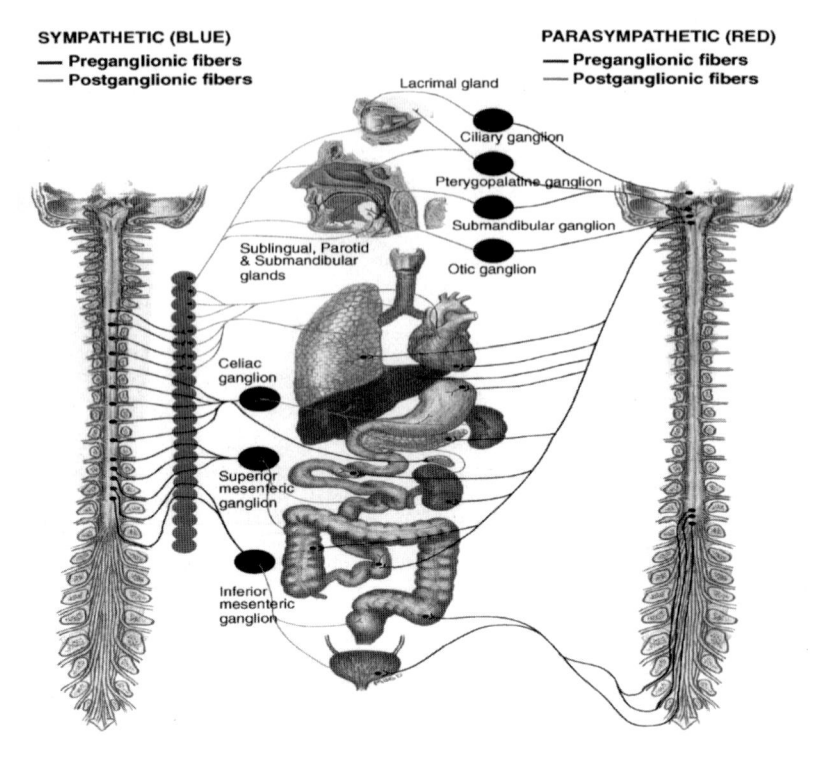

Fig. 7.14 Autonomic nervous system: left, sympathetic (thoracolumbar); right, parasympathetic (craniosacral) (From EM Marcus and S Jacobson, Integrated neuroscience, Kluwer, 2003)

that form the autonomic nervous system. In this section, we identify these neurons in the central and peripheral nervous systems and list the responses produced by the autonomic nervous system in each organ. Note that normally an antagonistic balance exists between the functions of the parasympathetic and sympathetic systems. However, this equilibrium might be modified by environmental stresses or disease processes. The somatic nervous system has its own receptors and effectors, which provide cutaneous and motor innervation to the skin and muscles in the head, neck, and body. The visceral, or autonomic, nervous system also has its own receptors and effectors, which provide dual motor and sensory innervation to the glands, smooth muscles, cardiac muscles, viscera, and blood vessels. These axons run part of the way with somatic fibers, but they also have separate pathways.

In the somatic nervous system, the efferent neuron is the ventral horn cell or motor cranial nerve nucleus. Its axon leaves the central nervous system to innervate a skeletal muscle, whereas in the visceral nervous system, two efferent neurons are found: (1) the preganglionic neuron and (2) the postganglionic neuron.

1. Location of Preganglionic Autonomic Neurons. These preganglionic cell bodies are found in the central nervous system at the following sites:
   Parasympathetic (craniosacral):

a. Cranial parasympathetic (cranial nerves III, VII, IX, and X)
   b. Sacral parasympathetic (sacral segments 2–4)

   Sympathetic (thoracolumbar): Origin in all thoracic segments, T1–T12, and from lumbar segments L1 and 2

2. Location of Postganglionic Autonomic Neurons. The postganglionic neuron, or the motor ganglion cell, is located outside the central nervous system and ends in the appropriate gland, smooth muscle, or cardiac muscle. The nerve fibers are further classified as being adrenergic or cholinergic, depending on which nerve transmitter is involved: epinephrine or acetylcholine.

   Postganglionic parasympathetic fibers release acetylcholine at their terminals. These axons end on sweat glands, whereas those glands receiving innervation from the thoracolumbar system are adrenergic, postganglionic. Pilomotor axons, which also receive thoracolumbar innervation, are cholinergic. The two systems provide dual innervation to glands and viscera and act to produce optimal balances of internal conditions under any environmental condition.

A. Enteric Nervous System

In the alimentary canal (esophagus, stomach, intestines, and rectum) there are about 100 million neurons (Gershon, 2003). One network of neurons, Meissner's (submucosal) plexus, lies in the connective tissue between the longitudinal muscle coat and the circular musculature coat in the alimentary canal. Another, Auerbach's (myenteric) plexus, lies in the connective tissue between the circular muscle coat and the muscularis mucosa. Each plexus consists of multipolar, bipolar, and unipolar neurons. The enteric nervous system includes glia that appears similar to the astrocytes in the central nervous system. The astrocytes enwrap the neurons and their processes and also form a barrier between the blood and the enteric nervous system. There are extensive connections within the enteric nervous system and projections onto the pancreas, gallbladder, and elsewhere. The unipolar and bipolar neurons appear to function as sensory neurons, with the multipolar cells forming many of the connections. The noradrenergic sympathetic postganglionic fibers project onto the submucosal and myenteric plexus, where they are inhibitory to the cholinergic neurons. (The cholinergic vagal input from the dorsal motor nucleus of the X to the submucosa and mucous membrane is excitatory.)

B. Parasympathetic System (Cranio-sacral) (Fig. 7.14)

The parasympathetic system contains cranial nerves III, VII, IX, and X.
   Cranial Nerve III. The fibers originate in the Edinger–Westphal nucleus in the midbrain and innervate the constrictors of the pupil via the ciliary ganglion.

Cranial Nerve VII. The axons originate in the superior salivatory nuclei of the pons and innervate glands in the nasal cavity and palate and in the lacrimal (pterygopalatine), submandibular, and sublingual (submandibular) glands.

Cranial Nerve IX. The fibers originate in the inferior salivatory nucleus in the medulla. The axons exit with nerve IX and end in the otic ganglion, which innervates the parotid gland.

Cranial Nerve X. The dorsal motor nucleus of the vagus nerve in the medulla provides preganglionic parasympathetic innervation to the pharynx, larynx, esophagus, lungs, heart, stomach, and intestines up to the transverse colon. The postganglionic fibers originate from ganglia in the respective organs, in the nodes of the submucosal plexus of Meissner, and in the myenteric plexus of Auerbach in the enteric nervous system associated with the alimentary tract.

Spinal Cord Parasymapthetic Segments S2–S4. The parasympathetic sacral plexus originates from sacral segments S2 and S4 and supplies the genitalia, sigmoid colon, rectum, bladder, and ureter. The ganglia are located in the organ and are named for the organ.

C. Sympathetic System (Fig. 7.14)

The preganglionic fibers originate in the intermediolateral column of gray matter in the spinal cord from T1 to L2; they exit in the white rami and enter the paravertebral sympathetic ganglionic chain. The ganglia form a continuous paravertebral trunk in the cervical, thoracic, lumbar, and sacral segments. The postganglionic fibers originate in the ganglia and exit via the gray communicating rami. The postganglionic fibers reach their destinations via plexuses enwrapping the arteries or by joining cranial and spinal nerves.

1. Cervical Sympathetic Ganglia

The cervical ganglia extend posterior to the great vessels from the level of the subclavian artery to the base of the skull. There are usually three ganglia in the cervical region: superior, middle, and inferior. The *superior cervical ganglion* is flat and the largest (25–35 mm). The internal carotid nerves leave the superior border of the superior cervical ganglion, enwrap the carotid artery, and enter the cranial cavity via the carotid canal. This ganglion provides postganglionic innervation to the eye, lacrimal gland, parotid gland (otic ganglia), submaxillary and sublingual glands (sublingual ganglia), and some branches to the heart. The *middle cervical ganglion* is small and variable and is found at the level of the cricoid cartilage. It provides postganglionic fibers to the heart. The *inferior cervical ganglion* vs usually fused with the first thoracic ganglion to form the stellate ganglion. It is also small and irregular in shape and lies posterior to the vertebral artery. The postganglionic fibers innervate the heart.

2. Thoracic Sympathetic Ganglia

The thoracic portion of the sympathetic trunk extends anterior to the heads of the first 10 ribs and then passes along the sides of the lower 2 thoracic vertebrae. The number of thoracic ganglia varies from 10 to 12. The ganglia of T1–T4 provide postganglionic fibers to the heart and lungs. The greater splanchnic nerve is formed by preganglionic rootlets from T5 to T10; the axons pass through the thoracic ganglia and end in the celiac ganglion. The lesser and least splanchnic nerves arise from spinal cord segments T11 and T12 and run through the paravertebral ganglion to the celiac and superior mesenteric ganglia. From there, the postganglionic axons run to the small intestine, ascending portion of the colon, and the kidneys.

3. Lumbar Sympathetic Ganglia

The lumbar sympathetic trunk consists of small ganglia that lie along the anterior border of the psoas muscle. The preganglionic fibers originate from the intermediate column of spinal cord, segments L1 and L2, pass through the lumbar ganglia, and end in the inferior mesenteric ganglion, where they innervate the descending colon, rectum, urinary bladder, ureter, and external genitalia.

Innervation in the pelvis and perineum is along the parasympathetic and sympathetic fibers. The *parasympathetic* fibers originate from segments S2–S4 and end in ganglia in the organs. Postganglionic fibers supply the uterus, vagina, testes, erectile tissue (penis and clitoris), sigmoid colon, rectum, and bladder. They cause contraction and emptying of the bladder and erection of the clitoris or penis. The *sympathetic* fibers originate from the lowest thoracic segments and segments L1–L3 and run in the hypogastric nerves to the target viscera, where the ganglia are found. They cause vasoconstriction and ejaculation of semen and inhibit peristalsis in the sigmoid colon and rectum.

Chapter 8
Cerebral Cortex Functional Localization

The cerebrum is the largest part of the brain and consists of the cerebral cortex and basal ganglia. The cerebral cortex consists of four lobes: the anteriorly placed frontal lobe, the parietal lobe in the middle, and the occipital lobe posteriorly with the inferiorly placed temporal lobe. The cortical neurons are found on the surface of the cerebrum covering the cortical white matter. Deep in the white matter is found the basal ganglia, which will be discussed in the motor system.

I. Anatomical Considerations

Cerebral Cortical Gray Matter

A firm knowledge of the structure and function of the cerebral cortex and the relationship of the cerebral cortex to the subcortical centers is of crucial importance for all who wish to understand the behavior and accomplishments of humans as opposed to other animal forms. The use of the thumb and fingers for fine manipulations with tools, the use of linguistic and mathematical symbols for communication in the auditory and visual spheres, and the capacity for postponement of gratification all reflect the evolution of the cerebral cortex.

The cerebral cortical gray matter consists of 16 billion nerve cells and is organized into a laminar arrangement, with an almost infinite number of synapses. Beaulieu and Colonnier (1966) have estimated that in a cubic millimeter of neocortex there are 2.78×10^8 synapses, 84% of which are type I (excitatory) and 16% are type II (inhibitory). The almost infinite number and variety of circuits present not only have provided the anatomical substrate for the recording of an infinite number of past experiences but have also allowed for a plasticity of future function projected both in time and in space. The attempt to relate functional differences to the differences in structure of the various areas of the cerebral cortex was a scientific outgrowth of the earlier philosophical arguments concerning the relationship of mind and body. We may indicate at the onset that, to a certain degree, cytoarchitectural differences do reflect functional differences:

1. Thickness: evident on simple visual inspection (average thickness is 2.5 mm; motor cortex: 4.5 mm; visual cortex: 1.45–2.0 mm.). The basis for the differences in overall thickness is related to differences in the thickness of the layers.

2. Density of the various cell types (evident in Nissl stains for cells or Golgi silver stains) to be discussed below (e.g., pyramidal vs. stellate): It must also be noted that within a given area of the cortex, the various layers of the cortex differ with regard to the relative distribution of these various types of cell (Fig. 8.1)
3. Density of horizontal fiber plexuses (stripes, evident on myelin and silver stains) and density of axodendritic and other synapses (evident on electron micrographs).

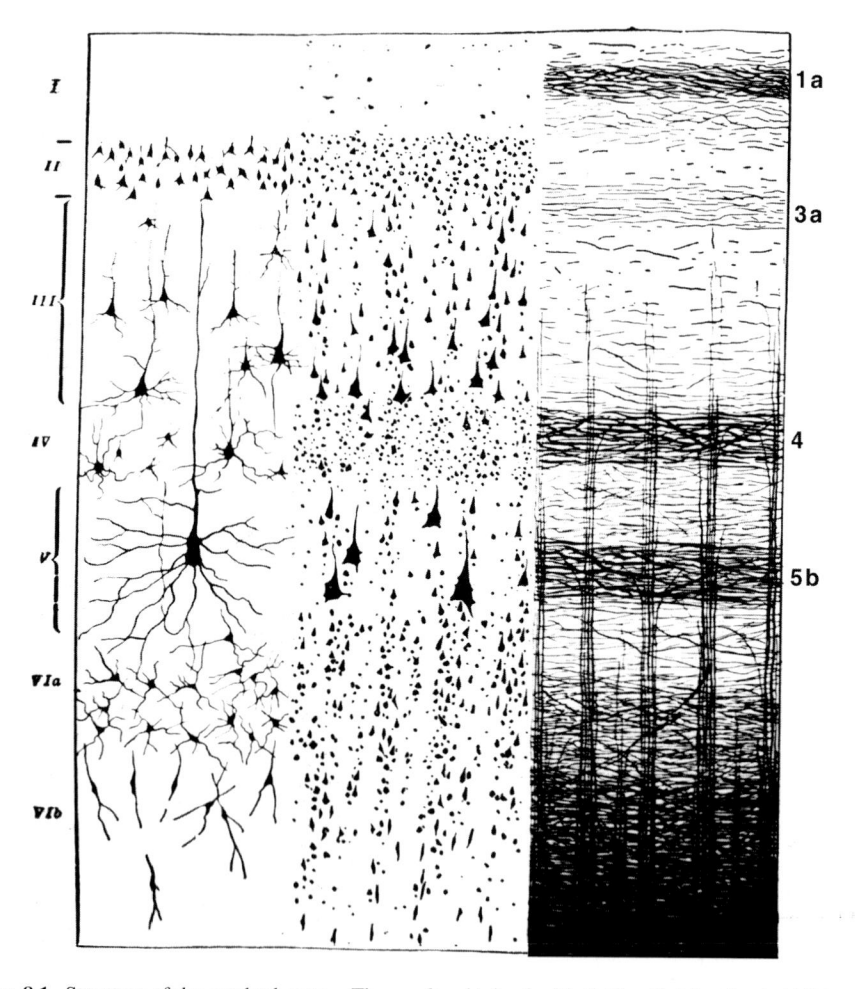

Fig. 8.1 Structure of the cerebral cortex. The results obtained with (left) a Golgi stain, (middle) a Nissl stain or other cellular stain, and (right) a myelin stain are contrasted. I = molecular layer; II = external granular layer; III = external pyramidal layer; IV = internal granular layer; V = large or giant. Pyramidal layer (ganglionic layer); VI = fusiform layer. The following features should be noted in the myelin stain: 1a = molecular layer; 3a1 = band of Kaes Bechterew; 4 = outer band of Baillarger; 5b = inner band of Baillarger. [After Brodmann (1909) from SW Ranson and SL Clark, The anatomy of the nervous system, Saunders, 1959, p. 350]

4. Degree of myelination of intracortical fiber systems (evident on myelin stain): To some extent, the various myelinated bands might be seen with the naked eye in freshly cut sections of the cerebral cortex. For example, Gennari in 1782 and Vicq d'Azyr in 1786 independently had noted a white line in the cortex near the calcarine fissure. The various features to be noted in the several types of stains are demonstrated in Figure 8.1

A. *Cytology*

From the standpoint of cytology, the neurons within the cerebral cortex can be classified into two major categories: pyramidal and stellate (Fig. 8.2). The axons of pyramidal cells (type I cells: neurons with long axons) form the majority of association, callosal, and subcortical projections, whereas the stellate cells (type II: neurons

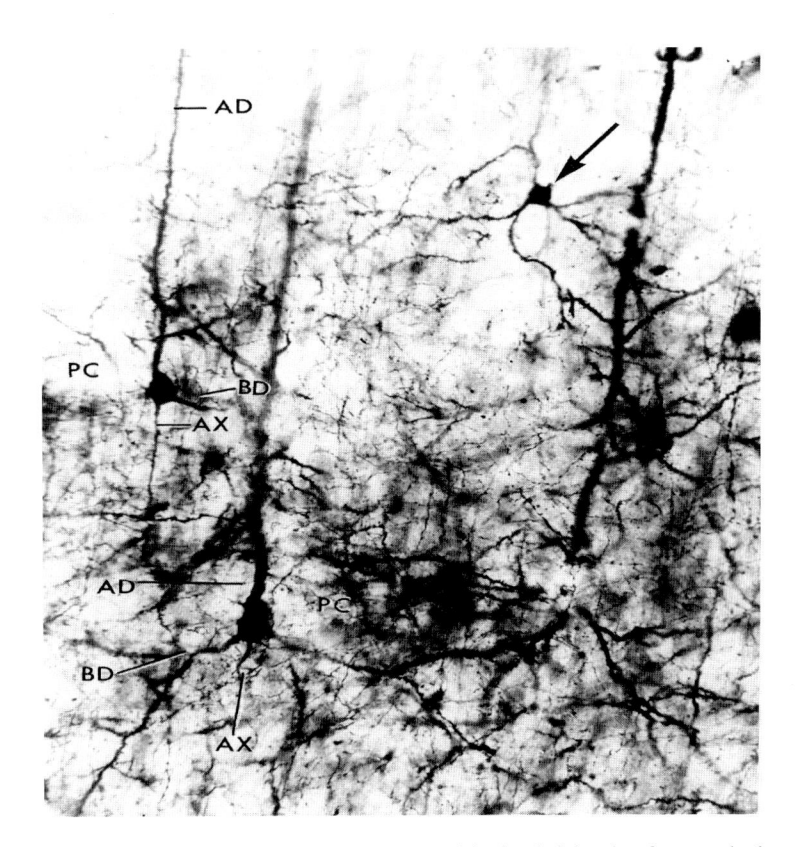

Fig. 8.2 Pyramidal and stellate cells as demonstrated in the Golgi stain of cat cerebral cortex. PC = pyramidal cells; arrow = basket type of stellate cell; AD = apical dendrite of pyramidal cell; BD = basal dendrite of pyramidal cells; AX = axons of pyramidal cell (From EM Marcus and S Jacobson, Integrated neuroscience, Kluwer, 2003)

with short axons) form the local circuits. Neurons can also be classified on the basis of whether the dendrites have spines.

From the standpoint of electrophysiology, the neurons can be classified as fast spiking, regular spiking, or bursting (repetitive discharge).

1. Spiny Neurons

Spiny neurons are neurons whose dendrites have spines and account for the majority of the neurons in cerebral cortex. These are excitatory and are of two types: pyramidal cells and spiny stellate cells

Pyramidal Cells

Pyramidal cells constitute two-thirds of all neurons in the neocortex. These cells are defined by a prominent apical dendrite extending through all layers of the cortex to layer. Pyramidal neurons have cell bodies of variable height: small, 1–12 µm; medium, 20–25 µm; large, 45–50 µm; giant, 70–100 µm. The giant pyramidal cells are characteristic of the motor cortex. The upper end of the pyramidal cell continues toward the surface as the apical dendrite. Numerous spines are found on the dendrite, providing a large surface area for synaptic contact. The pyramidal neurons have the largest dendritic trees and also contain axons that form associational, callosal, and projectional connections. In general, the larger the cell body, the larger the apical dendrite and the wider the spread in terms of ramifications of the terminal horizontal branches. The wider the spread of the apical dendrite, the greater the number of possible axodendritic synapses. The larger the cell body, the greater the number of possible axosomatic synapses. In addition to the apical dendrite, shorter basal dendrites arise from the base of the cell body and arborize in the vicinity of the cell body. The axon emerges from the base of the cell body and descends toward the deeper white matter. In general, the axons of small- and medium-sized pyramidal cells terminate as association fascicles within the cortex. The axons of large and giant pyramidal cells enter the deeper white matter as (1) association fibers to other cortical areas, (2) commissural fibers to the contralateral hemisphere, or (3) projection (efferent) fibers to subcortical, brain stem, and spinal cord areas. In addition, recurrent collateral association fibers might branch off from these axons within the cortex. In addition to functioning as the main efferent outflow of the cerebral cortex to subcortical areas (thalamus, basal ganglia, brain stem, and spinal cord) and to other cerebral cortical areas (ipsilateral and contralateral hemispheres), the pyramidal cells also provide a massive collateral network to other neurons with in the local area. Each pyramidal cell also receives multiple inputs onto its dendritic spines. The Betz cells, the giant pyramidal neuron of the motor cortex, have at least 10,000 spines, each with a type I excitatory synapse.

Spiny Stellate Neurons

In spiny stellate neurons the prominent apical dendrite is not present. Instead, dendrites of almost equal length radiate out from the cell body. These neurons are found primarily in layer IV of the sensory cortex. Their axons project only locally and not to distant sites.

Physiology: Most spiny neurons/pyramidal cells utilize glutamate as a synaptic transmitter and demonstrate "regular" spiking that is an adapting pattern of discharge in response to a constant current injection. These cells generally demonstrate a significant after-hyperpolarization (AHP). Some of the pyramidal cells located in deeper layers, primarily layer V, might demonstrate a bursting pattern at low threshold. These neurons respond to depolarization by generating repetitive spikes, usually three in sequence. The regular firing pyramidal cells are characterized by thin apical dendrites that do not branch extensively in layer I. In contrast, the burst firing pyramidal cells have thick apical dendrites and extensive branching in layer I.

2. Smooth Neurons

Smooth neuron without spines are also referred to as stellate cells. In these neurons, the prominent apical dendrite is not present. Instead, dendrites of almost equal length radiate out from the cell body. These neurons are found in all layers but predominantly in layer 4 of the sensory cortex. Their axons project locally within the area, not to distant sites. The most common type of stellate/granule cell is the basket cell, which has axons that form baskets around the soma of the pyramidal cells. There are additional types of stellate cell: (1) the horizontally oriented Retzius–Cajal neurons in layer I and (2) the vertically oriented Martinotti cell with a double dendritc bouquet and the fusiform cells in layer VI,.with an axon that ascends to layer I.

Physiology: The smooth cells are generally fast spiking and make type II inhibitory connections utilizing gamma-aminobutyric acid (GABA) as the transmitter. In contrast to pyramidal neurons, the action potentials of these fast-spiking cells are brief; the repolarization phase is rapid and followed by a significant undershoot. This has been correlated with very large, fast repolarizing potassium current. The interrelationship of neurons within the neocortex is demonstrated in Figure 8.2.

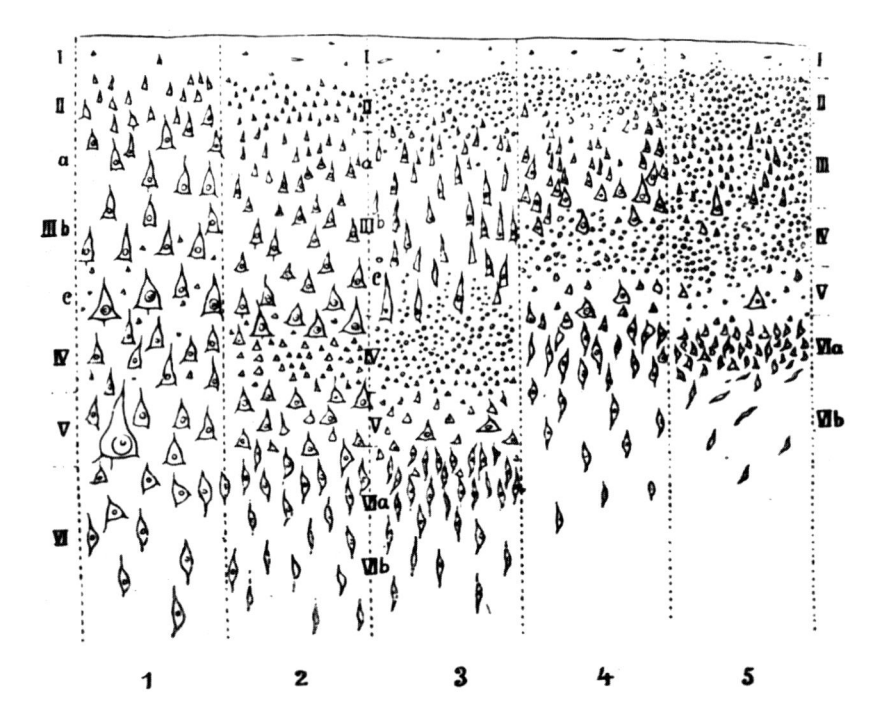

Fig. 8.3 The five fundamental types of neocortical areas. Type 1 = agranular motor cortex; type 2 = frontal granular; type 3 = parietal homotypical–associational type of cortex; type 4 = polar cortex area 18; type 5 = granular koniocortex of striate cortex. (From C Von Economo, The cytoarchitectonics of the human cerebral cortex, Oxford University Press, 1929, p. 16)

B. Basic Design and Functional Organization of Cerebral Cortex

The cerebral cortex is characterized by two essential design principles: (1) The neurons have a horizontal laminar arrangement and (2) connections occur in a vertical columnar manner. These vertical columns extend from the pial surface to white matter. A specific stimulus or a specific stimulus orientation activates each cylindrical column. Activation of a specific column also activates inhibition in adjacent columns: the concept of inhibitory surround. In the primary sensory, each column is concerned with a particular modality of sensation for a particular representation of the body or specific sector of a field of stimulation. This columnar organization was first described by Mountcastle for somatosensory cortex and has been subsequently found throughout the cortex.

1. Fundamental Types of Cerebral Cortex

Brodmann in 1903 (Brodmann, 1909) distinguished two fundamental types of cerebral cortex: homogenetic and heterogenetic.

1. *Homogenetic*. This type has a six-layer pattern at some time during development and is recognizable by the end of the third fetal month. This type of cortex is also called neocortex or isocortex, or neopallium or supra limbic.
2. *Heterogenetic*. This type of cortex does not have six layers at any time during development or during adult life. This type of cortex is also called allocortex and has three layers. There is also a transitional zone between the allocortex and the neocortex: the mesocortex that corresponds to the cingulate gyrus, the parahippocampal gyrus, the piriform area, and the anterior perforated substance (Fig. 8.4).

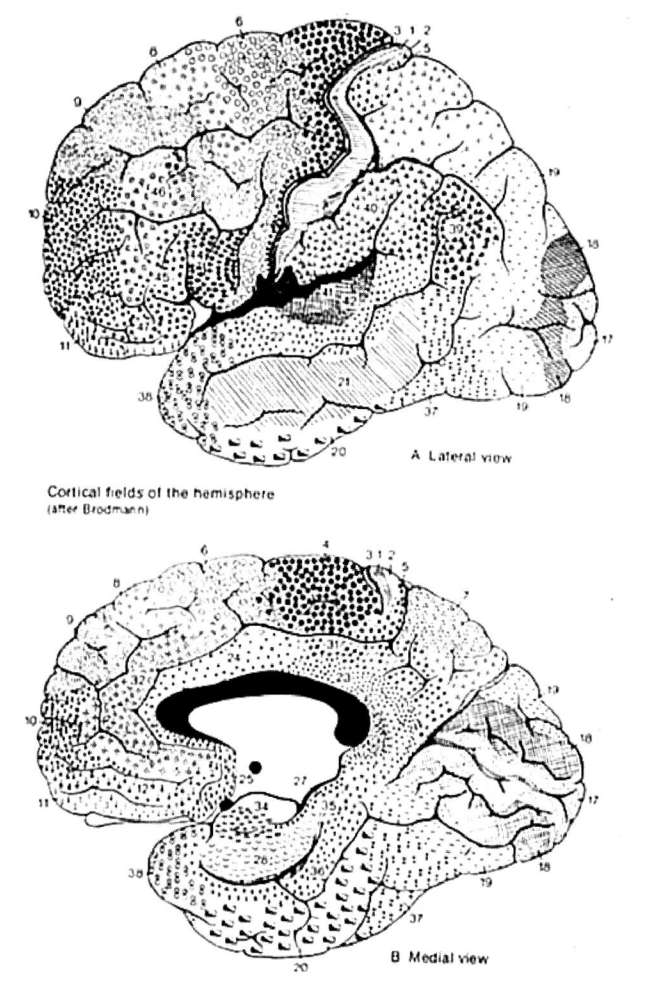

Fig. 8.4 Cytoarchitectural areas of cerebral cortex as designated by Brodmann. (From Curtis BA, S. Jacobson, and EM Marcus, An introduction to the neurosciences: Saunders, 1972. Modified from Ranson S and Clarks S. The Anatomy of the Nervous System. W.B. Saunders 1959)

Neocortex

Layer I: molecular or plexiform layers. This layer primarily consists of dendrites and axons from other cortical areas, deeper layers, and nonspecific thalamic input. A tangential plexus of fibers composed of the apical dendrites, ascending axons, and axon collaterals provides a dense collection of axodendritic synapses.

Layer II: external granular layer. This layer is a relatively densely packed layer of small stellate granule cells and small pyramidal cells whose apical dendrites terminate in the molecular layer and whose axons are sent to lower cortical layers.

Layer III: external pyramidal layer. This layer is composed of medium to large pyramidal cells whose apical dendrites extend to layer I. The axons of these cells function as association, commissural, or intracortical association fibers.

Layer IV: internal granular layer. This layer is densely packed with granular/stellate cells. A dense horizontal plexus of myelinated fibers, the external band of Baillarger is also present composed of the branches of the specific thalamo-cortical projection system. These fibers synapse with the stellate cells of this layer and the basilar dendrites of the pyramidal cells of layer III or with the apical dendrites of the pyramidal cells of layers V and VI.

Layer V: internal or large and giant pyramidal cells. This layer is also called the ganglionic layer. It layer serves as the major source of outflow fibers particularly to motor areas of the basal ganglia, brain stem, and spinal cord and possibly to the projection nuclei of the thalamus. Although cortico-cortical output arises primarily from the more superficial pyramidal cells of layer III, layers V and VI also provide such an output. Note that there is some overlap of the functions of pyramidal cell of layers V and VI. Collaterals of the axons also function as intracortical association fibers. In the deeper portion of this layer, a dense horizontal plexus of fibers is present—the internal band of Baillarger.

Layer VI: fusiform or spindle cell multiform layer. This layer is composed of a mixture of spindle-shaped cells, pyramidal cells, and stellate cells. Dendrites ascend to various cortical levels. The axons enter the white matter as short association fibers or ascend to other cortical layers. However, the cortico-thalamic outflow usually originates from pyramidal cells in this layer.

Summary of Cortical Circuitry: Associational fibers-arise in all layers, with the long running bundles from layer V.

Callosal: arise and interconnect layers II and III

Thalamocortical: from the many specific thalamic nuclei to layer IV; nonspecific to I.

Projectional: from layer V to brain stem and spinal cord and from layer VI to thalamus.

2. Classification of the Various Types of Neocortex

It is evident that, in the adult human, not all areas of neocortex have the same appearance and Brodmann found over 52 different areas and termed those areas of homogenetic neocortex that demonstrated the typical six-layered pattern seen in Figure 8.1

I Anatomical Considerations

as homotypical. There are other types of cortex, including the agranular motor cortex. This has many giant pyramidal cells present in layer V but with a virtual absence of an internal granular layer IV. At the opposite extreme is the granular primary sensory projection cortex. Layers II, III, and IV appear as an almost continuous granular layer. Layers of V and VI are thin and few large pyramidal cells are present. There have been other authors who have divided the cerebral cortex into areas. Von Economo (1929) also divided the cerebral cortex into categories (Fig. 8.3); however, in this text we are following the classification of Brodmann (Fig. 8.6).

3. How the Brodmann Areas Got Their Numbers

Numbers were initially assigned as histological distinct areas were identified in successive horizontal slices moving in anterior and posterior directions from the central sulcus. The crest of the postcentral gyrus would then appear in the first horizontal section and would be assigned the number 1. The primary motor cortex is assigned 4, the premotor cortex is assigned 6, and the frontal eye field is assigned 8. Thus, there is no logical reason why the numbers 6 and 8 are assigned to areas of the frontal lobe and 5 and 7 to areas in the parietal lobes. Nevertheless, some of the numbers are used with sufficient frequency in everyday neurological language that the student should commit these to memory. Before we review these areas with their correlated function, it is necessary to consider the methods employed in the study of functional localization.

C. *Correlation of Neocortical Cytoarchitecture and Function*

1. General Observation

Those neocortical areas with specialized cytoarchitectural deviating from the classical six-layered cortex have very specialized function (e.g., primary motor or sensory cortex). In contrast, the classical six-layered pattern tends to occur in association areas such as prefrontal, which have more complex and not such limited highly specialized functions. The Brodmann numbers are listed in the following outline with a brief note as to function. The list is not all inclusive and does not include the areas of the mesial temporal lobe.

Figures 8.5A and 8.5B demonstrates the lobes, gyri, and major sulci visible on the lateral surface of the cerebral hemisphere. Figures 8.6A and 8.6B demonstrate the lobes, gyri, and major sulci visible on the medial surface of the cerebral hemisphere.

2. Frontal Lobe (Table 8.1)

The following cortical areas according to Brodmann (Fig. 8.4) are noted in the frontal lobe: primary motor area 4, premotor areas 6 and 8, inferior frontal areas

(Broca's) 44, 45, and 47, frontal eye fields area 8, prefrontal/orbital areas 46, 9, 10, 11, 12, 13, and 14. It is customary to divide the frontal lobe into those more posterior areas devoted to motor functions and those more anterior areas devoted to nonmotor functions.

a. Motor Area 4

This area corresponds in general to the precentral gyrus and functions as the primary motor cortex. It is continued onto the medial surface as the paracentral lobule (Figs. 8.5A and 5B).

Stimulation Discrete repetitive focal movements are produced based on the area stimulated (e.g., repetitive jerks of the thumb or of the foot). A march of focal movements might occur (e.g., starting in the thumb, then spreading to hand, then to arm, then to face and leg, etc.)

Lesions An upper motor neuron type weakness occurs with a contralateral monoplegia or hemiparesis.

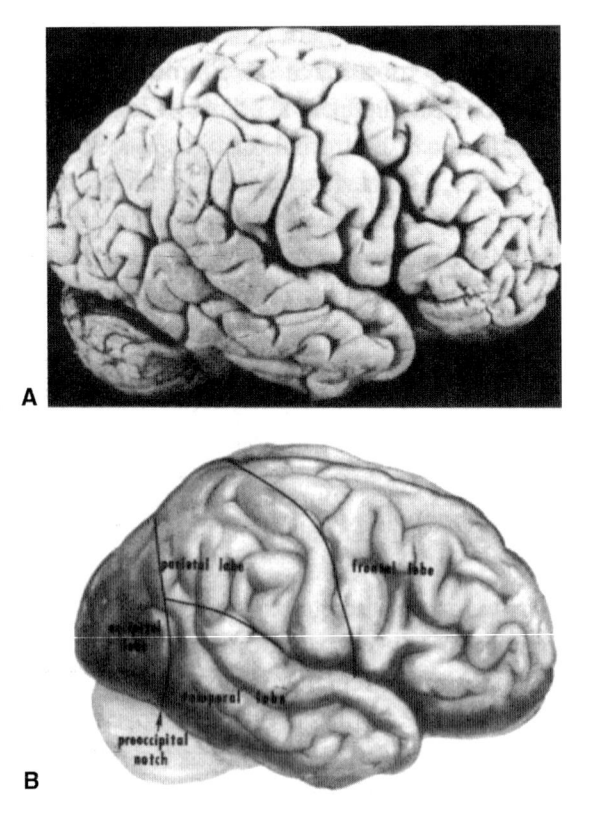

Fig. 8.5 (From EM Marcus and S Jacobson, Integrated neuroscience, Kluwer, 2003). **A** Gross brain, **B** lobes labeled, **C** Major sulci labeled, **D** Major gyri labeled. Lateral surface

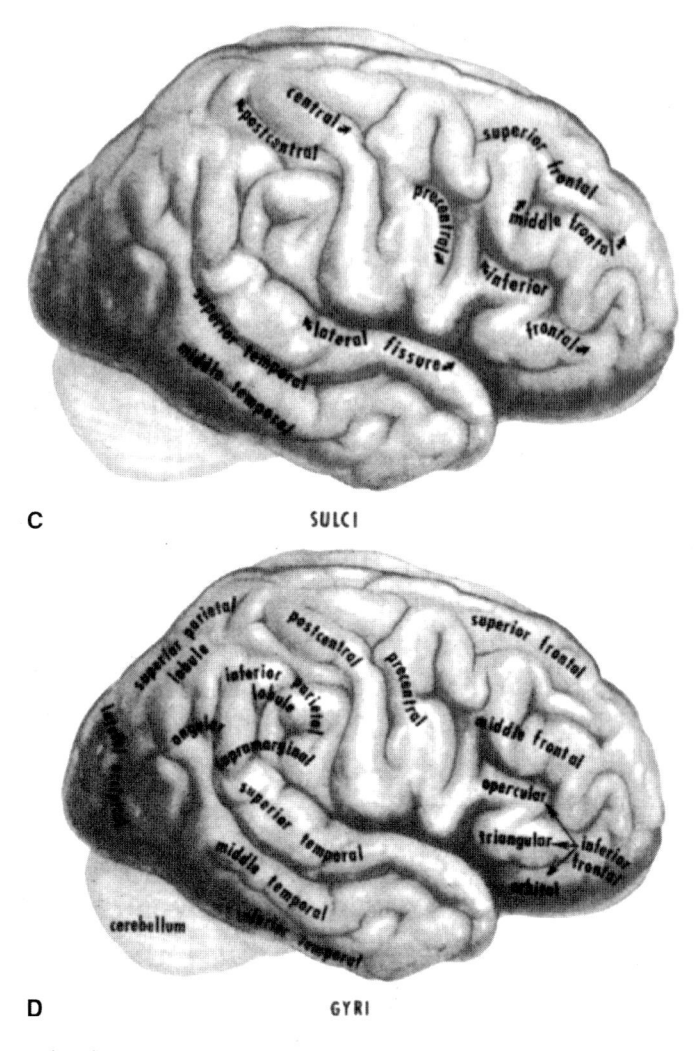

Fig. 8.5 (continued)

b. Area 6

This area is located anterior to area 4 and is thus referred to as the premotor cortex; it functions as a motor association or elaboration area (Fig. 8.5B). This area also functions to inhibit certain functions of the primary motor cortex. The supplementary motor cortex represents the continuation of this area onto the medial aspect of the hemisphere (Fig. 8.4).

Stimulation Patterns of movement occur (e.g., tonic rotation of head eyes and trunk to the opposite side associated with tonic abduction of the arm at the shoulder and flexion of the arm at elbow). Stimulation of the supplementary motor cortex produces similar complex patterns of tonic movement. In addition, there are bilateral movements of the trunk and lower extremities.

Lesions Transient release of "primitive" automatic reflexes mediated by the primary motor cortex such as instinctive grasp and suck occurs. A more persistent release of these instinctive reflexes follows the combined ablation of areas 6 and 8 and supplementary motor cortex. Such lesions also produce a significant gait apraxia (an impairment of the ability to walk without actual weakness or sensory deficit). Other deficits occur in the initiation of patterns of movement in relationship to somatosensory or visual stimuli.

Area 8 (Fig. 8.6B) is located anterior to area 6 in the middle frontal gyrus in the human and is often grouped with area 6 as the premotor cortex. This area functions in relation to conscious eye turning and is often referred to as the frontal eye field or frontal center for adversive or contraversive eye movement. This area is involved when the subject responds to the command "turn your eyes to the left (or right)." The effects are mediated by descending crossed connections to pontine lateral gaze center.

Stimulation A conscious repetitive conjugate eye movement to the opposite field occurs. More specific effects can also be achieved by discrete stimulation

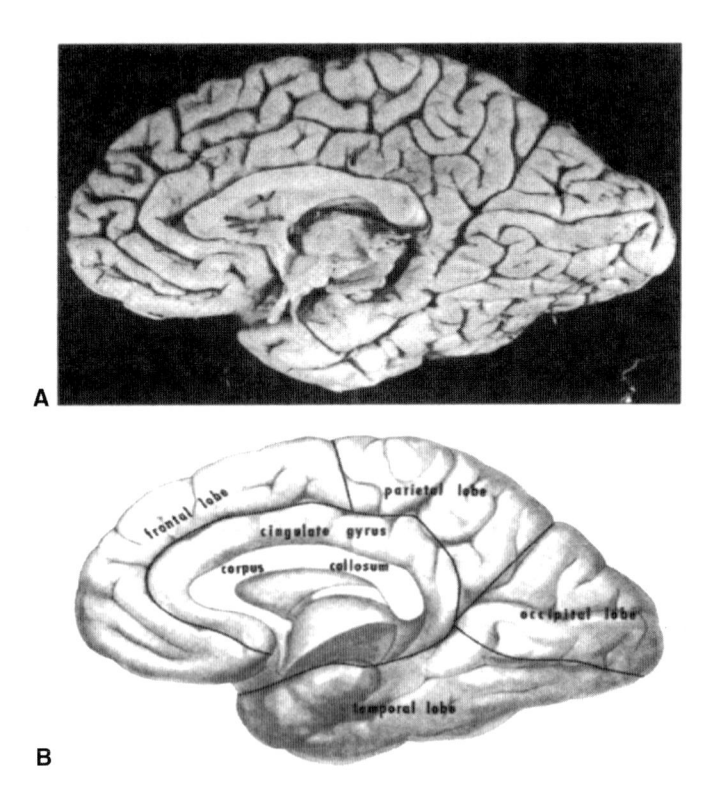

Fig. 8.6 Medial surface (From EM Marcus and S Jacobson, Integrated neuroscience, Kluwer, 2003). **A** Gross brain, **B** lobes labeled, **C** sulci labeled, and **D** Gyri labeled

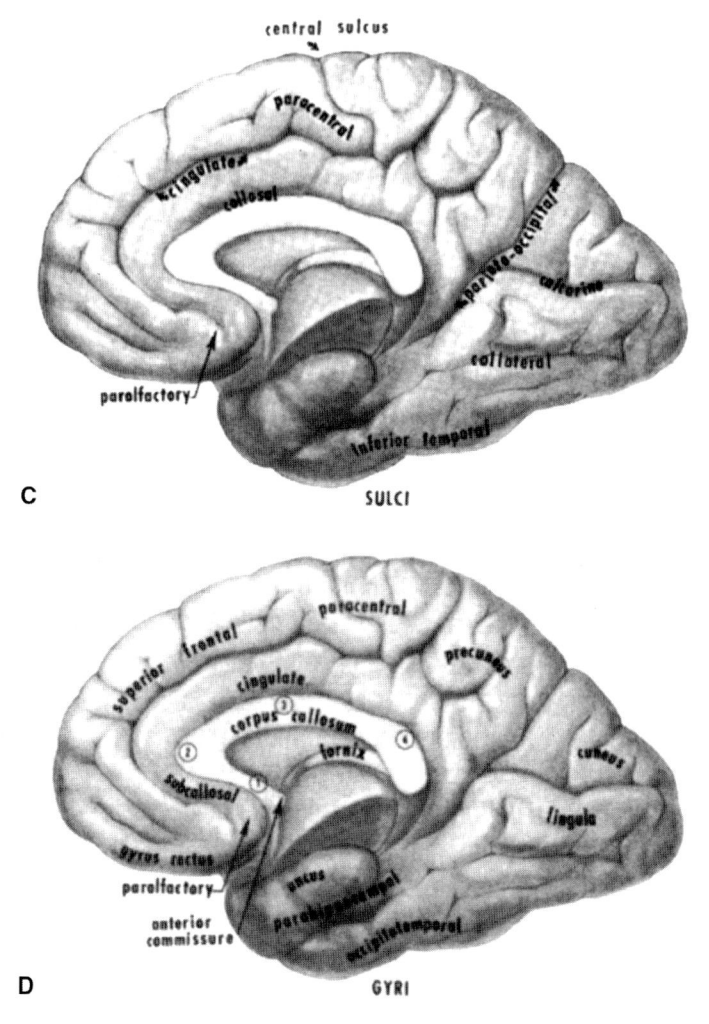

Fig. 8.6 (continued)

Table 8.1 Major gyri in the frontal lobe

Gyri	Major functions
Precentral	Origin of volitional motor pathways, the corticospinal and corticonuclear
Superior frontal	Premotor and prefrontal regions
Middle frontal	Prefrontal (executive functions) and frontal eye fields
Inferior frontal	Broca's motor and speech area
Orbital	Gyrus rectus/limbic cortex

within this area (e.g., eyelid opening, repetitive eyelid movements, pupillary dilatation) Bilateral stimulation will result in conjugate upward eye movements.

Lesions Transient paralysis of voluntary conjugate gaze to the contralateral visual field occurs. In general, in man, this paralysis of voluntary gaze does not occur in isolation but is associated with those processes, such as infarction, that have also produced a severe hemiparesis. The patient then lies in bed with the contralateral limbs in a hemiplegic posture and with the head and eyes deviated toward the intact arm and leg. This deviation most likely reflects the unbalanced effect of the adversive eye center of the opposite intact hemisphere. The effect is usually transient, clearing in a matter of days or weeks. A transient neglect syndrome primarily involving the contralateral visual field occurs although contralateral tactile and auditory stimuli might also be neglected. Bilateral ablations within this area in the monkey resulted in animals that had a bilateral neglect of the environment, remained apathetic, and continued to have a "wooden expression." Similar findings might occur in the human patient with degenerative disorders involving these frontal areas.

c. Areas 44 and 45

These areas correspond in general to the inferior frontal gyrus triangular and opercular portions. In the dominant (usually left) hemisphere, this constitutes Broca's motor speech center or the anterior speech center

Stimulation Arrest of speech occurs. Occasionally, simple vocalization occurs. A similar arrest of speech mighy also occur on stimulation of the other speech centers (posterior and superior).

Lesion A nonfluent aphasia occurs. The patient is mute. With limited lesions, considerable language might return. Although isolated, a lesion (e.g., embolic infarcts) involving this area might occur; more often this type of language defect is associated with a contralateral hemiparesis.

d. Areas 9, 10, 11, and 12

These prefrontal areas are concerned with executive functions, including emotional control and other aspects of cognitive function (Fig. 9.21 of Chapter 9). Areas 13 and 14 are also often included in this group.

Stimulation Complex partial seizures with alteration of personality, emotion, and behavior occur often accompanied by tonic motor components. Note that 25% of complex partial seizures originate in frontal lobe structures, but 75% originate in the temporal lobe.

Lesion An alteration in personality, affect and control of emotion might occur and/or there might be an alteration in cognitive/executive function. See Chapter 14.

3. Parietal Lobe

a. Areas 3, 1, and 2 (Postcentral Gyrus)

These areas correspond to the *postcentral gyrus* and function as the somatosensory projection areas (and also continue into the paracentral lobule on medial surface of a hemisphere).

Stimulation Episodes of localized tingling paresthesias occur that might spread as with focal motor seizures.

Lesion Deficits in cortical sensory (discriminative sensory) modalities will occur (affecting stereognosis, position sense, graphesthesia, and tactile localization).

b. Areas 5, 7, 39, and 40 (Parietal Lobules)

The intraparietal sulcus extend posteriorly, about halfway up the postcentral sulcus, dividing the parietal lobe into superior and inferior parietal lobules. The *inferior parietal lobule* consists of the supramarginal and angular gyri. The s*upramarginal gyrus* surrounds the posterior ascending limb of the lateral sulcus. The *angular gyrus* is behind the supramarginal gyrus and surrounds the posterior end of the superior temporal sulcus (Table 8.2).

c. Areas 39, 40

These areas correspond to the inferior parietal lobule (the angular and supramarginal gyri). These areas in the dominant hemisphere function in relation to reading and writing as higher integrative areas for language. This area is part of the posterior speech area. In the nondominant hemisphere, these areas relate to our concepts of visual space.

Stimulation In the dominant hemisphere, arrest of speech occurs.

Lesions In the dominant hemisphere, defects in reading, writing, and calculations occur. These deficits include difficulty in finger identification, and left/right

Table 8.2 Major gyri in the parietal lobe

Gyrus	Major functions
Postcentral areas 3-1-2	Somatosensory cortex
Superior parietal lobule	Sensory associational
Inferior parietal lobule	Dominant hemisphere, posterior sensory language areas
-Pars supramarginal	Nondominant hemisphere concepts of body image and visual space
-Pars angular	

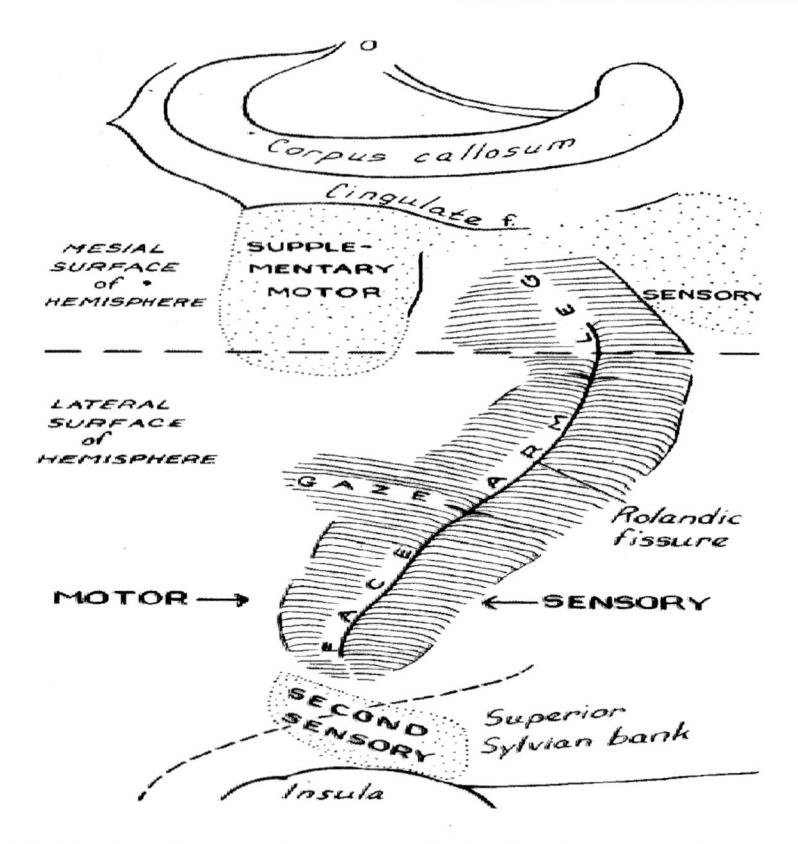

Fig. 8.7. Map of somatic motor and sensory areas. Note that the primary motor and sensory areas extend into the depth of the rolandic fissure/central sulcus and extend onto the medial surface of the hemisphere. Note that the premotor cortex, area 6 also extends onto the medial surface of the hemisphere as the supplementary motor cortex. Note the location of the adversive gaze field in relation to motor cortex. (From W Penfield and H Jasper, Epilepsy and the functional anatomy of the human brain, Little, Brown and Company, 1954, p.103)

confusion constitute aspects of the Gerstmann syndrome: a variety of fluent aphasia occurs. In the nondominant hemisphere, a denial or neglect syndrome occurs.

4. Temporal lobe

The temporal lobe is a complex structure that includes the six-layered neocortex and more primitive areas lacking six layers; allocortex, mesocortex, and a subcortical nucleus (the amygdala). At its posterior borders, it merges into the parietal and occipital lobes.

Major areas in the temporal lobe are presented in Figures 8.5B and 8.6B. They include the following: auditory (transverse temporal gyri of Heschel) 41; auditory

I Anatomical Considerations

associational 42 and 22; middle temporal 21; inferior temporal 20; posterior temporal 37; and entorhinal 27, 28, 35. The temporal lobe lies below the lateral sulcus and its lateral surface has a sequence of three anterior-to-posterior arranged gyri: the superior, middle, and inferior temporal gyri. The inferior temporal gyrus extends onto the ventral surface of the cerebrum. The superior temporal gyrus forms the temporal operculum. Near its posterior end, two gyri are seen running into the lateral fissure (Table 8.3). These are the transverse temporal gyri, consisting of the primary receptive auditory cortex, area 41. One of the first conclusive bits of evidence on the significance of anatomic asymmetry in the cerebrum was noted by Geschwind and Levitsky (1968) in the temporal lobe. They made a horizontal section through the lateral sulcus and removed the overlying parietal and occipital cortex. They called this exposed region of the superior temporal gyrus the *planum temporale*. They noted a larger planum temporale in the left temporal area than on the right side in right-handed humans (Wernicke's area). In left-handed individuals, the area was the same on both sides. Their observations on the anatomical basis of cerebral dominance have since been supported by radiological and other anatomical studies.

Areas 41 and 42

These areas are the transverse gyri of Heschl function and are primary auditory projection areas.

Stimulation Episodic tinnitus (ringing sensation) occurs. Usually, such seizures are not limited to this symptom in isolation, as other aspects of temporal lobe are involved to produce the more complex phenomena of the temporal lobe seizure.

Lesions Limited unilateral lesions might produce a disturbance in the ability to localize sounds. Bilateral lesions of a limited nature would be rare but could produce "cortical deafness."

Table 8.3 Gyri in the temporal lobe

Gyri	Major functions
Superior temporal	Auditory associational and Wernicke's sensory speech area
Transverse temporal gyrus of Heschel	Primary auditory cortex
Middle temporal	Limbic and facial recognition
Inferior temporal	Limbic
Occipitotemporal	Limbic
Parahippocampal (entorhinal)	Limbic
Hippocampal formation (Ammon's horn and dentate gyrus) allocortex	Limbic and memory
Amygdaloid nucleus	Emotional/arousal center

Area 22

This area corresponds to the superior temporal gyrus and surrounds areas 41 and 42. This is an auditory higher association center. In the dominant hemisphere, the posterior half of this area represents an auditory association area concerned with the reception and interpretation of spoken language. This is one component of the posterior speech area. The area is often referred to as Wernicke's receptive aphasia center. In the nondominant hemisphere, this area is more concerned with visual aspects of space.

Stimulation Stimulation of the posterior speech (Wernicke's) area (Fig.15.2 of Chapter 15) produces an arrest of speech. Seizures originating in the superior temporal gyrus are also characterized by "experiential" phenomena: distortions of auditory and visual perception alterations in the sense of time and in well-formed visual and auditory hallucinations (psychical seizures). Complex partial seizures might begin with these "simple" phenomena but they subsequently proceed to impairment of awareness (amnesia) and automatisms: (unconscious stereotyped patterns of movement). The progression to the complex partial seizure reflect involvement of or spread to the mesial temporal areas including the hippocampus and amygdala. Fear that might also accompany these seizures reflects involvement of the amygdala. Olfactory hallucinations that might also accompany these seizures reflects involvement of the mesial temporal structure, the uncus.

Lesions Damage to Wernicke's area (a part of the posterior speech area) produces a deficit in the comprehension of speech, a type of fluent aphasia. Suffice it to say that bilateral damage to the hippocampus will produce severe deficits in the ability to record new memories. Damage to the amygdala will alter emotional control

5. Occipital Lobe

The occipital gyri consist of the visual receptive cortex/calcarine-cortex found on the medial surface of the hemisphere and the visual associative cortex on the lateral surface of the hemisphere (Table 8.4)

a. Area 17

This area corresponds to the striate cortex bordering the calcarine sulcus and functions as the primary visual projection area. In modern physiological terms, the designation V1 is utilized.

Table 8.4 Gyri of the occipital lobe

Gyri	Major functions
Calcarine	Primary visual cortex: V1
Lingula	Superior visual field
Cuneus	Inferior visual field
Lateral occipital	Visual associational cortex: V2–V4

Stimulation Seizures originating in this area produce simple unformed visual hallucinations such as flashing lights, stars or jagged lines. The phenomena might be localized to the contralateral visual field or at times to the contralateral eye.

Lesions Homonymous visual deficits are produced (e.g., a homonymous hemianopsia or quadrantanopsia).

b. Areas 18 and 19

These areas form surrounding circular stripes around area 17. These areas function as visual association areas of varying complexity. The terms V2-V4 are utilized. These areas are also involved in the fixation and following of objects in the contralateral visual field. Note that visual areas V4 and V5 extend into the adjacent temporal-parietal cortex.

Stimulation: Some of the effects are similar to stimulation of area 17. In addition conjugate deviation of the eyes to the contralateral field will occur

Lesions: Selective lesions may produce deficits in some of the more complex visual activities. In addition defects in visual fixation and following may occur. (See Chapter 13)

II. Methods for Study of Functional Localization

A. *How Do We Study Function?*

In considering functional localization and the correlation with cytoarchitecture, it has been customary to consider the effects of two general categories: stimulation and lesions.

A. Stimulation

Various disease processes in man involving the cerebral cortex might produce a local area of excessive discharge (in a sense, a local area of stimulation resulting in focal seizures). Actually, our first understanding of cortical function was derived from the systematic study of such cases of focal discharge (termed *focal seizures, focal convulsions, focal epilepsy*, or *epileptiform seizures*). Hughlings Jackson was able to predict that an area existed in the cerebral cortex that governed isolated movements of the contralateral extremities. Such focal seizures are also referred to as partial seizures in the International Classification of seizures and Epilepsy 1985–1989. These partial seizures are subdivided into simple partial (simple motor or sensory or psychic symptoms) as contrasted to complex partial (complex motor sequences of automatic behavior and/or alterations in mental function with regard to perception, memory, and affect) accompanied by impairment of consciousness, confusion, and amnesia. Partial seizures or partial epilepsy must be distinguished from those seizures that are nonfocal [i.e., bilateral or generalized in their onset (generalized or idiopathic epilepsy in the

International Classifications)]. Other methods of stimulation such as electrical stimulation, of the cerebral cortex can be carried out in the human patient undergoing operative procedures on the cerebral cortex to identify specific cortical areas. Transcranial electrical and magnetic stimulation of the cerebral cortex has been developed as a clinical technique. In the interpretation of the effects of stimulation, consider (1) reproduction of the function of the area (e.g., focal motor seizure), (2) additional effects of spread to interconnected area, and (3) the effects on cortical function without the participation of the area simulated (e.g., stimulation of speech areas produces speech arrest and stimulation of the hippocampus produces interference with memory recording).

B. Evoked Potentials

Simulation of afferent pathways can be utilized to identify short latency responses evoked from specific cortical areas. When stimulating the lower extremity (or sciatic nerve), response is limited to that specific area of the contralateral postcentral gyrus devoted to the representation of the lower extremity, whereas stimulation of the upper extremity will elicit a similar response from a different portion of the contralateral postcentral gyrus: that devoted to representation of the upper extremity. With this approach, it is possible to prepare a map of the somatosensory projection cortex of the postcentral gyrus. It is possible to refine this technique at the response end by using microelectrode recordings from within or just outside neurons at particular depths of the cortex. It is also possible to define more carefully the parameters of stimulation as Hubel and Wiesel have done in their analysis of responses in the visual cortex to specific types of visual stimuli (see Chapter 13). The evoked potential method has been subsequently developed as a practical noninvasive clinical method to study conduction in the visual, auditory, and somatosensory systems. By stimulating particular thalamic nuclei, one can map out specific thalamocortical relationships. A modification of the technique is that of antidromic stimulation of an efferent pathway, such as the pyramidal tract, to map out the cortical nerve cells of origin of fibers in this tract

C. Lesions

In humans, many conditions can produce damage to the cerebrum, including interruption of the blood supply, tumors, or trauma. Clinical pathological correlation has produced considerable information.

1. The observed effects might indicate deficits directly related to a failure of a positive function normally subserved by the area destroyed; for example, after destruction of parietal somatosensory projection areas, deficits in stereognosis and position sense occur.
2. The effects might reflect the release of centers in other cortical areas or at other levels of the neural axis. This presumes that the area ablated usually inhibits

certain functions of these other centers. The end result then is an enhancement of certain functions (e.g., spasticity after ablation of the motor cortex).
3. A temporary loss of function of a lower center might follow acute ablation of higher centers. For example, following acute unilateral ablation of the motor cortex, a temporary loss or depression of deep tendon reflexes and of other stretch reflexes in the contralateral extremities will be observed. Initially, there will be a flaccid paralysis. This state (referred to as the "diaschisis of Von Monakow") is analogous to spinal shock or pyramidal shock. As recovery occurs, deep tendon reflexes return and become progressively more active. The flaccid paralysis is replaced by a spastic paresis.

B. How Do We Confirm the Location of the Pathology?

The initial method was postmortem correlation of lesion location with the specific symptoms and signs. Initially, data as to the actual location of the disease process and its nature (old scar from traumatic injury, old area of tissue damage due to ischemia, infection, local compression, or invasive brain tumor) were dependent on final examination of the brain at autopsy. With the development of neurosurgery, it became possible to define and to confirm, during life, the anatomical boundaries of the disease process, often allowing for treatment or cure of the basic disease process. The specialized techniques of electroencephalography, arteriography, pneumoencephalography, and radioactive brain scan were developed during the 1930s, 1940s, and 1950s to provide some information about localization prior to surgery. The subsequent development of the more precise noninvasive imaging techniques of computerized axial tomography, (CT scan), magnetic resonance imaging (MRI scan), and positron-emission tomography (PET scan) during the 1970s and 1980s now provides the closest correlation of clinical findings and anatomical location of the lesion in the living patient, supplanting the earlier radiological techniques of arteriography, pneumoencephalography, and radioactive brain scans.

C. Neurophysiology Correlates of Cortical Cytoarchitecture and the Basis of the EEG

The cerebral cortex of the human and of other mammals is characterized by continuous rhythmic sinusoidal electrical activity of variable frequency. This electrical activity can be recorded through the scalp in the form of the electroencephalogram (EEG) or directly from the cerebral cortex during surgery in the form of the electrocorticogram (ECG). The term EEG will be used during the subsequent discussion to refer to the activity found in either the EEG or the ECG. The EEG provides a physiologic correlate of the various states of consciousness.

1. Alpha Rhythm (Fig. 8.10)

The normal EEG in the awake adult who is recorded while resting with eyes lightly closed is characterized by a dominant activity in the posterior (parieto-occipital) recording areas of continuous 8–13-Hz sinusoidal waves of 25–75 µV. Amplitudes recorded directly from the cortex .are higher. This pattern was the first pattern discovered. The frequency of the basic background activity is also influenced by the age and level of consciousness of the patient.

2. Beta Rhythm

The second described pattern (the beta rhythm) is found predominantly in the frontal recording areas and consists of low-amplitude 14–30-Hz activities.

3. Slow Waves

Any pattern slower than alpha is referred to as composed of slow waves (theta = 4–7 Hz; delta = 0.5–3 Hz; Fig. 8.10). Many sedative medications will produce increased amounts of generalized beta activity and eventually generalized slow-wave activity. In general, the activity of the two cerebral hemispheres is relatively symmetrical and synchronous. The EEG recorded is not a result of axon potentials and does not represent the activity of individual neurons and synapses. Instead, EEG rhythms represent the summated electrical activity of a large number of synapses located in the more superficial cortical layers. Seizure discharges do not originate in single neurons but, instead, are generated in a group of neurons that manifest increased excitability and synchronization (the epileptic focus). Certain mutations involving the neuronal membrane and the synaptic receptor provide the basis for several types of genetically determined seizure disorder.

4. Abnormalities of the EEG. They are either Focal or Generalized.

a. Focal Abnormalities include the following:

1) Focal slow waves. these imply focal damage.
2) Focal spikes or sharp waves they imply focal excessive neural discharges as in focal or partial seizures.
3) Focal suppression of voltage implies severe focal damage on nonelectrical tissue such as hematoma below the electrodes.

b. Generalized Abnormalities include:

1) Generalized symmetrical and synchronized spikes or spike slow wave-correlates with generalized idiopathic epilepsy.

2) Generalized slow wave activity correlates with generalized depression of consciousness. This dysfunction may be structural or functional.
3) Generalized absence of activity-neocortical death.

III. Subcortical White Matter Afferents and Efferents

There are fundamentally three types of fiber system that occupy the subcortical white matter: (1) projection fibers, (2) commissural fibers, and (3) association fibers.

A. *Projection Fibers (Fig. 8.8)*

These fibers consist of the corticopetal afferent fibers (the thalamocortical radiations) and the corticofugal efferent fibers (the corticothalamic, corticospinal,

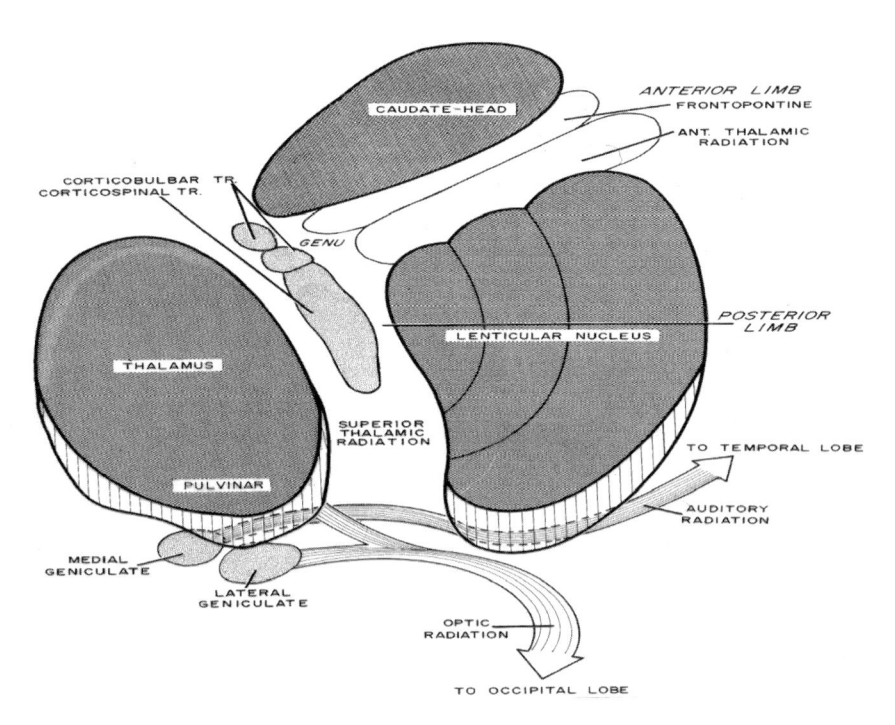

Fig. 8.8 Projection fiber systems passing through the internal capsule. The superior thalamic radiation includes fibers projecting from ventral thalamus to the sensory and motor cortex. (From EM Marcus and S Jacobson, Integrated neuroscience, Kluwer, 2003)

corticoreticular, corticopontine, and corticorubral tracts) (see also Figs. 9.5 and 9.6 of Chapter 9)

B. Commissural Fibers

The major commissural systems are the corpus callosum, anterior commissure, and hippocampal commissure.

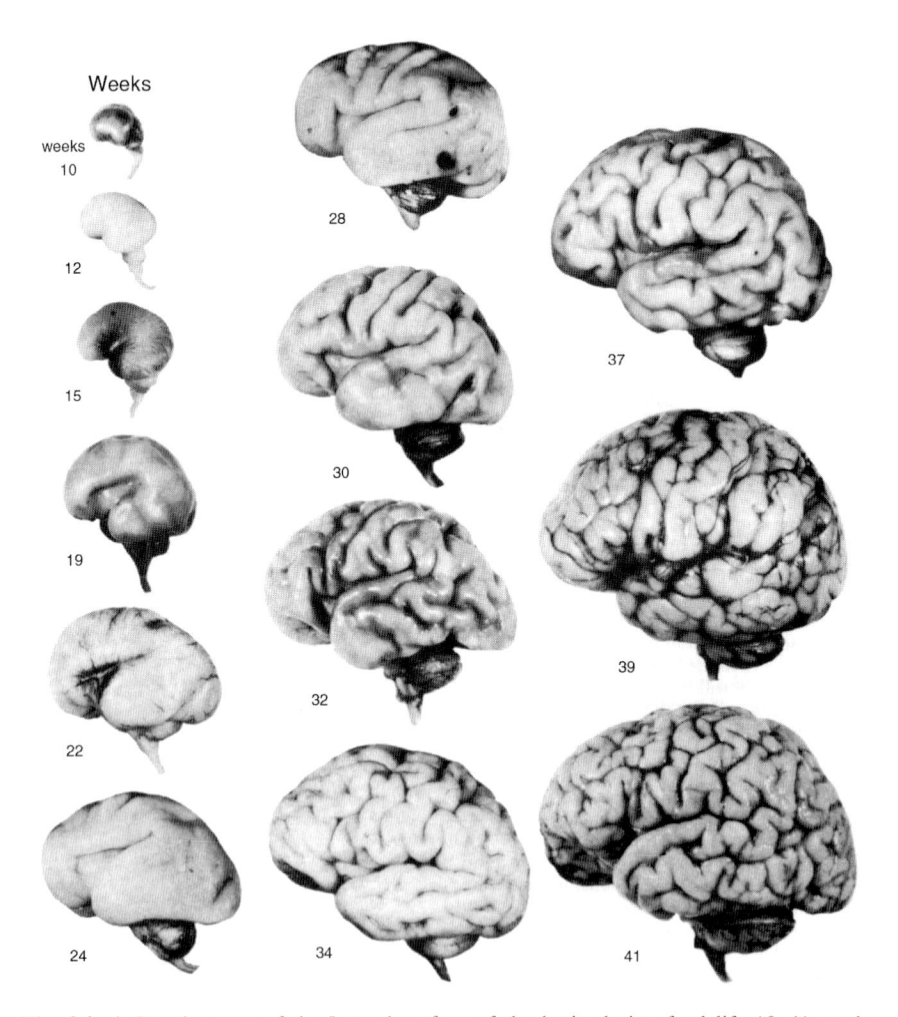

Fig. 8.9 A Development of the Lateral surface of the brain during fetal life 18–41 weeks (From J Larroche, The development of the CNS during inuterine life. In F Falkner (ed.) Human development, Saunders 1966). B Development of the Medial surface of the brain during fetal life 18–41 weeks (From J Larroche, The development of the CNS during inuterine life. In F Falkner (ed.) Human development, Saunders 1966)

Weeks 19

25

32

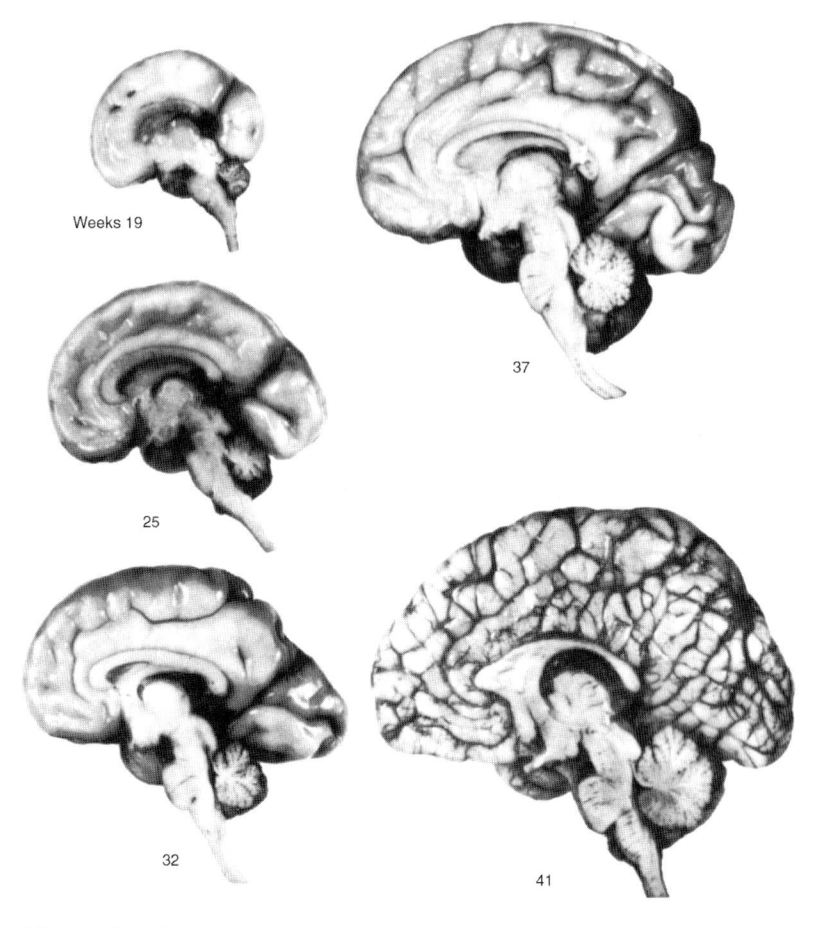

37

41

Fig. 8.9 (continued)

1. Callosal Connections

The corpus callosum is the major commissure for the neocortical areas (Figs. 8.6 A–D and 6B). In general, the homologous areas of the two hemispheres are interconnected. However, there are significant regional variations between homologous areas with regard to fiber density. Thus, there is a high density of callosal fibers connecting the premotor areas of the two cerebral hemispheres. On the other hand, the primary visual projection area (area 17) has almost no direct callosal connection to the contralateral hemisphere. Area 17 transmits to the adjacent area 18 that has a connection to contralateral area 18. Cortical areas also differ with regard to the spread of fibers to asymmetrical as well as to symmetrical points in the contralateral hemisphere. Thus, area 6 has widespread connections in the contralateral hemisphere,

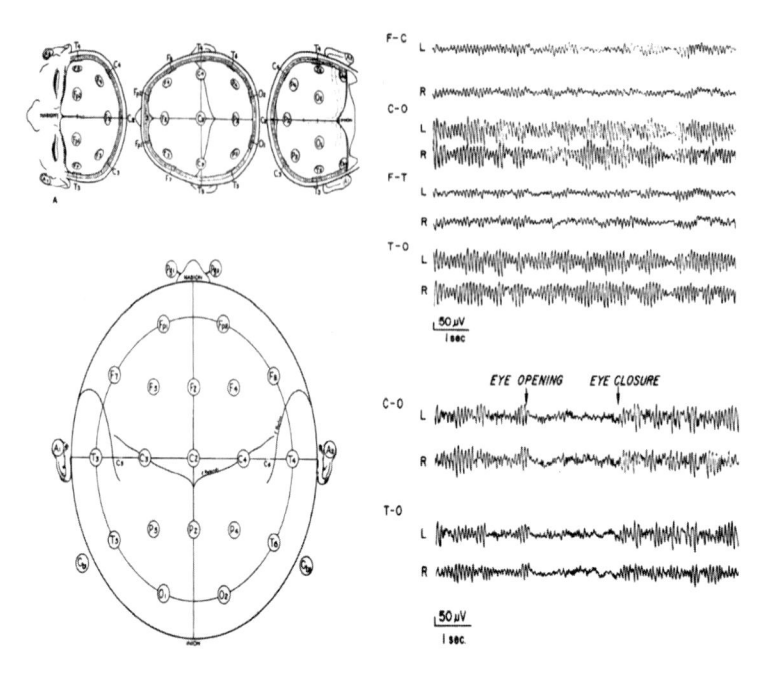

Fig. 8.10 The International Ten-Twenty Electrode Placement System. The patient is awake but in a resting recumbent state with eyes closed. 8BF = frontal, T = temporal, C = central and O = occipital. Bottom: Effects of eye opening and closure. (From EM Marcus and S Jacobson, Integrated neuroscience, Kluwer, 2003). The International 10–20 Electrode Placement System is described in Jasper, 1958

not only to area 6 but also to areas 4, 5, 7, and 39, whereas area 4 has a discrete contralateral connection only to the homotypical points.

2. Anterior Commissure

The anterior commissure interconnects the rostral portions of the superior, middle, and inferior temporal gyri, inferior–posterior orbital gyri, and paleocortex (parahippocampal gyrus).

3. Hippocampal Commissure

The hippocampal commissure interconnects the hippocampal formation and dentate gyri (archicortex) and the surrounding presubiculum, entorhinal, and adjacent inferior temporal gyri. These fibers are conveyed via the fimbria of the fornix and cross at the point beneath the splenium of the corpus callosum, where the posterior pillars of the fornix converge

C. Associational Fibers

Two types of subcortical association fibers can be distinguished: (1) short subcortical fibers, U fibers, that interconnect adjacent gyri and (2) long fiber bundles that reciprocally interconnect distant cortical areas. The following long fiber bundles have been identified:

1. The *superior longitudinal fasciculus* interconnects the superior and lateral frontal, parietal, temporal, and occipital areas (Fig. 15.5 of Chapter 15). The *arcuate* fasciculus is an extension of this fiber system into the temporal area, passing through the subcortical white matter of supramarginal and angular gyri. This fasciculus has considerable importance because of its role in connecting Wernicke's receptive language centers of the temporal lobe with Broca's expressive motor speech centers of the inferior frontal gyrus. Such a connection must be made if a sentence that has been heard is to be repeated.
2. The *inferior longitudinal fasciculus* interconnects the occipital and inferior temporal areas.
3. The *inferior frontal–occipital fasciculus* interconnects the frontal and occipital areas. It is often difficult to clearly differentiate this fiber system from the uncinate fasciculus.
4. The *uncinate fasciculus* passes deep to the lateral sulcus and interconnects the orbital, medial prefrontal, and anterior temporal areas.
5. The *cingulum* passes within the subcortical white matter on the medial surface of the cerebral hemisphere in the cingulate gyrus; it interconnects the subcallosal, medial-frontal, and orbital-frontal with the temporal lobe, occipital areas, and cingulate cortex. The cingulate bundle is also an important cortical link in the Papez circuit (see Chapter 14).

D. Afferent Inputs and Efferent Projections of Neocortex

The thalamus is the major source of afferents to the cortex with glutamate as the transmitter. With the exception of the olfactory system, all sensory information passes through the thalamus. The thalamus is composed of a series of relay nuclei that connect in a reciprocal manner with the cerebral cortex and the striatum (refer to Chapter 6). These nuclei can be subdivided into specific projection (or relay) nuclei and nonspecific or diffuse projection (or relay) nuclei.

The specific nuclei have defined afferent inputs and project in a relatively precise topographic manner to specific areas of the cerebral cortex. Experience or functional disuse might modify the precise cortical areas devoted to the topographic representation of a given sensory area. Examples of specific nuclei (VPL, MGN, LGN) and their cortical projections have been provided in Chapter 6. The densest input from the specific relay projection nuclei is to layer IV, but as indicated earlier, pyramidal cells in layers III and V might also receive direct or indirect inputs. There is a modality specific columnar arrangement.

The *nonspecific nuclei* have a more diffuse input and a more diffuse but not necessarily generalized cortical interaction. Examples include midline nuclei, central–medial and intralaminar nuclei, and reticular dorsal and medial dorsal (which is diffuse to frontal areas). Nonspecific nuclei project mainly to layer I.

The reticular nucleus of the thalamus is a thin-shell nucleus that is external to the external medullary lamina. It does not project directly to the cerebral cortex but receives innervation from the cerebral cortex. The subnuclei grouped as the thalamic reticular nucleus have specific reciprocal relationships with projection (relay) nuclei in the dorsal thalamus. The drive on the relay nuclei is therefore inhibitory.

E. Nonthalamic Sources of Input

– Non adserensic from lows cesvleus of brainstem.
– Seso torisesic from sophe nuclei of pons and medvila and postise reticular formation
– depaminesic from ventrral tegmental -sostoal. Mesescephalic rvelead groups
– Cholineagic from basal forabrain nucleus of megn
– ABA engic from basal forebrain, vestral fegentalaroe

F. Efferent Projections

Projections occur in a reciprocal manner with many subcortical areas (thalamus, basal ganglia, brain stem, and spinal cord). However, these projections are estimated to account for only 0.1–1% of all fibers in the white matter. Instead, most fibers are involved in intrahemispheric and interhemispheric connections.

IV. Development of the Cerebral Cortex

The telencephalon is part of the forebrain and consists of two regions: a cortical zone that becomes the cerebral hemisphere and a deeply placed nuclear region the corpus striatum that becomes the caudate, putamen, and globus pallidus. At 10 weeks *in utereo*, there are paired smooth telencephalic vesicles.

A. Primary Sulci

The first prominent features in the telencephalon are the primary sulci, the lateral sulcus (Fig. 8.9A), and the medial-placed Parieto-occipital sulcus (Fig. 8.9B). which appear at 19 weeks. By 24 weeks, the lateral sulcus is more pronouced and the calcarine sulcus I is now prominent on the medial surface. Shortly thereafter, the other

IV Development of the Cerebral Cortex   217

primary sulci (the central, callosal, and hippocampal) appear. After about 30 weeks, the primary sulci are well formed and there is an increase in the secondary sulci and the gyri that form. By term, formation of the gyri and sulci is complete. The cerebral cortex is formed by waves of migrating neuroblasts (migrating neurons). Formation of the cerebral cortex begins at the seventh week and is completed in the first year. The cortex forms from the inside out, with the first layer to form is layer VI and the last is layer II; the pyramidal cells mature before the stellate cells.

B. Myelination

All of the circuitry is in place before birth; however, myelination in the cortex begins in the last trimester and continues into the post natal period. Myelination begins in the spinal cord and spreads into the brain stem and diencephalon before birth and in the motor–sensory strip at 1–2 months postnatal and spreads into the adjacent associational regions by 2–4 months postnatal and, finally, into the polysensory areas: the prefrontal, inferior parietal, and posterior temporal gyri after 4 months. Myelination a misthes cerebral cortex continues into adult life.

Part II
The Systems within the Central Nervous System

Chapter 9
Motor System I: Movement and Motor Pathways

There are three major areas that are responsible for producing the fine motor movements in the human: the cerebral cortex, basal ganglia, and cerebellum. The motor cortex in the cerebral hemispheres and regions in the frontal lobe anterior to the motor–sensory strip provide voluntary control to all of the skeletal muscles in the body. Our understanding of localization in the motor system is based on studies following lesions at various levels of the neuroaxis.

I. Cerebral Cortical Motor Functions

In studying the motor system, we will consider reflex activity, the central generators of patterns of movement, voluntary movement, and learned movements. We will also consider two interrelated aspects: posture and movement. Under posture, we will be studying static or tonic reactions. In the following discussion, we will examine motor function at the level of the spinal cord, the brain stem, and cerebral cortex. It is important to realize that motor functions are represented at successively higher physiological and anatomical levels of the neural axis. As we go higher in the neural axis, we are utilizing and modifying mechanisms that have been integrated at a lower level of the neural axis—a concept first expressed in the modern era by Huglings Jackson (Taylor, 1931). Thus, pattern generators/centers make use of the motor mechanisms involved in reflexes without the necessity of afferent input. The terms utilized in defining a reflex are presented in Table 9.1.

In turn, voluntary and learned movements incorporate or impose a higher level of a more complex cortical control of these reflex and central pattern mechanisms. In subsequent sections, the role of basal ganglia and the cerebellum in modulating movement will be considered.

A. *Concept of Central Pattern Generators*

Specific neural circuits or neuronal clusters or centers might be responsible for the generation of simple and complex patterns of movement in the absence of afferent

Table 9.1 Definition of a reflex

Term	Example
Adequate stimulus	Nociceptive, proprioceptive, tactile, visual
Synapses involved	Monosynaptic in the stretch reflex.
	Polysynaptic in the nociceptive and flexion reflex
Segments involved	Segmental stretch reflex
	Intersegmental: interlimb reflexes
	Suprasegmental: brain stem tonic neck/labyrinthine
	Cerebral cortex:- placing and long loop reflexes
Movement or response	Flexion, extension, righting, response standing, walking, grasp, avoidance, placing
Aim or purpose	Avoidance of pain, escape from predator, acquisition of food

input. In actuality, such central patterns are modified or modulated by afferent input, utilize many of the reflex components to be discussed here, and are controlled and/or modified by descending control motor systems from higher centers. Such pattern generators have been described at the level of the spinal cord and the brain stem (mesencephalic locomotor system) for more complex behavior related to locomotion. The latter system is also involved in the motor components of emotional expression, as well as chewing, licking, and sucking behaviors. For the accurate performance of complex patterned movements involving the limbs, afferent input and reflex activities are essential. In the absence of such afferent input, the movements, although repetitive, are not as well coordinated (sSee Harris-Warrick and Johnson, 1989).

1. Spinal Pattern Generator

This is probably composed of bilateral clusters of interneurons in the intermediate gray matter at the base of the posterior horn. They are responsible for the locomotor movements from this level. These interneurons are the same interneurons involved in the flexion reflex to be described below. In the spinal cord preparation to be discussed below, the effects of dopaminergic and adrenergic agents in triggering behavior can be demonstrated.

2. Mesencephalic Locomotion Pattern Generator

This center is located in the periventricular tegmentum at the junction of midbrain and pons. Axons from this region descend to the medullary medial reticular formation. From this region, axons descend as the medial reticulospinal tract in the ventrolateral funiculus to the spinal locomotor system of the lumbar spinal cord. In the decerebrate preparation, glutaminergic effects can be demonstrated. Glutamate receptor antagonists will prevent the locomotion effects that occur on stimulation of the mesencephalic center.

3. Other Motor Pattern Centers

The premotor cortex for visually guided movements, the subthalamic nucleus, the pontine reticular formation, and the cerebellum will be discussed in Chapter 11.

B. Effects of Spinal, Brain Stem, and Cerebral Lesions on the Motor System

1. Acute Transection of the Spinal Cord in the Human

In humans, transection of the spinal cord reflects (1) the effects of spinal shock, (2) the prepotency of flexion reflexes as reflex recovery occurs, and (3) the slower recovery of monosynaptic stretch reflexes.

Spinal Shock

After the acute and sudden loss of suprasegmental control, a temporary expression of segmental reflex activities occurs. Deep tendon stretch reflexes are depressed. Cutaneous and autonomic reflexes might also be affected. The excitability of alpha and gamma motor neurons and of interneurons is depressed. In humans, the duration is usually of several weeks. In primates, spinal shock depends more on damage to the cortical spinal tracts. It is important to note that the depth and duration of spinal shock depends on the rapidity (acuity) and completeness of spinal cord section. Slowly evolving chronic spinal cord compressions are not characterized by spinal shock. Fever or infection will prolong the duration of spinal shock or might result in a recurrence of spinal shock once recovery has occurred. Immediately after an acute transection, the patient will demonstrate a depression of all the deep tendon stretch reflexes and, to a lesser degree, the flexion reflexes. The limbs will be flaccid and paralyzed. The bladder will be flaccid (i.e., hypotonic or atonic) due to depression of the bladder stretch reflex.

Pattern of Reflex Recovery

Within a short period of time (1–5 weeks), the flexion reflexes, including the sign of Babinski, will recover and remain prepotent. The flexion reflex (the withdrawal of the foot and leg on painful stimulation) is a polysynaptic reflex. The afferent component of the reflex arc is mediated by group II and III myelinated fibers and group IV unmyelinated fibers, from a variety of receptors in the skin, muscle, and joints. Initially in the human, the flexion reflex might be manifested as a mass reflex.

Mass Reflex

When fully developed, this mass reflex might consist of generalized reflex flexor spasms of the muscles of the trunk and lower extremities accompanied by sweating, emptying of the bladder, and, in males, penile erection and ejaculation. Initially, the receptive field for triggering this response is quite extensive, involving any tactile or nociceptive stimulus to the foot, leg, abdomen, or thorax. With the passage of time, the mass reflex becomes less prominent and a local sign develops. The more specific reflex movement then depends on the more specific location of the stimulus. For example, a painful stimulus to the outer border results in flexion, inversion, and adduction as opposed to a similar stimulus to the inner border of the foot; those results in flexion, eversion, and abduction. Over a number of weeks to months depending on the completeness of transection (longer if complete), stretch reflexes with other extensor reflexes and postures will return. The deep tendon stretch reflexes will then become hyperactive. There will be a spastic rather than a flaccid weakness. Bladder emptying and bowel evacuations return 3–4 weeks after injury.). With marked extensor tone, a partial capacity for brief periods of standing might return. Clearly, there is a phylogenetic factor; such extensor capabilities are more limited in the spinal human.

2. Transection of the Brain Stem: Decerebrate Preparation

Location of Lesion

The decerebrate preparation involves a transection of the brain stem at a level between the vestibular nuclei and the red nucleus, usually at the intercollicular midbrain level. In the human, the decerebrate state might reflect several pathological processes such as basilar artery thrombosis with brain stem infarction, temporal lobe herniation with midbrain compression, or massive destruction of both cerebral hemispheres. This acute state occurs in all mammals

Decerebrate Rigidity

Decerebrate rigidity develops almost immediately and can be defined as the exaggerated posture of extension of the antigravity muscles due to the enhancement of proprioceptive stretch reflexes. This posture is also referred to as opisthotonos. Sherrington (1947) has described the posture of decerebrate rigidity as "an exaggerated caricature of standing." This is most intense in those muscles that normally counteract the effect of gravity. In an animal, such as the sloth, which normally hangs upside down, it is the flexor posture that is exaggerated. It is important to note that although the extensor tone is sufficient to allow the animal to stand, the animal is pillarlike. The pillarlike limb seen in this extensor posture is indicative of the positive supporting reaction already seen in the "spinal" cat and dog. This is the tonic form of the extensor thrust reaction previously noted. The animal lacks righting reflexes and has no reactions to sudden displacement.

Modification of Decerebrate Rigidity

The posture of decerebrate rigidity can be modified by several influences: tonic neck reflexes, tonic labyrinthine reflexes, noxious stimuli, and time. Tonic neck reflexes depend on afferents from the cervical joints and are best studied after destruction of the labyrinths. In contrast labyrinth reflexes are best studied after section of the sensory/dorsal cervical roots or immobilizations of the head. The afferents are conveyed from the otoliths via the vestibular nerves to the vestibular nuclei.

1. Tonic neck reflexes result in the classical fereex's posture turn the hea to the right, the right as extends, the reflexes.
2. Noxious stimulus to a limb produces the classic interlimb reflex: The stimulated hind limb withdraws and the contralateral hind limb extends. The ipsilateral forelimb extends and the contralateral forelimb flexes.
3. Time: With time, 7–14 days in the cat and dog, rigidity decreases and righting reflexes begin to emerge.

The Anatomical Basis of Decerebrate Rigidity

The studies of Sherrington (1947) demonstrated that the corticospinal tracts (i.e., pyramidal tracts) were not involved in decerebrate rigidity because with hemisection of the brain stem at the intercollicular level, the unilateral rigidity that occurred was ipsilateral to the hemisection. Decerebrate rigidity was conceived to be a release phenomenon.

Bulbar Reticular Inhibitory Area

The studies of Magoun and Rhines (1946) demonstrated that the stimulation of the reticular formation in a ventromedial area of the medullary tegmentum produced a decrease in decerebrate rigidity. This inhibitory area of the medulla corresponds to the nucleus reticularis gigantocellularis, which is found in the lateral column of the spinal cord. In contrast, stimulation of a region of the reticular formation in the pons and midbrain resulted in an increase in the degree of decerebrate rigidity. This facilitory area extends into the subthalamus, the adjacent hypothalamus, and the intralaminar areas of the thalamus (e.g., centrum medianum). This area in the more lateral tegmental reticular formation in the pons and midbrain might be termed a brain stem (or bulbar reticular) facilitatory area. From an anatomical standpoint, the facilitatory area of the pons corresponds to the nucleus reticularis pontis oralis and caudalis. The neurons of this nucleus give rise to that medial reticulospinal tract found in the anterior column of the spinal cord. Classic decerebrate rigidity might be considered essentially "gamma rigidity."

3. Midbrain Preparation

In this preparation the brain stem is transected above the superior colliculus with the red nucleus and portions of subthalamic nucleus and posterior hypothalamus intact. In this preparation, proprioceptive reflexes are modified by contactual stimuli. In the dog or cat midbrain preparation, a sequence of righting reflexes occurs. The concept of righting reflex is easily visualized if one considers an intact cat dropped upside down. First, the head rights then the shoulders, followed by the upper body, and then the lower torso. The legs then right and all four limbs extend. The cat lands softly and walks away. In the midbrain cat preparation, kinetic reactions such as standing and walking might also occur. In contrast, the primate midbrain preparation is unable to stand. There are some fragments of a traction grasp.

4. Role of Hypothalamus

The motor responses for rage are represented at a brain stem level. Stimulation of the midbrain of the decerebrate preparation will produce all of the motor components of anger. The hypothalamus, however, is essential for the sham rage reaction of the decorticate animal.

5. Decorticate Preparation

The cat and dog preparations are similar to the midbrain animal. In the primate, the tonic grasp reflex might be present. The decorticate human or primate manifests a considerable degree of spasticity, often referred to as decorticate rigidity. The posture of the decorticate primate or human involves extension of the lower extremities and flexion of the upper extremities. In a sense, this is a double hemiplegic posture. Such a posture is often found in patients who have suffered anoxic encephalopathy after cardiac arrest or after anesthetic accidents. Contrast this posture to the four-limb extension found in the patient with decerebrate rigidity

6. Reactions Dependent on Cerebral Cortex

The cerebral cortex might be considered as providing for a complex reaction to the external environment. The cerebral cortex analyzes afferent information from many sources and utilizes reactions that have been integrated at lower levels of the neural axis. Thus, the presence of the cerebral cortex allows for the accurate projection of the limb in space and for the interaction of various reflex activities with visual and tactile stimuli. The following reflexes are associated with the cerebral cortex.

I Cerebral Cortical Motor Functions

a. Optical Righting Reflex

This response results in the righting of the head in relation to a visual stimulus. This response continues to occur after elimination of the labyrinths and dorsal roots from the upper cervical neck proprioceptors. The response can still be demonstrated after bilateral ablation of the motor and premotor areas in the primate. We should, of course, note that righting of the head when the head and body have been turned on the side is dependent not only on visual cues but also on labyrinthine cues and asymmetrical body contact stimuli.

b. Placing Reactions

A number of placing reactions are also noted in the intact animal: visual placing reaction and tactile placing reaction.

c. Grasp Reflex (Fig. 9.1)

This reflex consists of a stereotyped progressive forced closure of the subject's hand on the slowly moving stimulus of the examiner's fingers across the palm. Tactile pressure is the appropriate stimulus. Local anesthesia infiltration of the palm abolished the reflex.

1) Instinctive Grasp

The instinctive grasp is a more complex and variable response and is triggered by a tactile (or contactual) stimulus to the hand. The hand orients so as to grasp the object. Often, the hand follows a moving object, making moment-to-moment adjustments so as to grasp the object. This instinctive tactile grasping reaction allows for the exploration of space with the hand. The grasp reflex and instinctive tactile grasping reaction are, of course, seen normally when the intact individual attempts to grasp onto an object or to reach for an object. It is not surprising that intact motor and parietal cortex is required. The movements involve distal finger and thumb movement and require precise feedback information. At times, however, the reaction is seen to be released in abnormal form and degree and to occur in a context where voluntary grasping is inappropriate. Patients with release of the tactile grasp reaction usually have damage to the premotor cortex (areas 6 and 8) and the adjacent supplementary motor cortex and cingulated area on the medial surface of the hemisphere. The instinctive tactile grasp reaction, then, although requiring an intact motor and parietal cortex, is released in an abnormal form by lesions of the premotor, supplementary motor, and cingulate areas. This is a form of transcortical release. The same lesion, of course, also acts to release various subcortical areas.

AT BIRTH
TRACTION RESPONSE
Stimulus: Stretch shoulder adductors and flexors
Response: All joints flex

GRASP REFLEX
(INITIAL COMPONENT)   (FULLY FORMED)
S. Contact between thumb and index   S. Distally moving contact medial palm
R. Thumb and index adduct alone   R. All fingers flex

INSTINCTIVE GRASP REACTION
S. Contact radial or ulnar side   S. Contact hand (any part)   S. Contact hand (any part)
R. Hand orients   R. Hand gropes   R. Hand grasps

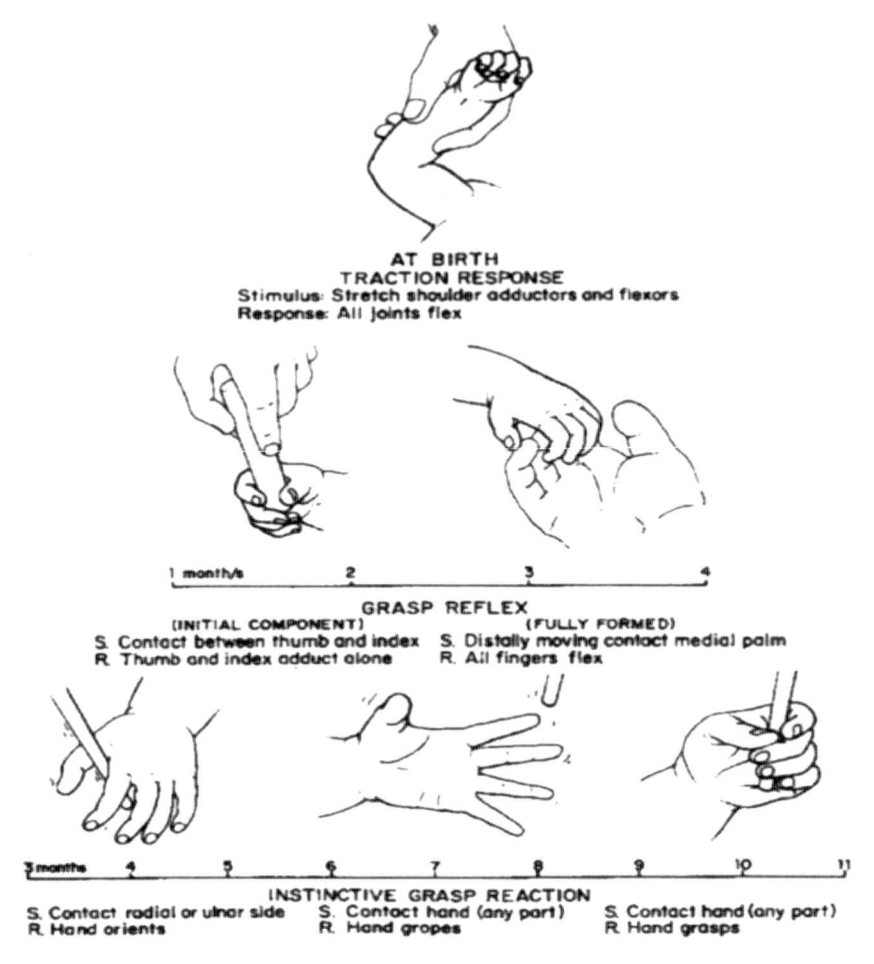

Fig. 9.1 Evolution of the automatic grasping responses of infants (From TE Twitchell, Neuropsychologia 3:251, 1965)

2) Instinctive Tactile Avoiding Reaction

The instinctive tactile grasp reaction is to be compared to the instinctive tactile avoiding reaction. The stimulus for this reaction is a distally moving tactile stimulus to the outer (ulnar) border of the hand, to the palmar surface of the hand, or to the dorsum of the hand. This stimulus leads to a nontonic orientation of the hand away from the stimulus. This is a more precise form of the tonic avoiding response noted in a decorticate preparation. The instinctive tactile avoiding response does not require the pyramidal system. The integrity of the cingulate area, area 8, and the supplementary motor cortex is required. The reaction is released in abnormal degree by damage to the parietal lobe (and adjacent motor cortex). The instinctive grasp and avoidance responses normally are in equilibrium from an anatomical and physiological standpoint.

3) Other Reflexes Associated with the Cerebral Cortex

In addition to the instinctive tactile grasp and avoiding responses, various visual triggered responses can be seen:

Visual Instinctive Grasp Reaction. The visual instinctive grasp reaction. requires the integrity of the posterior parietal areas, the visual mechanisms, and the motor cortex. The CT scan evidence suggested a medial frontal location of lesion responsible for such release.

Visual Instinctive Avoiding Reaction. This requires the integrity of the temporal lobe and visual mechanisms. The response is independent of the pyramidal system. Again, in the intact individual, equilibrium of grasp and avoidance is to be noted. Although our discussion has focused on the upper extremity, similar grasp and avoiding reactions can also be noted in the lower extremities. Lesions of the premotor cortex also affect gait, producing an apraxia in walking termed a frontal lobe gait apraxia. This will be discussed later in this chapter.

II. Postnatal Development of Motor Reflexes

Some reflexes are present at birth and disappear with time: the startle reflex, sucking, rooting, and the extensor plantar response. A sequence of reflex activities is noted in the developing human infant with regard to the use of hands and fingers. In addition, a sequence of changes occurs with regard to sitting, standing, and walking, as summarized in Table 9.2

III. Relationship of Primary Motor, Premotor and Prefrontal Cortex

In order to understand voluntary and learned behavior, it is necessary to explore the relationship of primary motor, premotor (including supplementary motor), and prefrontal areas. Voluntary and learned behaviors also reflect the motivational role of limbic structures to be considered in subsequent chapters.

A. Functional Overview

1. Primary Motor Cortex: Area 4 (Figs. 9.2 and 9.4)

The large and giant pyramidal cells in the primary motor cortex discharge, producing movements in specific directions (flexion, extension, abduction, adduction) that result from the contraction of muscles at a specific joint. Among other sources of

Table 9.2 Developmental changes in motor responses

Response	Postnatal time
Sucking and rooting reflexes	Present at birth
The Moro (startle) reflex	Present at birth and persisting until 2 months
Tonic neck reflexes	Might be present at birth, fragments at 1–2 months
Traction response grasp	Might be present at birth elicited by traction at shoulder
Grasp reflex: all fingers flex	Birth to 8 weeks briefly as unit
Instinctive grasp reaction	Begins at 4–5 months
Projected grasp (pincer).	Begins at 6–8 months reflex, then thumb-finger opposition; subsequently this is voluntary
Plantar grasp response	Present at birth; disappears at 9–12 months
Extensor plantar response	Present at birth; on lateral plantar stimulation disappears at 9–12 months (due to immature corticospinal pathway).
Plantar flexion on lateral plantar stimulus (dorsiflexion)	Present from 9–12 months and normal through life
Sitting	Infants usually sit unsupported at 6–7 months
Standing with support	At 8–9 months, pulls self to standing position At 10–12 months, holding onto object
Walking:	
Begins to walk with support	At 10–12 months.
Begins to walk independently	At 9–16 months

Note: We have not considered here the prenatal motor development of the human fetus. Wolff and Ferber (1979) reviewed many of the earlier studies (which demonstrated considerable capacity for swimming movements) and discussed the use of high-resolution ultrasound to study in a noninvasive manner movement *intrautero*. Such studies allow more precise establishment of actual fetal maturation. The suck reflex is actually present in the prenatal state

input, the primary motor cortex receives information from the somatosensory cortex of the postcentral gyrus and the ventral lateral nucleus of the thalamus. This input might allow the response of motor cortex to peripheral stimuli in what are termed *long-loop reflexes*.

2. Premotor Cortex: Areas 6 and 8

In contrast, the premotor cortex is concerned with patterns of movements (i.e., motor programs). Neurons in the premotor cortex discharge during the preparation phase prior to movement. The premotor cortex projects in parallel pathways to the motor cortex, to basal ganglia, and to the spinal cord. The premotor cortex receives projections, among other sources, from the unimodal sensory association cortex (somatosensory auditory and visual) as well as the primary somatosensory cortex and the ventral lateral nucleus of the thalamus.

3. Prefrontal Areas

The prefrontal areas have functions that are related to the limbic system and the motor system. The dorsal region of the prefrontal cortex is concerned with the executive functions involved in the planning, regulation, and sequencing of behavior and movement over time. In this regard, movement is simply the expression of behavior. This same area of prefrontal cortex functions as the anterior multimodal association cortex and receives projection from the posterior multimodal association cortex located at the intersection of the major sensory association areas. In turn, this dorsal prefrontal area projects to the premotor cortex.

B. *Primary Motor Cortex Area 4 (Figs. 9.2 and 9.4)*

Area 4 is distinguished by the presence of giant pyramidal cells in layer V. Occasionally giant pyramidal cells may also be found more anteriorly in area 6. In humans, much of area 4, particularly its lower half, is found on the anterior wall of the central sulcus. More superiorly, area 4 appears on the lateral precentral gyrus and continues onto the medial aspect of the hemisphere in the anterior half of the paracentral lobule.

1. Pyramidal Tract (Fig. 9.7)

Of the approximately 1 million fibers in the pyramidal tract, 31% arise from area 4, 29% arise from the premotor cortex of area 6, and 40% arise from areas 3, 1, 2, 5, and 7 of the parietal lobe. Fibers with a large diameter (9–22 µm) number about 30,000 and correspond roughly to the number of giant pyramidal cells. Most of these large fibers originate from area 4. It is also important to note that the same cortical areas that give rise to the pyramidal tract also give rise to other descending motor pathways: corticoreticular, corticorubral, as well as corticostriate and corticothalamic. From a phylogenetic standpoint, this is a logical arrangement because this relatively new control center, the motor cortex, has available a relatively new fast conducting pathway to the anterior horn cells. At the same time, the motor cortex also has available pathways to certain older brain stem centers, which formerly exercised primary control (e.g., the reticular formation and the red nucleus). Note that 70–90% of the fibers passing through the medullary pyramid will decussate and continue as the lateral corticospinal tract. These fibers originate primarily from those portions of area 4 representing the distal extremities and synapse directly on alpha motor neurons in the lateral sector of the ventral horn. In contrast, the remaining uncrossed fibers arise from area 6 and the portions of area 4 representing the trunk and neck areas These uncrossed fibers, as they descend through the brain stem, give off collaterals to the bilateral medial brain stem areas of the reticular formation. These uncrossed fibers continue as the anterior corticospinal tract and terminate bilaterally on interneurons, which, in turn, will synapse on neurons in the ventral medial cell column of motor neurons. The fibers originating in the parietal cortex synapse in relationship to interneurons, which, in turn, synapse in relationship to neurons of the dorsal horn.

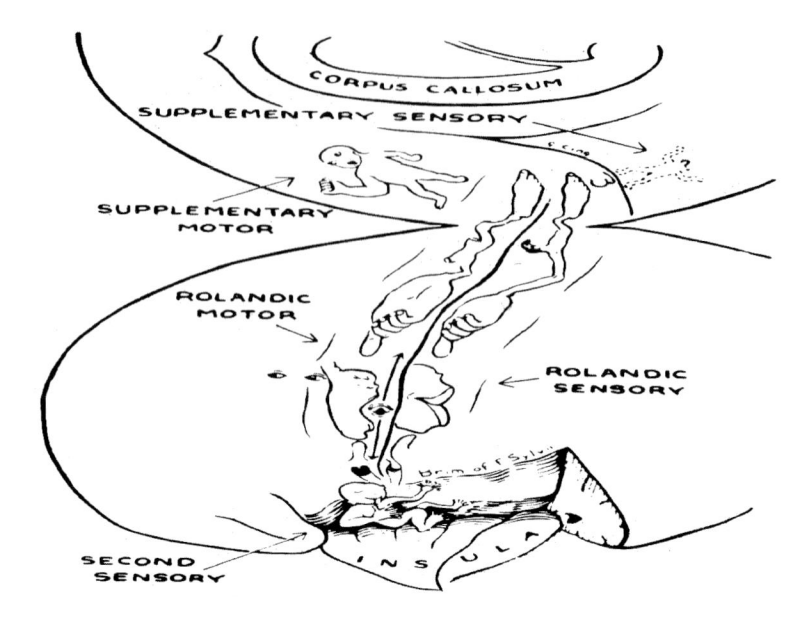

Fig. 9.2 Somatic figurines. The primary motor and sensory representation and the effects produced by supplementary motor and second sensory stimulation are demonstrated. (From W Penfield and H Jasper, Epilepsy and the functional anatomy of the human brain, Little Brown and Co., 1954)

2. Area 4: Stimulation (Fig. 9.2)

Electrical stimulation of the motor cortex at the threshold level produces isolated movements of the contralateral portions of the body. The movements can be described as a twitch or small jerk. With increased strength or increased frequency of stimulation, repetitive jerks occur; the movements involve a larger part of the contralateral side of the body and an orderly sequence of spread might occur. The repetitive jerks are similar to the movements of a focal motor seizure. The orderly sequence of spread is duplicated in the "Jacksonian March" of a focal motor seizure. This is, in a sense, inherent in the intracortical connections between the neurons representing the various muscles related to a given movement. In the conscious patient, these movements, which can be evoked by stimulation of the motor cortex during surgery, are not clearly the same as consciously willed voluntary movements. According to Penfield and Jasper (1954), the patient recognizes that these are not voluntary movements of his own and will often attribute them to the actions of the neurosurgeon. Although there is considerable overlap and plasticity in the representation of movements on the surface of the motor cortex, a certain sequence is apparent. In the classical homunculus, the representation of the pharyngeal and tongue movements is located close to the lateral sulcus. Unilateral stimulation of these areas results in bilateral movements. Next in upward sequence are movements of the lips, face, brow, and neck. In the middle third of the precentral gyrus, a large area is devoted to the representation of the contralateral thumb, fingers, and hand. In the upper third, the sequence of the contralateral wrist, elbow, shoulder, trunk, and hip is found. The remainder of the lower extremity in the sequence (knee, ankle, and toes) is represented on the medial surface of the hemisphere in the paracentral lobule. The representation of the anal and vesicle sphincters has been located lowest on the paracentral lobule. It is important to note how large the areas devoted to the thumb, fingers, and face (particularly the lips and tongue) are in the human, reflecting the evolution in use of these structures with regard to speech, writing, tool making, and manipulation.

It is also important to note that not all areas of the motor cortex have the same threshold for discharge. As studied in the baboon (Phillips and Porter, 1977), the thumb and index finger are the areas of lowest threshold. Slightly stronger shocks are required to produce a movement on stimulation of the great toe area; at slightly greater intensities of shock, movements of the face begin to occur on stimulation of the appropriate area. These findings might relate in part to the fact that those pyramidal fibers making direct connection to anterior horn cells are primarily those mediating cortical control of the distal finger and thumb muscles. It must also be noted that, at times, on stimulation of the postcentral gyrus, discrete movements will be produced similar to those elicited from the motor cortex. There is some evidence that such discrete movements are produced at a higher threshold than in the precentral cortex and that the ability to obtain such discrete movements is lost after ablation of the primary motor cortex. However, not all of these neurosurgical data on stimulation of postcentral gyrus in the human is consistent.

3. Area 4 Lesions

The effects of ablation of the motor cortex must be distinguished from the more selective effects of limited section of the pyramidal tract in the cerebral peduncle or medullary pyramid. Damage to or ablation of the motor cortex affects not only the pyramidal tracts but also the corticorubral, corticostriate, corticothalamic, and corticoreticular pathways

4. Modern Concepts of the Plasticity of the Primary Motor Cortex

More recent studies reviewed by Sanes and Donoghue (2000) indicate a considerable dynamic plasticity in the primate primary motor cortex. Thus, although modules for the head region, upper extremity, and lower extremity can be distinguished, there is, within each module, a considerable overlap and plasticity with less of the specific localization for each movement suggested by the earlier maps. Instead, a mosaic is present. The area devoted to specific movement representations might be modified by trauma to peripheral nerves, by central nervous system damage, and, most importantly, by motor skill learning and cognitive motor actions.

The following case history illustrates the effects of focal disease involving the motor cortex.

Case History 9.1 (Fig. 9.3)

Three months prior to admission, this 29-year-old, right-handed, white female manager developed an intermittent "pulling sensation" over the anterior aspect of the left thigh, lasting a few minutes and which recurred about twice a week. On the evening of admission, she again had the onset of this same sensation and then had numbness followed by "jerky movements" in her left foot, which progressed to

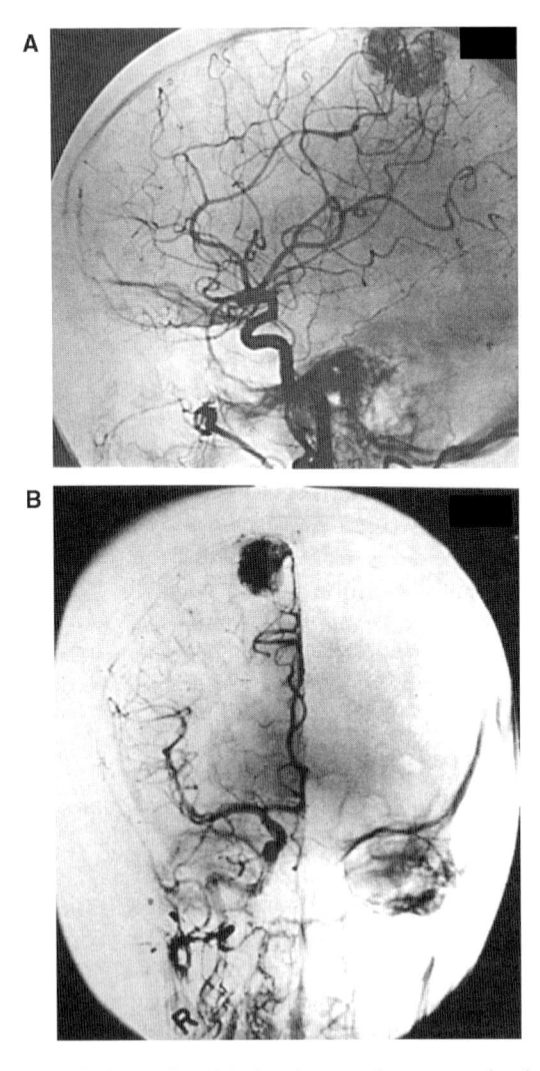

Fig. 9.3 Parasagittal meningioma. Case 9.1. Arteriograms demonstrate the characteristic features of a circumscribed vascular tumor in the parasagittal central sulcal area, supplied by the right anterior cerebral artery and meningeal branches of the external carotid artery. **A** Arterial phase lateral common carotid injection; **B** arterial phase view, common carotid injection. See text for details

involve the whole left lower extremity. She was then observed to have tonic extension and then clonus of all four limbs with loss of consciousness for a minute or less. She still appeared dazed and slightly confused for the next 5 minutes and was subsequently amnestic for the secondarily generalized seizure.

Physical examination: Normal except for bite marks on the left lateral aspect of the tongue.

Neurologic examination: Significant only for a slight asymmetry of deep tendon reflexes, left greater than right.

Clinical diagnosis: Focal seizures originating right parasagittal motor cortex (foot area), with question of parasagittal meningioma.

Laboratory data: Electroencephalogram demonstrated rare focal spikes, right central sulcus parasagittal consistent with the focal source of the secondarily generalized seizure. Arteriogram (Fig. 9.3) demonstrates an enhancing lesion in the right parasagittal region just anterior to the central sulcus. These findings were consistent with a right parasagittal meningioma.

Subsequent course: The patient did well after removal of a well-circumscribed right parasagittal meningioma by Dr. Bernard Stone. The patient received diphenylhydantoin (Phenytoin) to prevent additional seizures.

5. Corticorubral Spinal System

In the monkey and cat, if the pyramidal tracts are intact, destruction of the rubrospinal system produces no definite motor deficit. However, in monkeys and cats that have already had bilateral section of the pyramidal tract (a section that has produced very little qualitative change in limb movement), a complete loss of distal limb movement might occur after additional bilateral destruction of the rubrospinal tract. In the intact monkey, complex coordinated movements might be evoked by stimulation of the red nucleus. Flexor excitation predominates with extensor inhibition. The studies of Lawrence and Kuypers (1968) suggested that the rubrospinal tract is mainly concerned with steering movements of the extremities, particularly distal segments. On the other hand, the ventromedial brain stem reticulospinal tract is mainly concerned with steering movements and posture of head or body and synergistic movements of the extremities (Kuypers, 1987).

C. Areas 6 and 8: Premotor Cortex (Fig. 9.4)

These are widespread connections of these areas both within and between the hemispheres (see Pandya and Kuypers, 1969; Pandya and Vignolo 1971; Ward et al. 1946). Histologically, area 6 is similar to area 4 but lacks, in general, giant pyramidal cells. Area 8 has some similarity to area 6 but is also transitional to the granular prefrontal areas. There are considerable variations when the several cytoarchitectural maps are compared as to the primary motor cortex.

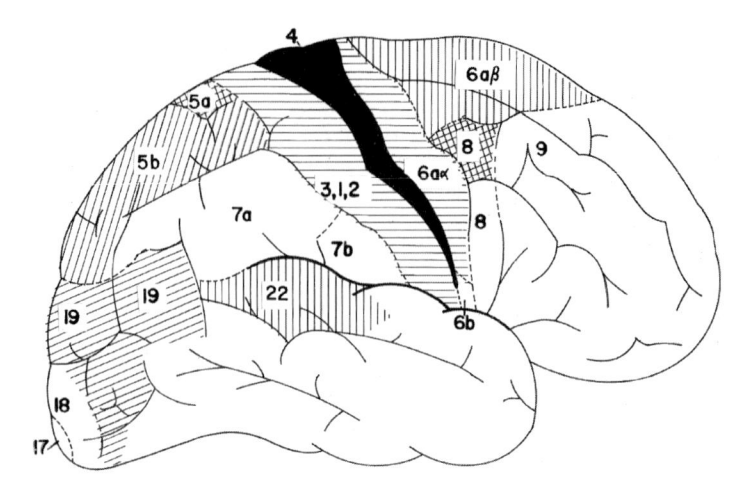

Fig. 9.4 The motor cortical fields of the human as determined by Foerster employing cortical stimulation. Stimulation of the area indicated in black produced discrete movements at low threshold and was designated pyramidal. Lines or cross-hatching indicates other areas producing movements and were designated in a broad sense as extrapyramidal. These movements were more complex synergistic and adversive movements and included the adversive responses obtained by stimulation of area 19; areas extend to the midline is continued on to the medial surface. (Modified from O Foerster, Brain 59:137, 1936) (From EM Marcus and S Jacobson, Integrated neuroscience, Kluwer, 2003)

Areas 44 and 45 in the inferior frontal convolution (Broca's motor speech or expressive aphasia centers in the dominant hemisphere) should also be included within this motor association or intermediate frontal grouping, from both the functional and cytoarchitectural standpoints. For simplicity of presentation, these areas will be considered later in relation to language function. Note that clinical lesions, such as vascular infarcts and tumors, usually do not respect cytoarchitectural borders. Thus, vascular lesions that involve the lateral premotor aspect of area 6 usually also involve the primary motor cortex. Tumors such as parasagittal meningiomas might involve both the lateral premotor and the supplementary motor cortex or might involve the prefrontal and the several premotor areas. With increasing research, there has come recognition that there are differential functions for the various sectors of these premotor areas (Freund and Hummel-Stein 1985; Krakauer et al., 2004; Wise et al., 1997).

The premotor sector of area 6 projects primarily to the medial pontomedullary areas. The spinal projections of this area consist of the ventromedial descending reticulospinal system and the vestibulospinal tracts. The red nucleus receives inputs from the rostral (proximal) arm and leg sectors of area 4 and the adjacent premotor cortex and projects as the rubrospinal system.

1. Supplementary Motor Area

The supplementary motor area (SMA) is concerned with the sequencing of movements that are initiated internally without triggering from external sensory stimuli.

The SMA is concerned with the performance of learned sequences of movement. The area just anterior to the SMA, the pre-SMA, is involved along with other prefrontal areas, such as area 46, in the actual process of learning the sequences (training of a motor skill). This same pre-SMA is also involved when the same complex movement sequence is imagined. Once a sequence has been fully and usually overlearned, control might be centered in the primary motor cortex.

The Readiness Potential (Bereitschafts Potential). In humans, surface electrodes placed on the scalp are able to record, 1–2 seconds prior to voluntary movements, a significant surface negative potential with maximum amplitude over the midline of the frontal area corresponding primarily to the supplementary cortex. The process of initiation of more complex actions enhances this potential. In contrast, when the same movements were triggered by external stimuli, there is no such readiness potential.

2. The Lateral Premotor Area

In contrast to the SMA, the lateral premotor area (PMA) is concerned with the sequencing of movements that are triggered by external stimuli. In this regard, there is a close relationship of the lateral PMA to the sensory association cortex of the superior parietal cortex (area 5) and the posterior parietal multimodal sensory areas, which are directly connected with the extra striate visual cortex. This relationship is complex, as demonstrated in reaching for and grasping precisely an object presented at a particular point in visual space. The lateral PMA has been divided into two major functional areas: the dorsal and the ventral.

Dorsal Lateral PMA

The dorsal lateral PMA neurons are involved in reaching for that object. This area is also involved in learning to associate specific stimuli that are particularly visual with the motor response of reaching for that stimulus. As opposed to the learning of a repetitive motor act (pressing a key) involving the supplementary motor cortex, this is a different type of learning, referred to as associative learning. Multiple bits of information must be integrated for this reaching activity: visual position of object, position of gaze, position of arm, and so forth; this involves information received from various posterior parietal areas adjacent to the extrastriate cortex discussed in greater detail by Wise et al. (1997).

Ventral Lateral PMA

Ventral lateral PMA is concerned with grasping. This area also must receive information from the extrastriate cortex regarding the size and shape of the object as well as other multimodal and sensory association data, so that an effective grasp of the object can occur.

3. Area 6 Stimulation: Stimulation of SMA and PMA

Essentially stimulation of PMA and SMA results in a pattern of movement characterized by tonic abduction of the contralateral arm at the shoulder with flexion of the elbow plus deviation of head and eyes to the contralateral field. In addition, with SMA stimulation, bilateral lower-extremity movements might occur. Lesions of these areas will result in a release of grasp reflex and an apraxia in limb movements. Bilateral lesions of the SMA might result in an akinetic mute state.

D. Area 8: Premotor

Lesions of area 8, due to infarction following occlusion of the blood supply, results in a transient paralysis of voluntary conjugate gaze to the contralateral visual field. In general, in humans this paralysis of voluntary gaze usually does not occur in isolation but is associated with a sufficient degree of infarction to have produced a severe hemiparesis. The patient then lies in bed with the contralateral limbs in a hemiplegic posture and with the head and eyes deviated toward the intact arm and leg. This deviation might reflect the unbalanced effect of the adversive eye center of the intact hemisphere. The effect is usually transient, clearing in a matter of days or weeks. These patients at times also appear to neglect objects introduced into the visual field contralateral to the lesion. The patients also neglect contralateral tactile and auditory stimuli. In the past such a unilateral neglect has been attributed to associated involvement of the inferior parietal area lesions of the inferior parietal area in the non-dominant hemisphere produce longer lasting neglect syndromes.

E. Suppressor Areas for Motor Activity (Negative Motor Response)

These "suppressor" areas cannot be distinguished on cytoarchitectural grounds from adjacent cortical areas. Similar effects have been elicited from stimulation of the caudate nucleus and of the bulbar medullary inhibitory center. Note must, of course, be made that a nonspecific interference with voluntary a movement may occur in relation to stimulation of motor cortex. Thus, the patient who is experiencing a focal motor seizure involving the hand cannot use that hand in the performance of voluntary skilled movements. Similarly, arrest of speech is frequently produced by stimulation of the various speech areas of the dominant hemisphere, whereas vocalization is only rarely produced. The areas of cortex include (1) the inferior frontal gyrus, immediately anterior to primary motor area for face, (2) less often, premotor cortex just anterior to the frontal eye field or the hand area, and (3) the supplementary motor cortex. The authors, Luders et al. (1988) speculate that these effects are produced because the stimulation interferes with the preparation for movement.

F. Prefrontal Cortex (Areas 9.14 and 46)

The term prefrontal refers to those portions of the frontal lobe anterior to the agranular motor and premotor areas. Three surfaces or divisions are usually delineated as (1) lateral–dorsal convexity, (2) medial, and (3) orbital or ventral. All three are concerned with executive function and relate to the mediodorsal thalamic nuclei. In general, the lateral–dorsal relates to motor association executive function; the orbital–ventral and the medial relate to the limbic–emotional control executive system. Recent studies have questioned this simplistic approach. Thus, certain functions previously considered to be associated with dorsolateral location have been now localized to a medial location (Alexander and Stuss 2000). In addition, there is evidence that dorsal lateral lesions also alter affect and motivation. Moreover, most pathological processes involve the functions associated with more than one division either directly or through pressure effects. Most studies of stimulation or of lesions involving the prefrontal areas have included, within the general meaning of the term prefrontal: areas beyond areas 9–12. The posterior orbital cortex (area 13) and the posteromedial orbital cortex (area 14) have been added to the orbital frontal group. These areas all share a common relationship to the medial dorsal nucleus of the thalamus. Areas 46 and 47, located in man on the lateral surface of the hemisphere, are also included. The anterior cingulate gyrus (area 24) is also included in the prefrontal area, although it might be considered part of the limbic system and relates to the anterior thalamic nuclei.

Our concern here is primarily with the dorsal prefrontal subdivision. The ventral/orbital and medial subdivisions also will be considered here but also in the limbic system chapter. We are concerned with the sequencing of behavior, the planning of motor function, and a very short type of memory referred to as working memory. (See Chapter 14 for a discussion on changes in personality following lesions in the prefrontal cortex).

Working memory provides for the temporary storage of information that is needed for ongoing behavior. In the human, aspects of prefrontal function are tested with a variety of tests. These include more complex forms of the delayed response and delayed alternation tests. The Luria motor sequences can be easily utilized in the office: The patient must reproduce a demonstrated sequence of hand movements (strike the thigh with the open hand, then with the ulnar aspect of the open hand, and then with the ulnar aspect of the closed fist). In the Wisconsin Card Sorting test, the patient must sort 60 cards based on color, symbol, or number of symbols. By changing the rules once the patient has developed a pattern set, it is possible to also determine how readily the set can be changed. Failure to change the set is termed *rigidity*. Activation of the ventrolateral region (inferior to the inferior frontal sulcus) has been correlated with the updating and maintenance of information. Activation of the dorsolateral region (dorsal to the inferior frontal sulcus) has been correlated with the selection/manipulation/monitoring of that information. Activation of the anterior prefrontal cortex (anterior to a line drawn vertically from the anterior edge of the inferior frontal sulcus) has been correlated with the selection of processes/subgoals.

IV. Disorders of Motor Development

Nonprogressive disorders of motor function or control recognized at birth or shortly after birth are referred to as cerebral palsy. The two types are spastic displegia (double hemiplegia with greater involvement of the legs) and a severe involvement of diorder–double the tosis. Other types of spastic cerebral palsy include: a. quadriplegia, and b. infantile hemiplegia. As reviewed by Paneth (1986), cerebral palsy affects at least 1 of 500 school-age children. More severely affected patients might die in infancy. Severely affected children may have subnormal intelligence (50%) and seizure disorders (25%). Little related the motor deficits to the circumstances of birth: premature delivery, breech presentations, or prolonged labor. Often, respiration was impaired at birth. Convulsions often occurred in the neonatal period. Little's hypothesis (1862) regarding causation has had serious medical and legal implications for the practice of obstetrics. Sigmund Freud (1897), prior to his investigations into hysteria and the psychoneuroses, undertook detailed clinical and neuropathological studies of these developmental motor syndromes. Freud unified the various disorders into a single syndrome and proposed the classification, which is presented above. He also proposed that the abnormalities of the birth process (noted by Little) were, in actuality, the consequence and not the cause of the perinatal pathology. The National Collaborative Perinatal Project followed 54,000 pregnancies and the subsequent offspring to age 7. The conclusions of that study (Nelson and Ellenberg 1986) tended to support Freud's hypothesis. Major prenatal malformations, such as microcephaly and prenatal rubella infection, were more common in the 1890s. In children with cerebral palsy, maternal mental retardation, and low birth weight below 2000 grams were also significant predictors of cerebral palsy, as was breech presentation but not breech delivery. In contrast, birth asphyxia did not add predictive power in the analysis.

V. Studies of Recovery of Motor Function in the Human

The sequence of events in the human following partial vascular lesions (infarcts) of the precentral cortex or of the motor pathways in the internal capsule has been studied by Twitchell (1951) as follows.

1. **Flaccid paralysis**. Particularly with a large lesion, there is a transient period of flaccid paralysis of the contralateral arm and leg and the lower half of the face, with a transient depression of stretch reflexes. The duration of this flaccid state (reflecting "cortical shock") is variable; it might be quite fleeting depending on the extent of the lesion (which is usually less than total). The sign of Babinski is present with an extensor plantar response on tactile or painful stimulation of the lateral border of the foot.

2. **Return of deep tendon reflexes/stretch reflexes**. After a variable period of time, the deep tendon reflexes in the affected limbs return and become hyperactive. There is increased resistance on passive motion at the joints (when there are severe lesions, even this stage might not be reached). This does not affect all movements equally but tends to have its greatest effect on the flexors of the upper extremity. A hemiplegic posture and gait is noted: the upper extremity is flexed at the elbow, wrist, and fingers and the leg is extended at the hip, knee, and ankle and externally rotated at the hip with circumduction in walking. Walking is possible with the hemiplegic limb as movement of the proximal limb begins to return. In general, recovery of function in the lower extremity is usually greater than in the upper extremity. Following the return of stretch reflexes, variation in the relative degree of spasticity in the flexor and extensor muscles in relation to tonic neck reflexes mightbe noted With the return of tonic neck reflexes one will begin to note proximal limb movements.
3. **Return of traction response**. Abduction of the arm and shoulder will result in stretch of the abductors. There will then be an adduction of the arm at the shoulder. In addition, synergistic flexion of the arm at the elbow, wrist, and fingers will occur, and objects can be grasped in this manner. This flexion synergy, however, results in a whole-hand grasp; all of the fingers flex together.
4. **Return of selective ability to grasp**. This stage of recovery involves a more selective grasp of the entire hand, albeit a tonic grasp. There is some dissociated synergy so that the hand alone tends to grasp. Finger and thumb opposition begin to return.
5. **Return of the instinctive tactile grasp reaction with the capacity for projectile movement.** This occurs with incomplete lesions. If a massive lesion is present, the instinctive grasp reflex does not return.
6. **Return of distal hand movement**. Depending on the extent of the lesion, there will be some return of distal hand movement, but such recovery is usually less than that which occurs for movements at more proximal joints (e.g., at the shoulder and hip). The logic of this pattern of recovery should be evident, as one sees in this sequence a recovery of spinal reflex activities, then of brain stem reflex activities, and, finally, those reflexes integrated at a cortical level.

VI. Cortical Control of Eye Movements

Eye movements are in the frontal (area 8) and parieto-occipital eye fields (Fig. 9.4). Vision is sharpest at the fovea, the center of the macular area of the retina. There are various mechanisms for moving the eyes so that points of interest are centered on the fovea of the two eyes. These might be listed as follows: saccadic eye movements, smooth eye movements, vergence eye movements, fixation system, vestibulo-ocular responses, and opticokinetic responses.

A. Saccadic Eye Movements

Rapid movements of gaze that shift the fovea rapidly to a point of interest are termed *saccades*. These movements might be triggered by attending to a visual point of interest, by memories, or by commands. The brain stem substrate for all horizontal saccades is the lateral gaze center of the paramedian pontine reticular formation, associated with the nucleus of cranial nerve VI. Several types of neurons are present in this center, Neurons referred to as medium-lead burst cells provide direct excitation to motor neurons in the ipsilateral abducens nucleus and to the interneurons in that nucleus, which give rise to the medial longitudinal fasciculus. Long-lead burst cells drive the medium-lead burst cells and receive excitatory input from the higher centers (discussed below); the medium-lead burst neurons also excite inhibitory burst neurons, which, in turn, suppress the discharge of contralateral abducens neurons. In addition, there are neurons located in the dorsal raphe nucleus of the pontomedullary area, which cease firing during saccades (omni pause neurons). These neurons, if stimulated during a saccade, will stop the saccadic movement. Additional neurons provide tonic discharges relevant to eye position and maintain the eyes in a position set by the saccade: the flocculus of the old vestibular portion of the cerebellum, the vestibular nucleus, and the nucleus prepositus hypoglossi. For vertical saccades, burst and tonic neurons are found in the rostral interstitial nucleus of the MLF in the mesencephalic reticular formation.

B. Central Control of Saccades

A complex system with redundant features is present. Essentially, two major cortical areas are involved:

1. The contralateral frontal eye field [area 8, frontal eye fields (FEF)]
2. The contralateral lateral intraparietal area of the posterior parietal area at the border of the extrastriate cortex

Area 8 provides the motor input for the response to the command (e.g., "look to the left or right field"). The frontal eye field has major interconnections in a topographic manner with areas that participate in the streams of information that flow out of the extrastriate visual cortex. The posterior parietal area, in contrast, is involved with visual attention, with representations of the location of objects of interest. There is a significant interconnection with area 8 with an exchange of information.

Both areas project to the intermediate and deep cell layers of the *superior colliculus*. The superior colliculus, in turn, projects to the mesencephalic and pontine reticular formation. The same areas of the superior colliculus also receive inputs from the prestrate areas of the posterior temporal gyri. The superficial layers of the superior colliculi receive input from both the striate cortex and the retina. The frontal eye field (area 8) projects not only to the superior colliculus but also directly to the pontine and mesencephalic reticular formation. The interaction of area 8 and the superior colliculus is even more complex. Area 8 projects to the caudate nucleus. The caudate nucleus inhibits the

S. nigra pars reticularis, which, in turn, acts to inhibit the superior colliculus. Other cortical areas are also involved in saccades: the SMA eye field and the dorsolateral frontal cortex. The former area has neurons that direct movement to part of a target. The latter contains neurons that discharge when the saccade is made to a remembered target.

Lesions of either area 8 or of the superior colliculus would produce a defect in saccades to the contralateral field and a transient visual neglect. The defect, however, would be temporary, because the alternative system would be intact. Lesions of the parietal cortex tend to produce longer lasting visual neglect syndromes.

C. Smooth Pursuit in Contrast to Saccade

This keeps the image of a moving target on the fovea. Eye movement velocity must match the velocity of the target. The medial vestibular nucleus, the nucleus prepositus hypoglossi, and the paramedian pontine reticular formation receive information from the vermis and flocculus of the cerebellum and project to the abducens nucleus and the oculomotor nucleus. The cortical inputs for smooth pursuit arise from area 8 (frontal eye field). In the monkey, stimulation of area 8 produces ipsilateral pursuit. Area 8 receives input from the posterior sectors of the superior and middle temporal gyri. This area, in turn, receives input from the visual cortex. In humans, the posterior parietal area may serve this same function.

D. Fixation System

The fixation system holds the eyes still during intentional gaze on an object. During fixation, saccades must be suppressed. The most rostral segment of the superior colliculus has a representation of the fovea, and neurons here are active during fixation. These neurons act to inhibit neurons in the more caudal superior colliculus. In addition, these neurons excite the omni pause neurons in the lateral gaze center, which, as discussed earlier, inhibit saccades.

E. Vergence Movements

Vergence movements (convergence and divergence) and accommodation are controlled by neurons within the third nerve complex of the midbrain.

F. Vestibulo-ocular Movements

These movements hold images stable on the retina during brief head movements. Input occurs from the vestibular system.

G. Opticokinetic Movements

These movements hold images stable during sustained head rotation. In nonprimates such as the rabbit, retinal neurons projects to the nucleus of the optic tract in the pretectum which then projects to the medial vestibular nucleus. This non cortical system responds to stimuli moving slowly in a temporal to nasal direction. In primates there is in addition a "cortical system" magnocellular layers of the LGN are also involved with striate cortex, and the posterior portions of the middle and superior temporal gyri. The noncortical system responds to stimuli moving slowly in a temporal to nasal direction while this "cortical system" can respond to stimuli moving with higher velocities in a nasal to temporal direction. This evolution in the system can reflect the changes that have occurred in the placement of the orbits in the skull when the primates are compared to the rabbit. Opticokinetic nystagmus occurs when a pattern of vertical black and white lines is moved slowly in front of the eyes. A similar nystagmus occurs as one sits in a moving train looking out the window at the series of telephone poles flashing by. A lesion in the cortical system produces defective opticokinetic nystagmus to visual stimuli moving toward the side of the lesion.

VII. Major Voluntary Motor Pathways

In the older literature, the motor pathways were divided into pyramidal (corticospinal and corticonuclear) and extrapyramidal (rubrospinal, tectospinal, etc.). These terms are no longer used, and each pathway is now usually described in terms of its function.

A. Basic Principles of Voluntary Motor System

The motor system consists of two portions: an upper and lower motor neurons.

Upper motor neurons are found in layer V of the motor sensory strip of the cerebral cortex. The pathways from the upper motor neurons, corticospinal or corticonuclear, descend and cross over to innervate the lower motor neurons in the brain stem or spinal cord. The axons of the upper motor neurons either synapse on interneurons or end directly on the lower motor neurons.

Lower motor neurons are the ventral horn cells in the spinal cord and motor cranial neurons in the brain stem. The axons from the *lower motor neurons* leave the central nervous system and form the motor division of the peripheral nervous system.

Voluntary movements of the muscles associated with the motor cranial nerves V, VII, IX, X, XI, and XII are controlled via the corticonuclear–bulbar pathway. Control of the cranial nerves that move the muscles (III, IV, and VI) of the eyes is through the corticomesencephalic pathway to the midbrain and pons, which coordinates movements under the control of cranial nerves III and VI interconnected by the ascending portion of the medial longitudinal fasciculus.

B. Corticospinal Tract: Voluntary Control of the Limbs, Thorax, and Abdomen (Fig. 9.5)

This tract innervates the motor neurons that control the skeletal muscles in the neck, thorax, abdomen, pelvis, and the extremities. It is essential for accurate voluntary movements and is the only direct tract from the cortex to the spinal motor neurons. This system originates from the pyramidal cells and giant cells of Betz in the upper two-thirds of the precentral gyrus, area 4, and to a lesser degree from area 6 and parts of the frontal, parietal, temporal, and cingulate cortices. The fibers pass through the genu of the internal capsule into the middle third of the cerebral peduncles and enter the pons, where they are broken into many fascicles and are covered by pontine gray and white matter. In the medulla, the fibers are again united and are found on the medial surface of the brain stem in the medullary pyramids.

About 75–90% of the corticospinal tract fibers cross at the medullospinal junction and are thereafter found in the lateral funiculus of the spinal cord. (The majority of fibers from the dominant hemisphere cross at the medullospinal junction.) Some corticospinal fibers remain uncrossed in the anterior funiculus but will slowly cross at cervical levels. About 50% of the corticospinal fibers end at the cervical levels—about 30% goes to the lumbosacral levels and the remainder to the thoracic levels. Most of the corticospinal fibers end on internuncial neurons in laminae 7 and 8 of the spinal cord, but in regions in the spinal cord where the digits are represented (in the cervical and lumbosacral enlargements), the corticospinal fibers sometimes end directly on the motor horn cells. The internuncial neurons are located in lamina 7of the spinal cord. Lesions in the corticospinal tract produce contralateral upper motor neuron symptoms, whereas lesions in the spinal motor neurons or rootlets cause lower motor neuron symptoms.

C. Corticonuclear/Corticobulbar System: Voluntary Control of the Muscles Controlled by Cranial nerves V, VII, and IX–XII (Fig. 9.6)

This system distributes to the motor nuclei of the cranial nerves V, VII, IX, X, XI, and XII and provides voluntary and involuntary control of the muscles and glands innervated by these nerves. In the older literature, this pathway was called corticobulbar because the medulla and pons containing these nuclei are collectively called the bulb. We prefer the term corticonuclear because it makes the student realize the

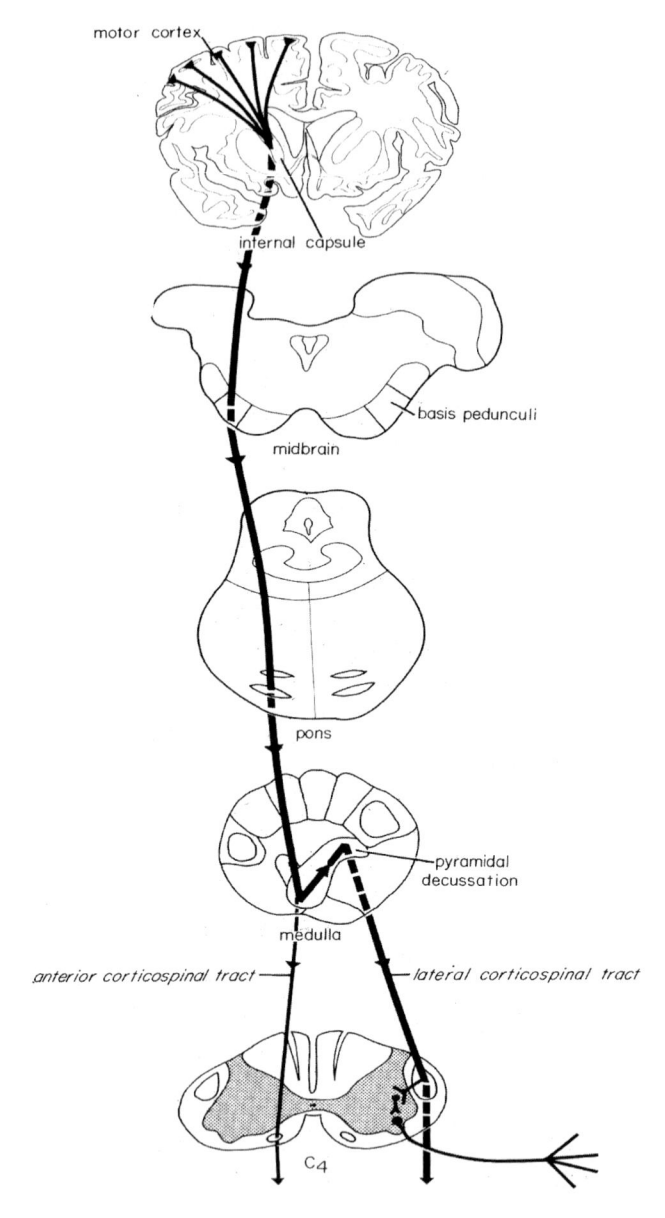

Fig. 9.5 Corticospinal pathway. voluntary movement of the muscles in the extremities, thorax and abdomen (From EM Marcus and S Jacobson, Integrated neuroscience, Kluwer, 2003)

key role of the cranial nerve nuclei in movement. These fibers are found anterior to the corticospinal fibers in the genu of the internal capsule and medial to the corticospinal tract in the cerebral peduncle, pons, and medullary pyramids. The fibers supplying the cranial nerves have the following cortical origin:

VII Major Voluntary Motor Pathways

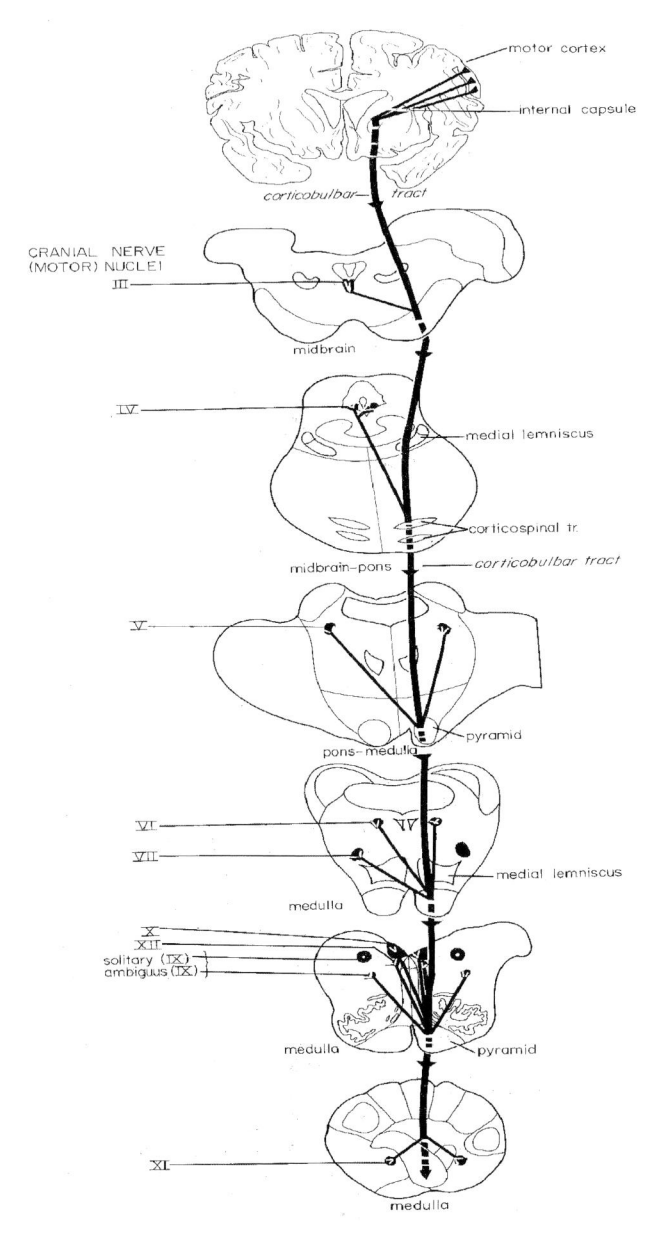

Fig. 9.6 Corticonuclear system: voluntary control of the muscles controlled by cranial nerves V, VII, and IX– XII (From EM Marcus and S Jacobson, Integrated neuroscience, Kluwer, 2003)

1. Muscles of facial expression (motor nerve VII), mastication (motor nerve V), and deglutition (ambiguous nuclei of nerves IX and X) originate from pyramidal cells in the inferior part of the precentral gyrus, area 4.

2. Muscles in the larynx are controlled from the inferior frontal gyrus and from the frontal operculum (the posterior part of the pars triangularis, area 44). It appears that many corticonuclear axons end on interneurons and not directly on the motor neuron of the cranial nerve.
3. The cortical innervation of the cranial nerves is bilateral, with the exception of the lower facial muscles, which are innervated by the contralateral cortex. Consequently, unilateral lesions in the corticonuclear system produce only weakness and not paralysis. Paralysis results only from bilateral lesions in the corticonuclear system.
4. Contrast lower motor versus upper motor neuron lesions. A lesion of the motor nucleus or rootlet produces a lower motor neuron syndrome, with paralysis, muscle atrophy, and decreased reflexes. Supranuclear lesions produce upper motor neuron symptoms. Associated with bilateral lesions of the corticonuclear system is pseudobulbar palsy with bilateral upper motor neuron signs and inappropriate behavior such as excessive cranial nerve expression of emotion such as excessive laughter or crying in response to a minor stimulus. This probably represents the release of cortical control via the corticonuclear innervation of the muscles controlled by the cranial nerves. Limbic control to these nuclei takes over, via descending autonomic pathways, and the emotional state are now undamped by the frontal lobes.

D. Corticomesencephalic System: Voluntary Control of Muscles Associated with Eye Movements (Cranial Nerves III, IV, and VI)

The fibers controlling eye movements originate from the frontal eye fields in the caudal part of the middle frontal gyrus area 44 and the adjacent inferior frontal gyrus (area 8). At upper midbrain levels, fibers to cranial nerves III and IV leave the cerebral peduncles and take a descending path through the tegmentum, usually in the medial lemniscus (corticomesencephalic tract). Fibers to cranial nerve VI also descend in this way to the parabducens and parapontine reticular nuclei (Fig. 9.6).

Chapter 10
Motor System II: Basal Ganglia

The term "basal ganglia" originally included the deep telencephalic nuclei: the caudate, putamen, globus pallidus, the claustrum, and nucleus accumbens) The globus pallidus and putamen are lens-shaped and are called lenticular nuclei. Collectively, the putamen and caudate are called the corpus striatum. Additional structures now included within this group are the substantia nigra, the subthalamic nuclei, the ventral tegmental area. and the ventral pallidum. One is very familiar with disease of the basal ganglia due to the widespread presence of Parkinson's disease with its tremor at rest.

I. Anatomy

The caudate and putamen have the same structure (Table 10.1) and are continuous anteriorly. The globus pallidus has two sectors: a medial or inner and a lateral or outer. The substantia nigra has two components: a ventral pars reticularis, which is identical in structure and function to the medial sector of the globus pallidus, and a dorsal darkly staining component—the pars compacta—which contains large dopamine- and melanin-containing neurons. The nuclei of the basal ganglia can be categorized as (1) input nuclei (caudate, putamen, and accumbens), (2) intrinsic nuclei (lateral segment of the globus pallidus, subthalamic nucleus, pars compacta of the substantia nigra, and the ventral tegmental area), and (3) output nuclei (medial segment of the globus pallidus, pars reticularis of the substantia nigra, and the ventral pallium). The consideration of the chemical and pharmacological anatomy of transmitters and circuits within the system will provide an understanding of function and dysfunction within this system. It should be noted that the nuclei of the basal ganglia, the circuits involving the basal ganglia, the cortical areas projecting to the basal ganglia, the cerebellar nuclei relating to the basal ganglia, and the reticular formation (which has connections with both the cortex and the basal ganglia) were once grouped together as an "extrapyramidal system." The term "extrapyramidal disorders" was used to refer to the effects of lesions within the basal ganglia system. However, the term "extrapyramidal" was also utilized to refer to the descending pathways: corticorubral spinal and corticoreticulospinal that were

Table 10.1 Anatomical divisions of corpus striatum

Basal ganglia: nuclei	Function
Neostriatum	Input nuclei: neocortical and substantia nigra., s. nigra compacta
Dorsal caudate/putamen	
Ventral nucleus accumbens	Input nucleus limbic system
Paleostriatum (globus pallidus)	
Dorsal portion:	Intrinsic nucleus
Lateral segment	Output nucleus
Medial segment	Output nucleus
Ventral portion	Limbich
Ventral pallidum	Limbich
Basal ganglia associated nuclei	
1. Subthalamic nucleus	Intrinsic nucleus
2. Substantia nigra of midbrain	
–Dorsal:pars compacta– (dopaminergic)	Intrinsic nucleus
Ventral: pars reticularis	Output nucleus

alternative descending systems to the pyramidal system. The term "extrapyramidal" then becomes very nonspecific and confusing and is best not utilized. It is important to note, moreover, that the cortical areas giving rise to this extrapyramidal system are also, in part, the same areas giving rise to the pyramidal system.

A. *Connections*

The connections and transmitters within this system are presented in Figure 10.1 and are summarized here. The essential pathways are the following: cerebral cortex (+ excitatory) to striatum (caudate/putamen) (– inhibitory) to globus pallidus (– inhibitory) to ventrolateral thalamus (+ excitatory to cerebral cortex). Additional circuits involve subthalamus and substantia nigra. Adding greater detail to the outline indicates the following sequence:

1. The major input into the basal ganglia is from the cerebral cortex to the striatum (caudate nucleus/putamen/accumbens). This input is excitatory, utilizing the transmitter glutamate, and arises primarily from the supplementary motor premotor and motor cortices (areas 8, 6, and 4). However, the projection from the neocortex occurs from many other areas. In general, frontal and parietal, lateral temporal, and occipital areas project to the caudate nucleus, supplementary motor, premotor and primary motor cortices, and primary somatosensory cortex project to the putamen. The hippocampal areas, medial and lateral temporal, project to the nucleus accumbens.
2. There are local circuits within the striatum, which involve the excitatory transmitter acetylcholine.
3. The striatum also receives a major dopaminergic input from the substantia nigra compacta. Based on the type of receptor, this input can be either excitatory (D1)

I Anatomy

```
              CEREBRAL CORTEX: PRIMARILY SMA, 6, 4, 8
                            (+)│ GLU
        S.NIGRA   DOPA     STRIATUM              DOPA  S.NIGRA
        COMPACTA  (+)→ D1 RECEPTORS  D2 RECEPTORS ←(-)  COMPACTA

                       (-)/  GABA        (-)\  GABA
                             substance P       enkephalin
                       DIRECT PATHWAY    INDIRECT PATHWAY

                                      LATERAL GLOBUS PALLIDUS
                                         (-)│ GABA
         MEDIAL GLOBUS PALLIDUS   GLU SUBTHALAMIC NUC.
         (And S. Nigra Reticularis)         (+)

              (-)│ GABA
         VENTRAL LATERAL NUCLEUS THALAMUS
         (And ventral anterior Nucleus of Thalamus)
                            (+)│ GLU
              CEREBRAL CORTEX: PRIMARILY SMA, 6, 4, 8
```

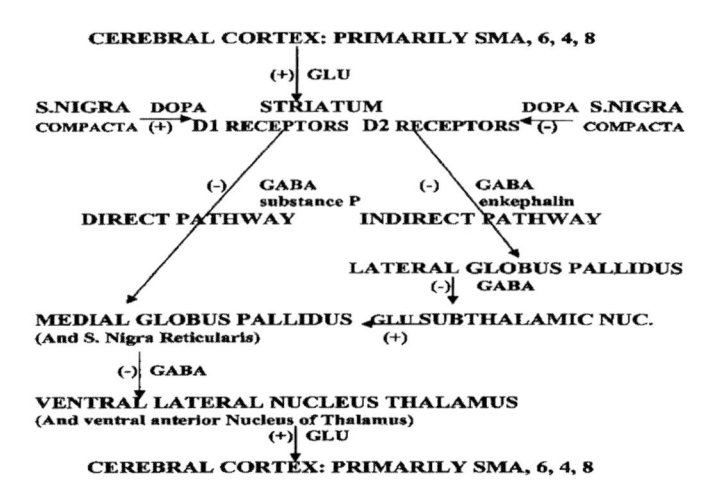

Fig. 10.1 Diagram of the connections of the basal ganglia with transmitters and major transmitter action (+) excitatory or (−) inhibitory. GABA = gamma-aminobutyric acid (−); GLU = glutamate (+); dopamine = Dopa (+). The action of acetylcholine within the striatum has been omitted. See text for details. (From EM Marcus and S Jacobson, Integrated neuroscience, Kluwer, 2003)

or inhibitory (D2). Dopamine receptors have been classified based on the dopamine-mediated effects on the enzyme adenylate cyclase (which is involved in the ATP to cyclic AMP transformation). D2 receptors are involved in the inhibition of the enzyme; D1 receptors in stimulation of the enzyme. Note that five dopamine receptors have now been identified

Because of this differential input, two pathways have been identified: the direct described below in item 4 and the indirect described in item 5.

4. The major direct outflow pathway arises from those striatal neurons receiving excitatory dopaminergic input (D1) and passes to the medial globus pallidus, which is mediated by the inhibitory transmitter gamma-aminobutyric acid (GABA). Substance P is also involved as a transmitter or modulator in this pathway. The major outflow is to the inner (medial) segment of the globus pallidus. There is a lesser outflow of fibers from the striatum to the substantia nigra reticularis, which has the same microscopic structure as the globus pallidus. The outflow of the neurons of these structures is to the ventral lateral and ventral anterior thalamic nuclei, mediated again by the inhibitory transmitter GABA. This outflow (Fig. 10.2) passes via two fiber systems: the fasciculus lenticularis and the ansa lenticularis. The fasciculus lenticularis penetrates through the posterior limb of the internal capsule; the ansa lenticularis loops around the undersurface of the most inferior part of the internal capsule and then passes upward to join the fasciculus lenticularis. These two fiber systems join at a point known as the Field H2 of Forel and then curve back toward the thalamus in the Field H1 of Forel, where they are known as the

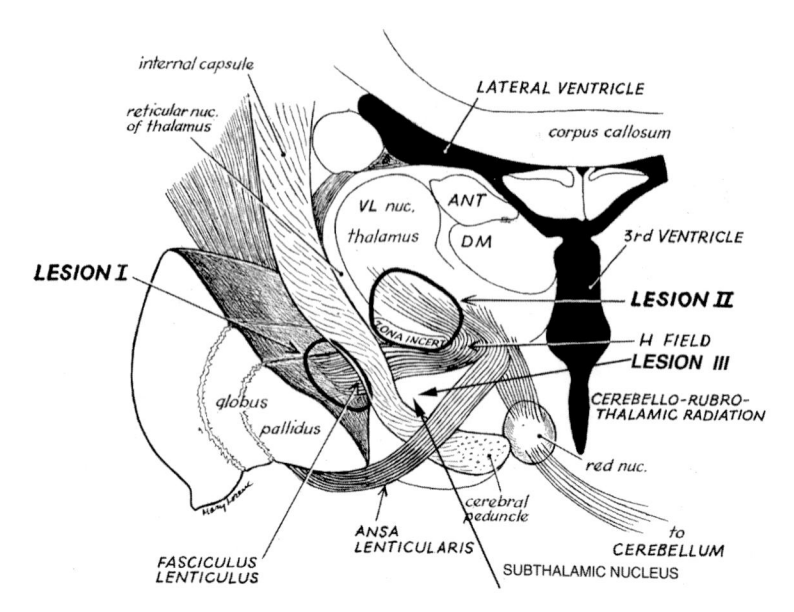

Fig. 10.2 Major connections of the basal ganglia with sites for surgical lesions or implantation of stimulators in Parkinson's disease. Lesion I: globus pallidus; lesion II: ventral lateral (VL) nucleus of the thalamus (the sector of VL involvement is termed VIM = nuc. ventral inferior medial); lesion III: subthalamic nucleus (Modified from FH Lin, S Okumura, and IS Cooper, Electroenceph. Clin. Neurophysiol. 13:633, 1961) (From EM Marcus and S Jacobson, Integrated neuroscience, Kluwer, 2003)

thalamic fasciculus. This fasciculus then enters the ventrolateral and ventral anterior nuclei of the thalamus. The outflow of these thalamic nuclei is excitatory to the motor areas of the cerebral cortex, predominantly to the premotor and supplementary motor areas.

5. The major indirect outflow pathway arises from those striatal neurons receiving inhibitory dopaminergic input (D2) and passes to the lateral (or external) sector of the globus pallidus mediated by the inhibitory transmitter GABA. Enkephalin is also involved as a transmitter or modulator in this pathway. The neurons of the lateral globus pallidus give rise to inhibitory fibers, again utilizing GABA, which connect to the subthalamic nucleus. This nucleus gives rise to excitatory fibers utilizing glutamate, which provide additional input to medial (inner) sector of the globus pallidus and the substantia.

As in the direct pathway, the subsequent connection of these structures is inhibitory to the ventral lateral and ventral anterior nuclei of the thalamus. The subsequent step in the sequence, as in the direct pathway, is excitatory to the motor areas of the cerebral cortex utilizing the transmitter glutamate. In the case of both the direct and the indirect pathways, a loop has been completed: cerebral cortex to striatum to globus pallidus to ventrolateral and ventral anterior thalamus back to cortical motor areas.

6. In addition to its major input from the cerebral cortex, the caudate–putamen also receives a lesser afferent input from the nucleus centrum medianum (C.M.) (excitatory transmitter presumed to be glutamate or aspartate) It should be noted that the C.M. represents a major interlaminar nucleus of the thalamus and is the main thalamic extension of the reticular formation. The C.M., however, also receives fibers from the globus pallidus (inhibitory transmitter: GABA). This again completes an additional loop, relating the basal ganglia to the reticular formation. It should also be noted that some fibers leave the ansa lenticularis, and rather than passing into the thalamus, they descend to the pedunculopontine nucleus within the tegmentum of the mesencephalon. This nucleus then projects back to the globus pallidus. There is also evidence for a direct excitatory input from motor and premotor areas to the subthalamic nucleus, providing an additional circuit for modulating motor activity.
7. The final effect of this system is on the thalamus. If one starts the analysis at the striatum, the end result of the direct system on the thalamus is excitatory. Inhibition of inhibitory drive is referred to as *disinhibition*. In contrast, following through the sequences of the indirect system, one finds that the end result is inhibitory. If one begins at the substantia nigra compacta and follows through the direct and indirect circuits, it is evident that both have the same end result with regard to the final effect at the thalamic and the cortical level.
8. Effects of decreased dopaminergic input on the thalamus and cortex = less excitation. The eventual thalamic and cortical effects of a decrease in dopaminergic input at the level of the striatum results in the thalamus and cortex receiving less excitation. The results are the same whether the direct or indirect pathway is considered. Movement then will be decreased or slowed down. The terms *akinesia* or *bradykinesia* are employed. On the other hand, if there is increased dopaminergic activity, the end results at the thalamic or cortical level will be an increase in motor activity irrespective of whether the direct or indirect pathway are considered. The terms *hyperkinesia* or *dyskinesia* are employed. The ultimate effect of dopamine then is to facilitate movement. (Cote and Crutcher, 1991; Bergman et al. 1990). The striatum also has GABA-mediated inhibitory outflow to substantia nigra compacta. Specific sites in the striatum project to the substantia nigra compacta (and receive afferent input from the Substantia nigra compacta).

B. *Microanatomy of the Striatum*

From histochemical and histological standpoints, distinct areas can be noted, referred to as patches within the larger background matrix. The **patch areas** have a high density of neurons; matrix areas are less dense. Patch zones contain little acetylcholine esterase; matrix areas are rich in acetylcholine esterase. Patch areas are associated with bundles of dopamine fibers early in development. Later in development there are less dense dopaminergic inputs to the matrix (Graybiel and Ragsdale, 1978). Neurons within the striatum can be additionally categorized on the basis of size and morphology of the dendrite and size as (1) medium-sized, spiny neurons and (2) large or small aspiny neurons. There is considerable evidence that the dopamine-containing terminals and their related synapses and the cholinergic synapses in the striatum have essentially antagonistic actions. Normally, however, they remain in a particular balance. We will discuss the problem of an imbalance when we come to discuss Parkinson's disease. As discussed above, both transmitters tend to be found in specific sites in the striatum (striasomes). These sites exist in a larger matrix of the striatum. Several additional loops and circuits must now be considered.

C. Overview of the Dopaminergic Systems

In addition to the dopaminergic nigral–striatal pathways, which are of major concern in our discussion of motor function, there are other major dopaminergic pathways. These other pathways, however, are of importance when one begins to utilize L-dopa (the precursor of dopamine) in the therapy of Parkinson's disease. In addition, these pathways are of importance when agents (such as neuroleptics employed in the treatment of psychoses or other psychiatric syndromes) are used to block the dopamine receptor or to prevent the storage of dopamine. In addition, the overall disease state in Parkinson's disease might also involve these pathways.

1. The mesolimbic system originates in the ventral tegmental area, medial and superior to the substantia nigra, and projects to the nucleus accumbens (ventral striatum), the stria terminalis septal nuclei, the amygdala, hippocampus, mesal frontal, anterior cingulate, and entorhinal cortex. A major role in emotion, memory, and perception is postulated for this system. Hallucinations are a side effect of a high dose of L-Dopa in the Parkinsonian patient.
2. The mesocortical system originates in the ventral tegmental area and projects to the neocortex, particularly the dorsal prefrontal cortex with a role in motivation, attention, and organization of behavior.

D. Overlap with the Cerebellar System

It is also important to remember that the major outflow from the cerebellum (via the superior cerebellar peduncle) terminates in relation to the same ventrolateral and ventral anterior nuclei of the thalamus. This thalamic nucleus receives a projection from both the cerebellum (predominantly the dentate nucleus) and the globus pallidus. However, the information from the cerebellum is projected in a more limited manner primarily to the VL nucleus of the thalamus and then to the motor areas of cerebral cortex, predominantly the premotor and motor cortices. The basal ganglia differ from the cerebellum in several important respects. In contrast to the more limited relationship of the cerebellum to VL nucleus of the thalamus (projection to motor areas of cortex and projections from motor areas of cortex via pontine nuclei to the cerebellum), the basal ganglia have widespread inputs from the cerebral cortex and via several thalamic nuclei back to multiple cortical areas. These cortical basal ganglia relationships are not diffuse or overlapping but, instead, are organized in parallel systems.

II. Clinical Symptoms and Signs of Dysfunction

A. General Overview

Based on the anatomical circuits considered above a variety of symptoms and signs may occur. See Table 10.2.

It should also be very clear then that patients with disease of the basal ganglia might have significant manifestations not only of motor dysfunction but also of cognitive and emotional dysfunction (see Kandel and Cooper 1991) and Starkstein et al. (1989) in relationship to Parkinson's disease and Laplane et al (1989) with regard to obsessive–compulsive behavior change). There is considerable overlap with frontal lobe, premotor, and supplementary motor dysfunction, as one might predict based on the anatomical connections discussed earlier. An understanding of the basic anatomical interrelationships will aid in an understanding of the effects of the various medical and surgical modalities of treatment in Parkinson's disease. Thus, when dysfunction develops in the circuit, lesions at specific critical points,

Table 10.2 Clinical symptoms and signs of basal ganglia disease

1. A lack of movement
   An inability to initiate movement or a slowness of movement because of excessive inhibition (a lack of excitation) so that the excitatory effect of ventral lateral nucleus of the thalamus on supplementary motor neurons fails to occur due to a decrease in dopamine. This lack of movement and slowness of movement is defined as akinesia and bradykinesia.
2. Elimination of inhibition
   At a critical point in the system, a release of disordered movement might result. The disordered movements are defined as dyskinesia and tremor.
3. Rigidity
   The relationship of muscle tone in antagonists and agonists might be altered so that excessive tone is present, defined as rigidity. Tremor superimposed on this rigidity produces cog wheel rigidity.
4. Dystonia
   A specific extreme of posture might be maintained, defined as dystonia.
5. Problems in the sequencing of movements as in walking or of posture as in standing might occur.
6. Note however that akinesia or bradykinesia is frequently seen in patients with disease of frontal premotor and supplementary motor areas.

such as the ventrolateral thalamic entry area or the subthalamic nucleus, might allow restoration of considerable function.

It should be evident that such a lesion in the ventral lateral nucleus of the thalamus would also be extremely effective in eliminating tremor originating in the cerebellum because such a lesion would prevent a disordered cerebellum from acting on the circuits passing through this nucleus and this disordered cerebellum would no longer influence motor activities originating at a cortical level. We must consider the circuits through the basal ganglia, modulators of cortical function, particularly with regard to movement. For voluntary movement, these circuits clearly provide complex modulation acting on the supplementary motor cortex. This modulation represents a focused equilibrium of opposing influences. When one of these influences acts to a disproportionate degree, dysfunction in the main circuit from the cortical motor cortex to the anterior horn cells and to the reticular formation will result. In general, a relative decrease in the action of dopamine produces a decrease in movement. Too much of dopamine action produces an excess of movement (a dyskinesia).

B. Parkinson's Disease and the Parkinsonian Syndrome

The most common disease involving the basal ganglia is Parkinson's disease, first described by James Parkinson in 1817. Parkinson's disease, after Alzheimer's disease, is the most common degenerative disorder of the central nervous system affecting approximately 1 million individuals in the United States. The basic pathology is the progressive loss of dopamine-producing neurons in the pars compacta of the substantia nigra. A certain mixture of symptoms then develops, each of which might vary in severity (Table 10.3).

(Recent reviews are provided by Olanow and Tatton, 1999, and Riley and Lang, 2000)

1. Cardinal Signs of Parkinson's Disease

The essential signs and symptoms consist of: tremor, rigidity, akinesia, and defects in postural and righting reflexes. Charcot essentially described this total picture and he named the disease after Parkinson (Goetz, 1986). The cardinal signs and symptoms are summarized in Table 10.3 Some cases, throughout their course, might continue to manifest primarily tremor as the major symptom and such cases tend to have a more favorable prognosis. Several types of Parkinson's disease and syndrome might be specified. The most common variety is the idiopathic or primary Parkinson's disease, a degenerative disease arising insidiously in the patient of age 40, 50, or 60. In a series of 1644 patients with Parkinsonism evaluated by Jankovic (1989) at a movement disorder clinic, 82% were found to have idiopathic Parkinson's disease. At times, there is a family history of the same disease.

II Clinical Symptoms and Signs of Dysfunction

Table 10.3 Clincal signs in Parkinson's disease

Clinical signs	Clinical appearance
Rhythmic tremor at rest: alternating type initially fine 4–5 Hz and then the disease progresses often coarse. The tremor disappears with movement but might remerge when a posture is maintained.	The tremor affects not only the extremities but also the eyelids, tongue, and voice. The alternate movements of thumb against opposing index finger are referred to as "pill rolling"
Clinical rigidity which must be differentiated from decerebrate rigidity, which is primarily spasticity with the jack-knife quality of splasticity (sudden resistance is encountered and with additional force, it is overcome)	Comparable to the resistance of a lead pipe bending under force in which the resistance is relatively constant throughout the range of motion. As with spasticity, it might reflect increased activity of the gamma system; tremor results in cogwheel
Akinesia: lack of spontaneous movement and difficulty in initiating movement. A corollary term is bradykinesia (a slowness of spotanteous movements)	Seen in the fixed faces of the Parkinsonian patient, lack of spontaneous blinking and movement, loss of associated movements (the swing of the arms in walking). Correlated with the reduction in dopaminergic input
Defects in posture; righting-reflexes are best noted as the patient stands, walks, and attempts to turn	Normally, as an individual turns, the eyes move initially in the direction of the turn, his head moves, then shoulders and arms, and the body then follows. The Parkinson's patient however, turns "en bloc." In walking, at times develops forward propulsion and seems to tilt forward and develop increasing speed, producing loss of balance with falls and injuries, and instability when turning

2. Pathology of Parkinsonian Lesions

The basic pathology involves a progressive loss of the pigmented neuromelanin-containing neurons of the pars compacta of the substantia nigra (Fig. 10.3). (The pigmented neurons of the locus ceruleus are also affected. The student will recall that both melanin and dopamine are steps in the metabolic pathway involving phenylalanine.) A long preclinical period estimated at 5 years occurs in this progressive disease. Various estimates suggest that when symptoms first occur, 70–80% of striatal dopamine is depleted. In terms of neurons in the pars compacta of the substantia nigra, 50% (compared to age-matched controls) to 70% (compared to young individuals) are already lost. Normally, the rate of cell loss in the nigra during normal aging is 5% per decade. In patients with Parkinson's disease, the rate of cell death is subsequently 45% per decade. On histologic examination (Fig. 10.4), there is a characteristic, specific finding of Lewy bodies: round, eosinophilic bodies surrounded by a clear halo within the cytoplasm of surviving neurons of the substantia nigra. In many cases, these Lewy bodies are found in other neurons (e.g., cerebral cortex). Immunocytochemistry has demonstrated that a small protein a-synuclein is a major component of the Lewy body. Under normal circumstances,

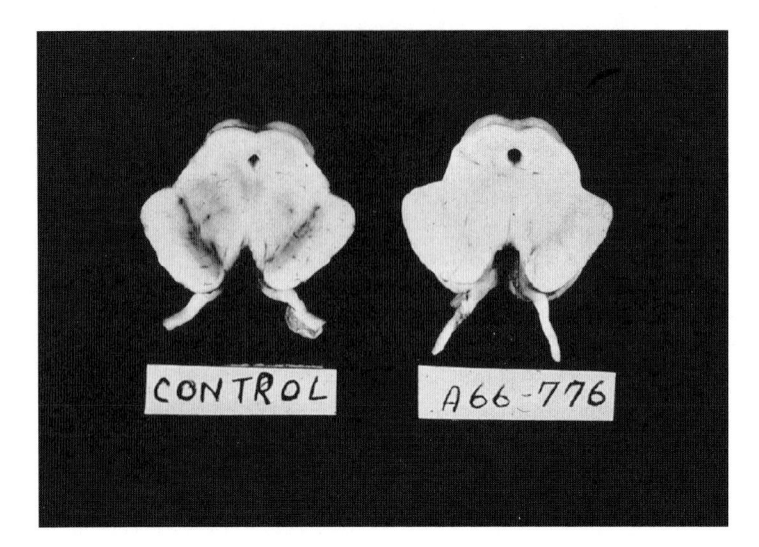

Fig. 10.3 Parkinson's disease. The substantia nigra in Parkinson's disease. Left: normal substantia nigra; right: similar region in the case of idiopathic Parkinson's disease. A marked loss of pigmentation is evident. (Courtesy of Dr. Thomas Smith, University of Massachusetts) (From EM Marcus and S Jacobson, Integrated neuroscience, Kluwer, 2003)

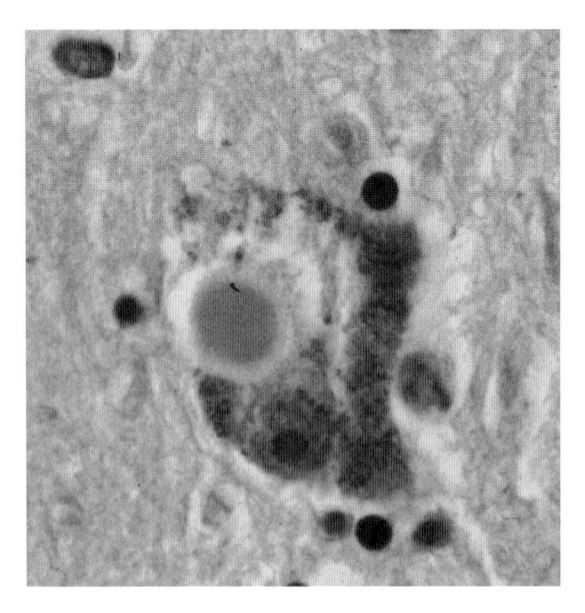

Fig. 10.4 Lewy body in a pigmented neuron of the substantia nigra. H&E stain, original at ×125. (Courtesy of Dr. Thomas Smith) (From EM Marcus and S Jacobson, Integrated neuroscience, Kluwer, 2003)

this protein is located in presynaptic nerve terminals. It is therefore likely that the Lewy body represents degenerated and aggregated neurofilaments. The Lewy body can also be found in diffusely in cerebral cortex.

3. Etiology of Parkinson Disease

Why do the neurons degenerate? The major etiologic factors have involved a genetic predisposition and possible toxic exposures The majority of cases with L-dopa responsive Parkinson's disease do not have a clear-cut genetic basis. However, mutations in the alpha synuclein gene have been found in several families of Italian or Greek origin with early-onset autosomal-dominant Parkinson's disease mapping to locus 4Q21-25 (Goedert et al. 1998; Duvoisin, 1998). Other families with autosomal recessive juvenile onset disease have implicated the parkin gene at locus 6q25.2-27. Other gene loci are discussed by Dunnett & Bjorklund, 1999.

4. Neuroleptic Agents

The use of neuroleptic agents might unmask an underlying genetic predisposition to Parkinson's disease. A significant percentage (15–60% depending on age) of patients who receive neuroleptic tranquilizers will develop symptoms of Parkinsonism. These agents include phenothiazides such as chlorpromazine, haloperidol, and reserpine. Reserpine interferes with the storage of dopamine within the nerve cells and, thus, depletes the brain of this neurotransmitter. Phenothiazides and related compounds, on the other hand, block the postsynaptic receptors at dopaminergic synapses. In general, of the symptoms produced by these agents, the akinesia, rigidity and tremor, will disappear when the agent has been cleared from the body. However, in some cases, the Parkinsonian features appear to remain as a permanent deficit. In such patients, one might find evidence in the family history of other cases of Parkinson's disease (Schmidt and Jarcho, 1966).

5. Toxic Agents

The major impetus for the study of possible toxic exposure has come from the study of cases of Parkinson's disease developing in miners exposed to manganese and more recently in cases exposed to the agent 1-methyl-4 phenyl-1,2,3,6,tetrahydro-pyridine (MPTP).

Manganese Poisoning

A relatively rare cause of a Parkinsonian syndrome is manganese poisoning. Manganese apparently accumulates in the melanin-containing neurons of the substantia nigra and interferes with the enzyme systems involved in the production of dopamine (Mena et al. 1970).

MPTP Toxicity

Another form of Parkinson's disease induced by the drug MPTP has provided major insights into the possible etiology of the degeneration of the dopamine-containing neurons in the substantia nigra. This agent is a by-product of the chemical process for the production of a reverse ester of meperidine (demerol): 1-methyl-4 propionoxypiperidine (MPPP). Meperidine is a controlled narcotic; MPPP is not controlled, has narcotic action, and can be produced in clandestine laboratories. MPTP had been recognized shortly after its legitimate synthesis in 1947 to produce a severe rigid akinetic state in monkeys. However, it was not until the summer of 1982, when MPPP was produced on a large scale in northern California for illicit mass distribution, that significant numbers of young drug abusers began appearing at emergency rooms with an acute Parkinsonian syndrome. The investigation of these cases has been presented in a series of papers by Langston et al. (1983, 1992) and Ballard et al. (1985). Essentially, the compound MPTP has selective toxicity on the dopaminergic neurons of the pars compacta of the substantia nigra, not only in human but also in subhuman primates, such as the Rhesus monkey, Macaca mulatta, and the squirrel monkey. Both in the human and the monkey, MPTP reproduces the clinical disease, the pathological findings, and the effects of pharmacological therapy of Parkinson's disease to a relatively complete degree, including, in older animals, the presence of Lewy bodies. With such an animal model, experimental medical and surgical therapy can be studied with a clear-cut correlation with the human disease (e.g., see Aziz et al, 1991; Langston and Ballard, 1984; Kopin, 1988). Moreover, because there were many patients exposed to the agent but not all exposed patients and not all acute cases developed the chronic Parkinsonian state, prospective studies can be performed as to the natural history and factors influencing the onset of the disease (Stone, 1992). In part because of MPTP, increased attention has been focused on other environmental exposures that might be involved in producing neurotoxic effects culminating in Parkinson's disease (Tanner and Langston, 1990; Olanow and Tatton, 1999). MPTP, is a mitochondrial neurotoxin. A recent report has implicated another mitochondrial neurotoxin, trichlorethylene (TCE) in this disorder. TCE has been commonly employed in industry and the military as a degreasing agent.

6. Other Pathological Processes

The following disorders also will produce some of the symptoms of Parkinson's disease without the presence of Lewy bodies (Gibb, 1988):

1. Bilateral necrosis of the globus pallidus might produce akinesia. Such a necrosis could occur in carbon monoxide poisoning, or anoxia, or in relation to infarction. Rigidity in flexion might also follow bilateral necrosis of the putamen or the globus pallidus.

II Clinical Symptoms and Signs of Dysfunction

2. A tremor at rest might occur with destructive lesions involving the ventromedial tegmentum of the midbrain. Such lesions, apparently, interrupt the pathway from the substantia nigra to the striatum. A degeneration of neurons in the substantia nigra and a decrease in dopamine in the corpus striatum occurs.
3. "Arteriosclerotic or vascular (multi-infarct)" variety in which bilateral infarcts occur in the putamen. Usually, this is the accompaniment of a lacunar state in which multiple small infarcts occur in the basal ganglia and internal capsule. The resultant syndrome of rigidity and akinesia affects the lower half of the body and thus can be clearly distinguished from the classical idiopathic Ll-dopa responsive Parkinson's disease.
4. Von Economo's encephalitis. A certain proportion of cases can be termed *postencephalitic*. These cases had a clear onset in relation to the epidemic of Von Economo's encephalitis lethargica in the years 1916 to 1926. Approximately 50% of the survivors of this devastating disease subsequently developed Parkinson's syndrome due to severe loss of neurons in the substantia nigra. In such cases, Lewy bodies are not found, but neurofibrillary tangles are present in surviving cells. In addition to the aforementioned Parkinsonian symptoms and signs, these patients presented other evidence of involvement of the central nervous system, including oculogyric crises (tonic vertical gaze), chorea, dystonia, and sleep disorders. The symptoms and signs indicated the involvement not only of the substantia nigra but also of multiple other sites, including the midbrain, hypothalamus, and hippocampus (see Gibb, 1988). The essential distribution of the pathology was in the periventricular areas about the aqueduct and third ventricle. A dramatic example of postencephalitic Parkinsonism and its treatment is presented in the movie "Awakenings," based on the book by Oliver Sacks.

7. Management (Refer to Lang & Lozano, 1998 and Riley and Lang, 2000)

When one considers that the basic pathology in Parkinson's disease represents a marked loss of dopamine-containing neurons in the substantia nigra and a marked decrease in dopamine concentration in the striatum, whereas other transmitters and neurons are affected to a much lesser degree (e.g., cholinergic system within corpus striatum now has a relatively unopposed action), then the possible types of treatment are very evident.

a. Early Medical Treatment

For many years, the standard treatment involved the use of anticholinergic compounds such as belladonna and atropine or of synthetic analogues. Trihexyphenidyl (Artane) and benztropine (Cogentin) are examples of such agents. The anticholinergic compounds affect primarily the tremor and, to some extent, the rigidity of Parkinson's disease and in many cases can produce limited improvement.

b. Early Approaches to Surgical Intervention

Various surgical procedures were attempted in the 1940s and 1950s, involving limited ablations of area 4 or area 6 or a section of the cerebral peduncle. In 1952, while attempting to section the cerebral peduncle, Cooper (1969) accidentally occluded the anterior choroidal artery, and the Parkinsonian patient showed a significant improvement in contralateral symptoms, presumably as a result of infarction of the inner section of the globus pallidus. Cooper soon came to employ stereotaxic techniques for the direct production of lesions in the globus pallidus (Fig. 10.2, lesion 1). The procedure produced a significant reduction in contralateral tremor and, to a lesser extent, a decrease in the contralateral rigidity. Cooper was able to demonstrate later that lesions in the ventrolateral thalamus were much more effective (Fig. 10.2, lesion 2). Initially, the stereotaxic lesions were produced by the injection of alcohol or coagulation. In later procedures, a freezing cannula was employed.

c. Transmitter Replacement

With the increasing knowledge of the neurotransmitters involved in the substantia nigra and corpus striatum and with reports of deficits of dopamine in these structures in patients with Parkinson's disease, the use of replacement therapy with L-dopa (dihydroxyphenylalanine) subsequently developed (Cotzias et al., 1967, 1969). As we have indicated, the anticholinergic drugs had a limited effect primarily on the tremor and rigidity; surgery affected primarily the tremor; the use of L-dopa, which was more a physiological technique, improved the severe akinesia in addition to decreasing the tremor and rigidity. The use of L-dopa required large amounts of medication because much of the administered dosage was decarboxylated to dopamine at peripheral systemic sites and never crossed the blood-brain barrier to be converted to dopamine (note that L-dopa crosses the blood-brain barrier; dopamine does not). There were significant side effects due to the peripheral actions of the drug. By combining L-dopa with a peripheral dopa decarboxylase inhibitor, carbidopa (in the form of Sinemet), a much lower dosage of L-dopa was required with fewer gastrointestinal and blood pressure side effects. Central side effects continued to occur (e.g., peak dose-related dyskinesias such as chorea and myoclonus). The short duration of action also continued to provide problems in terms of wearing off of dose. Other sudden on–off effects with sudden arrests of movement that were not entirely related to time of dose also occurred as the disease progressed. Individual titration of dosage specific to the particular patient is often required. The basic long term problem is that Parkinson's disease is progressive. Once the neurons in the substantia nigra have all died off or have become dysfunctional, there is no capability to convert L-dopa to dopamine.

d. Other Therapeutic Maneuvers

Other therapeutic maneuvers have been developed for such patients. Various dopamine agonists have been developed to bypass the problem of conversion of

L-dopa to dopamine. Additional surgical procedures involving the implantation of stimulators in deep brain structures such as the globus pallidus, thalamus, and subthalamic nucleus have also been developed (Fig. 10.2). Transplantation of fetal mesencephalic cells into the striatum has benefit in younger patients but might increase dyskinesias. The problems of management are reviewed in the following patient who was followed for 19 years.

Case History 10.1

A 48-year-old married left-handed white male schoolteacher and administrator, 3 months prior to evaluation in 1982, developed a sense of fatigue, stiffness, and lack of control in the left arm. After a period of prolonged writing, he would have to make a conscious effort to continue writing, Approximately 1 month prior to the evaluation; he first noted a slowing down in movements of the left leg. "I have to remember to lift it."

Neurological examination: The following positive findings were present:
Cranial nerves: He had a tremor of the closed eyelids and a positive glabellar sign (he was unable to suppress eyelid blinking on tap of forehead).
Motor system: In walking, there was a tendency to turn en bloc and a decreased swing of the left arm. There was a slight increase in resistance to passive motion at the left elbow, wrist, and knee (rigidity without definite cogwheel component). Although the patient was left-handed, there was a slowness of alternating finger movements of the left hand. No tremor was present. Handwriting was intact with no evidence of micrographia.
Clinical diagnosis: Early Parkinson's disease, predominantly unilateral.
Laboratory data: All studies were normal, including CT scan of the brain and thyroid studies.
Subsequent course: The patient did not wish to begin any treatment at that point in time. Reevaluation at 3 months indicated progression. One year after onset of symptoms, there was significant micrographia (Fig. 10.5) and cogwheel rigidity at the left shoulder, elbow, and wrist. Therapy with Sinemet (10–100 mg, 3 times per day.) was begun. Reevaluation 1 week later demonstrated a marked improvement in handwriting (Fig. 10.5B) and decreased cogwheel rigidity. As the dose of Sinemet was increased to 10–100 mg 5 times a day over the next week, occasional choreiform movements of the fingers of the left hand and occasional backward flinging (hemiballistic) movements of the left arm occurred at 60–90 minutes after later doses of Sinemet*. As the dose reached 10–100 mg 6 times a day, the patient reported that his handwriting now was back to normal. Examination confirmed the significant improvement in all findings. Ninety minutes after a Sinemet dose,

---

*This is a combination of 10 mg of carbidopa (dopa decarboxylase inhibitor) and 100 mg. of l-dopa. Higher dosage combinations are (1) 25 mg. of carbidopa with 100 mg l-dopa and (2) 25 mg carbidopa with 250 mg l-dopa. A sustained release form is also available (50 mg of carbidopa with 200 mg of L-dopa) and is marketed as Sinemet CR.

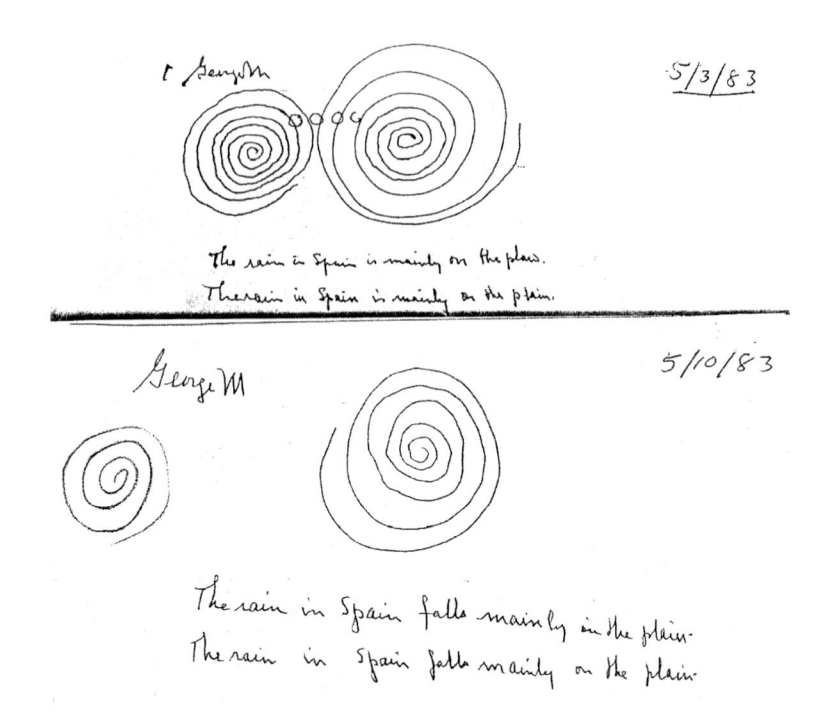

Fig. 10.5 Parkinson's disease. Case 10-1. See text. On 05/03/83 before therapy with L-dopa/carbidopa and on 05/10/83, 1 week after starting therapy. In each example, the larger circles have been drawn by the examiner. (From EM Marcus and S Jacobson, Integrated neuroscience, Kluwer, 2003)*

the patient did experience a tremor of the left shoulder and an exaggerated grasp of the left hand. Foot and toes on lifting the leg (possible form of dyskinetic or dystonic reaction). He was aware that some symptoms of his underlying disease would emerge 3–4 hours after a dose of Sinemet. Five years after onset of symptoms, despite a higher dosage of Sinemet of 25–100 mg. tablets, 7 or 8 times a day, a pill-rolling tremor of the left hand had emerged. (In actuality 75–100 mg of carbidopa is probably sufficient to saturate the peripheral dopa decarboxylase system). Moreover, choreiform (movements of the left arm) occurred 60–90 minutes after each dose of Sinemet. The choreiform movements disappeared when the dosage of Sinemet was reduced to 25–100 mg 6 times a day. The increased rigidity and tremor responded to the addition of a dopa agonist bromocriptine (Parlodel). Over the subsequent years, other agents were employed, but the disease continued to progress with all of the problems in management discussed above. Eventually, in 1997 he underwent mesencephalic fetal cell transplantation. Evaluation 3 years after the transplant demonstrated a little improvement but a marked increase in the dyskinesias. Some improvement in dyskinesias occurred after dopaminergic medications again were reduced.

C. Differential Diagnosis of Parkinson's Disease

The following other disorders are discussed in Marcus and Jacobson (2003).

1. In considering the akinesia and bradykinesia of Parkinson's disease, the major differential diagnosis in the older population are the syndromes of gait apraxia related to the frontal lobe, as discussed in Chapter 9.
2. Secondary Parkinsonism.
3. Parkinsonism plus syndromes.
4. Other neurological syndromes with rigidity and akinesia.
5. Parkinsonism–dementia–ALS complex found in the Chamorros of Guam (and several other Pacific groups). This is a probable toxic disease involving an excitatory amino acid (βN-methyl-amino L-alanine) derived from the cycad seed once used in the production of flour.
6. In considering, the tremor, the major differential is essential tremor-(see below)

D. Chorea, Hemichorea, Hemiballismus and Other Dyskinesias

A Committee of the Movement Disorder Society in 1981 formally defined chorea as "excessive spontaneous movements irregularly repeated randomly distributed and abrupt in character." These movements, although involuntary, consist of fragments or sequences of normal, coordinated movements. Face, mouth, head, proximal, or distal limbs might be involved. Two types of process must be distinguished: (1) lateralized hemichorea and (2) generalized chorea. The lateralized hemichorea merges into a remarkable disorder, hemiballismus, characterized by sudden flinging or ballistic movements at the proximal joints such as throwing or sudden swinging movements at the shoulder joint. Patients mightstart with hemichorea and evolve into hemiballismus. In discussing this topic, we will deal first with the focal problem and then with the generalized type. We will see in both varieties the abnormalities of the anatomic sequences discussed earlier in this chapter.

1. Hemichorea and Hemiballismus

Initial clinical and experimental studies correlated these disorders with lesions of the contralateral subthalamic nucleus. It was recognized that the underlying lesion could be a small infarct due to posterior cerebral penetrating branch involvement, often on a lacunar basis, in a patient with hypertension or diabetes mellitus. At times, the adjacent thalamic areas were involved as well (e.g., the VPL or VL).

At times, the responsible lesion was a small hemorrhage in the subthalamic nucleus (Fig. 10.6) and, rarely, a small metastatic tumor. With regard to the explanation for such contralateral excessive movement, hemichorea or hemiballismus, one needs only to consult Fig. 10.1. Decreasing the excitatory drive from the subthalamic

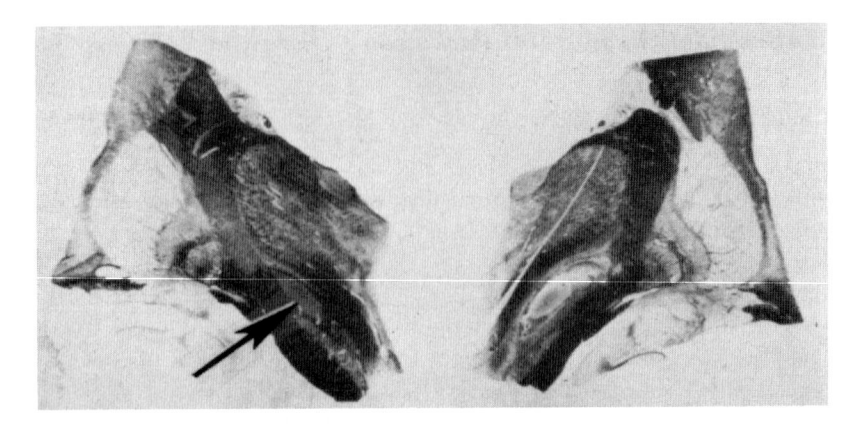

Fig. 10.6 Hemiballismus. Myelin stain of basal ganglia demonstrating a discrete hemorrhage into the right subthalamic nucleus (arrow points to the subthalmic nucleus on the unaffected side) (From LA Luhan, Neurology, Williams and Wilkins, 1968.) (From EM Marcus and S Jacobson, Integrated neuroscience, Kluwer, 2003)

nucleus acting on the medial globus pallidus and the substantia nigra reticularis would decrease the inhibitory drive from those structures acting in the thalamus. Therefore, the thalamic motor SMA motor cortex circuit would be more active and more movement would result.

2. Generalized Chorea

Bilateral generalized chorea might occur under the following circumstances:

a. Acute onset with an immunological basis (a) one aspect of rheumatic fever (plus or minus carditis and arthritis). (b) In relationship to various autoimmune disease such as lupus erythematosus.
b. As an acute complication of L-dopa therapy in patients with compromise of the dopaminergic system or with disease of the basal ganglia - as discussed above.
c. As an acute or chronic complication of various metabolic disorders: hepatic, renal, hypocalcemia, etc, rarely occurring in pregnancy.

Chronic Progressive Choreiform Disorder due to Degeneration.

Huntington's Chorea

This is a chronic progressive disorder (indicating a degenerative disease) affecting the striatum and other sectors of the basal ganglia and cerebral cortex. The most frequent disease in this category is Huntington's disease. Huntington's disease is of importance beyond its frequency in the population of 5–10/100,000. The disease was originally described in 1872 by George Huntington, based on clinical observations made by his grandfather, his father, and himself of an apparent autosomal-dominant syndrome of high penetrance that occurred in families living on Long Island in New York State. The syndrome was characterized by the development in midlife of a progressive psychological change (depression, paranoia, psychoses), dementia, and involuntary choreiform movements. The disease often affected multiple members of a family over a number of generations. Genetic studies have confirmed that the pattern of inheritance follows an autosomal-dominant pattern with extremely high penetrance. In large series, 50% of offspring will manifest the disease; that is, all who carry the affected gene will manifest the trait. Modern genetic analysis using recombinant DNA techniques and studying families with large numbers of affected individuals or large numbers at risk have allowed the localization of the gene defect. The expression of the gene can vary.

The mean age of onset in a large sample as defined by choreiform movements was 42 years. However, behavioral changes, including depression and suicide attempts, might precede the movement disorder by 10 years or more. Earlier onset of disease is associated with a more rapid course and with the earlier development of features of rigidity. Such patients, for unknown reasons, are more likely to have inherited the gene from the father. Late-onset cases (onset between 50 and 70 years of life) have a much more prolonged course and choreiform movements are often the predominant feature (see Sax and Vonsattel, 1992). At times, the term *senile chorea* has been utilized for these late-onset cases. The specific mutation has now been identified as an unstable trinucleotide repeat in the CAG series coding for polyglutamine tracts at the 4p16.3 locus on this chromosome. Normal individuals have 6–34 repeats; patients with Huntington's disease have 36–121 (Zoghi and Orr, 2000). The gene product is huntingtin. Patients with onset at an early age or with a more severe form of the disease have a greater number of repeats. Similar mutations involving an excessive number of CAG repeats have also been found in a number of other neurological disorders, including several of the spinocerebellar degenerations and spinobulbar muscular atrophy but these excessive repeats occur on chromosomes other than Chromosome 4. Other disorders such as myotonic dystrophy and Friedreich's ataxia have an excessive number of repeats in other trinucleotide sequences. Presymptomatic detection is possible, but this should be done under circumstances where a full gamut of genetic counseling is available. The underlying gross pathology has been well established (Fig.10.7). Marked atrophy of caudate nucleus and putamen are the pathologic and neuroimaging hallmarks. This gross pathological feature allows for diagnosis and subsequent monitoring of progression by means of CT scan and MRI scan (Fig. 10.8). In addition significant cortical atrophy also occurs as the disease progresses. The gene product huntingtin is expressed widely throughout the brain. There are high levels in large striatal interneurons and medium spiny neurons, as well as cortical pyramidal cells and cerebellar Purkinje cells. Additional analysis has indicated that not all cells and transmitters in the caudate and putamen are involved in the process of degeneration. The extensive studies have been reviewed by Albin et al. (1992), Martin and Gusella (1986), Penney and Young (1988), Young et al. (1988),

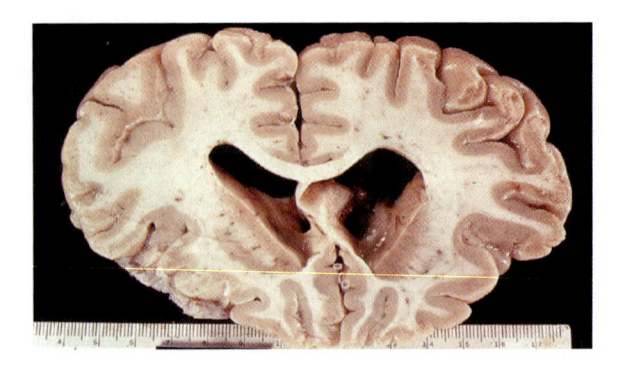

Fig. 10.7 Huntington's disease. Marked atrophy of the caudate and putamen with secondary dilatation of the lateral ventricles is evident. Cortical atrophy was less prominent in this case. (Courtesy of Dr. Emanuel Ross, Chicago) (From EM Marcus and S Jacobson, Integrated neuroscience, Kluwer, 2003)

and Wexler et al. (1992). The cells affected are classified as medium-sized spiny neurons that project to sites outside the striatum and constitute approximately 90% of all striatal neurons. All contain the transmitter GABA. However, different subtypes are distinguished based on the other cells with interneuron functions that are not affected: (a) the large, spiny specific associated neuropeptide that they contain (e.g., enkephalin, substance P, and/or dynorphin) and based on the projection destinations. Also on the other hand acetylcholine containing interneurons, and the small aspiny interneurons containing the neuropeptides somatostatin and substance Y are not affected. Early in the course of the disease there is a loss of those spiny neurons that contain the GABA and enkephalin transmitters and that project to the lateral globus pallidus. At this point in the disease, the loss of these striatal inhibitory inputs to the lateral globus pallidus results in increased activity (inhibitory) of the lateral globus pallidus neurons. Because the lateral globus pallidus is the main input to the subthalamic nucleus and this input is inhibitory, an increased inhibition of the subthalamic nucleus results. As a result, the excitatory output from the subthalamic nucleus to the medial globus pallidus (MGP) and substantia nigra reticularis (SNR) is reduced. The output of neurons in MGP and SNR is reduced. This output is inhibitory and destined for ventrolateral thalamus. The output of the ventrolateral thalamus is no longer inhibited and increased activity occurs in the ventrolateral to supplementary motor cortex circuit. Abnormal excessive movements due to increased activity of SMA then results as discussed earlier for hemichorea. The following case history provides an example of Huntington's disease.

Case History 10.2

A 54-year-old right-handed white female was referred for evaluation of a movement disorder. The patient lived alone and it was difficult to obtain much of any history from the patient. She denied any neurological or psychiatric disorders except for a problem with memory during the last year. She did indicate that she had worked for 10 years as a secretary/computer operator but had been fired 10 years ago because "she did not work fast enough." She had a past history of hypertension but had not been reliable in taking her prescribed medication. The patient's son was aware that changes in emotion, personality, speech, gait, and a movement disorder had been present for at least 3 years.

Family history: The patient denied any neurological family history except for a maternal uncle who had the "shakes" (actually a maternal aunt with essential tremor). Additional investigation revealed that her father clearly had Huntington's disease with onset of choreiform movements at age 30 and of irritability, paranoia, and gait disturbance in his early fifties. The patient's eldest daughter (age 34) had been depressed for 5 years and had problems with speech and walking for a number of years.

Neurological examination: Mental status: The patient often avoided eye contact. Affect was usually inappropriate, with laughter as the response to many questions. The overall minimental status score was 27/30. The only deficit related to delayed recall portion of the test 0/3 objects.

Cranial nerves: The patient had a hyperactive jaw jerk. She had frequent facial grimacing.

Motor system: The patient had decreased tone at wrists and elbows. As she sat, she had frequent restlessness and, at times, choreiform movements of hands and feet. Gait was often "bizarre" in terms of occasional sudden movements of a leg as she moved down the hall and came to a stop. These same movements occurred as she remained standing. In addition, in standing or walking, there were intermittent dystonic postures of either left or right arm.

Sensory system: Intact

Reflexes: Deep tendon stretch reflexes were everywhere hyperactive but plantar responses were flexor.

Clinical diagnosis: Huntington's disease.

Laboratory data: MRI scans of head (Fig. 10.8): The heads of the caudate nuclei were very atrophic. Cortical sulci were wide consistent with cortical atrophy. The lateral and third ventricles were secondarily dilated. Polymerase chain reaction testing for the CAG trinucleotide repeats of the huntingtin gene on chromosome 4 indicated one allele normal at 17 and the second excessive at 42 repeats.

Subsequent course: The patient refused any treatment for the movement disorder. According to the son, the patient's neurological status worsened to a moderate degree over the subsequent 3 years with regard to speech, unsteadiness of gait, and mood swings.

3. Other Movement Disorders Associated with Diseases of the Basal Ganglia Movement Disorders Induced by Dopamine-Blocking Agents (Neuroleptics)

These agents produce frequent acute and chronic complications in psychiatric patients (Sethi, 2001)

Fig. 10.8 Huntington's disease. Case 10.2. Marked atrophy of cerebral cortex and of caudate nucleus in this patient with a familial history of the disorder and a significant increase in the CAG trinucleotide repeats MRI, T_2. (From EM Marcus and S Jacobson, Integrated neuroscience, Kluwer, 2003)

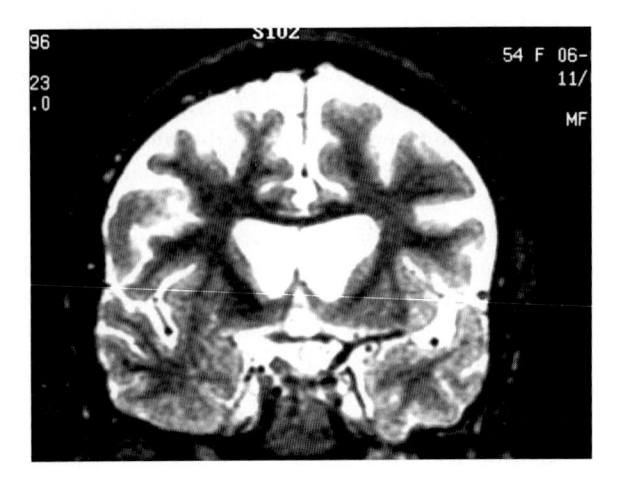

1. Acute dystonic reactions, which occur in 2–5% of patients within hours or days after onset of therapy. Note that this complication occurs not only in psychiatric patients but also in patients receiving agents of this class as antiemetics.
2. Acute akathisia, acute restlessness and an inability to sit still, occurs in 20% of patients receiving dopamine-blocking agents.
3. Acute drug-induced Parkinsonian syndrome discussed earlier. Although symptoms might appear in any patient given a sufficient amount of dopamine-blocking agents to block 80% of receptors, there does appear to be a predisposition to develop symptoms in patients with a family history of Parkinsonism.

4. Tardive Dyskinesia and Other Tardive Reactions

In this context, tardive has been defined as 3 months of exposure to the agent (1 month for patients over the age of 60 years). Usually these are seen in the older patient and involves the mouth, tongue, lips, and face (oral buccolingual facial masticatory syndrome). Symptoms usually develop after 6–24 months of drug therapy, and in many patients, the symptoms decrease after drug withdrawal. In another group of patients, symptoms develop only after neuroleptic drugs have been withdrawn or dosage has been reduced again after prolonged use.

5. Tics

Motor tics are sudden, brief involuntary movements resembling jerks or gestures. The patient might experience a poorly defined sensation of an irresistible urge to move prior to the movement. These movements might be simple: an eye blink, a head jerk, or a facial grimace. In some cases, the movement is more complex and patterned. resembling a compulsive act. In some cases (syndrome of Gilles de la Tourette, which has autosomal-dominant transmission), multiple tics occur accompanied by vocalizations. The vocalization might consist of a simple sound or involve the use of obscene four-letter words referring to defecation, genitalia, or sexual acts (coprolalia). Then a flurry of tics might occur after the suppression. Tics might continue to occur in sleep. The more usual idiopathic tics are common in children and might persist into adult life. Lees and Tolosa (1988) indicated that 1 of 10 schoolboys will have idiopathic tics. Males are affected three times as often as females. No specific neuropathology has been established.

6. Other Disorders

The following disorders are discussed in Marcus and Jacobson (2003):

1. Hepatolenticular degeneration (Wilson's disease)
2. Double athetosis (a variety of cerebral palsy)
3. Dystonias: segmental, focal, generalized and hemi dystonia
4. Familial paroxysmal dyskinesias.

Chapter 11
Motor Systems III: Cerebellum and Movement

The cerebellum forms the tectum of the medulla and pons. It lies in the posterior cranial fossa under the tentorium of the dura. It is a layered cortical region and projects onto subcortical nuclei that, in turn, project onto the brain stem, spinal cord, and diencephalon. It functions as an unconscious center for coordination of motor movements. It is commonly affected by drugs and vascular and other pathological lesions.

I. Anatomy

A number of schemes for dividing the cerebellum into various lobes and lobules have been devised. From a functional standpoint, it is perhaps best for the student to visualize the cerebellum as composed of various longitudinal and transverse divisions. In order to visualize these subdivisions, it is necessary to unfold and to flatten the cerebellum, as shown in Figure 11.1.

A. *Longitudinal Divisions (Table 11.1)*

The major longitudinal divisions are the median (vermal cortex), the paramedian (paravermal), and the lateral (remainder of the cerebellar hemispheres).

The major projections and functional correlations are outlined in Table 11.1. In general, the median (vermal) region is concerned with the medial descending systems and with axial control. The lateral and paramedian regions are concerned with the lateral descending systems and with the appendages (the limbs) (midline to axis; lateral to appendages).

B. *Transverse Divisions (Table 11.2)*

The transverse subdivisions are essentially phylogenetic divisions.: archicerebellum, paleocerebellum, and neocerebellum.

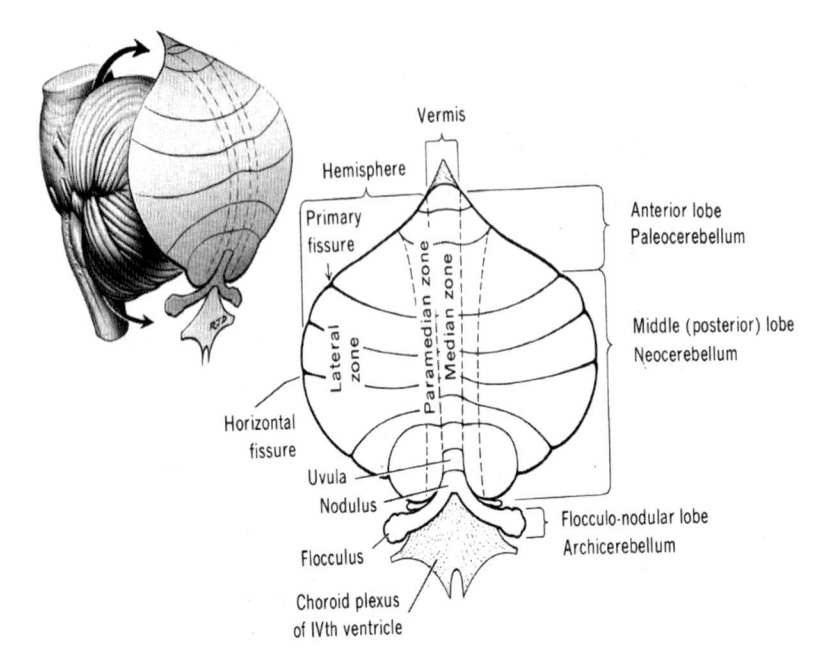

Fig. 11.1 Major transverse and longitudinal subdivisions of the cerebellum. The surface has been unfolded and laid out flat. (From CR Noback. et al,. The human nervous system. 4th ed, Lea & Febiger, 1991, p. 282)

Table 11.1 Longitudinal subdivisions of the cerebellum

Median	Paramedian	Lateral
Vermal Cortex	Paravermal	Remainder of cerebellar hemisphere
Projects to: fastigial nucleus (globus and emboliform)	Projects to: interpositus nuclei	Projects to: dentate nucleus
Role in: Posture & movements of axialbody	Role in: discrete ipsilateral appendicular movements	Role with cerebral cortex, thalamus, and red nucleus. Initiating, planning, and timing of movements.

1. The **archicerebellum**, flocculus and nodulus, is related primarily to the vestibular nerve and vestibular nuclei. Reflecting this anatomical relation, the archicerebellum has a role in the control of equilibrium (balance), axial posture, and eye movements.
2. The **paleocerebellum**, anterior lobe, relates primarily to the spinocerebellar system and to its analogue for the upper extremities—the cuneocerebellar pathway (the lateral cuneate nucleus is the analogue of the dorsal nucleus of Clarke). Its major function, however, appears more related to the lower extremities than

Table 11.2 Transverse divisions of the cerebellum

Archicerebellum: vestibulocerebellum–flocculonodular lobe	Paleocerebellum: anterior lobe	Neocerebellum: Middle or posterior lobe
Input: 1. Cranial nerve VIII, lateral vestibular nucleus, labyrinth (semicircular canals and otolith organs), and auditory trigeminal and visual 2. Vestibulospinal, MLF, extraocular motor nuclei, spinocerebellar, and cuneocerebellar from body and proximal limbs	Input: 1. Dorsal spinocerebellar and cuneocerebellar 2. Deep cerebellar nuclei: interpositus (globose and emboliform)	Input: 1. Motor and prmotor cerebral cortex (areas 4 and 6) and parietal cortex (areas 1–3 and 5) to pontine nuclei to lateral neocerebellum (after decussation) via the middle cerebellar peduncle 2. Deep cerebellar nucleus: dentate
Connections: 1. Via juxtarestiform body and central tegmental system to pontine, medullary reticular nuclei (origin of medial and lateral reticulospinal tracts) and to vestibular nuclei (origin of vestibulospinal tract) 2. Via superior cerebellar peduncle (decussated) to ventral nucleus of thalamus and onto motor cortex to anterior corticospinal tract	Connections: Lateral descending motor system, superior cerebellar peduncle decussates to magnocellular sector of the red nucleus (origin of rubrospinal system) and to VL nucleus of thalamus to primary motor area 4 and area 6 Damage produces cerebellar dysarthria that affects prosoady, harmony, and rhytmn of speech	Connections: 1. Superior cerebellar peduncle to contralateral red nucleus (parvocellular) and subsequent rubro-olivary fibers 2. VL of thalamus to motor and premotor cerebral cortex, the origins of corticospinal, corticorubrospinal, and coticoreticulospinal
Role: Axial equilibrium, trunk primarily) posture and equilibirium at trunk, muscle tone, vestibular reflexes, eye movements	Role: Equilibrium, posture, muscle tone in lower extremities, coordinated movements lower extremities (e.g. heel to shin), speech	Role: Limb coordination, phasic posture movements of upper and lower extremities, possible role in higher executive functions and emotions

to the upper extremities. The primary fissure separates the anterior lobe from the middle posterior lobe.

3. The **neocerebellum**, middle and posterior lobes, relates primarily to the neocortex. The function of the neocerebellum is primarily the coordination of discrete movements of the upper and lower extremities (see Table 11.2).

C. Cytoarchitecture of the Cerebellum

In contrast to the cerebral cortex, all areas of the cerebellar cortex are relatively thin and have the same basic three-layered cytoarchitectural pattern plus a correlated group of deep nuclei.

1. Outer molecular layer
2. Purkinje cell layer
3. Inner granule cell layer with underlying medullary layer of white matter, and
4. Deep cerebellar nuclei, from medial to lateral; dentate, emboliform, globose, and fastigial.

The arrangement of cells and the basic synaptic connections are indicated in Figure 11.2.

D. Cerebellar Fibers (Table 11.3)

1. Mossy Fibers

These fibers are so named because each terminates in a series of mosslike glomeruli, where axodendritic synaptic contacts are made with granule cells. The mossy fibers have already been encountered as spinocerebellar, cuneocerebellar, corticopontocerebellar, vestibulocerebellar, or reticulocerebellar pathways.

Table 11.3 Cerebellar fibers

INPUT/AFFERENTS:
A. Inferior cerebellar peduncle (restiform body)
   1. Uncrossed: from dorsal spinocerebellar and cuneocerebellar. Reticulocerebellar (from lateral reticular and paramedian nuclei)
   2. Crossed olivocerebellar
B. Juxta-restiform body
   Uncrossed from vestibular nuclei predominantly to the floccular nodular lobe, the archicerebellum, and the midline vermis. and note that some fibers from the vestibular nerve bypass the vestibular nuclei and directly enter the cerebellum
C. Middle cerebellar peduncle (brachium pontis)
   The corticopontocerebellar input (via the pontine nuclei) crossed from opposite cerebral hemisphere to the neocerebellum of the lateral hemisphere
D. Superior cerebellar peduncle (brachium conjunctivum)
   Crossed ventral spinocerebellar is primarily to the anterior lobe (paleocerebellum) and the intermediate area of the paraflocculus. Its major role is as the efferent pathway.

OUTFLOW/EFFERENTS:
A. Superior cerebellar peduncle (brachium conjunctivum)
   1. From dentate nuclei and interpositus (globus and emboliform) nuclei, crossed to VA and VL in thalamus and to the red nucleus (dentatorubrothalamic pathway.)
   2. Descending impulses to the paramedian reticular nuclei
B. Inferior cerebellar peduncle (restiform body)
   1. Floccular nodular lobe via fastigial nucleus to the lateral vestibular nucleus and reticular areas of the brain stem
   2. Uncinate fasciculus: fastigioreticular and fastigiovestibular pathways hook around the superior cerebellar peduncle and also project into juxtarestiform body
C. Middle cerebellar peduncle (brachium pontis). This peduncle only has efferent connections from the cerebrum onto the pontine nuclei.

2. Parallel Fibers

The granule cell axon is sent into the molecular layer where it divides into two long branches. These branches travel parallel to the long axis of the cerebellar folium and are designated as parallel fibers. These fibers make excitatory synaptic contact with the extensive dendritic arborizations of Purkinje cells, as well as with stellate cells, Golgi cells, and basket cells. The basket cells inhibit the Purkinje cells via axosomatic synapses, as these projections are inhibitory in nature The outflow of the Purkinje cells is inhibitory in nature The target of the axons of the Purkinje cells is the neurons of the deep cerebellar nuclei. The axons of the neurons in these deep cerebellar nuclei are the major outflow from the cerebellum to the brain stem and thalamus. The parallel fibers also have excitatory synapses in the molecular layer, but they synapse with the dendrites of the Golgi cells. The Golgi cell, in turn, is inhibitory to the granule cell with a synapse within areas of the molecular layer referred to as glomeruli. The parallel fibers also excite other interneurons (small stellate cells) in the molecular layer, which, in turn, are inhibitory to the dendrites of Purkinje cells.

3. Climbing Fibers

Climbing fibers are excitatory to the Purkinje cell dendrites. These fibers originate in the contralateral *inferior olivary nuclei* and represent a long latency pathway from the spinal cord or cerebral cortex to the cerebellum (olivocerebellar). These fibers are so named because they climb and wrap around the dendrites and body of the Purkinje cell neurons, making several hundred synaptic contacts. Stimulation of a single climbing fiber results in large postsynaptic potentials and in large high-frequency bursts of discharges. GABA (gamma aminobutyric acid) is the inhibitory transmitter for the Purkinje, basket, stellate, and Golgi cells. Because the outflow from the cerebellar cortex occurs via the axons of Purkinje cells and this outflow is inhibitory to the deep cerebellar nuclei, it is logical to ask what excitatory influences drive the neurons of these deep nuclei. The answer is that collaterals of the mossy fibers and climbing fibers serve this function.

II. Functions of the Cerebellum Topographic Patterns of Representation in Cerebellar Cortex

Stimulation of tactile receptors or of proprioceptors results in an evoked response in the cerebellar cortex. (Fig. 11.3).

Visual and auditory stimuli evoke responses primarily in the midline vermis. Stimulation of the specific areas of the cerebral cortex also evokes responses from the cerebellar cortex in an appropriate topographic manner. Moreover, direct

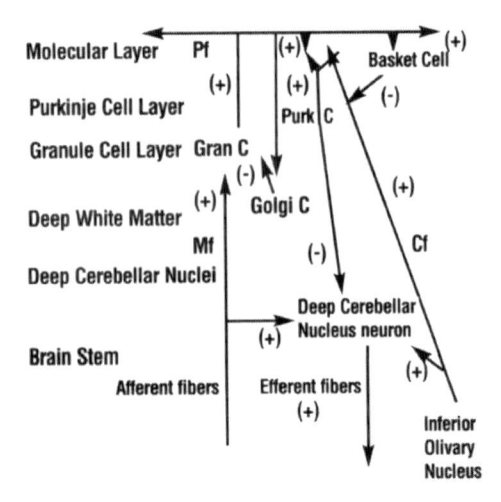

Fig. 11.2 The most significant cells, connections, and the afferent and efferent fibers in the cerebellar cortex. Arrows show direction of axonal conduction. (+) = excitatory;(−) = inhibitory. Cf = climbing fibers; pf = parallel fiber; mf = mossy fiber. Refer to text. (From EM Marcus and S Jacobson, Integrated neuroscience, Kluwer, 2003)

stimulation of the cerebellar cortex in the decerebrate animal will produce, in a topographic manner, movement or changes in tone of flexors or extensors. The general pattern of representation is consistent with the patterns of sensory and cortical representation previously noted. A somatotrophic representation in the primate cerebellum is shown in Figure 11.3. It is important to recall that there is no conscious perception of stimuli arriving at the cerebellar level.

Vestibular stimulation evokes responses not only in the floccular nodular regions but also in the superior and inferior midline vermis, the ventral lateral nucleus, and ventral anterior nucleus of the thalamus and to the red nucleus (dentatorubral–thalamic pathway).

The cerebellum acts as a servomechanism; that is, it acts as a feedback loop that dampens movements and motor power to prevent overshoot and oscillation (i.e., tremor). In short, it acts to maintain stability of movement and posture. More recently, a role in higher motor function, such as the initiation of planning, the timing of movement, and motor learning, has been suggested for the cerebellum. Table 11.4 reviews the functional correlations in each lobe.

III. Effects of Disease on the Cerebellum

A. *Overview*

The cerebellum can be viewed as a machine processing information from many sources (sensory receptors, vestibular nuclei, reticular formation, and cerebral cortex) and then acting to smooth out the resultant movements or postures. Sudden displacements

III Effects of Disease on the Cerebellum

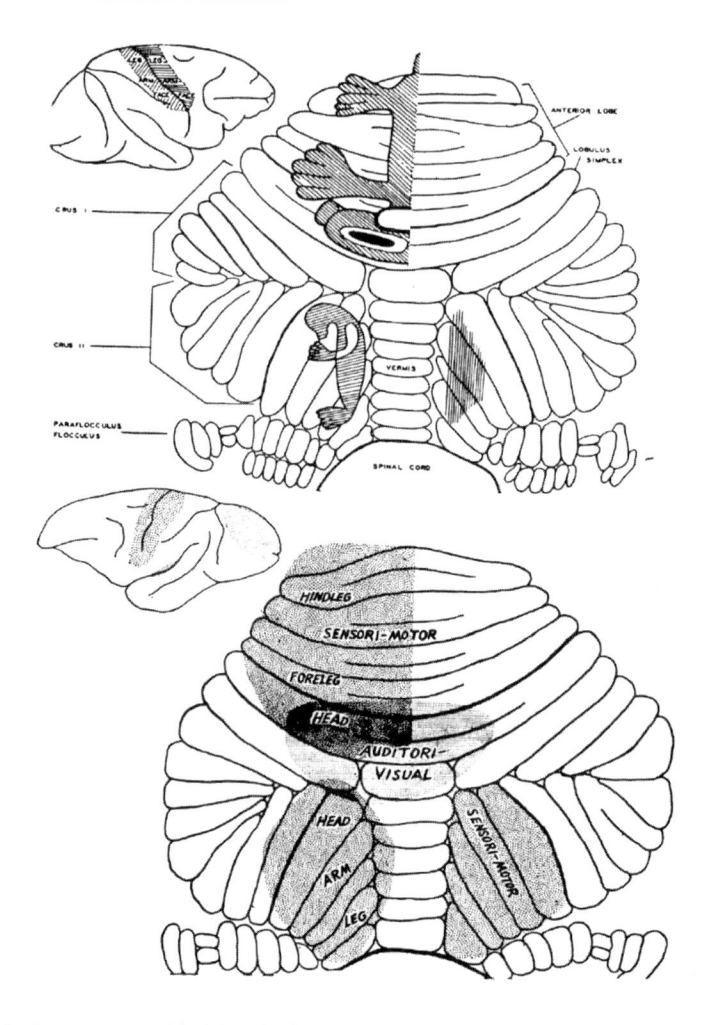

Fig. 11.3 Somat-topographical localization in the cerebellum of the monkey: **A** summary of projections from sensory area of the motor cortex; **B** summary of corticocerebellar projections. Note that there is a unilateral representation on the dosal surface and arepresentation within each paramedial lobule. (From RS Snider, Arch. Neurol. Psych. 64:204, 1950)

or lurches are prevented. The large number of inhibitory feedback circuits in the cerebellum (the major influence of Purkinje cells on deep cerebellar nuclei being inhibitory) qualifies the cerebellum for this role. Lesions might then result in undamped oscillation. As a result, tremor perpendicular to the line of movement might occur, called intention or action tremor. Lesions might also impair the ability to respond to sudden displacements, resulting in instability of posture and gait (ataxia). Release of function of ventral lateral nucleus of thalamus might also occur. The cerebellum is not designed for the direct, fast-conduction control of movements.

Table 11.4 Regional functional correlations

Vestibular Reflexes and Eye Movements—Flocculonodular lobe
  Input-vestibular
  Deep nucleus lateral vestibular > medial descending systems (vestibulospinal, MLF)
Posture and Equilibrium of the Trunk— Flocculonodular lobe (vestibulocerebellum)
Distal Motor Control and Speech/Voice control—Intermediate lobe, paravermal or paramedian lobe spinocereellar
  Input: spinocerebellar and cuneocerebellar
  Deep Nucleus interpositus → superior cerebellar peduncle to magnocellular red nucleus and to VL → areas 4and 6 and SMA → corticospinal pathway
Speech— Paravermal lobe more left than right
  Input: spinocerebellar.
Initiation, Planning and Timing of Movement—Lateral cerebellar hemisphere
  Input: Spinocerebellar and cuneocerebellar
  Nucleus dentate → superior cerebellar peduncle to VL → motor and premotor cortex → corticospinal, rubrospinal,vestibulospinal, and tectospinal tract.

In humans, extensive destruction of cerebellum might be present with little obvious deficit. Moreover, cerebellar symptoms, if present in a "static" disease process (such as cerebrovascular accidents), often disappear with time. However, more specific tests might still reveal a minor deficit in speed and coordination. More recent studies suggest a role for the cerebellum in classical conditioning and in motor learning, particularly with regard to the memory aspect for the timing of movements (Raymond et al., 1996). In addition, a role of the cerebellum in higher-order cognitive and emotional functions has been suggested by the studies of Schmahmann and Sherman, (1998) Levinsohn et al. (2000), and Riva and Giogi (2000). In patients with lesions of the posterior lobe and vermis, there was an impairment of executive functions such as planning, set shifting, verbal fluency, abstract reasoning, working memory, as well as visual spatial organization and memory. Higher-language disturbances also occurred. In addition, these patients had a personality change characterized by blunting of affect or disinhibition and inappropriate behavior. There are cognitive and affective syndrome suggested a disruption of the modulatory effects of cerebellum on the prefrontal, posterior parietal and limbic circuits discussed in Chapter 14. These changes did not occur with lesions of the anterior lobe. The effects on cognitive and emotional function in patients with prenatal cerebellar hypoplasia might be even more serious and include autism (Courchesne et al., 1994; Allin et al., 2001)

B. Major Cerebellar Syndromes

Although we will discuss three anatomically distinct syndromes of cerebellar disease (the floccular nodular lobe, the anterior lobe, and the lateral hemisphere), it should be pointed out that these strict anatomic borders do not confine many diseases that affect the cerebellum. Moreover, the cerebellum is positioned in a relatively tight compartment with bony walls and a relatively rigid tentorium cerebelli above. An expanding lesion in the posterior fossa, then, might produce a generalized compression of the cerebellum. In such clinical situations, it might be possible to differentiate only between midline (vermal) involvement and lateral (cerebellar hemisphere) involvement. *Midline lesions produce disorders of equilibrium and axial ataxia; lateral and lateralized paramedian lesions produce appendicular ataxia and tremor.*

1. Syndromes of the Floccular Nodular Lobe and Other Midline Cerebellar Tumors

The major findings in the human, monkey, cat, and dog are a loss of equilibrium and an ataxia (unsteadiness) of trunk, gait, and station. Thus, the patient, when standing on a narrow base with eyes open, has a tendency to fall forward, backward, or to one side. The patient might be unable to sit or stand. The patient walks on a broad base, often reeling from side to side, and often falling. Despite the loss of equilibrium, the patient usually does not complain of a rotational vertigo. When recumbent in bed, the patient often fails to show any ataxia or tremor of limbs. Thus, the finger-to-nose and heel-to-shin tests are performed without difficulty. With unilateral disease of the floccular nodular lobe, a head tilt might be present. Spontaneous horizontal nystagmus might be present as a transitory phenomenon.

In humans, the most common cause of this syndrome is neoplastic. The type of neoplasm depends on the age of the patient. The most likely cause in an infant or child is a medulloblastoma, a tumor arising in nests of external granular cells in the nodulus forming the roof of the fourth ventricle. In older children and young adults, the ependymoma, a tumor arising from ependymal cells in the floor of the fourth ventricle, might cause this syndrome by pressing upward against the nodulus. In adults, it is more appropriate to speak of midline tumors of the vermis because the tumors do not selectively involve the floccular nodular region. In middle-aged adults, the most common cause is probably the midline hemangioblastoma. In older adults, metastatic tumors might produce this syndrome. At all ages, rare arteriovenous malformations might produce the syndrome.

Midline Tumor in a Child

The course of a medulloblastoma is indicated in the following case history

Case History 11.1

A 27-month-old white female had been unsteady in gait, with poor balance, falling frequently since the age of 13 months, when she began to walk. Three months prior to admission, in relation to a viral infection manifested by fever and diarrhea, the patient developed increasing anorexia and lethargy. One month prior to admission, vomiting increased and the patient also became more irritable. Perinatal history had been normal and head circumference had remained normal.

Neurologic examination: Mental status: The child was irritable but cooperative.
Cranial nerves: The fundi could not be visualized. Bobbing of the head was present in the sitting position.
Motor system: A significant ataxia of the trunk was present when sitting or standing. Gait was broad-based and ataxic, requiring assistance. No ataxia or tremor of the extremities was present.
Clinical diagnosis: Midline cerebellar tumor, most likely, based on age, a medulloblastoma.
Hospital course: Doctor Peter Carney performed a suboccipital craniotomy, which confirmed the presence of a medulloblastoma with many malignant cells present in the cerebrospinal fluid obtained at surgery. Radiation therapy to the entire central nervous axis was begun 1 week after surgery. There was initial improvement but the patient deteriorated 3e months after the surgery and expired 1 month later despite chemotherapy At present, recommended therapy includes wide resection followed by radiotherapy to the entire central nervous system axis. Dissemination of tumor cells throughout the cerebrospinal fluid is frequent, because of the friable, cellular, nonstromal nature of the lesion. At present, 5-year survival has been increased to 60%. Recurrences can be treated with chemotherapy, although the combination of radiotherapy and chemotherapy might produce significant pathologic alterations in the white matter.

Midline Cerebellar Tumor in the Adult

The course of a midline cerebellar tumor in the adult is detailed in Figure 11.4: Case History 11.2

Case History 11.2

A 67-year-old white female weeks prior to admission developed intermittent vertigo, and then 2 weeks prior to admission, she had a progressive ataxia of gait followed by headaches and vomiting. She had a past history of a poorly differentiated highly malignant infiltrating ductal adenocarcinoma of the left breast for which she had undergone lumpectomy 6 years earlier and a radical mastectomy followed by radiation therapy 3 years prior to admission. On neurological examination (in the emergency room), she had difficulty standing, tending to fall to the right. She was unable to walk because of severe ataxia. There was no limb dysmetria. Computed tomography (CT) scan demonstrated that a large (3–4 cm) enhancing tumor was present involving the midline vermis and right paramedian cerebellum and compressing the fourth ventricle, subsequently confirmed by magnetic resonance imaging (MRI) scan of the brain (Fig. 11.5). A CT scan of the abdomen demonstrated two possible metastatic lesions in the liver. With dexamethasone, she had

III Effects of Disease on the Cerebellum

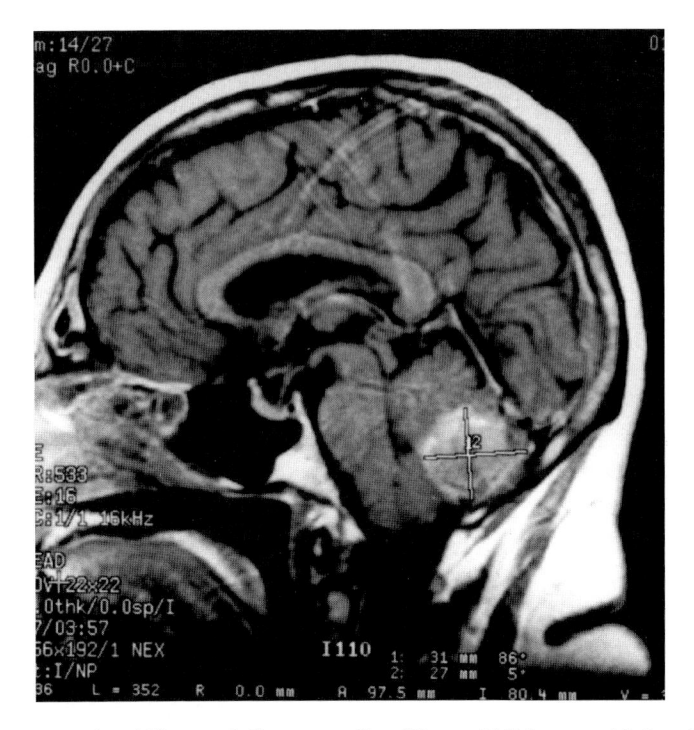

Fig. 11.4 Metastatic midline cerebellar tumor. Case History 11.2 (presented below. From EM Marcus and S. Jacobson, Integrated Neuroscience, Kluwer 2003)

significant improvement within 12 hours. At the insistence of the patient, Dr. Gerald McGullicuddy resected the solitary central nervous system metastatic lesion. She then received 3000 cGy (rads) whole-brain radiation and was begun on the antitumor agent tamoxifen. She expired at home, 4 months after surgery of systemic complications of the malignancy.

2. Syndromes of the Anterior Lobe: Anterior Superior Vermis

The major symptoms are an ataxia of gait with a marked side-to-side ataxia of the lower extremities as tested in the heel-to-shin test. Muscle tone is not changed. The upper extremities, in contrast, are affected to only a minor degree. This area corresponds to the representation of the lower extremities and to the area of projection of the spinocerebellar pathways. **Typical of this syndrome are cases of alcoholic cerebellar degeneration,** where a severe loss of Purkinje cells occurs in the anterior lobe (Fig. 11.5). Although this degeneration occurs primarily in alcoholics, the basic etiology is usually a nutritional deficiency (the specific factor is most likely thiamine). The major findings of a broad-based, staggering gait with truncal ataxia and heel-to-shin ataxia can be related to the anterior lobe of the cerebellum. Because symptoms improve or do not progress after the discontinuation of alcohol

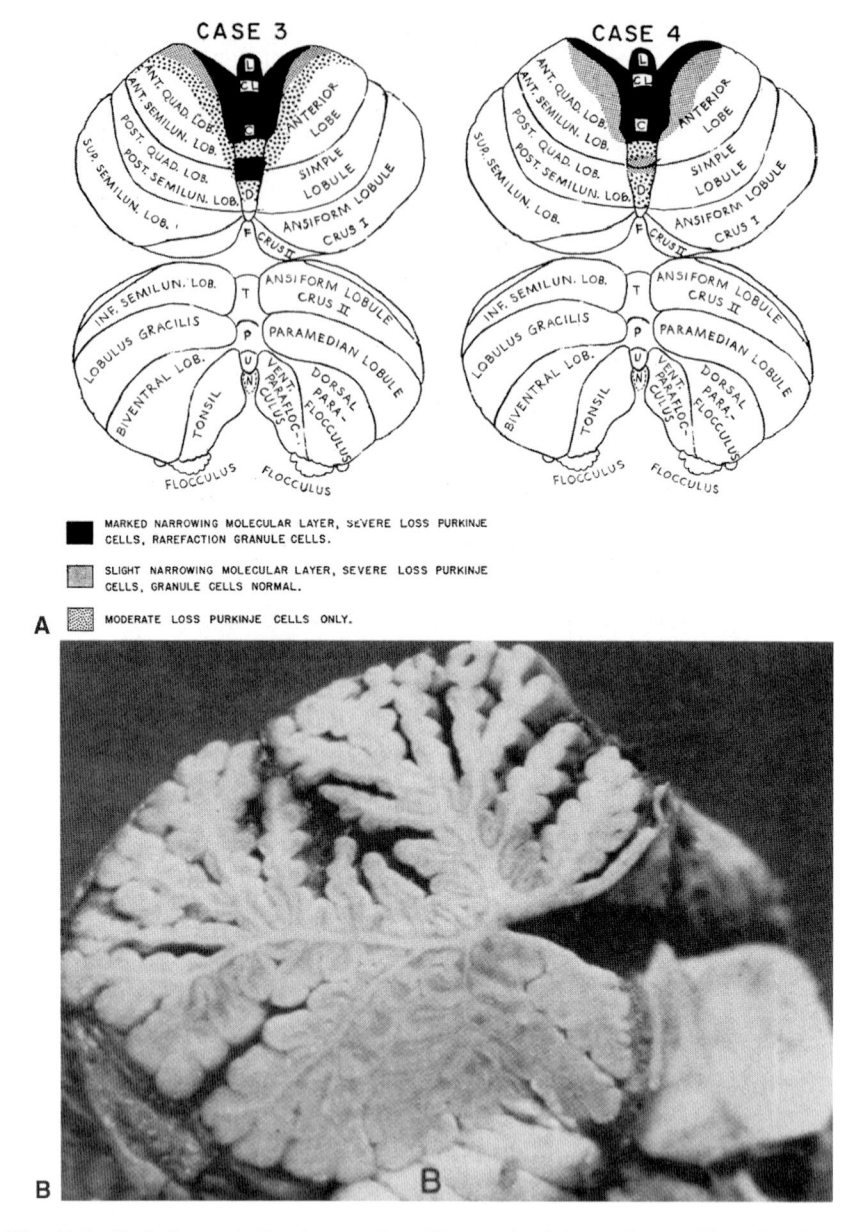

Fig. 11.5 Alcoholic cerebellar degeneration. The anterior lobe syndrome. There is a loss of Purkinje cells, atrophy of the cerebellar folia with relatively selective involvement of the anterior lobe. **A** Schematic representation of the neuronal loss; **B** atrophy of the anterior superior vermis in the sagittal sac (From M Victor, RD Adams, and EL Mancall, Arch. Neurol. 1:599–688, 1959)

and the administration of multiple B vitamins, a toxic or nutritional factor might be postulated. (The reliability of the history of alcohol intake is always open to question. The amount stated by the patient is usually assumed to be the minimum amount.)

Currently, the diagnosis of cerebellar atrophy can be confirmed by MRI (midline sagittal section) or by CT scan. As a result, evidence of cerebellar atrophy is now found in many patients presenting with other aspects of chronic alcoholism (Fig. 11.6) or other nutritional disease.

Wernicke–Korsakoff's Syndrome

This syndrome is discussed in chapter 14. This syndrome includes opthalmoplegia, obtundation, amnestic syndrome and a cerebellar ataxia. The cerebellar ataxia is similar to that seen with alcoholic cerebellar degeneration from both a clinical and pathological standpoint.

Other Causes of Cerebellar Atrophy

In other cases, this syndrome is the result of a chronic degenerative disease with a genetic basis. As in alcoholic cerebellar degeneration, the predominant histologic change is the loss of Purkinje cells in the anterior lobe. In contrast to alcoholic cerebellar degeneration, the pathology is usually more widespread and not strictly limited to the anterior lobe.

Prolonged high fever, meningitis, a postinfectious syndrome, and repeated grand mal seizures might all be associated with the development of an ataxia that suggests major involvement of the anterior superior vermis. In some of these cases, there is found more widespread atrophy on CT scan or MRI. At times, other residual neurologic findings might also be present. The topography of cerebellar atrophy, particularly with regard to midline structures, can best be studied with MRI.

Figure 11.6. demonstrates the findings in a patient with gait ataxia following high fever and meningitis, in which atrophy, although widespread, was most prominent in the anterior superior vermis. When the ataxia is of acute or subacute onset in childhood, the possibility of a cerebellitis must be considered. A varicella or other viral infection often precedes this entity.

Localization of ataxia of Posture

Diener and Dichgans (1992) discussed the localization of ataxia of standing body posture. Their conclusions were as follows:

1. Lesions of the spinocerebellar portion of the anterior lobe (mainly observed in chronic alcoholics) result in body sway along the anterior/posterior axis with a frequency of three per second. This sway, or tremor, is provoked by eye closure. Visual stabilization of posture occurs. Because the direction of sway of the trunk

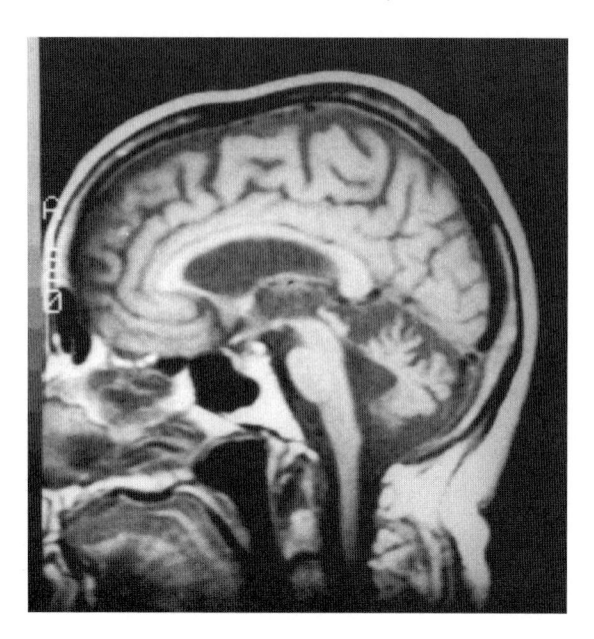

Fig. 11.6 Cerebellar atrophy most prominent in anterior superior vermis with predominant anterior lobe syndrome. MRI. This 70-year-old female 23 years previously had prolonged marked elevated temperature (104 °F, possibly as high as 110 °F) related to meningitis, coma, and convulsions with a residual ataxia of stance and gait. She was unable to walk a tandem gait but had no impairment on finger-tonose test. The patient also had episodes of tinnitus, vertigo, olfactory hallucinations, déjà vu, and musical hallucinations followed by loss of contact (temporal lobe seizures) (From EM Marcus and S Jacobson, Integrated neuroscience, Kluwer, 2003)

is opposite that of the head and legs, the center of gravity is only minimally shifted and the patient does not fall.
2. In contrast, the lesions of the vestibulocerebellum (floccular nodular lobe and lower vermis), which are primarily mass lesions, produce a postural instability of head and trunk during sitting, standing, and walking. Postural sway occurs in all directions often at a frequency of less than one per second. Visual stabilization is limited; falls, therefore, are more frequent.
3. Spinal ataxia (as in tabes dorsalis and combined system disease) results predominantly in lateral body sway. Visual stabilization is prominent

3. Syndromes of the Lateral Cerebellar Hemispheres (Neocerebellar or Middle–Posterior Lobe Syndrome) (Fig. 11.7)

Major findings in ipsilateral limb from lateral hemisphere lesions:

a. **Intention** Tremor in Upper Extremity, Disorganization of Alternating Movements (Dysdiadochokinesia), Dysmetria, Ataxia, and Overshoot on Rebound

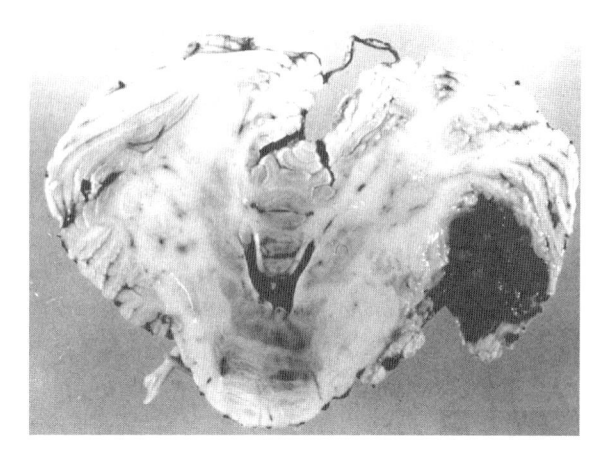

Fig. 11.7 Syndrome of the lateral cerebellar hemisphere (hemiangioblastoma). This 72-year-old had a 2-year history of vertex headaches (worse on coughing), blurring of vison, diploplia, papliodema, and a lateral ipsilateral tremor on finger-to-nose testing (Courtesy of Dr. Jose Segarra) (From EM Marcus and S Jacobson, Integrated neuroscience, Kluwer, 2003)

The intention tremor may be described as a tremor perpendicular to the line of motion; often becoming more prominent on slower movements and as the target is approached. This intention tremor may be noted in the finger-to-nose test. As with the other signs of cerebellar disease, intention tremor is usually ipsilateral to the hemisphere involved. The intention tremor represents a defect in the ability to dampen oscillations and to dampen overshoot. In the monkey, cerebellar tremor occurs despite deafferentation(section of the dorsal roots.), a finding consistent with the concept that there is a central generator of oscillations. (However, other theories reviewed by Diner and Dichgans(1992) suggest the oscillations arise from dysfunction in long-latency tarnscortical reflexes.

b. Intention Tremor in Lower Extremity
  One might also demonstrate a similar tremor in the lower extremities as the patient attempts to raise the foot off the bed to touch the examiner's finger or in the heel-to-shin test. As with the other signs of cerebellar disease, intention tremor is usually ipsilateral to the hemisphere involved.
c. At times, in addition to an intention tremor, a sustained postural "tremor" might be evident, more particularly if the superior cerebellar peduncle or its midbrain connections are involved.
d. Brief jerks at the onset of movement (intention myoclonus) might occur when the dentate nucleus or superior cerebellar peduncle is involved.
e. In disease of the cerebellar hemisphere, a marked disorganization of repetitive and alternating movement occurs. This is almost always ipsilateral to the hemisphere involved. Having the patient slap the thigh above the knee with the hand

at a particular rhythm tests repetitive movements. One can also test for repetitive movement in the lower extremities by having the patient tap toes or heel on the floor. Alternating movements are tested by having the patient slap the hand on the thigh or on the opposite hand, alternating between the palmar and dorsal surface of the hand. These patients are also said to demonstrate dysmetria. This can be defined as a defect in the estimation of the force and rate of movement necessary for an extremity to reach a target. The result is that the patient fails the finger-to-nose test. The finger might overshoot its goal or fail to reach the goal. Note that in disease involving the older regions of the cerebellum, eye movements might also be described as dysmetric.

f. The intention tremor represents a defect in the ability to dampen oscillations and to dampen overshoot. In the monkey, cerebellar tremor occurs despite deafferentation (section of the dorsal roots), a finding consistent with the concept that there is a central generator of oscillations. [However, other theories reviewed by Diener and Dichgans (1992) suggested that the oscillations arise from dysfunction in long-latency transcortical reflexes.]

g. In addition, there is often a failure to properly adjust force when pulling with the arm against an opposing force. As this opposing force suddenly gives way, the patient fails to apply a brake to the action of the arm and tends to have a rebound overshoot. The arm might even strike the body or face.

h. Finally, patients with the cerebellar hemisphere disease are said to have a dyssynergia (disturbance in the synergy of movements), a defect of the coordination of multiple sets of agonists and antagonistic muscles when reaching for an object, producing a decomposition of movement.

i. Cerebellar dysartria might also be present and is characterized by scanning (hesiation) that affects the pattern and intonation of speech. Prosody (rhythm and harmony) is affected rather than word sequence. At times, a tremor of speech is also apparent. Dysarthria might occur in cerebellar atrophy or in focal processes involving the hemisphere or the superior vermis. The crucial area of involvement is the superior paravermal area. The left hemisphere is more often involved than the right.

j. The patient with disease of the cerebellar hemisphere also has a disturbance of gait. This tends to be an ipsilateral disturbance of gait, with a tendency to fall toward the side of the lesion. As discussed earlier, in lesions of the hemisphere, balance is often well maintained, in contrast to lesions of the flocculo nodular lobe, where a disturbances of gait and of sitting balance also occur.

4. Syndromes of the Cerebellar Peduncles (Orioli and Mettler, 1958; Carrera and Mettler, 1947)

Some patients demonstrating cerebellar symptoms and findings actually have damage to the cerebellar peduncle rather than direct cerebellar involvement.

1. Lesions of the inferior cerebellar peduncle as in the lateral medullary syndac, result in an ipsilateral ataxia of the extremities, with falling toward the side of the lesion and nystagmus.
2. Lesions of the superior cerebellar peduncle produce a unilateral ataxia of the limbs, with intention tremor and hypotonia. The symptoms are essentially those of hemisphere lesions. The clinical symptomatology is ipsilateral if the lesion occurs between the dentate nucleus and the decussation of the brachium conjunctivum. If the lesion is above this decussation, the tremor is usually contralateral and often is more than a pure intention tremor. These are components of a sustained postural type of tremor.
3. Lesions of the middle cerebellar peduncle. The effects of isolated lesions of the middle cerebellar peduncles are not certain. There is some evidence that an incoordination of fine limb movements and an ataxia of gait might result. In experimental lesions, circling might result.

5. Vascular Syndromes of the Cerebellum: Vertebral Basilar Disorders

The development of CT and MRI scans has resulted in an increased recognition of infarcts, hemorrhages, and arteriovenous malformations involving the cerebellum. Many have been clinically silent or previously attributed to brain stem or labyrinthine pathology.

The cerebellum is the site of hypertensive hemorrhage in 8–10% of all cases. Before the era of CT and MRI, the precise diagnosis was often not established. Small hemorrhages might have only nonspecific symptoms of headache and dizziness. Essentially, three circumferential arteries of the vertebral–basilar circulation supply the several surfaces of the cerebellum, the arteries being named for the surface they supply:

1. **The posterior inferior cerebellar artery (PICA)** usually originates from the vertebral artery. An initial medial branch of this vessel supplies the lateral tegmental area of the medulla, and the medial and lateral branches supply the cerebellum. For a case involving the PICA, see Fig. 11.8.
2. **The anterior inferior cerebellar artery (AICA)** originates from the lower third of the basilar artery. The initial branches of this vessel supply the lateral tegmental area of the lower pons.
3. **The superior cerebellar artery** originates from the rostral basilar artery. Proximal branches supply the lateral tegmental area of the rostral pons and caudal mesencephalon.

    All three vessels have extensive leptomeningeal anastomoses over the surface of the cerebellum that are similar to the leptomeningeal anastomoses of the arteries supplying the cortical surface of the cerebral hemispheres. As a consequence, vertebral artery occlusion might result in a lateral medullary infarct and syndrome, but infarction of the cerebellum might be absent or limited. On the other hand, embolic occlusion is implicated in cerebellar vascular syndromes. Additional discussion of vascualr syndromes will be found in Chapter 17.

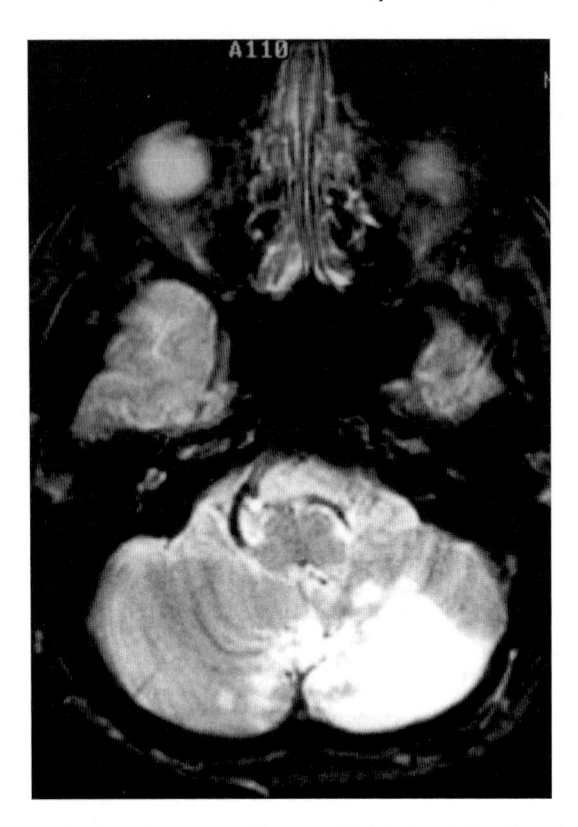

Fig. 11.8 PICA, Cerebellar infarct. This 64-year-old right-handed male woke at 3 AM with a severe posterior headache, vertigo with projectile vomiting, and slurred speech. Shortly thereafter, he had left facial paraesthesias, clumsiness of the left extremity, and diplopia. Two weeks previously he had been seen for a transient episode of difficulty in controlling the right arm and impairment of speech. Past history was significant for hypertension, coronary bypass, 6 years previously episodes of transient weakness, and numbness of the right arm with transient aphasia. He had been placed on long-term anticoagulation. After intensive rehabilitation, 6 months later he showed only slowness of alternating hand movements. T2-weighted axial sections at level of upper medulla. (From EM Marcus and S Jacobson, Integrated neuroscience, Kluwer, 2003)

6. Spinocerebellar Degenerations

The most frequent disorder is Fredreich's ataxia, which accounts for 50% of progressive hereditary ataxias. This disorder is due to an excessive number of GAA repeats on chromosome 9. It is an autosomal recessive disorder of childhood which primarily affects the spinal cord.

This extensive topic has been reviewed in Marcus and Jacobson (2003).

7. An Overview of Tremors

Tremor is defined as an involuntary movement of a body part characterized by rhythmic oscillation that develops when there is a synchronized discharge of many motor units. Before concluding our discussion of the motor system, we will briefly outline the various types of tremor. This is necessary because many common postural tremors (physiologic and essential) are mistakenly attributed to more serious disease of the basal ganglia or cerebellum.

a. Rest Tremor

This is a tremor that occurs in a body part that is not voluntarily activated and is completely supported against gravity. This tremor, particularly when a pill-rolling component is present, is almost always indicative of Parkinson's disease or of related disease of the basal ganglia.

b. Action Tremors

This is a tremor that is produced by voluntary contraction of muscle. Within this category, several subtypes may be specified:

c. Postural Tremors

These are by far the most common tremor and include the following:

1. **Physiologic Tremor** is usually fine and rapid (6–12 Hz) and occurs when at attempting to sustain a posture. This tremor is significantly increased or "enhanced" by anxiety, excessive caffeine intake, exercise, and fatigue. It is also increased by thyrotoxicosis, hypoglycemia, alcohol withdrawal, and β-adrenergic drugs, such as those used in the treatment of asthma, and by drugs commonly employed in psychiatry, such as lithium, neuroleptic, or tricyclic medications. Beta-adrenergic blockers such as propranolol are often effective against this tremor if specific etiologic factors cannot be corrected (use in asthmatics, however, is contraindicated).
2. **Essential Tremor** (also called familial, benign, or senile tremor). Aside from physiologic tremor, essential tremor (from 4 to 12 Hz) is the most common tremor encountered by the physician. Prevalence studies suggest a high frequency, particularly in the over-40 age group; 5% of the population might be affected. When familial, the genetic pattern appears to be autosomal dominant. The hands are most commonly affected, but in some patients, the head is primarily involved (side-to-side movement). The underlying pathology of essential tremor is unknown. Functional MRI studies suggest an abnormal bilateral overactivity of cerebellar connections.

d. Kinetic tremors usually approximately 5 Hz are tremors occurring during any voluntary movement. Several types have been specified:

1. *Tremor during target directed movements: intention tremor.* Amplitude increases during visually guided movements toward a target, particularly at the termination of movement. This characteristic cerebellar tremor might be decreased by lesions of the contralateral VL thalamic nucleus.
2. *Task-specific kinetic tremor.* These appear or becomes exacerbated during specific activities related to occupation or writing. Primary writing tremor and occupational tremors of musicians represent a variant of essential tremor.

Chapter 12
Somatosensory Function and the Parietal Lobe

The parietal lobe is the middle lobe of the cerebrum and is found on the lateral surface of the hemisphere; it consists of the postcentral gyrus and superior and inferior parietal lobules. In this region of the cerebrum, one finds major functional differences between the left hemisphere, which is dominant for language, and the right hemisphere, which is important for body imagery. The pain pathway has been discussed in Chapters 3 (spinal cord), 4 (brain stem), and 6 (diencephalon). In this chapter we will cover the pathways subserving tactile information for (1) the extremities, thorax, and abdomen, the posterior columns and (2) the trigeminal systems subserving tactile information from the head. The importance of these pathways is apparent when one walks in the dark or drinks or eats.

I. Postcentral Gyrus: Somatic Sensory Cortex [Primary Sensory S-I]

The postcentral gyrus is not, from a histological standpoint, homogeneous. Four subtypes can be distinguished areas 3a, 3b, 1 and 2.

A. *Organization of the Postcentral Gyrus*

Within the postcentral gyrus, four representations of the body and limb surfaces are present in a parallel manner, with each representation relatively modality specific:
  The response of the cells in the postcentral gyrus is as follows:
  Area 3a responds primarily to muscle stretch receptors.
  Area 3b responds to rapidly and slowly adapting skin receptors (as in movement of a hair or skin indentation).
  Area 1 responds to rapidly adapting skin receptors.
  Area 2 esponds to deep pressure and joint position receptors.
  Within areas 1 and 2, there are additional neurons that do not receive direct thalamic input. These neurons respond, instead, to more complex properties of the

stimulus, such as the specific direction of movement, and have been labeled complex cells area 5. Studies of cortical potentials evoked by tactile stimulation in the monkey have suggested that there is a secondary somatic sensory projection area (S-II) in addition to the classic postcentral contralateral projection area. This second area has a bilateral representation and is found partially buried in the sylvian fissure at the lower end of the central sulcus. A similar second area of representation has been reported by Penfield and Jasper (1954), Luders et al. (1985), and Blume et al. (1992) in those seizure patients in whom an abdominal sensation (aura) was followed by a sensation of paraesthesias in both sides of the mouth and in both hands. Note also that, at times, stimulation of the precentral gyrus by the neurosurgeon during surgery has produced contralateral tingling or numbness, in addition to the more frequent motor responses (Penfield and Jasper, 1954).

Areas 3a and 3b (Fig. 8.4) receive the major projection form the ventral posterior medial nucleus (VPM) of the thalamus. Some of the neurons in areas 1 and 2 receive direct input from the ventral posterior thalamic nuclei, whereas other neurons are dependent on collaterals from area 3.

B. *Postcentral Gyrus Stimulation*

Stimulation of the postcentral gyrus produces a sensation over the contralateral side of the face, arm, hand, leg, or trunk described by the patient as a tingling or numbness and labeled *paresthesias*. Less often, a sense of movement is experienced. The patient does not describe the sensation as painful. These various phenomena, occurring at the onset of a focal seizure, can be described as a somatic sensory aura. There is in the postcentral gyrus a sequence of sensory representations that, in general, is similar to that noted in the precentral gyrus for motor function (Fig.12.1). The representation of the face occupies the lower 40%; the representation of the hand occupies the middle upper 40%; and the representation of the foot occupies the paracentral lobule. As in the precentral gyrus, certain areas of the body have a disproportionate area of representation (i.e., the thumb, fingers, lips, and tongue). Those areas of the skin surface that are most sensitive to touch have not only the greatest area of cortical representation but also the greatest number and density of receptors projecting to the postcentral gyrus. The peripheral field of each of these receptors is also very small (compared, e.g., to the less sensitive skin areas of the trunk). In addition, at the lower end of the postcentral gyrus, extending into the sylvian fissure (the parietal operculum), there is found a representation of the alimentary tract, including taste. Gustatory hallucinations might arise from seizure foci in this area (Hausser-Hauw and Bancaud, 1987). There is also a representation of the genitalia in the paracentral lobule, and a rare patient with seizures beginning in this area had reported paroxysmal sexual emotions, including orgasm and nymphomania (Calleuja et al. 1989). Microelectrode techniques for recording from single cells in the cerebral cortex of the monkey and cat have allowed a considerable elaboration of the functional localization within the sensory cortex. The studies of Mountcastle (1957) indicated a vertical columnar

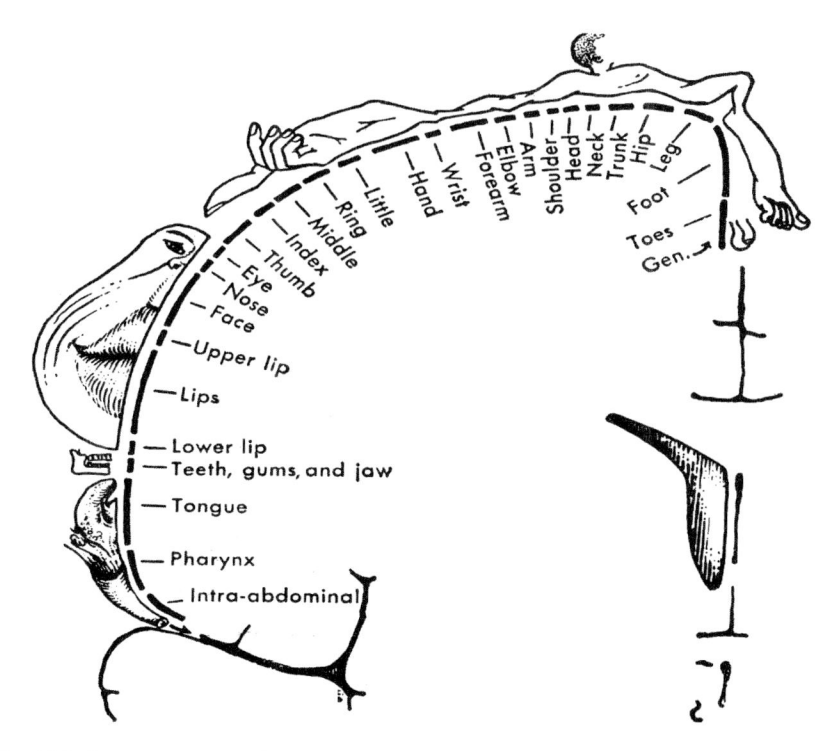

Fig. 12.1 Sensory representation as determined by stimulation studies on the human cerebral cortex at surgery. Note the relatively large area devoted to lips, thumb, and fingers. (From W Penfield and T. Rasmussen, The cerebral cortex of man, Macmillan, 1955, p. 214)

organization from the cortical surface to the white matter. Although each column is modality-specific, each neuron in a particular column is activated by that specific sensory modality (e.g., touch, movement of a hair, deep pressure, joint position). Jones et al. (1982), Kasdon and Jacobson (1979), Jones and Powell (1970), and Killackey and Ebner (1973) discussed the thalamocortical relationships.

C. Postcentral Gyrus Lesions

Immediately following complete destruction of the postcentral gyrus, there will often be found an almost total loss of awareness of all sensory modalities on the contralateral side of the body. Within a short time, there is usually a return of some appreciation of painful stimuli. The patient will, however, often continue to note that the quality of the painful stimulus differs from that on the intact side. An awareness of gross pressure, touch, and temperature also returns. Vibratory sensation might return to a certain degree. Certain modalities of sensation, however, never return or return only to a minor degree. (In partial lesions of the postcentral gyrus, however, these various modalities often return to a variable degree). These modalities are often referred to as the cortical modalities of sensation or as discriminative modalities of sensation. In contrast, the sensory modalities of pain, gross touch, pressure, temperature, and vibration are referred to as primary modalities of sensation. Perception of these modalities continues to occur after ablation of the postcentral gyrus, although some alteration in quality of sensation is noted. It has been assumed that the anatomical substrate for such awareness must exist at the thalamic level. In terms of more specific localization of specific modalities of cortical sensation, the studies of Randolph and Semmes (1974) suggest that in the monkey, small lesions restricted to area 3b (hand region) produce deficits in discrimination of texture, size, and shape. Lesions in area 1 interfere with the ability to discriminate texture only; lesions in area 2 interfere with the ability to discriminate size and shape.

The following types of sensory awareness usually included in this category are summarized in Table 12.1.

The following case histories demonstrate the types of sensory phenomenon found in disease involving the postcentral gyrus.

Table 12.1 Modalities of sensation in postcentral gyrus

Cortical discriminative sensory modalities	Type of stimulus
1. Position sense	Ability to perceive movement and the direction of movement when the finger or toe is moved passively at the interphalangeal joint, the hand at the wrist, or the foot at the ankle.
2. Tactile Localization	Ability to accurately localize the specific point on the body or extremity that has been stimulated.
3. Two-point discrimination	Ability to perceive that a double stimulus with a small separation in space has touched a given area on the body, limb, or hand.
4. Stereognosis	Ability to distinguish the shape of an object and thus to recognize objects based on their three-dimensional tactile form.
5. Graphesthesia	Ability to recognize numbers or letters that have been drawn on the fingers, hand, face, or leg.
6. Weight discrimination	Ability to recognize differences in weight placed on the hand or foot.
7. Perception of simultaneous stimuli	Ability to perceive that both sides of the body have been simultaneously stimulated. When bilateral stimuli are presented but only a unilateral stimulus is perceived, "extinction" is said to have occurred.
8. Perception of texture	Ability to perceive the pattern of texture surface stimuli encountered by the moving tactile receptors.

Case History 12.1 (Fig. 12.2)

One week prior to evaluation, a 40-year-old right-handed married white male had the onset of repeated 15–21-minute duration episodes of focal numbness (tingling or paresthesias) involving the left side of the face, head, ear, and posterior neck and occasionally spreading into the left hand and fingers and rarely subsequently spreading into the left leg. Biting or eating would trigger the episodes. There was no associated pain or headache or weakness or associated focal motor phenomena.

Initial neurological examination: Totally intact with regard to mental status, cranial nerves, motor system, reflexes, and sensory system

Clinical diagnosis: Focal seizures originating in the lower third postcentral gyrus, of uncertain etiology.

Laboratory data: Computed tomography (CT) scan: The enhanced study demonstrated a small enhancing lesion just above the right sylvian fissure.

Magnetic resonance imaging (MRI) (Fig.12.2): Demonstrated the more extensive nature of this process involving the operculum (above the right sylvian fissure

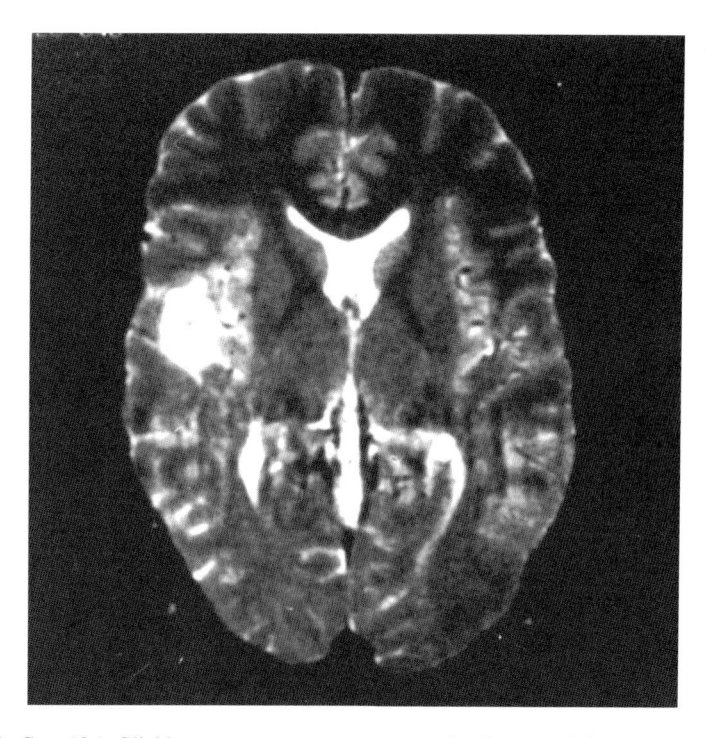

Fig. 12.2 Case 12.1. Glioblastoma, somatosensory cortex; focal sensory left facial seizures. MRI scans demonstrating the more extensive involvement of the cortex and white matter: horizontal (T2) (From EM Marcus and S Jacobson, Integrated neuroscience, Kluwer, 2003)

including the lower end of the post central gyrus, the area of representation of the face). A probable infiltrating tumor was suggested as the most likely diagnosis.

Arteriogram: Demonstrated a tumor blush at the right sylvian area, consistent with a glioblastoma, a rapidly growing infiltrating tumor originating from astrocytes.

Subsequent course: Treatment with an anticonvulsant reduced the frequency of the episodes. Neurosurgical consultation suggested an additional observation period and periodic CT scans. Three months after the onset of symptoms, the patient developed difficulty in coordination of the left arm. Examination now demonstrated left central facial weakness, mild weakness of the left hand, and, in the left hand, a loss of stereognosis and graphesthesia with a relative alteration in pinprick perception. The CT scan showed significant enlargement of the previous lesion. Dr. Bernard Stone performed a subtotal resection of a glioblastoma, in the right temporal parietal area. Radiation therapy was administered following surgery, 4000 rads total to the whole brain and 5210 rads total to the tumor area. Reevaluation 3 months after surgery indicated only rare focal seizures involving the face. The neurological examination was normal. Repeat CT scans, however, continued to demonstrate a large area of enhancing tumor in the right temporal–frontal–parietal area. Despite the use of dexamethasone and arterial chemotherapy with cis-platinum, he continued to progress with the development of a left field defect, left hemiparesis, and a loss of all modalities of sensation on the left side. Subsequently, he developed significant changes in memory and cognition. The CT scan demonstrated progression with extensive involvement of the right side of the brain and spread to the left hemisphere. Death occurred 33 months after onset of symptoms. At postmortem examination of the brain, almost the entire right hemisphere was replaced by a necrotic infiltrating tumor. The tumor now had spread into the corpus callosum and temporal and occipital lobes. The internal capsule, thalamus, hypothalamus, and basal ganglia were destroyed.

The following case presents a patient with episodic pain from a tumor in the left postcentral gyrus.

Case History 12.2

A 62-year-old white, right-handed housewife, 6 years prior to admission had undergone a left radical mastectomy for carcinoma of the breast. Five months prior to admission, the patient developed a persistent cough, left pleuritic pain, and a collection of fluid in the left pleural space (pleural effusion). She now had the onset of daily headaches in the right orbit, occasionally awakening her from sleep. Four months prior to admission, over a 3–4-week period, the patient developed progressive "weakness" and difficulty in control of the right lower extremity. Several weeks later, the patient now experienced, aching pain in the right index finger in addition to a progressive deficit in the use of the right hand. This was more a "stiffness and in coordination" than any actual weakness. One month prior to admission, the patient noted episodes of pain in the toes of the right foot and numbness of the right foot occurred, lasting 2–3 days at a time. She subsequently developed difficulty in memory and some minor language deficit, suggesting a nominal aphasia, prompting hospital admission.

Neurological examination: Mental status: Slowness in naming objects was present, although she missed only one item of six. Delayed recall was slightly reduced to three out of five in 5 minutes.

Cranial nerves: Early papilledema was noted: absence of venous pulsations, indistinctness of disc margins, and minimal elevation of vessels as they passed over the disc margin. A right central facial weakness was present.

Motor system: Mild weakness of the right upper limb was present. A more marked weakness was present in the lower limb (most marked distally). Spasticity was present at the right elbow and knee. In walking, the right leg was circumducted; the right arm was held in a flexed posture.

Reflexes: Deep tendon reflexes were increased on the right. A right Babinski response was present.

Sensation: An ill-defined alteration in pain perception was present in the right arm and leg—more of a relative difference in quality of the pain than any actual deficit. Repeated stimulation of the right lower extremity (pain or touch) produced dysesthesias (painful sensation). Position sense was markedly defective in the right fingers, with errors in perception of fine- and medium-amplitude movement in the right toes. Simultaneous stimulation resulted in extinction in the right lower extremity. Graphesthesia (identification of numbers, e.g., 8, 5, 4, drawn on cutaneous surface) was absent in the right hand and fingers and was poor in the right leg. Tactile localization and two-point discrimination were decreased on the right side.

Clinical diagnosis: Metastatic tumor to the left post central and precentral gyri with a cortical pain syndrome (pseudo-thalamic pain syndrome).

Laboratory data: Chest X-ray: Several metastatic nodules were present in the left lung. Imaging studies were consistent with a focal metastatic lesion in the parasagittal upper left parietal area.

Subsequent course: Because there was evidence in this case that the disease had spread to multiple organs, radiation and hormonal therapy were administered rather than any attempt at surgical removal of the left parietal metastatic lesion. Temporary improvement occurred in motor function in the arm, but the patient eventually expired 5 months after admission. Autopsy performed by Dr. Humphrey Lloyd of the Beverly Hospital disclosed extensive metastatic disease in the lungs, liver, and lymph nodes and a single necrotic metastatic brain lesion in the upper left post central of the gyrus parietal area, 1.0 cm below the pial surface and measuring $1.2 \times 1.0 \times 0.6$ cm. The occurrence of episodic pain in the involved arms and legs along with the production of an experienced painful sensation on repetitive tactile stimulation (dysesthesias) in patients with sensory pathway lesions is sometimes referred to as a "thalamic" or "pseudo thalamic" syndrome.

Comments: In this case, the anatomical locus for the pseudo-thalamic syndrome is apparent. The effects of deficits in cortical sensation on the total sensory motor function of a limb are apparent in this case. The actual disability and disuse of the right arm and leg were far out of proportion to any actual weakness. Such an extremity is often referred to as a "useless limb." The actual weakness that was present undoubtedly reflected pressure effects on the precentral gyrus and the descending motor fibers in adjacent white matter.

Destructive lesions of the postcentral gyrus during infancy or early childhood often produce a retardation of skeletal growth on the contralateral side of the body. Such a patient examined as an adolescent or adult will be found to have not only cortical sensory deficits in the contralateral arm and leg but also a relative smallness of these extremities (shorter arm or leg, smaller hand and glove size, smaller shoe size).

II. Superior and Inferior Parietal Lobules

The superior parietal lobule is composed of Brodmann's cytoarchitectural areas 5 and 7, the inferior parietal lobule of areas 40 and 39 (the supramarginal and angular gyri). All of these areas can be classified as varieties of the homotypical cerebral cortex. The major afferent input of the superior parietal lobule area 5 in the monkey is from the primary sensory areas of the postcentral gyrus. Area 7 of the monkey cortex receives indirect cortico-cortical connections. The superior parietal lobule receives projections from the posterior lateral nuclei of the thalamus; the inferior parietal lobule receives projections from the pulvinar (see Chapter 6). These secondary sensory areas project to an adjacent tertiary sensory area and then to a multimodal sensory association area at the temporal parietal junctional area. This latter area in the posterior parietal cortex then projects to another multimodal sensory association area in the frontal premotor and frontal eye fields (discussed earlier in relationship to premotor and prefrontal motor function). Area 7 has connections to the limbic cortex: cingulate gyrus (Mesulum et al. 1977). The implications of these connections for the integration of complex movements are clear. It is not surprising then that single-cell studies as reviewed by Darian-Smith et al. (1979) demonstrate responses of neurons in area 5 to manipulation of joints, hand manipulation as in grasping and manipulation objects, or in projecting the hand to obtain a specific object associate with reward. When damage to parietal cortex occurs, the patient manifests a variety of spatial deficits. A neglect occurs in relation to multiple sensory modalities but is most prominent with regard to contralateral visual space, as we will discuss below. There are corresponding deficits in the generation of spatially directed actions.

A. *Stimulation*

The threshold of the superior and inferior parietal lobules for discharge is relatively high. Although Foerster (1936) reported the occurrence of some contralateral paraesthesias on stimulation in man; Penfield did not confirm these results. Stimulation in the inferior parietal areas of the dominant hemisphere did produce arrest of speech, but this is a nonspecific effect, occurring on stimulation of any of the speech areas of the dominant hemisphere. It must, of course, be noted that space-occupying lesions in the parietal lobules might produce sensory or motor seizures by virtue of their pressure effects on the lower threshold postcentral and precentral gyri.

B. Lesions

Darian-Smith et al. (1979) have summarized the effects of selective lesions of area 5 and or area 7 in the monkey. The most selective studies were those of Stein (1977)using reversible cooling lesions. Cooling of area 5 resulted in a clumsiness of the contralateral arm and hand so that the animal was unable to search for a small object. Cooling of area 7 produced a much more complex clumsiness, which was apparent only when the arm was moved into the contralateral visual field. Movements of the hand to the mouth were intact. Earlier clinical studies in humans had suggested possible sensory deficits in relation to ablation of the inferior or superior parietal areas. The detailed studies of Corkin et al. (1964) on patients subjected to limited cortical ablations of the inferior or superior parietal areas (in the treatment of focal epilepsy) clearly indicated that no significant sensory deficits occurred. The postcentral gyrus rather than the parietal lobules is critical for somatic sensory discrimination. The more recent studies of Pause et al. (1989) confirmed that the anterior parietal (postcentral lesions) resulted in somatosensory disturbances, including surface sensibility, two-point discrimination, position sense, as well as more complex tactile recognition. In contrast, in posterior parietal lesions, there was a preferential impairment of complex somatosensory and motor functions involving exploration and manipulation by the fingers; this was not explained by any sensory deficit. There was a deficit in the conception and execution in the spatial and temporal patterns of movement. The end result is an impairment of purposive movement as discussed earlier. The response to visual stimuli in terms of attending to and reaching for objects is impaired (Nagel-Leiby et al. 1990). Lesions involving the white matter deep to the inferior parietal lobule might damage the superior portion of the optic (geniculocalcarine) radiation, producing a defect in the inferior half of the contralateral visual field, an inferior quadrantanopsia.

C. Parietal Lobules in the Dominant Hemisphere

Destructive lesions of the parietal lobules produce additional effects on more complex cortical functions. Those lesions involving particularly the supramarginal and angular gyri of the dominant (usually the left hemisphere) inferior parietal lobule might produce one or more of a complex of symptoms and signs known as Gerstmann's syndrome:

(a) *Dysgraphia* (a deficit in writing in the presence of intact motor and sensory function in the upper extremities)
(b) *Dyscalculia* (deficits in the performance of calculations)
(c) Left-right confusion
(d) Errors in finger recognition (e.g., middle finger, index finger, ring finger) in the presence of intact sensation (finger agnosia).

In addition, disturbances in the capacity for reading might be present. Some patients might also manifest problems in performing skilled movements on command (an apraxia) at a time when strength, sensation, and coordination are intact. Usually, only partial forms of the syndrome are present. The problem of the dominant parietal lobe in language function will be considered in greater detail and illustrated in the section on language and aphasia. The reader should refer to Chapter 15 for an illustration.

D. Parietal Lobules in the Nondominant Hemisphere

1. Disturbance in concept of body image (neglect) and denial of illness

Patients with involvement of the nondominant parietal lobe, particularly the inferior parietal lobule, often demonstrate additional abnormalities in their concepts of body image, in their perception of external space, and in their capacity to construct drawings.

These patients might have the following deficits: (1) a lack of awareness of the left side of the body, with a neglect of the left side of the body in dressing, undressing, and washing; (2) despite relatively intact cortical and primary sensation, the patient might fail to recognize his arm or leg when this is passively brought into his field of vision; and (3) yhe patient might have a lack of awareness of a hemiparesis despite a relative preservation of cortical sensation (anosognosia) or might have a total *denial of illness*. At times, this might be carried to the point of attempting to leave the hospital because as far as the patient is concerned; there is no justification for hospitalization. The denial of illness undoubtedly involves more than perception of body image. The disturbance in perception of external space can take several forms: N*eglect* of the left visual field and of objects, writing, or pictures in the left visual field. At times, this is associated with a dense defect in vision in the left visual field; at other times, there may be no definite defect for single objects in the left visual field, but extinction occurs when objects are presented simultaneously in both visual fields. Again, the problem of possible involvement of occipital cortex or of subcortical optic radiation should be considered as previously noted.

2. Inability to Interpret Drawings

The patient has difficulty with a map or in picking out objects from a complex figure. The patient is confused in figuring background relationships and is disoriented in attempting to locate objects in a room. When asked to locate cities on an outline map of the United States, the patient manifests disorientation as to the west and east coasts and as to the relationship of one city to the next. Chicago and New Orleans might be placed in the Pacific Ocean, Boston on the Florida peninsula, and New York City somewhere west of the Great Lakes. The disturbance in capacity for the construction of drawings has been termed a constructional apraxia or dyspraxia. An apraxia can be defined as an inability to perform a previously well-performed act at a time when voluntary movement, sensation, coordination, and understand are otherwise all intact. The following deficits might be present: The patient might be unable to draw a house or the face of a clock, or to copy a complex figure such as a three-dimensional cube, a locomotive, and so forth; in severe disturbances, the patient might be unable to copy even a simple square, circle, or triangle. The following case demonstrates many of these features.

Case History 12.3

A 70-year-old, single, white female, right-handed, underwent a left radical mastectomy for carcinoma of the breast, 3 years prior to admission. Four months prior to evaluation, the patient became unsteady with a sensation of rocking as though on a boat. She no longer attended to her housekeeping and to dressing. Over a 3-week period, prior to evaluation, a relatively rapid progression occurred, with deterioration of recent memory. A perseveration occurred in motor activities and speech. The patient was incontinent but was no longer concerned with urinary and fecal incontinence. For 2 weeks, right temporal headaches had been present. During this time, her sister noted the patient to be neglecting the left side of her body. She would fail to put on the left shoe when dressing. In undressing, the stocking on the left would be only half removed.

Family history: The patient's mother died of metastatic carcinoma of the breast.

Neurological examination: Mental status: Intact except for the following features: The patient often wandered in her conversation. She often asked irrelevant questions and was often impersistent in motor activities. There was marked disorganization in the drawing of a house or of a clock. A similar marked disorganization was noted in attempts at copying the picture of a railroad engine (Fig. 12.3). There was a marked neglect of the left side of space and of the left side of the body. The patient failed to read the left half of a page. When she put her glasses on, she did not put the left bow over the ear. When getting into bed, she did not move the left leg into bed. She had slipped off her dress on the right side, but was lying in bed with the dress still covering the left side. The patient had been reluctant to come for neurological consultation. Although she complained of headache and nausea, she denied any other deficits. Her relatives provided information concerning these problems. Much additional persuasion over a 2-week period was required before the patient would agree to be hospitalized.

Cranial nerves: A dense left homonymous hemianopsia was present. When reading, the patient left off the left side of a page. She bisected a line markedly off center.

Fig. 12.3 Case 12.3. Nondominant parietal constructional apraxia. The patient's attempts to draw a house are shown on the upper half of the page. Her attempts to copy a drawing of a railroad engine are shown on the lower half of the page (B and C). The examiner's (Dr. Leon Menzer) original is designated (A). (From EM Marcus and S Jacobson, Integrated neuroscience, Kluwer, 2003)

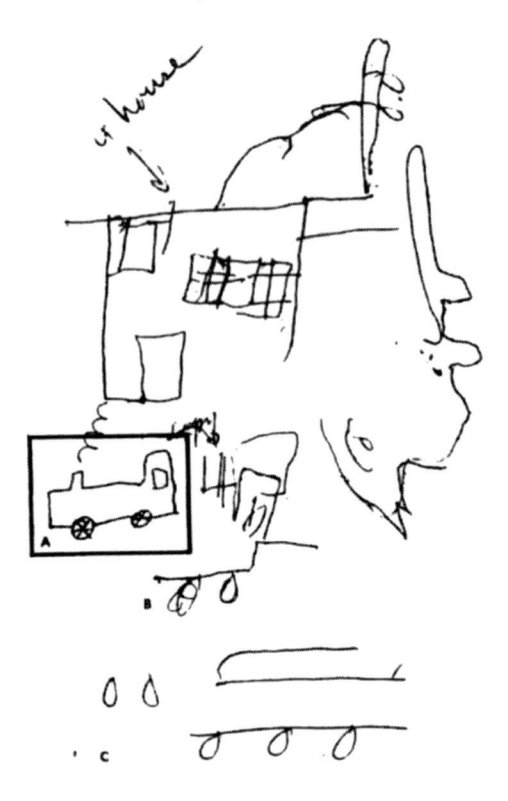

Disc margins were blurred and venous pulsations were absent, indicating papilledema. The right pupil was slightly larger than the left. A minimal left central facial weakness was present.

Motor system: Although strength was intact, there was little spontaneous movement of the left arm and leg. The patient was ataxic on a narrow base with eyes open with a tendency to fall to the left, and she was unable to stand with eyes closed even on a broad base.

Sensation: Although pain, touch, and vibration were intact, at times there was a decreased awareness of stimuli on the left side. Errors were made in position sense at toes and fingers on the left. Tactile localization was poor over the left arm and leg. With double simultaneous stimulation, the patient neglected stimuli on the left face, arm, and leg.

Clinical diagnosis: Metastatic breast tumor to right nondominant parietal cortex.

Laboratory data: Chest X-ray indicated a possible metastatic lesion at the right hilum. EEG was abnormal because of frequent focal 3–4 cps slow waves in the right temporal and parietal areas, suggesting focal damage in these areas.

Subsequent course: Treatment with steroids (dexamethasone and estrogens) resulted in temporary improvement. The patient refused surgery. Her condition soon deteriorated with increasing obtundation of consciousness. She expired 2 months following her initial neurological consultation.

Imaging Studies. A CT scan from a more recent case demonstrating many aspects of this syndrome is illustrated in Figure 12.4. In many cases, the location of a lesion might appear to be predominantly posterior temporal. Such large posterior temporal lesions would certainly compromise the cortex and subcortical white matter of the adjacent inferior parietal area. In this case, marked deficits in perception of the cortical modalities of sensation were present. In other cases, as in the case demonstrated in Figure 12.4, such involvement is much less marked. In some cases, involvement of the motor cortex is evident with an actual left hemiparesis accompanied by an increase in deep tendon reflexes and an extensor plantar response. At times, in patients with neglect syndromes, there might be several indications in the clinical examination and in the laboratory studies that the involvement of the frontal lobe areas is more prominent than the parietal involvement. We have already indicated that the neglect components of this syndrome might also be noted in lesions of the anterior premotor area (area 8). The prefrontal and premotor areas as discussed earlier and in Chapter 10 receive projection fibers from the multimodal area of the posterior parietal area. It is possible that in some cases the posterior temporal–inferior parietal location of the lesion might also be critical in interrupting these association fibers.

For these several reasons, it is perhaps more appropriate to use the term, syndrome of the nondominant hemisphere, rather than the more localized designation, nondominant inferior parietal syndrome. Duffner et al. (1990) have presented the concept of a network for directed attention—with right frontal lesions leading to left hemispatial neglect only for tasks that emphasize exploratory-motor components of directed attention, whereas parietal lesions emphasize the perceptual-sensory

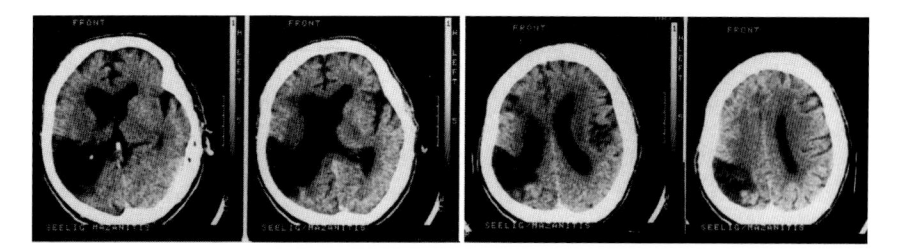

Fig. 12.4 Nondominant parietal lobe syndrome. CT scans. This 68-year-old left-handed male with diabetes mellitus and hypertension had the sudden onset of left arm paralysis, loss of speech, and ability to read and write, all of which recovered rapidly. On examination 9 months after the acute episode, he continued to have the following selective deficits: (1) he was vague in recalling his left hemiparesis; (2) in dressing, he reversed trousers and failed to cover himself on the left side, (3) he had extinction in the left visual field on bilateral simultaneous visual stimulation and over the left arm and leg on bilateral simultaneous tactile stimulation; (4) a left Babinski sign was present. CT scan now demonstrated an old cystic area of infarction in the right posterior temporal-parietal (territory of the inferior division of the right middle cerebral artery). (From EM Marcus and S Jacobson, Integrated neuroscience, Kluwer, 2003)

aspects of neglect. With lesions of the nondominant hemisphere, there is a significant alteration of the patient's awareness of his environment. The behavior of an individual is in part determined by his own particular perception of the environment If that perception is altered or disorganized, the behavioral responses of the patient might appear inappropriate to others. Obviously, not all individuals will respond in the same manner to a given environmental situation; part of the response will be determined by the past experience and personality of the individual. Thus, given the same lesion, one individual might be unaware of a hemiparesis, another might deny the hemiparesis but agree an illness is present, and a third might claim to be healthy and claim that people are conspiring to keep him in the hospital.

III. Parietal Lobe and Tactile Sensation from the Body

A. Basic Principle of Sensory System

All sensory systems have 3 levels of neuronal organisation:
First-order neurons found in the periphery: the dorsal root ganglion or ganglion of cranial nerve V, VII, IX, or X. Its axons are ipsilateral (Table 12.2).
Second-order neurons within the central nervous system in the spinal cord or brain stem; its axon crosses to the contralateral side.
Third-order neurons within the thalamus form the final subcortical neurons in the sensory systems and its axons project and synapse in the cerebral cortex.

B. Tactile Sensation from the Body – Medial Lemnsicus (Fig. 12.5)

The ascending sensory fibers from the neck, trunk and extremeties for touch (posterior columns), from the face for touch and pain (trigeminothalamic), and from the viscera for general and special sensation ascend in the medial lemniscus to the thalamus.

Table 12.2 Mechanicoreceptor

Modality	Receptor
Sound	Cochlea in inner ear of temporal bone transforms mechanical into neural impulses
Light touch and vibration	Encapsulated endings: Meissner's and Pacinian corpuscles
Proprioception: Encapsulated sensory endings	Muscle spindles and Golgi tendon organs in joints, Meissner's and Pacinian corpuscles, Merkel's tactile discs
Pain and temperature	Free nerve endings, end bulbs of Krausse and Golgi–Mason.

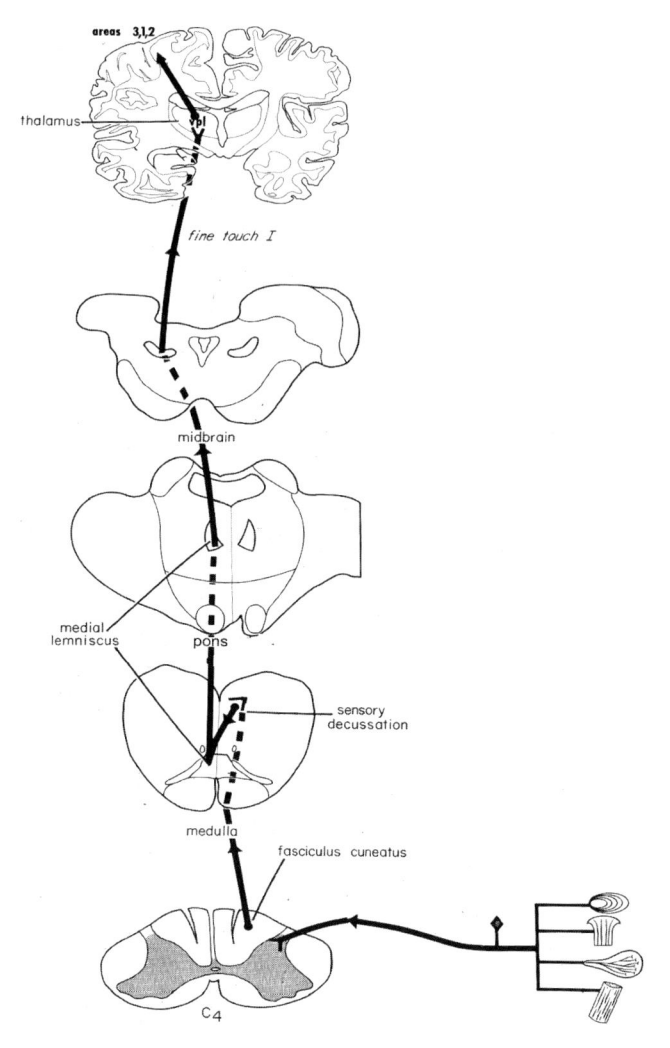

Fig. 12.5 Tactile sensation from the upper extremity, fasciculus cuneatus. Posterior columns → medial lemniscus →VPL → postcentral gyrus

1. Posterior Columns: Tactile Sensation from the Neck, Trunk and Extremities (Fasciculus Gracilis and Cuneatus)

The posterior columns—the fasciculus gracilis, and the fasciculus cuneatus (Fig. 12.5)—conduct proprioception (position sense), vibration sensation, tactile discrimination, object recognition, deep touch (pressure) awareness, and two-point discrimination from the neck, thorax, abdomen, pelvis, and extremities. The sensory receptors for the system are the Golgi tendon organs, muscle spindles, proprioceptors, tactile discs, and Pacinian corpuscles (deep touch or pressure). The primary cell body is located in the dorsal root ganglion. The well-myelinated fibers of this system enter the spinal cord as the medial division of the dorsal root and bifurcate into ascending and descending portions, which enter the dorsal column. In the spinal cord, the posterior columns are uncrossed and divided into the medial gracile fasciculus (lower extremity) and the lateral cuneate fasciculus (upper extremity). The fasciculus gracilis contains fibers from the sacral, lumbar, and lower thoracic levels, whereas the fasciculus cuneatus contains fibers from the upper thoracic and cervical levels. Fibers from the sacral levels lie most medial, followed by lumbar, thoracic, and, finally, cervical fibers. Fibers from the upper extremity form 50% of the posterior columns, with the lower extremity 25% and the remainder from the thorax and abdomen. The primary axons ascend in the dorsal columns of the spinal cord to the secondary cell body of this system located in the nucleus gracilis and the nucleus cuneatus in the narrow medulla and the spinomedullary junction. The secondary fibers cross in the sensory decussation and form the bulk of the medial lemniscus along with trigeminal and other fibers and ascend in the contralateral medial lemniscus to the ventral posterior lateral nucleus (VPL) in the thalamus. From this thalamic nucleus, these fibers are projected to the postcentral gyrus (areas 1, 2, and 3). Fibers also descend in the dorsal columns, but their functional significance is unknown. The fibers responsible for proprioception cross in the medial lemniscus. The fibers for vibration sensation and tactile discrimination ascend bilaterally in the medial lemniscus to the ventral posterior lateral nucleus. Consequently, a unilateral lesion can abolish proprioception, but tactile discrimination and vibration sensation will not be entirely lost.

2. Clinical Lesions

Injury to the posterior column appears not to affect pressure sense, but vibration sense, two-point discrimination, and tactile discrimination are diminished or abolished, depending on the extent of the lesion. Interruption of the medial fibers (cervical-hand region) affect the ability to recognize differences in the shape and weight of objects placed in the hand. Because the extremities are more sensitive to these modalities than any other body regions, position sense is impaired more severely in the extremities than elsewhere, and the person has trouble identifying small passive movements of the limbs. Consequently, performance of voluntary acts is impaired and movements are clumsy (sensory ataxia). Lesions in lateral fibers of the posterior column, gracilis, might be devastating due to the interruption of one of the most important sensory mechanisms: the ability to detect the sole of the foot. This is a major handicap in walking in a dimly lit or a dark room, in driving a car, and so forth.

C. Tactile Sensation from the Head (Fig. 12.6)

Cranial nerve V is the largest cranial nerve in the brain stem and it has three sensory nuclei (chief, spinal, and mesencephalic) and three divisions: ophthalmic (V1),

III Parietal Lobe and Tactile Sensation from the Body

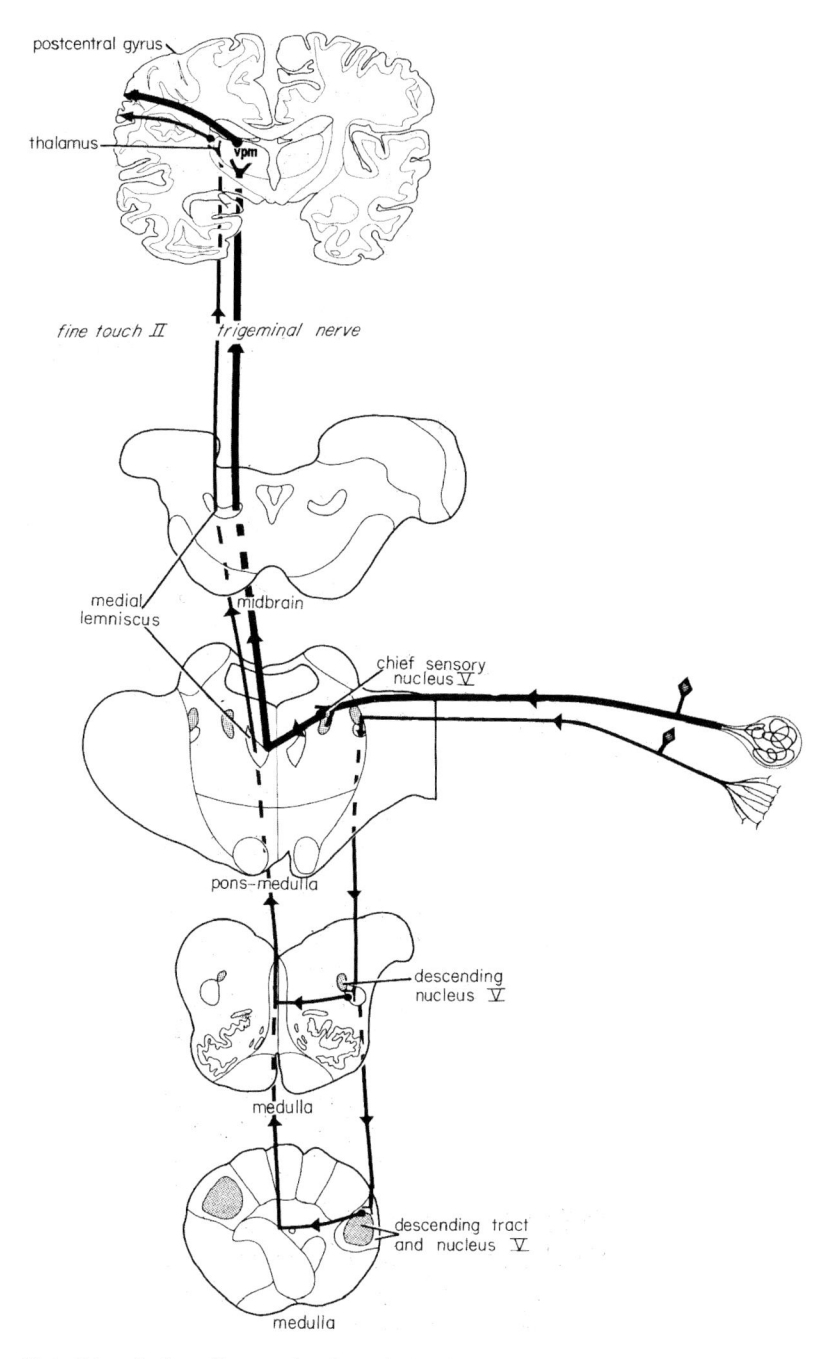

Fig. 12.6 Trigeminal: tactile sensation from the head: trigeminal lemniscus → medial lemniscus → VPM → postcentral gyrus

maxillary (V2), and mandibular (V3). The primary cells of the trigeminal nerve are located in the trigeminal ganglion in Meckel's cave in the middle cranial fossa. The second-order nuclei are in the brain stem and upper cervical spinal cord. Each of the three divisions bring in the sensation of pain, temperature, touch, and pressure from receptors in the skin, muscles, and sinuses that they innervate.

1. Proprioception from the Head: Mesencephalic Nucleus

These unique primary cell bodies are located not only in the trigeminal ganglion in the middle cranial fossa but also along the nerve rootlet within the pons and midbrain, where they form the mesencephalic nucleus of nerve V. The proprioceptive fibers arise from the muscles in the face, including the muscles of mastication, facial expression, ears, and the muscles of the eye The primary axons project to the motor nucleus of nerve V in the pons and into the reticular formation. Axons are also projected to the cerebellum and inferior olive.

2. Pain and Temperature: Spinal Descending Nucleus of Nerve V

The origin of the second-order neuron is the descending nucleus of nerve V. The secondary axons ascend in the dorsal portion of the medial lemniscus to the ventral posterior medial nucleus of the thalamus. (See Chapter 4.)

3. Tactile Sensation: Chief/Main Nucleus of Nerve V

The primary tactile fibers synapse within the chief sensory nucleus of nerve V the second-order nucleus of nerve V in the pons. The primary tactile and proprioceptive axons synapse in the chief nucleus and from this second-order neuron; the information ascends in the dorsalmost portion of the medial lemniscus, the trigeminal lemniscus, and synapse in the ventral posterior medial nucleus of the thalamus This information is then sent onto the lower third of the postcentral gyrus, the region that contains the somatotopic representation for the head and neck, including the face (Fig. 12.1). From this second-order nucleus axons also enter the brain stem, ascend bilaterally through the pons, and synapse in the midbrain on the mesencephalic nucleus of nerve V, from which fibers then descend onto the motor nucleus of nerve V in the pons for the "jaw jerk."

Chapter 13
Visual System and Occipital Lobe

The eye is the organ of vision and it is the only receptor that is actually part of the central nervous. The optic nerve grows out from the brain and enters the retina. The muscles that move the eye are attached to the outer surface of the eye. Of all our senses, vision is the most important: We perceive the world mostly through our eyes. Even though light intensity varies by a factor of 10 million between the brightest snowy day and a starlit night, our eyes and visual system adapt to these intensity changes. We can discriminate between thousands of hues and shades of color. Our eyes are set in our heads in such a way that each eye sees almost the same visual field, making depth perception possible. In the visual system, the primary, secondary, and tertiary neurons are in the retina and are all part of the central nervous system. The right field of vision projects to the left cerebral hemisphere; the left field of vision projects to the right cerebral hemisphere.

I. Structure of the Eye

A. Anatomy of the Eye

The anatomy of the receptor organ, the eye is shown in Figure 13.1. It has three layers, or tunics: (1) The outer fibrous tunic (cornea and sclera), (2) the middle vascular and pigmented tunic (choroids, ciliary body, iris, and pupil), and (3) the inner neural tunic (retina with pigmented epithelium and neuronal layers).

1. Outer Fibrous Tunic

This layer consists of two parts: the anterior transparent cornea and the posterior fibrous white sclera. The recti and oblique muscles that move the eye are attached to the sclera.

The cornea is the window to the world; it allows light rays to enter the eye. Most of the refraction needed to focus light rays on the retina occurs at the air–cornea junction. The sclera forms the white of the eye and the rest of the outer covering.

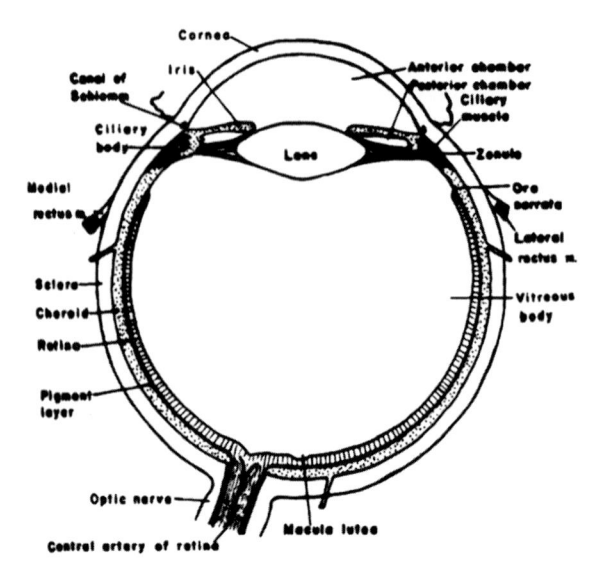

Fig. 13.1 Horizontal meridional section of the eye. (From Leeson and Lesson, Histology, Saunders)

2. Middle Tunic (Vascular and Pigmented)

This layer contains the choroid, the ciliary body, and the iris. The middle tunic is richly vascular and provides oxygen and nutrients to the inner, photoreceptor layer and the retina.

a. Choroid

The choroid is vascular and pigmented and forms the posterior portion of this tunic. Its inner portion is attached to the pigmented layer of the retina.

b. Ciliary Body

The ciliary body is found in the anterior portion of the middle tunic and consists of a vascular tunic and the ciliary muscle. The ciliary body surrounds the lens and consists of a vascular tunic (the ciliary muscle) and the suspensory ligaments (the zonule), which suspend the lens (Fig. 13.1). The ciliary muscles consist of meridional and circular fibers. The meridional fibers are external to the circular fibers. The ciliary muscles are the muscles used in accommodation, focusing on near objects.

c. Lens

The lens separates the anterior chamber from the vitreous body and completes the refraction of the entering light. A fibrous network, the zonule, suspends the lens.

I Structure of the Eye

For distance vision, the fibers are taut and the anterior surface of the lens is pulled flat. For close vision, the ciliary muscle contracts and the dilator pupillae dilate. When the circular muscles, the sphincter pupillae, contract, the iris is drawn together and the pupil constricts, such as purse strings close the mouth of a purse. The second set of muscles, the dilator pupillae, are radial and draw the iris back toward the sclera.

d. Pupillary Muscles

Two sets of muscles in the iris (rainbow) control the size of the pupil: sphincter and dilator pupillae.

The *sphincter pupillae* are supplied by the parasympathetic nervous system via the ciliary nerves, the fibers of which run together with nerve III. Transmission by this pathway is cholinergic.

The *dilator pupillae* are supplied by the sympathetic nervous system via the superior cervical ganglion. Neurotransmission in this pathway is alpha-adrenergic via the superior cervical ganglion and this is an alpha-adrenergic system. The pars ciliaris and iridica is primarily a pigmented region on respectively the ciliary body and iris. The pars optica contains the photoreceptor cells and will be the focus of our discussion.

The neural pathway arises in the occipital lobe and the final efferent fibers run together with nerve III. In looking at distant objects, the pupil dilates, and in focusing on nearby objects, the pupil constricts.

e. Pupillary Reflexes

When light is flashed into either eye, the pupil constricts the *light reflex*. When the eye focuses on close objects, accommodation occurs and the pupil constricts. Two sets of muscles in the iris control the size of the pupil: sphincter and dilator of the pupil.

Light Reflex. In dim light, the aperture of the pupil increases (dilator muscle), whereas in bright light, the aperture of the pupil decreases (sphincter muscle). The afferent fibers arise in the retina and travel with the optic nerve and optic tract; these fibers pass through the lateral geniculate and synapse in the pretectal region of the midbrain. Neurons of the pretectal nuclei project bilaterally to the Edinger–Westphal nucleus in the tegmentum of the midbrain.

Accommodation. This reflex changes the refractive power of the lens by the ciliary muscle contracting and decreasing the force on the suspensory ligament of the lens; the lens assumes a more rounded appearance with a shorter focal length.

Fixed Pupil. In the absence of trauma to the eye, a dilated pupil that does not respond to light (fixed) is usually a sign of pressure on the third nerve. This is typically from supratentorial mass lesions that have produced herniation of the mesial segments of the temporal lobe through the tentorium. The fibers go slack, and the anterior surface becomes more convex; the eye accommodates. Compression of the third nerve and of the midbrain occurs. Progressive rostral–caudal damage to the brain stem then evolves. This problem, when first detected, must be treated as a neurosurgical emergency.

Visual Acuity. A decreased ability to focus on close objects is a normal consequence of aging. The term "visual acuity" refers to the resolving power of the eye in terms of the ability to focus on near or distant objects. For example, 23/40 means the patient's eye sees at 23 feet what the normal eye sees at 40 feet. The pigmented iris on the anterior portion of the vascular tunic divides the space between the lens and the cornea into an anterior chamber (between the lens and the cornea) and a narrow posterior chamber (between the suspensory ligaments of the lens and the iris). The iris is a circular structure that acts as a diaphragm to control the amount of light falling on the retina. The opening is analogous to the f-stop of a camera. The eye has a range equivalent to the range from f2 to f22.

3. Inner Tunic

The inner tunic contains pigmented cells, photoreceptor cells, and nerve cells. They are organized so that photoreceptor cells containing the photo pigments are closet to the sclera and the nerve cells are above them. The nerve cells send fibers to the optic nerve through a pigment-free area, the optic disc.

Pars optica

The pars optica consists of the following 10 layers (Fig. 13.2A):

1. Outer pigmented epithelium (most external)
2. Receptor cell layer: rods and cones
3. External limiting membrane
4. Outer nuclear layer, containing the nuclei of rods and cones
5. Outer plexiform layer, containing the synapses of rods and cones with the dendrites of bipolar and horizontal cells and the cell bodies of horizontal cells
6. Inner nuclear layer, containing the cell bodies of bipolar and amacrine cells
7. Inner plexiform layer, containing the synapses of bipolar and amacrine cells with ganglion cells
8. Ganglion cell layer,
9. Optic nerve layer
10. Internal limiting membrane

Photoreceptor Layer: Rods and Cones

There are two types of photoreceptor cells: rods and cones.
 Rods: Vision in Dim Light and Night Vision. The *rods* are sensitive in dim light as they contain more of the photosensitive pigment rhodopsin than the cones.

I Structure of the Eye

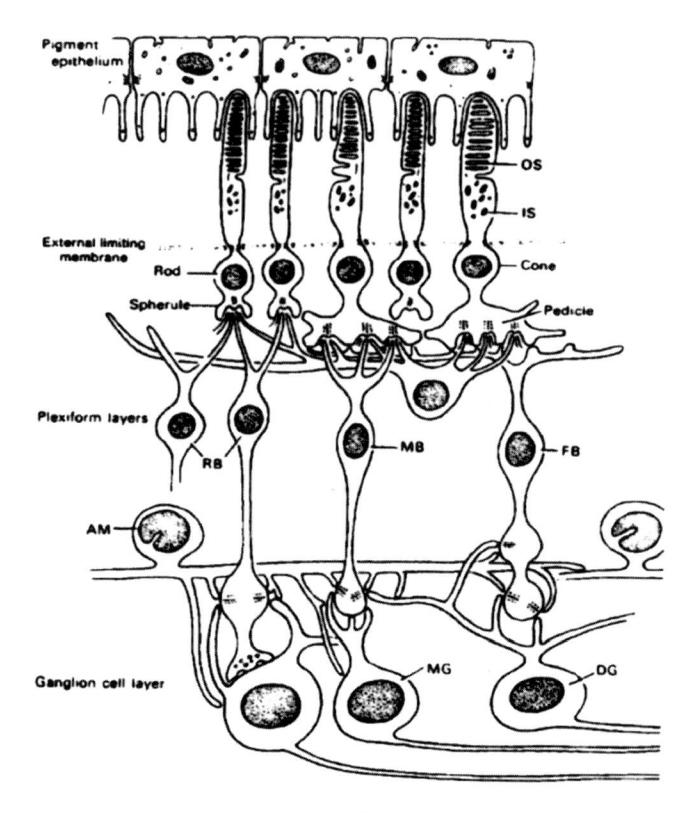

Fig. 13.2 Schematic diagram of the ultrastructural organization of the retina. Rods and cones are composed of outer (OS) and inner segments (IS), cell bodies, and synaptic bases. Photo pigments are present in laminated discs in the outer segments. The synaptic base of a rod is called a spherule; the synaptic base of a cone is called a pedicle. Abbreviations: RB = bipolar cells, MB = midget bipolar cell, FB = flat bipolar cell, AM = amacrine cell, MG = midget ganglion cell, DG = diffuse ganglion cell. (Modified from M.B. Carpenter, Core text of neuroanatomy, Williams and Wilkins)

This system has poor resolution; newspaper headlines are the smallest letters that can be recognized. The rods permit *scotopic* (dark vision) vision in the dark-adapted eye. Rods contain a single pigment, *rhodopsin*, which is related to vitamin A_1 (retinol). The spectral sensitivity of night or scotopic vision is identical to the absorption spectra of rhodopsin. Light that can be absorbed by rhodopsin is seen; light that cannot be absorbed, such as red light, is not seen at these low light intensities. Scotopic vision has quite poor definition; two light sources must be quite far apart to be distinguished as two sources rather than one. Peripheral vision has progressively poor definition. Several modern inventions, such as night goggles with infrared sensors, are designed to compensate for these limitations. The visual pigment, rhodopsin, absorbs photons. Each rhodopsin molecule consists of retinal and opsin. Retinal absorbs the light and opsin is the protein in the plasma membrane of the rods. The photon converts retinal from the 11-*cis* form to the all-*trans* form.

The all-*trans* retinal has to be re isomerized to 11-*cis* retinal to reform rhodopsin and to begin the process again. The photons result in hyperpolarization of the plasma membrane, which is the light-dependent part of visual excitation. The photoreceptors are depolarized in the dark due to the sodium channels remaining open, called the dark *currents*. The dark current is turned off in light. Daylight prevents the regeneration of rhodopsin in the rods. After 5–10 minutes in the dark, scotopic vision begins to return as the rhodopsin is regenerated, and it reaches maximal sensitivity after 15–23 minutes. Because rhodopsin does not absorb red light, using red goggles can preserve night vision.

Cones: Color Vision Cones. The *cones* are sensitive to color. There are three separate groups of cones, each of which contains photo pigments that are primarily sensitive (1) to blue, the short wavelengths, (2) to green, the middle wavelengths, or (3) to red, the longer wavelengths. The rods and cones are not uniformly distributed in the retina. Most of the 6 million cone cells are located in an area 2 mm in diameter—the macula lutea (Fig. 13.1), which can be seen through the ophthalmoscope. In the center of the macula lutea lies a zone of pure cones—the fovea (Fig. 13.2B). The rest of the macula lutea is composed of both rods and cones, and most of the peripheral retina contains only rods. There are about 123 million rods. Color vision requires much higher light intensities and occurs primarily when the image is focused upon the macula. Each cone contains one of the three color pigments. The visual pigment of the cones consists of opsin and retinal. The cones are divided into three separate groups that contain photo pigments primarily sensitive to a different part of the visible spectrum as follows:

1. Blue: 423 nm* (short wavelengths)
2. Green: 530 nm (middle wave lengths)
3. Red: 560 nm (longer wavelengths)

Color vision requires light levels greater than bright moonlight and has high resolution so that fine detail can be seen. Any color that does not excite two pigments, such as deep purple (400 nm) and deep red (650 nm), will be hard to discriminate. For this reason, a color of 660 nm in bright light can mimic a color of 640 nm in weaker light. To distinguish the color orange, 600 nm, the visual system compares the relative absorption by two visual pigments—in this case, red and green. Comparing the absorption by one visual pigment to the overall brightness apparently makes these color discriminations. For this reason, a color of 660 nm in bright light, mimics a color of 640 nm in weaker light. Only when the two types of cone are excited can color and brightness is determined independently.

B. *Optic Nerve*

The axons from the ganglion cells traverse the inner surface of the retina to the *optic disk*, where they form the optic nerve, which runs for about 3 cm before entering the optic foramen on its way toward the diencephalon. Electrical recordings from optic nerve fibers also show the types of data reduction occurring in the retina. First, there is convergence; many receptors must fire one bipolar cell in the fovea. Consequently, the "grain" of the visual image depends on the location on the retina. Second, brightness can be detected. Third, there is a center-surround contrast enhancement. Finally, there is a dynamic or motion detection. A spot of light on the retina will be coded in terms of its brightness and its edges. As it is turned on and off, the dynamic receptors fire to alert higher centers. The spot of light will receive much more attention in the optic nerve if it is moved around, because each tiny movement will activate a new set of dynamic or movement receptors and these receptors make up about half of the total optic nerve fibers. For example, the rotating flashing lights of emergency vehicles attracts more attention than a fixed light.

C. Blind Spot

The point of exit of the optic nerves (the papilla) *contains no rod or cone receptors* but will include retinal arteries and veins. This area is noted when visual fields are plotted. When swelling of the papilla occurs due to increased intracranial pressures (papilledema), the blind spot is enlarged because the receptors surrounding the point of exit will be covered by the edema.

II. Visual Pathway (Fig. 13.3)

A. *Retina and Visual Fields (Fig. 13.3)*

1. Retina

Each retina is divided into a temporal or nasal half by a vertical line passing through the *fovea centralis*. This also serves to divide the retina of each eye into a left half and right half. A horizontal line passing through the fovea would further divide each half of the retina and macula into an upper and a lower quadrant. Each area of the retina corresponds to a particular sector of the visual fields, which are named as viewed by the patient.

2. Visual Fields (Fig. 13.3)

For each eye tested alone, there is a left and right visual field corresponding to a nasal and temporal field. Because of the effect of the lens, the patient's left visual field projects onto the nasal retina of the left eye and the temporal retina of the right eye. Similarly, the upper visual quadrants project to the inferior retinal quadrants. At the chiasm, the fibers in the optic nerve from the left nasal retina cross the midline

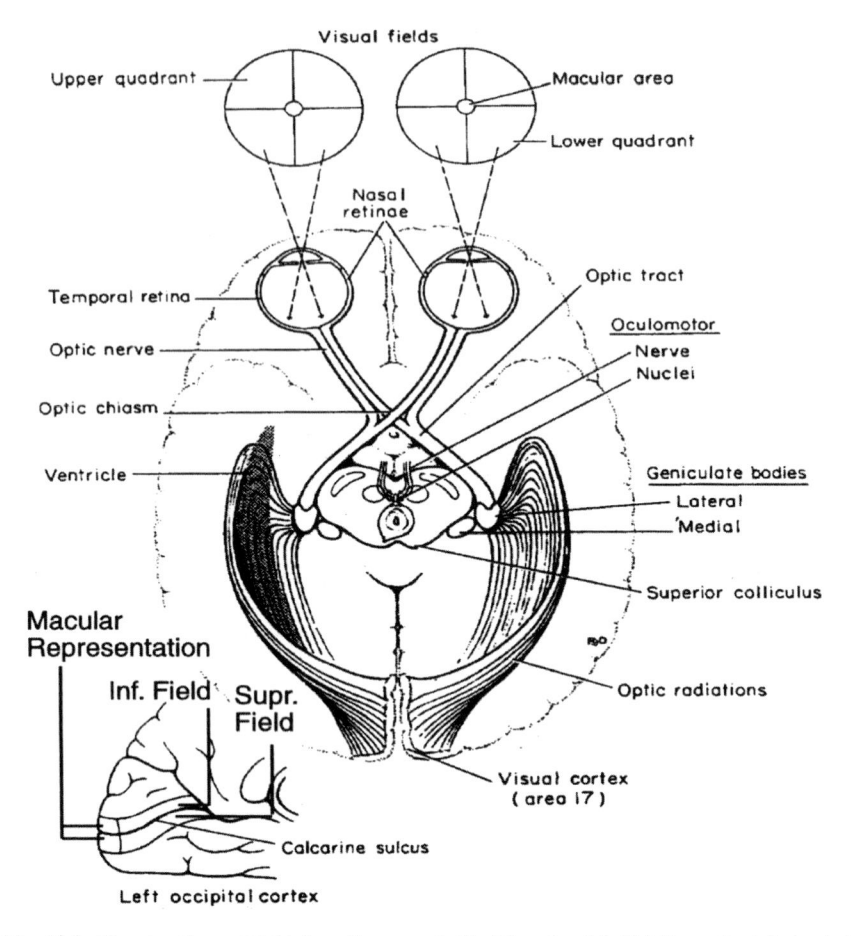

Fig. 13.3 Visual pathway. Light from the upper half of the visual field falls on the inferior half of the retina. Image from the temporal half of the visual field falls on the nasal half of the retina; image from the nasal half of the visual field falls on the temporal half of the retina. The visual pathways from the retina to the striate cortex are shown. The plane of the visual fields has been rotated 90° toward the reader (From EM Marcus and S Jacobson Integrated Neuroscience, Kluwer, 2003)

and join with the fibers from the right temporal retina to form the right optic tract. The fibers in the optic nerve from the right nasal retina cross the midline and join with the fibers from the left temporal retina to form the left optic tract. This dense bundle of over a million fibers runs across the base of the cerebrum to the lateral geniculate nucleus (LGN) of the thalamus. Fibers from the inferior nasal retina (the superior temporal visual field) cross in the inferior portion of the chiasm, which usually lies just above the dorsum sellae. By this crossing, all of the information from the left visual field is brought to the right LGN and, subsequently, to the right calcarine cortex. The fibers of the right nasal retina also cross in the chiasm to join the fibers from the left temporal retina. Thus, the visual pathway repeats the pattern

II Visual Pathway

in other sensory systems; the right side of the visual field is represented on the left cerebrum. Note that it is the information from each temporal visual field represented in each nasal retina that crosses the midline in the optic chiasm.

B. *Visual Pathway: Overview*

The pathway is divided as follows: optic nerve → optic chiasm → optic tract → LGN → visual cortex (Fig. 13.3).

1. Cranial nerve II originates from the ganglion cell layer in the retina.
2. Optic nerve goes up to the chiasm.
3. Optic chiasm where temporal half of fibers cross over and enter the optic tract while nasal half of fibers remain uncrossed, and enter optic tract.
4. Optic tract continues until it synapses in the LGN of the thalamus.
5. Visual radiation from the LGN projects to the striate (visual) cortex. In nonhuman primates and probably in the human, there are five visual areas involved in parallel processing of the visual images:

    a. Area 17 (V1), the area with initial processing in the visual cortex of the occipital lobe.
    b. Area 18 (V2 and V3), the next levels of processing in the visual cortex), with area 17 projecting to these visual association cortex.
    c. Area 19 (V3 and V4), the final level of processing in the occipital lobe.

6. Information then projects to the multimodal sensory association area, to the posterior multimodal areas in the temporal and parietal cortices, and also to the prefrontal (anterior) multimodal sensory association area in both hemispheres.

1. Optic Nerve Termination in LGN

The optic nerve fibers from the *temporal visual field*, the crossed fibers, end in layers 1, 4, and 6. The fibers from the *nasal visual field*, the uncrossed fibers, end in layers 2, 3, and 5. Each optic nerve fiber ends in one layer on five or six cells. The total number of cells and fibers is roughly equal. LGN cells respond well to small spots of light and to short, narrow bars of light, as well as to on–off stimuli. Diffuse light is a very poor stimulus.

The LGN is a rare site for a lesion. The LGN is a six-layer horseshoe shaped structure. Visual processing begins in the large and small cells in the LGN. The small cells are in layers 1–4, and the large cells are in layers 5 and 6. Each LGN cell receives direct input from a specific area of the retina, and. each area of the LGN is driven from a specific area of the visual field. Each receptive field has a center that is either on or off and a surround that has the opposite polarity. Some fibers might end in the LGN and enter the inferior pulvinar nucleus to reach extrastriate cortex.

2. Lateral Geniculate Nucleus

The LGN is primarily a relay station between the optic nerve and the visual cortex. Two major streams of visual information leave the retina to the LGN:

1. The magnocellular stream starts in the large cells of the retinal ganglion (Y-cells) that project to the *magnocellular layers* (layers 1 and 2), which subsequently project to the striate cortex into layer IV superficial to the paravocellular projection. The cortical neurons of the magnocellular stream respond to movements, orientation, and contrast, but they respond poorly to color.
2. The *parvocellular stream* begins with small cells of the retinal ganglion layer (X-cells) that project to parvocellular layer of LGN layers 3 and 6 and then projects onto layer IV of area 17. These cells respond either to shape and orientation or color: red/green and blue/yellow.

3. Light Reflexes

The optic fibers for papillary light reflexes bypass the LGN and descend into the pretectal region of the midbrain. Others, for coordinating eye and head or neck movements, descend to the superior colliculus. Also, some of the retinal fibers synapse in the suprachiasmatic nucleus of the hypothalamus and then project to the pineal by way of the sympathetic fibers, where they influence circadian rhythms and endocrine functions. Under visual stimuli, the cells in the LGN grow rapidly in the first 6–12 months after birth. Visual sensory deprivation during this period leads to atrophy and life-long blindness (amblyopia).

Figure 13.3 Inset: The projection of the quadrants of the right visual field on the left calcarine (striate) cortex. The macular area of the retina is represented nearest the occipital pole.

Pupillary light reflex: Fibers leave the optic tract, bypass the LGN and enter the pretectal region; fibers from the pretectal nucleus relay signals bilaterally to the visceral nuclei (Edinger–Westphal) of the oculomotor complex.

III. Occipital Lobe

A. *Areas in Occipital Lobe-17, 18, 19 (V1–V5)*

The occipital lobe in Brodmann's numerical scheme (Fig. 8.4) consists of areas 17, 18, and 19. From a cytoarchitectural standpoint, area 17 represents a classic example of specialized granular cortex, or koniocortex and receives the information conveyed by the optic radiation. Areas 18 and 19 represent progressive modifications from koniocortex toward homotypical cortex (Refer to chapter 8 for a discussion of cytoarchitectural).

B. Parallel Processing in the Visual Cortex

Most of our understanding of occipital lobe function comes from simian studies. In non-human primates, five visual areas are involved in parallel processing of the visual images:

V1:- corresponding to area 17 in humans
V2 and V3: corresponding to area 18
V4 and V5: corresponding to area 19

1. Striate Cortex Area 17 (V1)

The striate cortex, the principal visual projection area in humans, is found primarily on the medial surface of the hemisphere, occupying those portions of the cuneus (above) and lingual gyrus (below) that border the calcarine sulcus. For this reason, it is often termed the *calcarine cortex*. Much of this cortex is located on the walls and in the depths of this sulcus.

2. Extrastriate Visual Cortex Areas 18 (V2 and V3) and 19 (V4 and V5)

These areas form concentric bands about area 17 and is found on both the medial and lateral surfaces. Area 19 in humans extends onto the adjacent mid and inferior temporal gyri.

3. Intracortical Associations Between Striate and Nonstriate Cortex

Area 17 receives fibers from and sends fibers to area 18 but does not have any direct callosal or long-association-fiber connections to other cortical areas. Area 17 is more mature at birth and has the most precise map. The other visual areas develop through maturation and experience, although experience modifies area 17 as well. Area 18 has extensive connections with areas 18 and 19 of the ipsilateral and contralateral hemispheres, which were demonstrated by both the earlier strychnine neuronography studies and the more recent horseradish peroxidase studies.

4. Callosal Fibers

Callosal fibers enter the opposite hemisphere, and association fibers communicate with the premotor and inferior temporal areas and area 7 of the adjacent parietal cortex.

5. Termination of Optic Radiation

Area 17 receives the termination of the geniculocalcarine (or optic) radiation. This projection is arranged in a topographic manner, with the superior quadrant of the contralateral visual field represented on the inferior bank of the calcarine fissure and the inferior quadrant of the contralateral visual field represented on the superior bank. The macula has a large area of representation, which occupies the posterior third of the calcarine cortex and extends onto the occipital pole. The optic (geniculocalcarine) radiation terminates on the cells in the calcarine cortex, which respond like retinal ganglion cells. Their receptive fields consist of an excitatory region surrounded by an inhibitory region. Fields with the opposite pattern are also seen in higher mammals. The fibers form the LGN terminate in layer IV with the fibers from the magnocellular neurons termination more superficial than that of the parvicellular.

6. Ocular Dominance Columns

Ocular dominance columns are seen in the striate cortex. They are seen as an alternating series of parallel stripes that represent a column of neurons in the striate cortex innervated by either the ipsilateral or contralateral eye. Ocular dominance columns extend in alternating bands through all cortical areas and layers and are absent in only the cortical region, representing the blind spot, and the cortical area, representing the monocular temporal crescent of the visual field. The mosaic appearance of these ocular dominance columns is demonstrated with autoradiography using 2-deoxyglucose and stimulation of only one eye. Simple cells are driven by one eye, whereas the complex and hypercomplex cells are stimulated from both eyes.

7. Retinal Representation in the Occipital Cortex

Zeki (1992) has identified parallel pathways processing various aspects of visual information—color (wavelength), motion, stereognosis—thatform (line orientation) and extend from the retina to the LGN, striate cortex, and, finally, extrastriate cortex.

8. Response of Neurons in Striate Cortex: Simple, Complex, and Hypercomplex Cells

The groundbreaking work by Hubel and Wiesel (1977) identified these cells by shining light directly onto the retina, exciting the cells using an annulus (donut) of light. Hubel and Wiesel identified three types of cell in the visual cortex.

a. Simple Cell. This cell type responds best to a bar of light with a critical orientation and location. The excitatory field of each simple cell is bordered on one or both sides by an inhibitory field. The small portion of the visual field drives thousands of simple cells, each with a discrete location and orientation.

III Occipital Lobe

Each simple cell appears to receive input from a number of ganglion cells, which have their excitatory and inhibitory fields in a straight line. Every simple cell can be driven by input from either eye, but they usually show a preference for one eye or the other. For example, a simple cell might be strongly excited by a bar seen with the right eye and only weakly stimulated by the same bar seen with the left eye. This differential sensitivity might form the basis of binocular vision.

b. Complex Cell. A number of simple cells having the same orientation activate a complex cell. The complex cell has a definite orientation preference but a much larger receptive field. Any horizontal line of any length within the outlined visual field excites a specific complex cell. Lines with different orientations in the same visual field will drive other complex cells.

c. Hypercomplex cell. The final cell type is a hypercomplex cell. It has characteristics very similar to the complex cell except it discriminates lines of different lengths. Corners and angles also excite these cells. About half a million optic-tract fibers enter each occipital lobe. It should be clear that there are many more cells processing this information, and, indeed, the banks of the calcarine cortex contain many millions of cells. The response to a large rectangle of light varies with the type of cell. Only if the edge falls on the center of a cell's receptive field will it be excited.

9. Perceptual Pathways

Zeki (1992) identified four perceptual pathways (color, form with color, dynamic form, and motion) in specific visual areas:

1. Areas V1 (area 17) and V2 (area 18): columnar organization. The distribution of the information in V1 has columnar and intercolumnar areas. The columns stain heavily for the energy-related enzyme cytochrome oxidase. The neurons within the columns are wavelength-sensitive (I.e., color-sensitive), tend to concentrate in layers 2 and 3, and receive input primarily from the parvocellular layers of the LGN. Information concerning color is then projected either directly or through thin-column stripes in areas V2 (19) to V4 (19).
2. Areas V3 and V3A (area 18): selective for form. The cells of the adjacent areas V3 and V3A (area 18) are selective for form (line orientation) but are indifferent to wavelength. By contrast, form-selective neurons related to color are located in the intercolumnar areas in V1. Form related to color derives from the parvocellular neuron layers of the LGN, which project first to the intercolumnar areas of V1 and then to V4. The second form system is independent of color and depends on inputs from the magnocellular layers of the LGN to layer 4B of V1 projecting to V3. Extrastriate areas send projections back to the dorsal LGN and the pulvinar of the thalamus.

3. Area V5 (area 19): selective for motion and directionality. The neurons of area V5 (area 19) are responsive to motion and are directionally selective, but they are nonselective for color. In contrast, the majority of cells in V4 (area 19) are selective for specific wavelengths of light (color-sensitive), and many are selective for form (line orientation) as well. The motion-sensitive system consists of inputs from the magnocellular layers of te LGN to layer 4B of area V1, which then projects to area V5, either directly or through area V2. Each of the prestriate areas V5, V4, and V3 sends information back to V1 and V2, as well as to the parietal and temporal areas. This provides for more integrated visual perception.

10. Summary

Let us consider the example of a moving colored ball. Visual information is fractionated into the modalities of form, color, and motion in the calcarine cortex. That information is reassembled elsewhere and we are just beginning to understand where and how this parallel processing is done. If the object is the letter A, then further processing takes place in the dominant (usually left) lateral occipital cortex (Brodmann's areas 18 and 19), the visual language association area. Letters seen in the left visual field and represented in the right calcarine cortex must project across the posterior corpus callosum before they can be recognized in the left visual association cortex. Information needed to reconstruct complex forms flows through the inferior occipital/temporal region, primarily on the nondominant side. Located here are cells that can discriminate between human faces, for example. Many of these cells are sensitive to gaze, being strongly stimulated by direct eye contact but only weakly stimulated when the eyes are averted. Extraoccipital higher-level visual processing then occurs in area 7 of the parietal lobe and areas 23 and 21 of the temporal lobe.

C. *Effects of Stimulation of Areas 17, 18, and 19*

Direct electrical stimulation of areas 17, 18, and 19 in conscious people produces visual sensations (Pollen, 1975). These images are not elaborate hallucinations (as in complex partial temporal seizures) but, rather, are described as flickering lights, stars, lines, spots, and so forth. Often the images are described as colored or moving around. The images are usually localized to the contralateral field and at times to the contralateral eye. See Case History 13.4. At times the patient cannot determine laterality. In addition, stimulation of areas 18 and 19 (and sometimes 17) produces conjugate deviation of the eyes to the contralateral field and, at times, vertical conjugate movements. Discharge of the occipital cortex might be followed by transient visual defects similar to the transient post-ictal hemiparesis that might follow focal seizures beginning in the motor cortex (see Aldrich et al., 1989). A more complete discussion of seizures beginning in occipital cortex can be found in Salanova et al. (1993) and Williamson et al. (1992).

D. Effects of Lesions in the Occipital Visual Areas

1. Complete Unilateral Ablation of Area 17 (V1)

This produces a complete homonymous hemianopsia in the contralateral visual field (the same half-field in each eye has no conscious visual perception). Such patients might have some crude visual function, such as spatial localization, presumably because of optic-tract connections to the superior colliculus. Visual stimulus and its direction of motion can continue to occur after selective V1 lesions have been made, due to activation of V5 (possibly by subcortical input to V5). Vision is clearly abnormal in this blind field. V1 to V5 input is the dominant input in intact individuals.

2. Partial Lesions in Area 17

For example, if only the superior bank of the calcarine fissure is involved, the visual field defect is limited to the inferior quadrant of the contralateral field. A bilateral infarct of either the upper or lower bank of the calcarine fissure might produce a homonymous altitudinal hemianopsia in which apparent blindness exists in either the entire lower or upper field of vision (as appropriate). With such calcarine cortex lesions, the field defects are usually similar (congruous) in both eyes.

3. Vascular Lesions in Area 17

In particular, occlusion of the posterior cerebral artery often results in a homonymous hemianopsia with macular sparing; that is, vision in the macular area of the involved field remains intact. Such preservation of macular vision probably occurs because the macular area has a large representation in the most posterior third of the calcarine cortex and is the area nearest the occipital pole. This area, then, is best situated to receive leptomeningeal anastomotic blood supply from the middle cerebral and anterior cerebral arteries. Occasionally, with occipital infarcts, there might be preservation of vision in a small peripheral unpaired portion of the visual field, called the temporal crescent (see Benton and Swash, 1980). In vascular insufficiency or occlusion of the basilar artery, infarction might occur in the distribution of both posterior cerebral arteries, producing a bilateral homonymous hemianopia and a syndrome of "cortical blindness." Such patients lose all visual sensation, but pupillary constriction in response to light is preserved.

4. Lesions in Extrastriate Areas 18 and 19

Lesions in these areas produce deficits in visual association, including defects in visual recognition and reading. The problem with humans with a unilateral lesion is that one is almost never dealing with disease limited to areas 18 and 19. Rather, one finds that adjacent portions of either the inferior temporal areas or of the inferior parietal lobule (the angular and supramarginal gyri) are involved or that the lesion extends into the deeper white matter of the lateral occipital area and involves association and callosal fibers. Such more extensive lesions will, of course, produce the deficits previously noted in the discussion of parietal function.

5. Unilateral Lesion of Extrastriate Areas 18 and 19

A limited unilateral ablation might produce defects in visual following, as tested by evoking opticokinetic nystagmus (e.g., moving vertical black lines on a white background or looking at telephone poles from a moving train).

6. Bilateral Lesions Limited to Areas 18 and 19

These lesions occur only rarely in humans. In humans, such a lesion would deprive the speech areas of the dominant hemispheres of all visual information. The patient would presumably see objects but be unable to recognize them or to place visual sensations in the context of previous experience. Moreover, the patient would be unable to relate these visual stimuli to tactile and auditory stimuli. Horton and Hoyt (1991) have reported that unilateral lesions specific to the extrastriate V2/V3 (area 18) cortex produce quadrantal visual field defects. As discussed by Zeki (1992), rare patients with restricted lesions in V4 present with achromatopsia (deficiency in color perception). The patients see the world only in shades of gray, but perception of form, depth, and motion are intact. Rare patients with lesions limited to V5 have akinetopia; stationary objects are perceived, but if they are in motion, the objects appear to vanish. Plant and co-workers (1993) have reviewed such impaired motion perception. Selective deficits in form perception are even less frequent. Destruction of both form systems and therefore of V3 and V4 (areas 18 and 19) would be required and, as discussed earlier, such a lesion would also destroy V1 (area 17), resulting in total blindness.

E. Occipital Lobe and Eye Movements (See Also Chapter 9)

Areas 18 and 19 send fibers to the tectal area of the superior colliculus. Such connections are necessary for visual fixation and accurate following of a moving object. The slow and smooth conjugate eye movements that occur when the eyes are following a moving visual stimulus (pursuit movements) should be distinguished from the independent phenomenon of voluntary (saccadic) conjugate eye movements. Saccadic movements are rapid and shorter in latency and duration. They do not require a visual stimulus and do not depend on any connections to area 18. Rather, they are dependent on area 8 in the premotor cortex, which sends fibers to the lateral gaze center of the pons. Both areas send some fibers to the superior lliculus, and ultimately, both the saccadic and pursuit movements involve the pontine gaze centers. In addition, vestibular and cerebellar influences also determine eye movement. The following case illustrates the effects of a space-occupying lesion in the occipital lobe. One should compare these findings to those reported in Case History 13.1, a lesion in the visual system anterior to the optic chiasm.

IV. Visual Field Deficits Produced by Lesions in the Optic Pathway

A. *Overview of Localized Lesions in the Visual System (Fig. 13.4)*

1. Figure 13.4, #1. **Normal condition** with no defects in visual fields.
2. Figure 13.4, #2. **Monocular blindness.** Lesion in the retina or the optic nerve before the optic chiasm; there is no sight in that eye (*monocular blindness*). Incomplete lesions produce a *monocular scotoma*, as in optic neuritis.
3. Figure 13.4, #3. **Bitemporal hemianopsia.** Lesion in the optic chiasm; the fibers from both temporal visual fields are cut and the result is *bitemporal hemianopsia*.
4. Figure 13.4, #4. **Homonymous hemianopsia.** Lesions behind the optic chiasm produce blindness in one visual field, a *homonymous hemianopsia* from injury to the optic tract, the lateral geniculate body, or the geniculocalcarine radiation in Figure 13.4, #4 and #5).
5. Figure 13.4, #5. **Quadrantanopia.** Due to Partial lesions of the geniculocalcarine radiations. Lesions in the upper or lower bank of the calcarine cortex also produce a *quadrantanopia*.
6. Figure 13.4, #6. **Homonymous hemianopsia** with macular sparing is seen with lesions in the visual cortex following calcarine artery infarcts. The macula is represented close to the occipital pole and this area receives anastomotic flow from the middle cerebral artery.

B. *Case Histories from a lesion on the Visual System: Optic Nerve · Optic Chiasm · Visual Radiations · Striate Cortex*

1. Lesion in the Optic Nerve Before the Chiasm. Result: Monocular Blindness

Case History 13.1 (Fig. 13.5)

This 53-year-old, white, right-handed housewife had a progressive 23-year history of right-sided supraorbital headache with decreasing acuity and now almost total

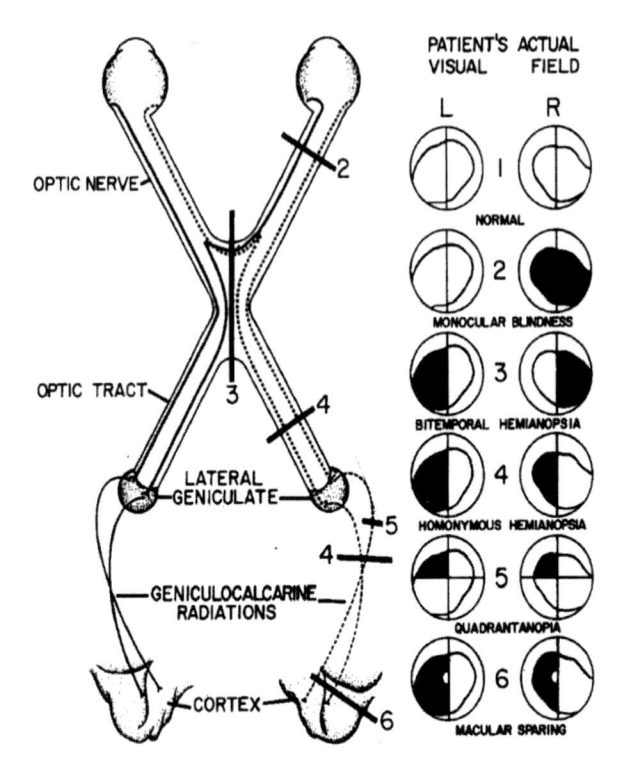

Fig. 13.4 Common lesions of the optic tract: (1) normal; (2) optic nerve; (3) optic chiasm; (4) optic tract or complete geniculate, complete geniculocalcarine, radiation or complete cortical; (5) if lesion in optic radiation temporal segment (Meyer's loop), it produces a superior quadrantanopia; if it is in the parietal segment of the radiation, it produces an inferior quadrantanopia; (6) calcarine or posterior cerebral artery occlusion producing a homonymous hemianopia with macular sparing. (From EM Marcus and S Jacobson, Integrated neuroscience, Kluwer, 2003)

blindness in the right eye. During the 3 years before admission, intermittent tingling paresthesia had been noted in the left face, arm, or leg. About 1 year before admission, the patient had a sudden loss of consciousness and was amnestic for the events of the next 48 hours. No explanation for the episode was clearly established. The cerebrospinal fluid protein was reported to be elevated (230 mg/dl). The patient and her family reported some personality changes over a period of several years, including a loss of spontaneity and increasing apathy.

General physical examination: There was a minor degree of proptosis (downward protrusion of the right eye).

Neurologic examination: Mental status: The patient was, in general, alert but at times would become lethargic. Her affect was flat. At times, she would laugh or joke in an inappropriate manner.

Cranial nerves: There was anosmia for odors, such as cloves, on the right, and a reduced sensitivity on the left. Marked papilledema (increased intracranial pressure with elevation of the optic disk and venous engorgement was present in the left eye.

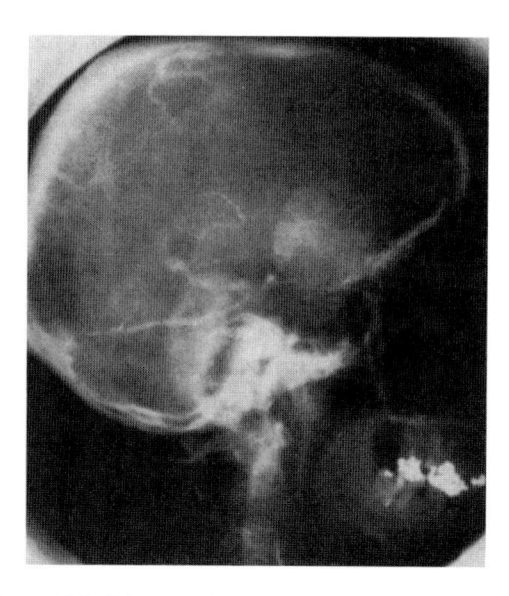

Fig. 13.5 Case History 13.1. Sphenoid-wing meningioma producing compression of the right optic and olfactory nerves. Refer to text. Right carotid arteriogram, venous phase, demonstrating tumor blush in the subfrontal area of the anterior fossa, extending into the middle fossa. (Courtesy of Dr. Samuel Wolpert. New England Medical Center Hospitals.) (From EM Marcus and S Jacobson, Integrated neuroscience, Kluwer, 2003)

In contrast, there was pallor of the right optic disk, indicating optic atrophy. Visual acuity in the right eye was markedly reduced. This combination of funduscopic findings is termed the Foster Kennedy syndrome. The patient had only a small crescent of vision in the temporal field of the right eye; only vague outlines of objects could be seen. A slight left central facial weakness was present.

Motor system: Movements on the left side were slow.

Reflexes: A release of grasp reflex was present on the left side.

Clinical diagnosis: Subfrontal meningioma rising from the inner third sphenoid wing. Alternative location would be olfactory groove.

Laboratory data: Imaging studies (Fig. 13.5) demonstrated a tumor blush in the right subfrontal region extending back to the right optic nerve groove, consistent with a meningioma arising from the olfactory groove or inner third sphenoid wing.

Hospital course: A bifrontal craniotomy performed by Dr. Sam Brendler exposed a well-encapsulated smooth tumor a meningioma attached to the inner third of the sphenoid wing. Approximately 90–95% of the tumor was removed, exposing the right optic nerve. Examination 4 months after surgery indicated that right anosmia and right optic atrophy were present.

2. Lesion at the Optic Chiasm. Result: Bitemporal Hemianopsia

In the optic chiasm, fibers from the temporal visual fields cross and unite with fibers from the contralateral nasal field to form the optic tract. Lesions here produce a bitemporal defect results (Fig. 13.6). This might also produce a bitemporal hemianopsia or an incomplete bitemporal field defect. The usual cause is a pituitary adenoma (Fig. 13.6) or a suprasellar tumor such as a craniopharyngioma.

Case History 13.2 (Fig. 13.7)

A patient of Dr. Martha Fehr, a 44-year-old man had an 18-month history of a progressive alteration in vision. He was unable to see objects in the right or left periphery of vision. He was concerned that in the process of driving, he might hit pedestrians stepping off the sidewalks in the periphery of vision. He had also experienced, over 6–8 months, progressive headaches and loss of energy and libido.

Neurologic exam: Normal except for a bitemporal hemianopia.
Clinical diagnosis: Pituitary adenoma compressing the optic chiasm.
Laboratory data: Endocrine studies: #ll were normal.

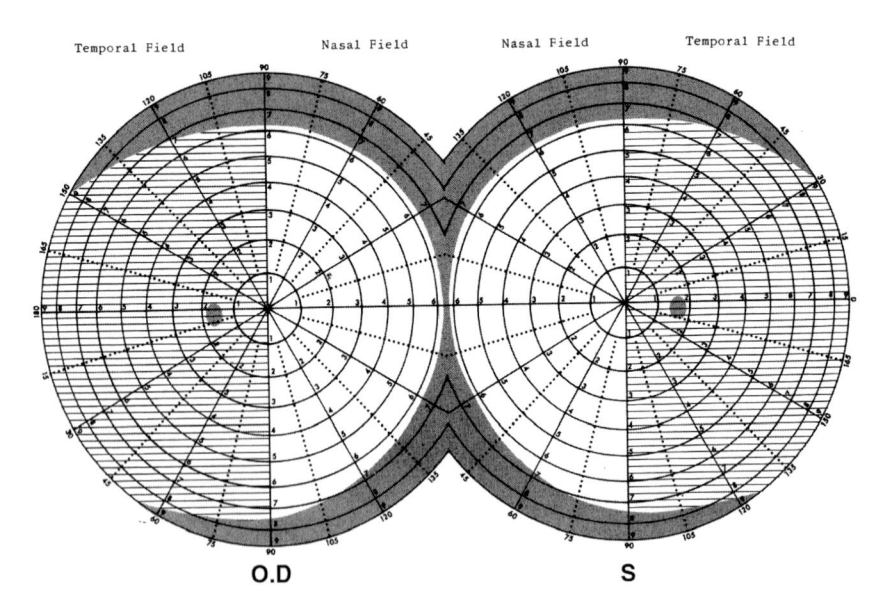

Fig. 13.6 Pituitary adenoma. Case History 13.2. Visual fields demonstrating a bitemporal hemianopia. S. = left eye; O.D. = right eye. This 51-year-old obese male with large puffy hands and a prominent jaw, with declining sexual interest for 18 years, an 8-year history of progressive loss of visual acuity, a 6–7-month history of diplopia due to a bilateral medial rectus palsy, and headaches, had stopped driving because he was unable to see the sides of the road. Urinary adrenal and gonadal steroids and thyroid functions were low, with no follicle-stimulating hormone. Imaging studies demonstrated a large pituitary adenoma with significant suprasellar extent. (From EM Marcus and S Jacobson, Integrated neuroscience, Kluwer, 2003)

IV Visual Field Deficits Produced by Lesions in the Optic Pathway   331

MRI: A large macroadenoma measuring 3 cm in diameter extended outside of the sella turcica to compress the optic chiasm (Fig. 13.7B).

Subsequent course: Dr. Gerald McGuillicuddy performed a gross total transsphenoidal resection of the tumor, with a significant improvement in vision. Three years later, the patient again had visual symptoms and eye pain. MRI scans indicated regrowth of tumor with possible impression on the right optic nerve. He was treated with radiotherapy. Note that males are more likely to present with large pituitary adenoma compressing the optic chiasm. Women are more likely to present with microadenomas or small macroadenomas because they are more likely to be seen at an earlier stage due to an initial complaint of amenorrhea.

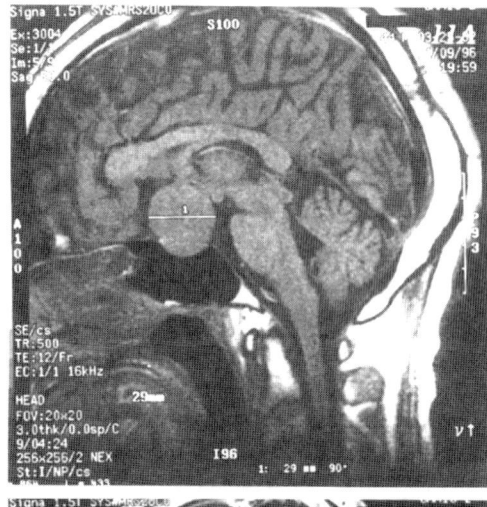

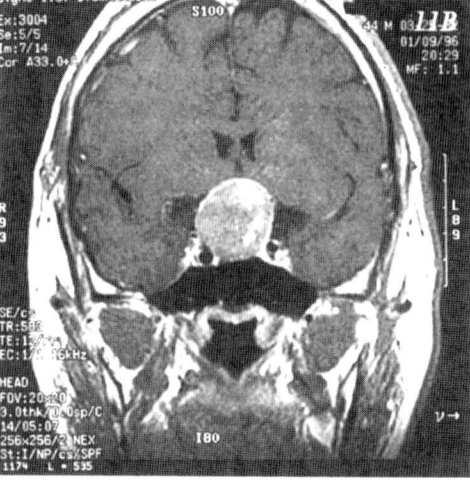

Fig. 13.7 Case 13.2. A large pituitary adenoma in a 44-year old man with a bitemproal hemianopsia. Significant extrasellar extension compresses the optci chiasm. MRI scans: (**A**) midline sagittal section; (**B**) coronal section. (From EM Marcus and S Jacobson, Integrated neuroscience, Kluwer, 2003)

3. Lesions in the Optic Radiation; Result Noncongruous Homonymous Hemianopsia or Quadrantanopia

The fibers to the visual cortex leave the LGN and form the optic (geniculocalcarine) radiation. The visual radiation is so extensive that not all the fibers can pass directly posterior to the calcarine cortex. The most dorsally placed fibers pass back deep into the parietal lobe. The more ventrally placed fibers pass forward (Myer's loop) and deep into the temporal lobe, swing posterior around the anterior portion of the inferior horn of the lateral ventricle, and continue posterior lateral to the ventricle wall to the calcarine cortex. In contrast to the densely packed optic tract, the optic radiation fans out widely on its passage through the parietal and temporal lobes. As a result, lesions might produce noncongruous or quadrantic visual field defects. However, similar types of defect might also occur, with lesions involving the optic radiation deep to the occipital lobe white matter. Note that lesions involving Myer's loop of the geniculo-calcarine radiation in temporal lobe white matter will produce a superior field quadrantanopia and those involving the deep parietal component of the geniculo-calcarine radiation will produce an inferior field quadrantanopia.

Case History 13.3 (Fig. 13.8)

A 47-year-old white, right-handed, real estate salesman developed a cough with some blood present in the sputum (hemoptysis), 1 month before admission. Five days before admission, the patient developed a generalized headache, which was precipitated by coughing or straining and which awakened him or prevented him from sleeping. Two days before admission, the patient noted blurring in the left inferior quadrant of his field of vision. The day before admission, he noted complete loss of vision in this quadrant. On the day of admission, the headache increased, and the patient was unable to see anything in the left visual field. Past history was significant. The patient had multiple pulmonary infections, treated with antibiotics.

Neurologic examination: Cranial nerves: Early papilledema was present. A noncongruous left homonymous hemianopia was present (Fig. 13.8A)

Clinical diagnosis: Mass lesion right occipital? Abscess, differential tumor.

Laboratory data: Imaging demonstrated an enhancing lesion in the posterior and medial aspect of the right hemisphere (occipital and adjacent parietal lobes). EEG was abnormal with frequent focal 3–4-cps slow waves in the right occipital area, and to a lesser degree, the right posterior parietal area. Cerebrospinal fluid (CSF) pressure was elevated to 210 mm (N = 110 mm).

Subsequent course: The patient was treated with antibiotics (cephalothin sodium, penicillin, and streptomycin), and visual fields, EEG, and spinal fluid findings improved. Within 3 weeks, the field defect had resolved to a noncongruous left inferior quadrantanopia (Fig. 13.8B).

Ten days later, the patient was readmitted to the hospital with a 3-day history of right eye pain, sweats, and chills. Neurologic examination now revealed a recurrence of blurred optic disk margins, a left homonymous hemianopia, and a slight increase

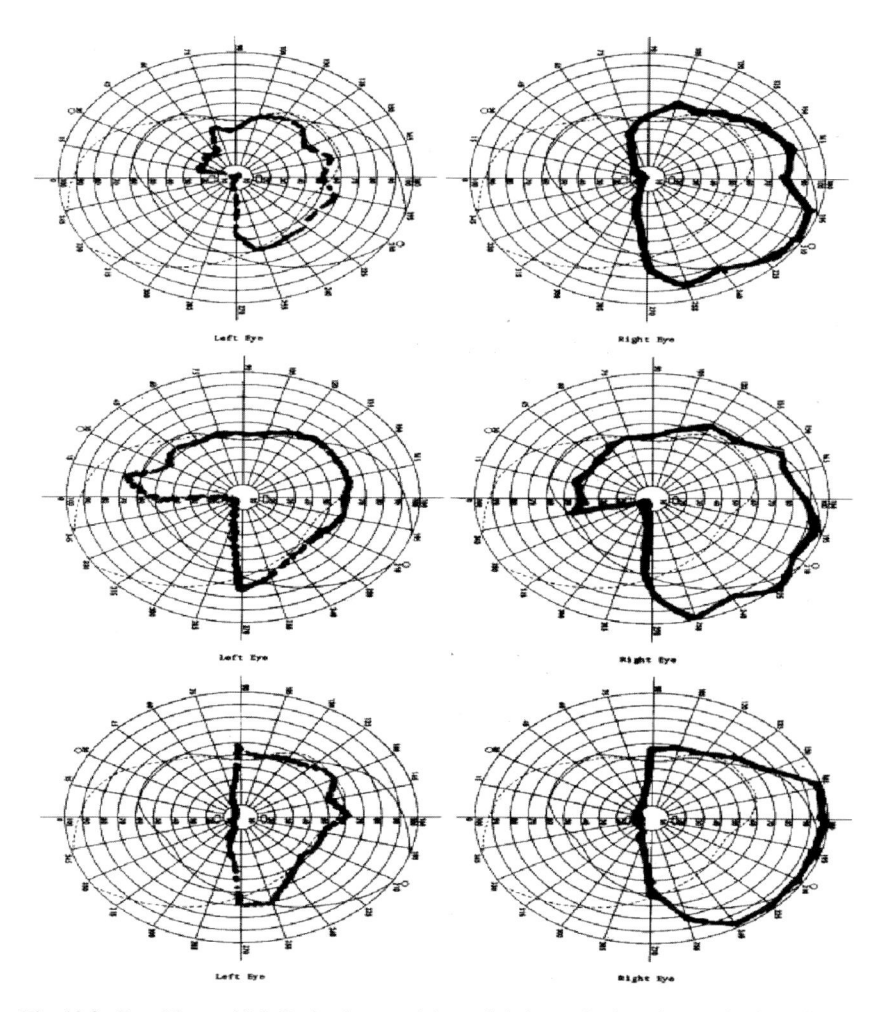

Fig. 13.8 Case History 13.3. Brain abscess, right occipital area. Perimetric examination of visual fields. Top row: Initial examination demonstrated a somewhat asymmetrical (noncongruous) *left homonymous hemianopsia*, less in the left eye than the right. The fields are shown from the patient's point of view. Middle row: 16 days later; with antibiotic therapy, an improvement had occurred. A *noncongruous quadrantanopia* is now present. Bottom row: Fields after an additional 24 days. A relatively *complete homonymous hemianopia*, present at the time of admission, persisted following surgery. (From EM Marcus and S Jacobson, Integrated neuroscience, Kluwer, 2003)

in deep tendon reflexes on the left. Imaging studies now revealed a large space-occupying lesion in the right parietal/occipital region, displacing the right lateral ventricle forward and downward. After 10 days of treatment with antibiotics (penicillin and streptomycin), a craniotomy was performed by Dr. Bertram Selverstone, revealing an abscess. The abscess and a large surrounding area of hard granulomatous cortex and white matter were removed. The etiologic organism was subsequently found to be a microaerophilic streptococcus. Follow-up examination 6 months after surgery was normal, except for the left homonymous hemianopia (Fig. 13.8C).

The case that follows is an example of the effects of a lesion in the occipital lobe with very different consequences than those seen in the previous case.

4. Lesion in Occipital Cortex. Result: Focal Seizures with a Noncongruous Homonymous Quadrantanopia

Case History 13.4 (Fig. 13.9)

An 18-year-old left-handed, single, white female had the onset of her seizures at age 14 when she had a sequence of five seizures in less than 12 hours. Each began with flashing lights, "like Christmas tree lights," all over her visual field, plus the sensation "those peoples' faces were moving." She then would have an apparent generalized convulsive seizure. Neurologic examination reported as normal. A CT scan done in Arizona had been reported as essentially within normal limits. An electroencephalogram reported occasional focal sharp and slow waves in the left hemisphere. The patient was treated with carbamazepine (Tegretol), with no additional seizures. She apparently had done quite well in the interim with no additional grand mal seizures. She had rare episodes of "fear attacks," which would last 23–25 minutes. The last attack had occurred 2 years ago, but as recently as 2 months ago, she had had one episode of flashing lights. A recent EEG was normal.

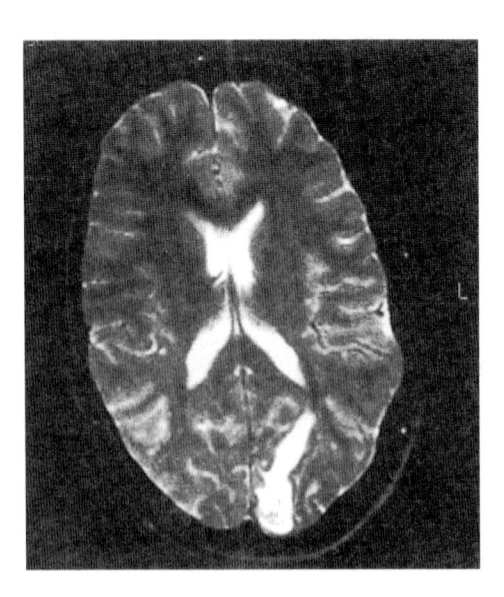

Fig. 13.9 Case History 13.4. Focal visual seizures characterized by "flashing lights" and the sensation of movement of the visual field with secondary generalization beginning at age 14 due to metastatic (thyroid) tumor of the left occipital lobe. MRI, T2-weighted, nonenhanced demonstrated a small tumor at the left occipital pole with surrounding edema. (From EM Marcus and S Jacobson, Integrated neuroscience, Kluwer, 2003)

Neurologic examination: Normal.

Clinical diagnosis: Seizures of focal origin left occipital lobe with secondary activation of temporal lobe and secondary generalization.

Subsequent course: The patient did well for 3 years and then had a recurrence of a generalized convulsive seizure possibly related to omission of medication. Six weeks later, she reported two additional episodes characterized by flashing lights, then movement of the lights away from a center circle, then dizziness, with a sensation of unreality. She also had other episodes of feeling unreal accompanied by fear The neurologic examination showed a minor right central facial weakness not present previously.

Laboratory data: EEG: normal. MRI scan now revealed a small tumor at the left occipital pole with surrounding edema (Fig. 13.9). On CT scan, this tumor appeared partially calcified but did show enhancement. Review of the CT scan obtained in Arizona at age 14 suggested similar findings. Angiograms suggested tumor vascularization of a type seen with meningioma. Dr. Bernard Stone removed a discrete encapsulated tumor that grossly, at surgery, appeared to be a meningioma. Subsequent microscopic examination, however, indicated a rare type of indolent follicular adenocarcinoma of the thyroid, which sometimes spreads as a single lesion to the brain and remains quiescent for many years. In the postoperative period, a noncongruous right-inferior-field defect was present (partial quadrantanopia). The blood level of thyroid-stimulating hormone (TSH) was elevated. A thyroid nodule was found. A thyroidectomy was performed, and thyroid replacement medication was prescribed. No additional seizures were observed over the next 18 months. The patient subsequently developed a lumbar vertebral metastasis.

Comment: Note that although the seizures began in the occipital cortex, with flashing lights and the sensation that objects were moving, she also had phenomena associated with temporal lobe seizures such as fear and déjà vu. It is not unusual for an occipital focus to spread and activate focal ictal phenomena associated with the temporal lobe.

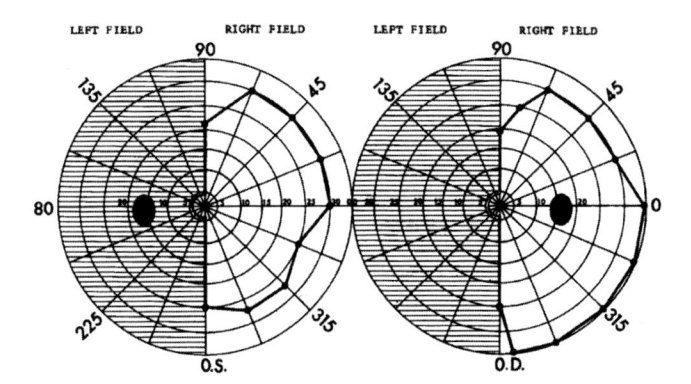

Fig. 13.10 Occlusion of right posterior cerebral artery. Tangent screen examination of visual fields 3 months after the acute event. This 75-year-old woman had the acute onset of headache, bilateral blindness confusion, vomiting, mild ataxia, and bilateral Babinski signs. All findings cleared except a homonymous hemianopia with macular sparing. O.S. = left eye; O.D. = right eye. (From EM Marcus and S Jacobson, Integrated neuroscience, Kluwer, 2003)

Fig. 13.11 CT scan of a total infarct, presumably embolic in the right posterior cerebral artery (PCA) cortical territory (occipital and posterior temporal lobes) obtained 5 days after the acute onset of confusion and possible visual hallucinations in an 86-year-old right-handed male. Dense left homonymous hemianopia with no evidence of macular sparing. In addition, a left field deficit is present As confusion cleared, examination indicated the patient would look to the right and to the midline but would not follow objects to the left. (From EM Marcus and S Jacobson Interpreted Neuroscience Kluwer, 2003)

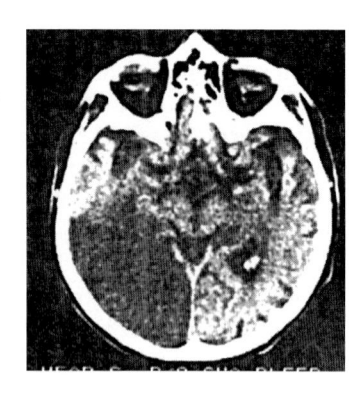

5. Vascular Lesions Within the Calcarine Cortex

These infarcts usually reflect embolic events. As discussed earlier, macular sparing reflects the leptomeningeal anastomotic flow from middle and anterior cerebral sources to the occipital pole, where there is some macular representation. Various types of visual field defect might occur, including a homonymous hemianopsia, with macular sparing (Fig. 13.10) or without macular sparing (13.11).

Chapter 14
Limbic System, the Temporal Lobe, and Prefrontal Cortex

Cure her of that: Canst thou not minister to a mind diseased; pluck from the memory a rooted sorrow; raze out the written troubles of the brain; and with some sweet oblivious antidote cleanse the stuff'd bosom of that perilous stuff which weighs upon the heart." —Macbeth V:iii.

I. Limbic System

The neurologist Paul Broca in the later half of the 19th century initially designated all of the structures on the medial surface of the cerebral hemisphere "the great limbic lobe." This region, due to its strong olfactory input, was also designated the rhinencephalon. The olfactory portion of the brain (rhinencephalon, or archipallium) comprises much of the telencephalon in fish, amphibians, and most mammals. In mammals the presence of a large olfactory lobe adjacent to the hippocampus was once considered to be evidence of the important olfactory functions of these regions. However, when comparative anatomists examined the brains of sea mammals that had rudimentary olfactory apparatus (e.g., dolphins and whales), the presence of a large hippocampus suggested other than olfactory functions for this region.

Papez Circuit. In 1937, Papez proposed that olfactory input was not the prime input for this region, and the experiments of Kluver and Bucy (1937, 1939) and Kluver (1952, 1958) demonstrated the behavioral deficits seen after lesions in this zone. More recently, it has been shown that in primates, only a small portion of the limbic lobe is purely olfactory: the olfactory bulb, olfactory tract, olfactory tubercle, pyriform cortex of the uncus, and corticoamygdaloid nuclei (Fig. 14.1). The other portions—hippocampal formation, fornix, parahippocampal gyrus, and cingulate gyrus—are now known to be the cortical regions of the limbic system (Fulton, 1953; Green, 1958; Papez, 1958; Scheer, 1961; Isaacson, 1982).

Since the initial observations of Kluver and Bucy (1937) and Papez (1937), which localized emotions in the telencephalon, many investigators have added information concerning the localization of behavior. We now know that many cortical and subcortical regions are incorporated in the **"emotional brain."** Different investigators have coined different terms to succinctly describe the limbic system, particularly the visceral, vital, or emotional brain. The term "visceral brain" would seem appropriate because much of our emotional response is characterized by specific responses in the viscera (Fulton, 1953). On the other hand, the importance of the emotional response for the self-preservation of the individual and the perpetuation of the species has led other investigators to call this region the vital brain (MacLean, 1955). We can separate the entire central nervous system into a "somatic brain," which controls the external environment through the skeletal muscles, and a " visceral brain," which controls the internal environment through the control of cardiac, smooth muscles and glands The term used most commonly by investigators and the one used in this chapter and throughout the volume is limbic lobe (limbus = margin) because the involved regions are located on the medial margin of the cerebrum and surrounds the brain stem as it enters the diencephalon (Fig. 14.1A–14.1C). The regions in the Limbic System are divided into cortical and subcortical zones that are interconnected.

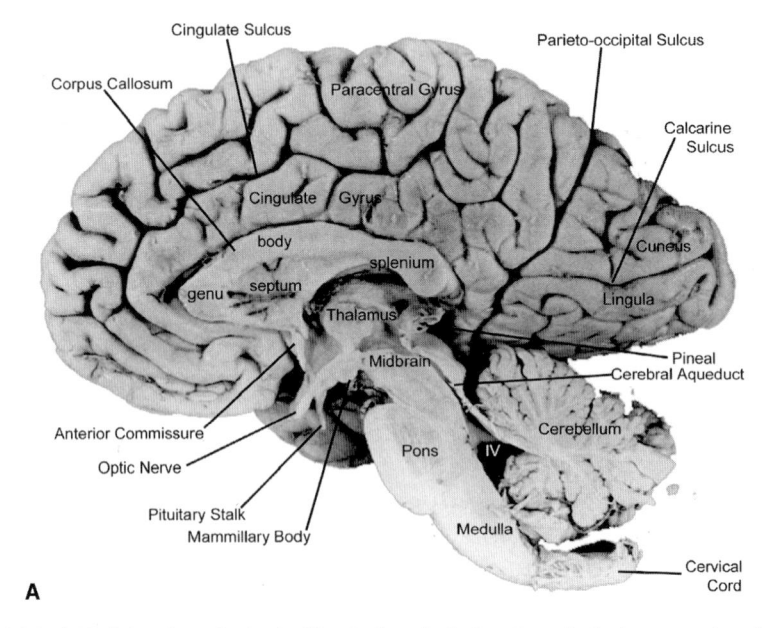

Fig. 14.1 A Medial surface of a cerebral hemisphere, including the entire brain stem and cerebellum (From EM Marcus and S Jacobson, Integrated neuroscience, Kluwer, 2003) **B** Medial surface with the brain stem and cerebellum now removed (From EM Marcus and S Jacobson, Integrated neuroscience, Kluwer, 2003) **C** Medial surface of the cerebrum with thalamus removed to demonstrate relationship between fornix and hippocampus (From EM Marcus and S Jacobson, Integrated neuroscience, Kluwer, 2003)

I Limbic System

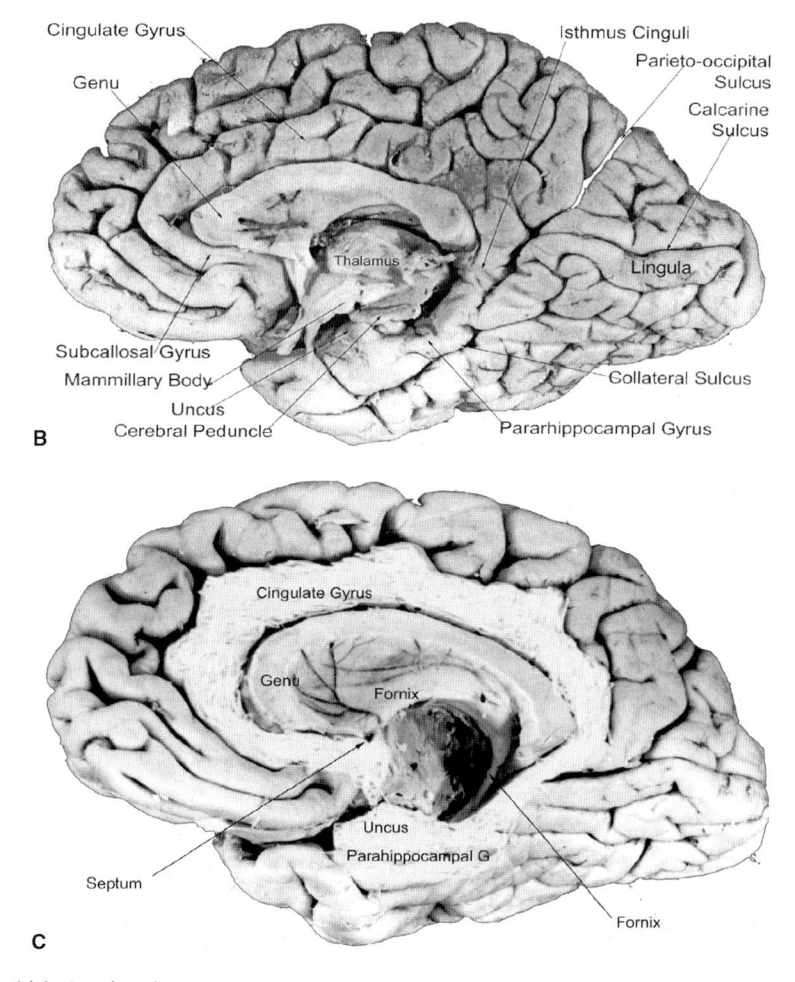

Fig. 14.1 (continued)

A. Subcortical Structures (Table 14.1A)

1. Reticular Formation of the Brain Stem and Spinal Cord

The spinal and cranial nerve roots are the first-order neurons for sensory information to reach the somatic and visceral brain. Peripheral spinal and cranial nerves send much sensory information via axon collaterals into the reticular formation. The reticular formation is organized longitudinally, with the lateral area being the receptor zone and the medial area being the effector. In the medial zone are the ascending and descending multisynaptic fiber tracts of the reticular formation: the central

Table 14.1 Regions of the limbic system

A. Subcortical Regions
   1. Reticular formation of the brain stem and spinal cord
      Midbrain limbic nuclei (interpeduncular nucleus)
      Paramedian nucleus: rostral portion of ventral tegmental area, ventral half of the periaqueductal gray
   2. Hypothalamus (preoptic, lateral, mammillary nuclei, longitudinal stria
   3. Thalamic nuclei: midline, intralaminar, anterior, dorsomedial
   4. Epithalamus
   5. Septum
   6. Nucleus accumbens
B. Cortical Regions
   1. Temporal Lobe: amygdala, hippocampal formation (hippocampus, dentate gyrus, and parahippocampal gyrus), occipitotemporal gyrus, inferior temporal gyrus
   2. Frontal Lobe: Prefrontalassociation areas: supracallosal gyrus and subcallosal gyrus, orbital frontal cortex
   3. Cingulate gyrus and cingulate isthmus

tegmental tract. There are important limbic nuclear groupings in the reticular formation: (1) the especially important mesencephalic reticular formation that provides a direct reciprocal pathway to the hypothalamus, thalamus, and septum, and (2) the locus ceruleus of the upper pons and the raphe of the midbrain that provide ascending serotoninergic and adrenergic systems onto the diencephalon and telencephalon.

2. Interpeduncular Nucleus

The interpeduncular nucleus is in the posterior perforated substance on the anterior surface of the midbrain in the interpeduncular fossa extending from the posterior end of the mammillary body to the anterior end of the pons. It receives fibers from the habenular nuclei (habenulopeduncular tract) and has reciprocal connections with the hypothalamus and the midbrain limbic region. Amygdaloidal information reaches this region through connections via the stria terminalis to the septum and then from the septum to the interpeduncular nucleus.

3. Hypothalamus (See Chapter 7)

The hypothalamus is the highest subcortical center of the "visceral brain." The basic function of this region is to maintain internal homeostasis (body temperature, appetite, water balance, and pituitary functions) and to establish emotional content. It is a most potent subcortical center due to its control of the autonomic nervous system

I Limbic System

4. Thalamus (See Chapter 6)

The anterior, dorsomedial, midline, and intralaminar dorsal thalamic nuclei receive input from the ascending nociceptive pathways, hypothalamus, reticular system (especially the midbrain reticular formation) and project onto the cingulate, and frontal association cortex.

5. Epithalamus (See Chapter 6)

The habenular nuclei of the epithalamus give origin to the habenulopeduncular tract which projects to the midbrain tegmentum and interpeduncular nucleus.

6. Septum

The septum forms the medial wall of the frontal horn and separates the lateral ventricle and consists of (1) the septum pellucidum with its lower portion containing the septal nuclei (dorsal, lateral and medial) and (2) the caudal glial velum interpositum. The septum pellucidum is rostral to the interventricular foramen. This paired glial membrane, along with the fornix, separates the bodies of the lateral ventricles.

The septum receives strong input from the following:

1. Amygdala via the stria terminalis and hypothalamus
2. Hypothalamus, interpeduncular nucleus, and the midbrain tegmentum via the medial forebrain bundle
3. Habenula via the stria medullaris
4. Basolateral amygdaloidal nuclei through the diagonal band

Destruction of the septum in cats causes docile animals to become fearful or aggressive, but only for a short time. Complete destruction of the septum might produce coma, probably because it destroys the strong connections between the septum and hypothalamus.

7. Nucleus Accumbens (See Chapter 10)

This nucleus lies below the caudate and receives fibers from the amygdala via the ventral amygdalofugal pathway and from the basal ganglia and thus provides a major link between the limbic and basal nuclei. This nucleus has a high content of acetylcholine. In Alzheimer's disease, there is a significant loss of cholinergic neurons in this nucleus.

B. Cortical Structures in the Limbic System (Table 14.1B)

Figures 14.1A–14.1C demonstrate the location of the limbic cortical areas on the medial surface of the cerebrum. In Fig. 14.1, A the brain stem is present, and in

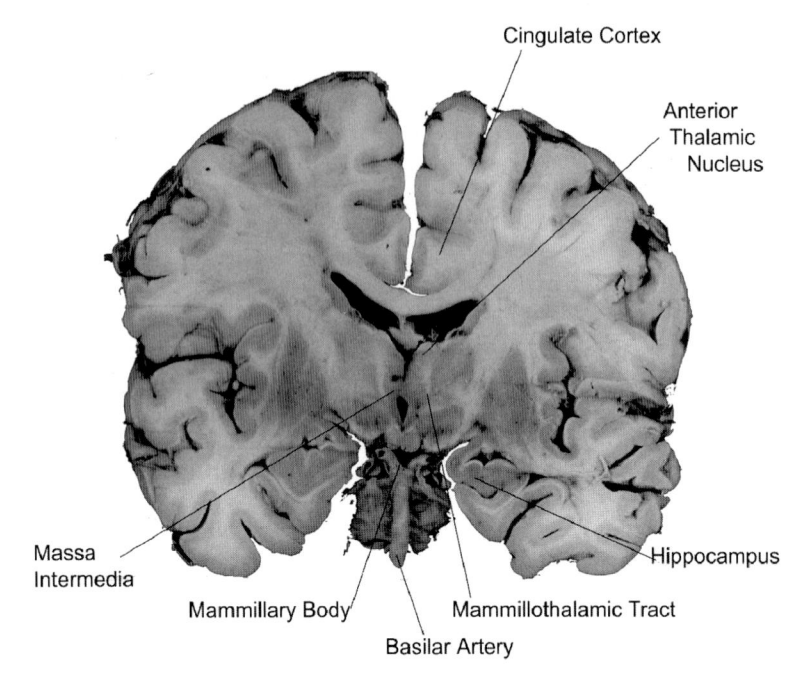

Fig. 14.2 Coronal section through diencephalon showing hippocampus in medial temporal lobe and mammillothalamic tract leaving the mammillary body. (From EM Marcus and S Jacobson, Integrated neuroscience, Kluwer, 2003)

Fig. 14.1B, the brain stem and cerebellum has been removed and the entire cingulate gyrus is now revealed. Finally, in Fig. 14.1C, the diencephalon has been removed and now most of the course of the fornix can be seen. Figure 14.2 is a coronal section through the diencephalon showing the hippocampus in the medial temporal lobe and the mammillothalamic tract leaving the mammillary bodies mammillary bodies (also note the basilar artery on the anterior surface of the brain stem).

1. Parahippocampal Gyrus of the Hippocampal Formation

The uncus (hook) forms the anteriormost regions in the parahippocampal gyrus on the medial surface of the temporal lobe. Its surface is the olfactory cortex and the principal nucleus of the amygdala lies internally (Figs. 14.1 and 14.2).

2. Amygdala (Figs. 14.3 and 14.4)

The amygdala is uniquely located to provide the intersection between the primary motivational drives of the hypothalamus and septum and the associative learning that occurs at the hippocampal and neocortical levels. The amygdala (Fig. 14.2)

I Limbic System

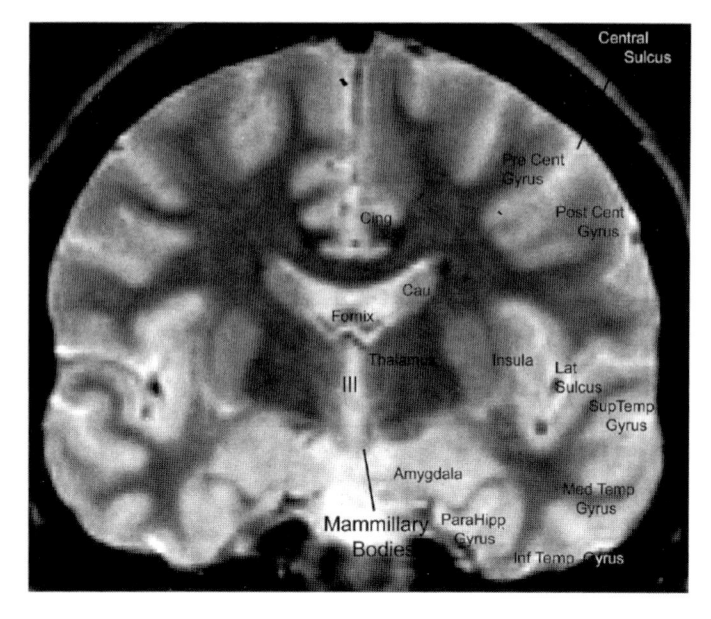

Fig. 14.3 Coronal section through mammillary bodies and amygdala. MRI-T2 (From EM Marcus and S Jacobson, Integrated neuroscience, Kluwer, 2003)

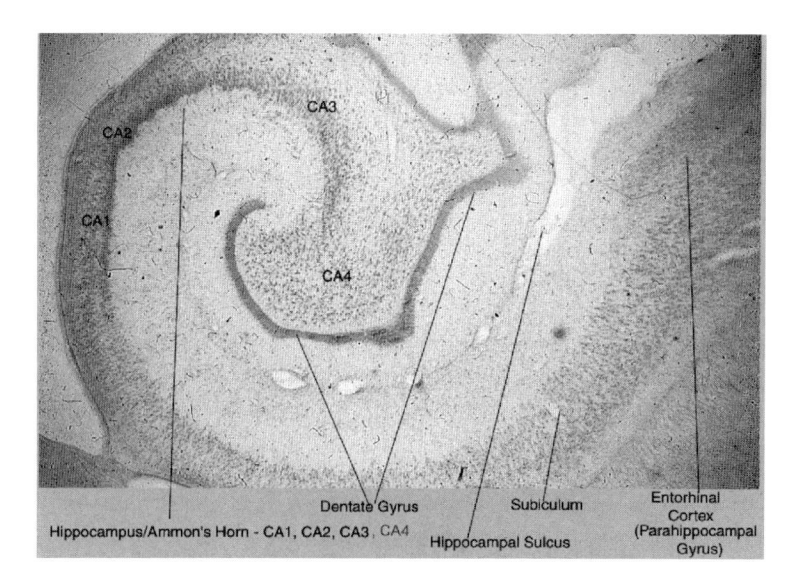

Fig. 14.4 The sectors in the hippocampus of the human. The dentate gyrus, subiculum, and related structures are demonstrated. Nissl stain. (From EM Marcus and S Jacobson, Integrated neuroscience, Kluwer, 2003)

Role of amygdala in fear & anger!

consists of three main groupings of nuclei: corticomedial, basolateral, and central. In addition to these principal nuclei, there are extratemporal neurons, including the nucleus of the stria terminalis and the sublenticular *substantia innominata*. The amygdala receives extensive projections from many cortical areas (the input from the prefrontal region is a very important inhibtiory circuit) and, in turn, sends projections to these areas. From the standpoint of the role of the amygdala in emotion and instinctive behavior, there are important connections with the basal forebrain, medial thalamus (medial dorsal nucleus), hypothalamus (preoptic, anterior and ventromedial and lateral areas), and the tegmentum of the midbrain, pons and medulla. In primates, as one expects in comparison to rodents, there has been a significant increase in the projections from and to the isocortex (neocortex) as opposed to the allocortex and mesocortex. These projections originate from (1) the multimodality sensory areas, (2) tertiary unimodal sensory association areas, (3) visual association areas that are particularly prominent in the primates, and (4) first- and second-order central sensory neurons of the olfactory system.

a. Connections of Amygdala

The more specific connections of the amygdaloidal nuclei are as follows:

1) Olfactory Cortex. The corticomedial group receives olfactory information from the lateral olfactory stria and interconnects with the contralateral corticomedial nuclei (via anterior commissure) and ipsilateral basolateral nuclei. The primary efferent pathway of the corticomedial nucleus is the stria terminalis, which projects to the septum, to the medial hypothalamus, including the preoptic nucleus of the hypothalamus, and to the corticomedial nucleus in the opposite hemisphere.
2) Limbic Nuclei. The central and basolateral nuclear grouping is associated with the limbic brain and has connections with the parahippocampal cortex, temporal pole, frontal lobe, orbital frontal gyri, cingulated lobe, thalamus (especially dorsomedial nucleus), catecholamine-containing nuclei of the reticular formation, and substantia nigra.
3) The ventral amygdalofugal pathway projects from the central nucleus to the brain stem and to the septum, the preoptic, lateral, and ventral hypothalamus, and—in the dorsal thalamus—to the dorsomedial, intralaminar, and midline thalamus.

b. Stimulation of the Amygdaloid Region

In monkeys, cats, and rats, stimulation produces aggressive behavior. The stimulated cats have a sympathetic response of dilated pupils, increased heart beat, extension of claws, piloerection, and attack behavior. When the stimulus stops, they

I Limbic System

[handwritten margin note: Lesions do not necessarily reduce aggression!]

become friendly. Animals will even fight when the amygdala is stimulated and stop fighting if the stimulus is off. Eating, sniffing, licking, biting, chewing, and gagging might also be stimulated here. In contrast, studies in the cat indicate that stimulation of the prefrontal areas will prevent aggressive behavior. In humans, stimulation of the amygdala produces feelings of fear or anger ((an amygdala prefrontal circuit) Cendes et al.,1994). A role in sexual behavior has also been postulated, although this might be more prominent in the female.

c. Lesions of Amygdala

In the studies of Amaral, selective bilateral lesions of the amygdala in the adult monkey significantly decreased the fear response to inanimate objects such as an artificial toy snake. The social interactions of the lesioned monkey with other members of a colony were significantly altered. The lesioned animals were more sociable, with more sexual and nonsexual friendly contacts with other members of the colony. They were described as socially uninhibited. In the male, aggression was decreased, but, occasionally, females had an increase in aggressive behavior. When the bilateral lesions were produced at 2 weeks of age, there was a decrease in fear responses to inanimate objects but an increase in fear responses to other monkeys. This latter effect interfered with their social integration into the colony.

[handwritten margin note: → not the centre of fear?]

Human Studies

In the human, bilateral lesions of the amygdala have been produced for control of aggression. In patients with bilateral lesions of the amygdale, there is an impairment of the ability to interpret the emotional aspects of facial expression (Adolph et al., 1998; Anderson and Phelps, 2000). Patients with high functioning autism have a similar disorder (Adolph et al., 2001).

When a quantitative MRI is performed in such autistic patients, there is significant enlargement of the volume of the amygdala. Howard et al. (2000) suggested that these results might indicate that a developmental malformation of the amygdala (possibly an incomplete neuronal pruning) might underlie the social, cognitive impairments of the autistic patient.

3. Hippocampal Formation (Figs. 14.4, 14.5, and 14.13)

This region includes the hippocampus, parahippocampus (entorhinal area and posterior segment of the piriform region), and dentate gyrus. This region of the limbic system has been of critical importance in our understanding of both the

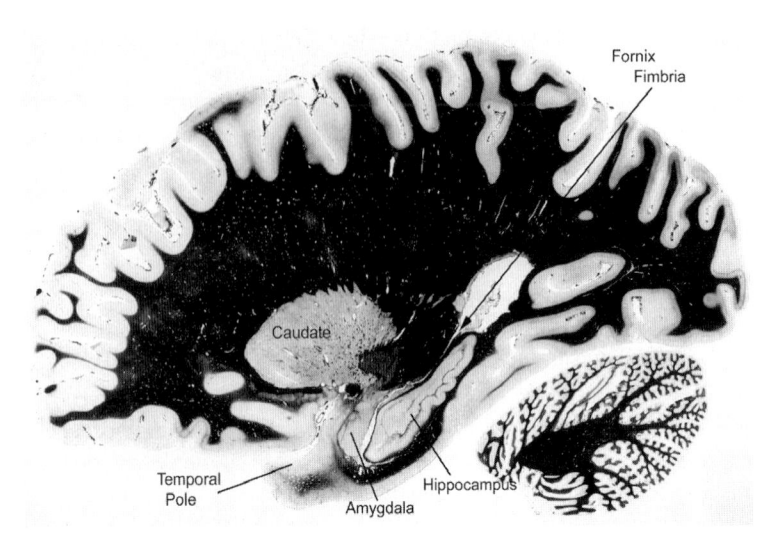

Fig. 14.5 Sagittal section. Temporal lobe with amygdala and hippocampus. Myelin stain (From EM Marcus and S Jacobson, Integrated neuroscience, Kluwer, 2003)

clinical aspects and underlying biological substrate of memory and of complex partial epilepsy.

Anatomical Correlates

From an anatomical standpoint, compared to other lobes of the brain, the temporal lobe is a complex structure. It contains four diverse regions:

1. Neocortex (six layers): superior, middle, and inferior temporal gyri (Chapter 9)
2. Paleocortex (three layers): the olfactory cortex, and the archi cortex-hippocampal formation and subiculum
3. Mesocortex [a transitional type of six-layer cortex (transitional between neocortex and allocortex]: entorhinal/parahippocampal gyrus (the large posterior segment of the pyriform region), presubiculum, and para subiculum
4. Cortical nucleus: the amygdala (discussed earlier).

The hippocampus is phylogenetically the older part of the cerebral cortex termed the allocortex (different from the six-layered cortex) and consists of three layers: polymorphic, pyramidal, and molecular. [The most primitive cortex is the paleocortex (old) of the olfactory bulb.] The cortex adjacent to the hippocampus changes from three layers to six layers and is classified as the transitional mesocortex and includes the parahippocampal gyrus (medial to the collateral sulcus), including the entorhinal cortex. The piriform (pear-shaped) lobe consists of the lateral olfactory stria, uncus, and the anterior part of the parahippocampal gyrus (Figs. 14.1B and 14.1C).

I Limbic System

Hippocampal Sectors (CA1, CA2, CA3, CA4)

The hippocampus is divided into sectors based on cytoarchitectural differences. (In the older literature, the hippocampus was designated as Ammon's (also corpus Ammon's) (from the ram's horn = CA) CA1, the sector closest and continuous, with the subiculum is referred to as the Summer's sector. This sector (Fig. 14.4) is most severely affected by cell loss following hypoglycemia, anoxia, and status epilepticus (Fig. 14.4). However, this selective vulnerability might also involve CA3 and CA4, with relative sparing of CA2 and the dentate granule cells. The subsequent gliosis (mesial sclerosis) of the hippocampus is the pathology found at surgery or autopsy in 75% of cases of complex partial seizures arising in the hippocampus.

The hippocampal regions are interconnected by a commissure, the hippocampal commissure. In the primate, the dorsal hippocampal commissure originates in the entorhinal cortex and presubiculum. The ventral commissure originates in the CA3 sector and the dentate hilus primarily in their more rostral areas (e.g., in relation to the uncus). CA4 is the end folium or end blade that merges with the hilus of the dentate gyrus.

Cytoarchitecture of the Hippocampus

In the hippocampus, the three layers are as follows:

1. Molecular cell layer, in which the apical dendrites of the pyramidal cells arborize. This layer is usually divided into a more external stratum lacunosum (moleculare) and a more internal stratum radiatum. It is continuous with the molecular layer of dentate gyrus and adjacent temporal neocortex.
2. Pyramidal cell layer (stratum pyramidal) contains pyramidal cells that are the principal cells of the hippocampus. Dendrites extend into the molecular layer and Schaffer axon collaterals arise from pyramidal neurons and synapse in the molecular layer on dendrites of other pyramidal cells.
3. Polymorphic layer (stratum oriens), in which the basilar dendrites of the pyramidal cells are found. It contains axons, dendrites, and interneurons. This layer in CA4 is continuous with the hilus of the dentate gyrus. Only the hippocampus sends axons outside the hippocampal formation.

Dentate Gyrus

Anatomy. The dentate gyrus Fig. 14.4 fits inside the hippocampus and, is similar to the hippocampus with three layers: molecular, granular, and polymorphic. The fibers of the dentate gyrus are confined to the hippocampal formation, whereas the hippocampal fibers leave the hippocampal formation through the fornix and project to either the septum or to the mammillary bodies and mesencephalic tegmentum.

Cytoarchitecture. The following three layers are found:

1. Molecular layer. This contains dendrites of granule cells.
2. Granule cell layer. Small neurons replace the pyramidal cell layer of the hippocampus. The granule cells are unipolar; all of the dendrites emerge from the apical end of the cell into the molecular layer. Efferent neurons from the granule cells are mossy fibers that synapse only with cells of hippocampal areas CA2 and CA3.
3. Polymorphic cell. Also referred to as the hilus, this layer is continuous with CA4 of hippocampus. The pyramidal and granule cell neurons are excitatory, utilizing glutamine as the transmitter; in addition, there are inhibitory interneurons (GABA-ergic), basket cells, in the polymorphic layer of both the hippocampus and dentate gyrus. There are also mossy cells in the polymorph layer, probably excitatory interneurons. There are also scattered interneurons in the molecular layers

Entorhinal Region (Fig. 14.4)

The entorhinal/parahippocampal gyrus forms the large posterior segment of the piriform region. This is Brodman's area 28 and constitutes the bulk of the parahippocampal gyrus. This area has extensive interconnection with the higher association cortex throughout the neocortex and also receives olfactory information from the olfactory stria. This is the major pathway for relating neocortex to the limbic cortex of the hippocampus.

The connections of the hippocampal formation are summarized in Table 14.2

Selective Vulnerability of Hippocampus

This occurs in diseases such as anoxia and hypoglycemia. In addition, the hippocampus appears to be significantly involved in most seizures of temporal lobe origin.

a. *Why Does This Selective Vulnerability Occur?* The CA1 region is rich in NMDA receptors. The dentate hilus and the CA3 sector are rich in kainate receptors. Activation of these receptors by glutamate would allow a considerable entry of calcium ions into the pyramidal neurons, beginning a virtual cascade. The pyramidal neurons of these sectors compared to CA2 and the granular cells of the dentate gyrus contain very little calcium-buffering protein and repeated activation of these pyramidal neurons could result in cell death.
b. *Seizure Activity.* Most seizures beginning after age 15 are classified as partial, and the majority of these are classified as complex partial. Approximately 75% of the complex partial seizures arise in the temporal lobe; the remainder arise in the frontal lobe. Although these seizures might arise in the temporal neocortex, the majority arise in the mesial temporal structures, particularly the

Table 14.2 Connections of hippocampal formation (Fig. 14.6)

Afferent Input
1. Perforant pathway from adjacent lateral entorhinal cortex onto granule cells of dentate gyrus and the alvear pathway from the medial entorhinal cortex. The entorhinal areas, in turn, receive their input from many of the highest levels of the associational cortex in the frontal, orbital, temporal (amygdala), parietal, and cingulate cortex through the cingulum bundle.
2. Septum through the stria terminalis.
3. Contralateral hippocampus, via the hippocampal commissure.

Efferent Output. The axons of pyramidal neurons in the hippocampal and subiculum to:
1. Lateral mammillary nuclei, habenular nuclei, anterior midline and intralaminar thalamic nuclei, lateral hypothalamic nuclei, midbrain tegmentum, and periaqueductal gray via a column of the fornix.
2. Septum, preoptic region, par olfactory. and cingulated cortex via the precommissural fornix and supracallosal.

hippocampus. The hippocampus has a low threshold for seizure discharge; consequently, stimulation of any region that supplies hippocampal afferents or stimulation of the hippocampus itself might produce seizures. Hippocampal stimulation produces respiratory and cardiovascular changes, as well as automatisms (stereotyped movements) involving the face, limb, and trunk.

c. *Why Do Not All of the Various Pathological Processes in the Hippocampus Produce Seizures?* There appears to be a critical age during infancy and early childhood for the acquisition of the pathology that is associated with seizures originating in the mesial temporal lobe. There might be an age-related remodeling of intrinsic hippocampal connections. Whether a single episode of epileptic status during infancy is sufficient to produce these changes or whether multiple episodes might be required is still under discussion.

d. *Is the Hippocampal Pathology Alone Sufficient to Explain the Complex Partial Seizures of Temporal Lobe Origin?* As reviewed by Gloor (1997), many of the specimens examined after surgery or at autopsy also demonstrate extensive changes in the amygdala as well as mesial and lateral isocortex. [Overall. however, in the autopsy studies of Margerison and Corsellis (1966), the most frequent site of damage and cell loss was in the hippocampus.]

4. Other Cortical Regions of the Limbic System

Cingulate Cortex

This region (Figs. 14.1A and 14.1B) receives reciprocal innervation from the anterior thalamic nuclei, contralateral and ipsilateral cingulate cortex, and temporal lobe via the cingulum bundle, as well as projecting to the corpus striatum and most of the subcortical limbic nuclei. The cingulate cortex is continuous with the parahippocampal gyrus at the isthmus behind the splenium of the corpus callosum.

Stimulation of the cingulate cortex also produces respiratory, vascular, and visceral changes, but these changes are less than those produced by hypothalamic stimulation. Interruption of the cingulum bundle, which lies deep to the cingulate cortex and the parahippocampal gyrus, has been proposed as a less devastating way to produce the effects of prefrontal lobotomy without a major reduction in intellectual capacity. (For additional discussion of the effects of stimulation and of lesions of the anterior cingulate area on autonomic function and behavior, refer to Devinsky et al. (1995).

Prefrontal Areas

This region will be discussed below.

II. Principal Pathways of the Limbic System

An understanding of the circuits in the limbic system is critical for our appreciating the important function role of the limbic system.

A. *Fornix (Fig. 14.6)*

Note that the fornix is the efferent pathway from the hippocampus and subiculum and is connected to the hypothalamus, septum, and midbrain. This tract takes a rather circuitous pathway to reach the hypothalamus. The fornix originates from the medial surface of the temporal lobe and runs in the medial wall of the inferior horn of the lateral ventricle, passing onto the undersurface of the corpus callosum at the junction of the inferior horn and body of the ventricles, and running in the medial wall of the body of the lateral ventricle suspended from the corpus callosum. The fornix finally enters the substance of the hypothalamus at the level of the interventricular foramen.

Different portions of the fornix have specific names:

1. Portio fimbria (fringe) found on the medial surface of the hippocampus and consists of fibers from the fornix and hippocampal commissure
2. Portio-alveus (groove), fibers covering ventricular surface of the hippocampus
3. Portio tenia (tape), connecting the hippocampus to the corpus callosum
4. Portio corpus body, underneath the corpus callosum and enters the hypothalamus
5. Portio columnaris, in the substance of the hypothalamus
6. Precommissural portion (in front of the anterior commissure), that enters the septum (from the hippocampus), and

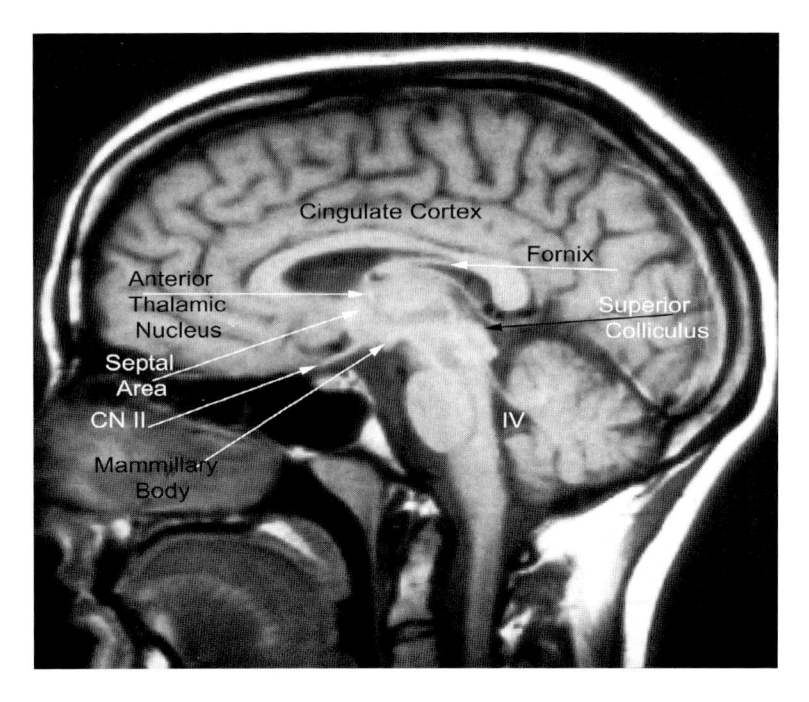

Fig. 14.6 Sagittal section showing the medial surface of the cerebrum demonstrating the limbic structures that surround the brain stem and are on the medial surface of the cerebrum, with portions of the Papez circuit labeled: fornix, mammillary bodies, and anterior thalamic nucleus cingulate cortex

7. The postcommissural portion (behind the anterior commissure), that distributes in the mammillary bodies and midbrain (from subiculum).

B. Circuits in Limbic Emotional Brain

1. Papez Circuit (Fig. 14.6): The identification of this fiber system by Papez in 1937 was one of the important landmarks in understanding the circuitry in the emotional brain the Papez, circuit includes the following:

a. Origin of fornix many hippocampal pyramidal cells synapse on the pyramidal cells of the adjacent subiculum. The pyramidal cells of the subiculum constitute the origin of most of the fibers in the fornix.
b. Fornix projects primarily to the mammillary bodies of the hypothalamus and septum.
c. Mammillary bodies then project via the mammillothalamic tract to the anterior nuclei of the thalamus.
d. Anterior thalamic nuclei project to rostral cingulate gyrus.

e. Cingulate gyrus then projects via the fibers in the cingulum bundle to the parahippocampal/entorhinal cortex, which subsequently projects to the hippocampus, completing the reverberating system.

2. Entorhinal Reverberating Circuit/Perforant Pathway (Fig. 14.7)

This circuit is an adjunct to the Papez circuitry just described:

a. Perforant pathway from adjacent lateral entorhinal cortex onto granule cells of dentate gyrus.
b. The alvear pathway from the medial entorhinal.
c. The CA3 pyramidal cells project via Schaffer collaterals to CA1 pyramids.
d. CA1 pyramidal cells collaterals project to the subiculum and entorhinal cortex.
e. The entorhinal cortex projects back to each of the neocortical polysensory areas.

3. Stria Terminalis

The pathway of the stria terminalis (the efferent fiber tract of the amygdala) parallels the fornix, but it is found adjacent to the body and tail of the caudate nucleus on its medial surface. The stria terminalis interconnects the medial corticoamygdaloid nuclei, as well as connecting the amygdale to the hypothalamus and septum.

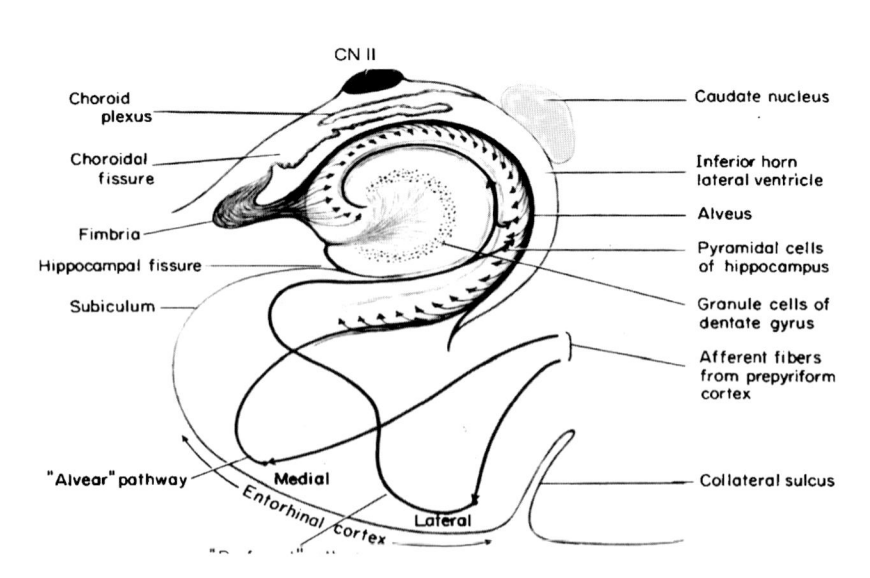

Fig. 14.7 Perforant pathway. (Modified from M.B. Carpenter (1970), Core text of neuroanatomy, William and Wilkins) (From EM Marcus and S Jacobson, Integrated neuroscience, Kluwer, 2003)

4. Ventral Amygdalofugal Pathway

This fiber pathway originates primarily from the basolateral amygdaloid nuclei and, to a lesser degree, from the olfactory cortex, spreads beneath the lentiform nuclei, and enters the lateral hypothalamus and preoptic region, the septum, and the diagonal band nucleus. Some of these fibers bypass the hypothalamus and terminate on the magnocellular portions of the dorsomedial thalamic nuclei.

5. Cingulum

This fiber bundle is found on the medial surface of the hemisphere and interconnects primarily the medial cortical limbic areas, especially the parahippocampal formation, with one another.

6. Intracortical Association Fiber System

The superior and inferior longitudinal fasciculus and uncinate fasciculus. Through these associational fiber systems, the limbic regions on the lateral surface of the hemisphere have strong interconnections with polysensory and third-order sensory cortical regions throughout the cerebral cortex.

7. Limbic System and the Corticospinal and Corticobulbar Pathway

The bulk of the pyramidal pathway originates in the motor/sensory strip. However, the planning for most movements begin with a "thought" in the prefrontal association areas. A possible example of release of the bulbar areas for emotional expression from frontal-lobe control is seen in pseudobulbar palsy. Such patients exhibit inappropriate responses to situations because the interrupted corticonuclear/corticobulbar pathway no longer dampens the strong descending autonomic/limbic input to the brain stem cranial nuclei. The resultant inappropriate behavior is termed *emotional liability*. A sad story might trigger excessive crying; a funny story might trigger excessive laughter.

III. Temporal Lobe

The temporal lobe includes (a) the allocortex of the hippocampal formation, (b) the transitional mesocortex bordering this area, (c) the neocortex then occupies all of the lateral surface and the inferior temporal areas, and (d) the amygdale, already discussed earlier.

A. Auditory and Auditory Association

Area 41, the primary auditory cortex, is located on the more anterior of the transverse gyri of Heschl (Fig. 8.4 in Chapter 8). This area is a primary special sensory region and has a pronounced layer IV (granular or koniocortex) similar to but considerably thicker than areas 17 and 3. Area 41 receives the main projection from the medial geniculate nucleus. Its tonal organization is organized with the lowest frequencies projecting to the more rostral areas. Some investigators limit the term "Heschl's gyrus" to the more anterior of the transverse gyri and localize the primary auditory cortex to the posterior aspects of that gyrus. The remaining neocortical areas of the temporal lobe are homotypical, with well-defined six cortical layers.

Areas 42 surrounds area 41 and receives association fibers from that area. Area 22, in turn, surrounds area 42 and communicates with areas 41 and 42. Both areas 42 and 22 are auditory association areas, and in the dominant hemisphere, they are important in understanding speech, as Wernicke's area.

B. Visual Perceptions

The inferior temporal areas have significant connections with area 18 of the occipital lobe, providing a pathway by which visual perceptions, processed at the cortical level, can then be related to the limbic areas.

C. Symptoms of Disease Involving the Temporal Lobe

When one considers disease processes affecting the temporal lobe, signs and symptoms obviously do not follow the precise subdivisions outlined earlier. Lesions in the more posterior temporal areas often involve the adjacent posterior parietal areas (i.e., the inferior parietal lobule). In contrast, anteriorly placed lesions of the temporal lobe often involve the adjacent inferior frontal gyrus. Lesions spreading into the deeper white matter of the temporal lobe (particularly in its middle and posterior thirds) often involve part or all of the optic radiation.

1. Symptoms Following Stimulation of the Temporal Lobe

In patients with seizures beginning after the age of 15 years, the seizures of focal onset are much more common than seizures that are not of focal origin. The latter are classified as primary generalized or idiopathic. *The most common*

III Temporal Lobe

site of origin of focal seizures is the temporal lobe. Depending on the origin of the seizure, a variety of symptoms may be produced. These may be classified as simple partial if awareness is retained or complex partial if awareness is not retained and the patient is amnestic for the symptoms. Patients with simple partial seizures do not have automatisms; those with complex partial seizures often do have automatisms. A simple partial seizure might progress to a complex partial seizure and subsequently secondarily generalize. As we have indicated, there is mixed symptomatology during partial seizures of temporal lobe origin. Brain tumors involving the temporal lobe often produce seizures, as illustrated below. However, the most common cause of complex partial seizures involving the temporal lobe is a pathologic process referred to as *mesial temporal sclerosis*. Note that 75% of patients with complex partial epilepsy have a pathology involving the temporal lobe and 25% have a pathology located in prefrontal areas. Seizures involving the temporal lobe are frequent and produce a variety of symptoms. These symptoms are outlined in Table 14.3 with the most likely anatomical correlate. Refer to Marcus and Jacobson (2003) for a more detailed discussion.

Table 14.3 Anatomical localization of symptoms from temporal lobe seizures or stimulation of the temporal lobe

Symptoms	Anatomical region
Autonomic phenomena	Amygdala. L
Fear, less often anger or other emotion	Amgydala
Crude auditory sensation, tinnitus	Heschl's transverse gyrus (primary auditory projection)
Vestibular sensations, dizziness vertigo	Superior temporal gyrus dizziness/vertigo posterior to auditory cortex
Arrest of speech	Wernicke's area and posterior speech area dominant hemisphere
Olfactory hallucinations Uncinate epilepsy of Jackson	Olfactory cortex of the uncus (termination of lateral olfactory stria)
Experiential phenomena: complex illusions such as déjà vu, déjà vécu, jamais vu, and illusions of recognition, visual and auditory hallucinations	Lateral temporal cortex: primarily superior temporal gyrus. However, ictal illusions is abolished by lesions limited to lateral temporal neocortex.
Automatisms = repetitive simple or complex often stereotyped motor acts, most commonly involving mouth, lips limbs, etc.	Primary or secondary bilateral involvement of amygdale, hippocampus. Invariably accompanied by confusion and amnesia for the acts
Defects in memory recording followed by amnesia for the event	Hippocampus

Note: Déjà vu =sensation of familiarity, jamais vu = sensation of strangeness, déjà vécu = perception is dream like partial epilepsy with temporal structures

2. Symptoms from Ablation of or Damage to the Temporal Lobe

a. Effects on Hearing

Unilateral lesion of auditory projection area may result in an inability to localize a sound. Bilateral lesions may produce cortical deafness.

b. Aphasia

Destruction of area 22 in the dominant hemisphere produces a Wernicke's receptive aphasia. Such a patient not only has difficulty in interpreting speech but also, in a sense, has lost the ability to use previous auditory associations. Lesions that deprive the receptive aphasia area of Wernicke in the dominant temporal lobe (area 22) of information from the auditory projection areas of the right and left hemispheres result in pure word deafness. Such a patient can hear sounds and words but is unable to interpret them.

c. Visual Defects

Unilateral lesions of the temporal lobe that involve the subcortical white matter often produce damage to the geniculocalcarine radiations. Since the most inferior fibers of the radiation that swing forward around the temporal horn (representing the inferior parts of the retina and referred to as "Meyer's loop") are often the first to be involved, the initial field defect may be a contralateral superior field quadrantanopia.

d. Klüver-Bucy Syndrome

This syndrome results from bilateral ablation of the temporal pole, amygdaloid nuclei, and hippocampus in the monkey (Klüver and Bucy, 1937). The animals could see and find objects, but they could not identify objects (visual agnosia). The animal showed marked deficits in visual discrimination, particularly with regard to visual stimuli related to various motivations. They also had a release of very strong oral automatisms and compulsively placed objects in their mouths, which, if not edible, were dropped. They had a tendency to mouth and touch all visible objects and manifested indiscriminate sexual practice. They willingly ate food not normally a part of their diet (e.g. cornbeef sandwiches). They showed a lack of response to aversive stimuli and had no recollection or judgment. The animals also lost their fear (release phenomenon) and, in the case of wild monkeys, became tame and docile creatures. They had a marked absence of the fight-or-flight response and also lost fears of unknown objects or objects that previously had frightened them. (See amygdale above). The problems in visual discrimination might reflect a disconnection of the visual association areas from the amygdala/hippocampal areas.

e. Memory

Bilateral damage to the hippocampus produces a marked impairment of the ability to form new associations, an inability to establish new memories at a time when remote memory is well preserved.

f. Unilateral Effects on Memory

Patients with resection of the left temporal lobe might lose the ability to retain verbally related material but gain the ability to retain visually related material relevant to the right hemisphere. The reverse is true for right temporal lobectomy patients.

g. Psychiatric Disturbances

Waxman and Geschwind (1975) have described an interictal personality disorder characterized by hyposexuality, hyperreligiosity, hypergraphia, and a so-called stickiness or viscosity in interpersonal relationships. In some patients, a psychosis might be apparent in periods between seizures (the interictal period) that is often most severe during periods when the seizure disorder is well under control. In other patients, a transient psychosis might be related to the actual seizure discharge or the postictal period. There is controversy regarding this issue.

h. Aggressive Behavior

Aggressive behavior and episodic dyscontrol might be related to dysfunction of the amygdala or to damage to prefrontal areas, which inhibit the amygdala. This remains an area of considerable controversy (Mark and Ervin, 1970; Geschwind, 1983; Ferguson et al., 1986).

i. Complex Partial Seizures

Complex partial seizures, as discussed earlier, might follow mesial temporal sclerosis. The following case history illustrates many of the points just discussed regarding simple and complex seizures of temporal lobe origin.

Case History 14.1: Seizures from Left Temporal Lobe

Three months before admission, this 56-year-old, right-handed, male baker had the onset of 4-minute episodes of vertigo and tinnitus, unrelated to position, followed by increasing forgetfulness. Later that month, he had a generalized convulsive seizure that occurred without any warning. The patient then developed episodes of confusion and unresponsiveness, followed by a left frontal headache. One month before admission, the patient began to have minor episodes, characterized by lip smacking and a vertiginous sensation, during which he reported seeing several well-formed, colorful scenes. At times, he had hallucinations of "loaves of bread being laid out on the wall." In addition, he would have a perceptual disturbance (e.g., objects would appear larger than normal). He also had colorful visions and terrifying nightmares: "terrifying dreams, crazy things".

Neurological examination: Seizures observed: The patient had frequent transient episodes of distress characterized by saying, "Oh, oh, oh, my" and at times accompanied by the automatisms: fluttering of the eyelids, smacking of the lips, and repetitive picking at bedclothes with his right hand. Consciousness was not completely impaired during these episodes, which lasted from 30 seconds to 3 minutes. The patient reported afterward that at the onset of the seizure, he had seen loaves of bread on the wall and smelled a poorly described unpleasant odor. At other times, the olfactory hallucination was described as pleasant, resembling the aroma of freshly baked bread.

Mental status: The patient was disoriented to time, could not recall his street address, and could not pronounce the name of the hospital

Cranial nerves: A possible deficit in the periphery of the right visual field and a minor right central facial weakness were present.

Reflexes: A right Babinski sign was present.

Clinical diagnosis: Simple and complex partial seizures originating in left temporal lobe probably involving at various times left lateral superior temporal gyrus, uncus, amygdale, and hippocampus, with tumor the most likely etiology in view of age and the focal neurological findings.

Laboratory data: EEG: Frequent focal spike discharge was present throughout the left temporal and parietal areas consistent with a focal seizure disorder.

Imaging studies: Left carotid arteriogram indicated a large space-occupying tumor of the left temporal lobe and a possible avascular mass lesion in the left posterior lateral frontal.

Subsequent course: These episodes were eventually controlled with anticonvulsant medication. Seizures symptoms recurred 7 months after onset. An aura of unpleasant odor was followed by a generalized convulsion followed by four or five subsequent seizures of a somewhat different character (deviation of the head and eyes to the right, then tonic and clonic movements of the right hand spreading to the arm, foot, and leg lasting approximately 1,2 minutes, followed by a post-ictal right hemiparesis). He also experienced minor seizures characterized by sensory phenomena on the right side of the body. Neurologic examination now indicated progression with a marked expressive aphasia, with little spontaneous speech and difficulty in naming objects. There was a dense right homonymous hemianopia, a flattening of the right nasolabial fold, and a right hemiparesis, with a right Babinski sign. The symptoms and findings suggested that the basic disease process might well have spread to involve the adjacent areas across the sylvian fissure—the speech areas of the inferior frontal convolution, premotor areas, and sensory motor

IV Role of the Limbic System in Memory

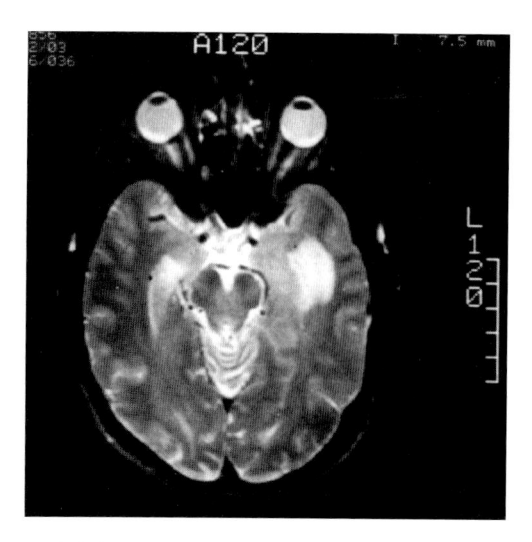

Fig. 14.8 Glioblastoma in left temporal lobe. **Case History 14.2.** Complex partial seizures with secondary generalization in a 47-year-old. MRI .T2 nonenhanced. Horizontal (From EM Marcus and S Jacobson, Integrated neuroscience, Kluwer, 2003)

cortex. At surgery a necrotic glioblastoma was found involving the superior temporal gyrus, the deeper temporal and extending superficially under the sylvian fissure to involve the adjacent posterior portion of the inferior frontal gyrus. A temporal lobectomy was performed (from the anterior temporal pole posterior for a distance of 6 cm).

Today, CT scan and MRI would be employed for early diagnosis in patients with this type of seizure disorder, as illustrate in Figure 14.8.

IV. Role of the Limbic System in Memory

A. Anatomical Substrate of Learning in Humans

There are two broad categories of learning: declarative and nondeclarative:

1. Declarative Learning (Explicit). This refers to the conscious recollection of fact or events. This system is rapid and one trial might be sufficient. There are essentially three major stages.

    a. Immediate or short-term working memory. This happens in a matter of less than 10 seconds with a capacity of about 12 items. It is best exemplified in simple digit repetition or the immediate repetition of three or five objects. Monkeys with prefrontal lesions are unable to consistently perform the task. The neural substrate involves reception in the primary sensory area and relay to the adjacent sensory association area (18, 22, and 5/7), then relay to the prefrontal area for very short-term storage, then relay to the motor association cortex, and then to the motor cortex so one can respond to the questioner.

   b. Long-term memory, labile stage. This is a stage of transcription and transduction. It has duration of about 20–180 minutes. RNA and protein synthesis are involved in this stage. Also, it occurs in the entorhinal, parahippocampal gyrus, hippocampus, and anterior and dorsomedial thalamus. These structures as previously discussed are inter-related in a circuit as the limbic system. From a clinical standpoint, we assess this stage with the delayed recall test. A list of three or five objects previously learned in the initial phase of the test are to be recalled at 5 minutes. For individuals who are nonfluent, the actual objects might be presented among a larger group of objects, with the patient making a selection. The ability to learn and retain new experiences is also referred to as retentive memory. In anterograde amnesia, this stage of memory is defective; following a traumatic or other brain injury, the patient cannot learn new material or recall events since the injury. In contrast, in retrograde amnesia, the patient cannot recall events that occurred prior to the injury. This latter problem might relate to interference with memories that were still in a labile stage of recording and not yet fully consolidated in the long-term remote memory. The initial extension backward in time often exceeds the expected duration of this labile stage. However, with time, there is usually a shrinking of this retrograde defect.

   Note that this labile stage of memory is also referred to as recent memory. Note also that in the past, this labile stage has also been referred to as short-term memory resulting with confusion, with the first stage of memory referred to as immediate or short-term working memory

   c. Long-term memory stage of remote memory. This stage is not discretely localized, rather it represents a diffuse storage throughout the cerebral cortex. Although not discretely localized, it must be remembered that remote memories can be triggered by stimulation of the lateral temporal lobe. The studies of Milner (1972) have shown that bilateral removal or damage of the hippocampus produces great difficulty in learning new information, a condition called anterograde amnesia. The excitatory amino acids l-glutamate and l-aspartame are important in learning and memory through the mechanism of long-term ostentation. Longterm potentation is a long-lasting facilitation that occurs after repeated activation of excitatory amino acid pathways and is most pronounced in the hippocampus and might be related to the initiation of the memory trace the amygdala and its connections are associated with the conditioning of the emotional response of fear [refer to review of Le Doux (2000)].

2. Nondeclarative (Implicit or Reflexive). This refers to a nonconscious alteration of behavior by experience. This type of learning is slow and requires multiple trials and includes motor habit and skilled learning. The anatomical substrate for motor habit and skilled learning includes motor cortex, striatum and cerebellum. The anatomical substrate for classical and operant conditioning includes (amygdala for emotional responses and cerebellum for motor response.

B. Disorders of Recent Memory; the Amnestic Confabulatory Syndrome of Diencephalic Origin; Wernicke–Korsakoff's

Wernicke, in 1881, described a common syndrome of relatively acute onset occurring in alcoholics or nutritionally deficient patients and consisting of a triad: mental disturbance (confusion and drowsiness), paralysis of eye movements, and an ataxia of gait of cerebellar origin. This syndrome now carries the label *Wernicke's encephalopathy*. The basic cause of the syndrome is a deficiency of the vitamin thiamine. This deficiency of thiamine and the other B-complex vitamins also produces a peripheral neuropathy, in some but not all of the patients. The basic pathological process consists of a necrosis of neural parenchyma and a prominence of blood vessels due to a proliferation of adventitial cells and endothelium with patchy hemorrhages about these vessels. The pathological findings involve the gray matter surrounding the third ventricle aqueduct and fourth ventricle. Lesions, in general, are most prominent in the mammillary bodies and the medial thalamic areas. With treatment with intravenous thiamine, the extraocular findings and the drowsiness will usually rapidly clear. The cerebellar findings might persist as an alcoholic cerebellar degeneration. In approximately 75% of cases; memory problems persist with an inability to record new events. An anterograde amnesia will be present as well as a retrograde amnesia for events surrounding the acute illness. A significant disorientation for time and a confabulation might persist. Korsakoff had encountered patients in state mental hospitals with the more persistent memory problems; such patients are now labeled with the term *Korsakoff's psychosis*. In the studies of Victor and his associates of the patients with a persistent memory deficits summarized in 1972, pathological examination of the brain revealed persistent lesions in the dorsomedial and anterior nuclei of the thalamus. The following case history illustrates the Wernicke–Korsakoff syndrome.

Case History 14.3. Wernicke–Korsakoff Syndrome (Patient of Dr. John Sullivan and Doctor John Hills)

A 62-year-old, white, right-handed male had been a known heavy alcoholic spree drinker for many years. The patient would drink large quantities of wine for 6–8 weeks at a time. Two years previously, the patient had been admitted to the Boston City Hospital because of delirium tremens (tremor and visual hallucinations). Two months prior to admission shortly following the death of a brother-in-law, the patient began his most recent drinking spree. He unaccountably found himself in Florida, not knowing where he was and why he was there. Apparently, he had drifted aimlessly for 5 weeks with no definite food intake for a month. The patient was brought back by his family and hospitalized at his local community hospital with diplopia, ataxia, and marked impairment of memory. The patient had complaints of numbness of his finger tips and unsteadiness of gait. The patient was shortly thereafter referred to the Neurology Service at the New England Medical Center.

General physical examination: The liver was enlarged with the edge palpated approximately 2-1/2 finger breadths below the costal margin.

Neurological examination: Mental status: The patient was markedly disoriented for time and place. At times, the patient thought that he was in New Jersey; at other times, he stated correctly that he was in Boston. The patient, with suggestion, would recall his apparent travels that day to various other locations within and outside Boston (he was actually in the hospital). Confabulation was also evident when it was suggested that he had recently seen various fictitious persons. The patient was unable to state his age. At times, the patient often indicated to visitors that his mother and father were still alive, although both parents had been dead for over 20 years. He appeared to have little insight for his disorientation in time. He had no insight as to his condition or for the reason for his hospitalization.

The patient could name various objects correctly when these were presented to him and yet he was unable to retain any memory of which objects had been presented to him 5 minutes previously. The patient was able to provide his birth date correctly. However, he could not give his address or telephone number or select the President of the United States on a multiple choice test. The patient was able to do two- and three-figure additions and multiplications without difficulty. He was able to do the initial subtractions in the serial 7 test but then lost track of the number to be subtracted.

Cranial nerves: There was horizontal diplopia on right lateral gaze. The minor degree of separation of images, however, did not allow precise identification of the muscle involved. A minor weakness of the right lateral rectus was suspected. Horizontal nystagmus was present on lateral gaze; bilaterally and vertical nystagmus was present on vertical gaze.

Motor system: Strength was intact except for a minor degree of weakness in the distal portions of the lower extremities evident on ankle dorsiflexion and toe dorsiflexion. The patient walked on a narrow base with eyes open and showed no evidence of an ataxia of gait. On a narrow base with eyes closed, the patient had a positive Romberg test.

Reflexes: Patellar and Achilles deep tendon reflexes were absent (0) even with reinforcement.

Sensory system: Pain and touch were decreased in the lower extremities below the midcalf. Vibratory sensation was absent at the toes and decreased over the tibia to a marked degree and to a lesser degree over the knees. There was, to a lesser extent, a decrease in the upper extremities at the finger tips and wrists. Position sense was decreased at fingers and toes.

Clinical diagnosis: Wernicke' encephalopathy plus nutritional poly neuropathy.

Laboratory data: Normal and EEG was normal.

Hospital course: The patient was treated with thiamine, 50 mg daily. There was a significant improvement in extraocular functions. There was no significant change in his mental condition or peripheral neuropathy. Evaluation 3 months later indicated persistent disorientation for time and place and severe selective deficits in memory (delayed recall was still grossly defective). At that point the additional diagnosis of Korsakoff's syndrome was appropriate.

C. Other Lesions of the Diencephalon and Adjacent Regions Producing the Amnestic Confabulatory Disorder Seen in the Korsakoff Syndrome

The following are a list of other lesions

1. Infarcts of the medial or anterior but not the posterior thalamic areas will produce amnesia (refer to von Cramon et al., 1988). Graff-Radford et al. (1990) concluded that small infarcts strategically located so as to interfere both with the mammillothalamic tract (hippocampal related neural structures) and ventral amygdalofugal pathway (which is adjacent to the mamillothalamic tract) produced diencephalic amnesia. Damage to the anterior thalamic area, the mammillary bodies, and their connections impair the neural systems related to hippocampus. Damage to the dorsomedial thalamic nuclei and their connections impair systems related to the amygdala and frontal systems. Isolated lesions of one system alone do not result in amnesia; combined lesions are required. In a recent MRI study by Tatemichi et al. (1992), of paramedian thalamopeduncular infarction, persistent amnesia was observed only when the dominant anterior thalamic nucleus or mammillothalamic tract was damaged.
2. Tumors of the posterior but not the anterior hypothalamus might be associated with this syndrome.
3. Lesions in the fornix might also be associated with significant problems in memory recording. Such damage is likely to occur with colloid cysts or with the surgical procedure necessary to remove this potentially life-threatening nonmalignant tumor The effects are most prominent if bilateral but might also occur with unilateral left-sided damage. Recall that the fornix is the major outflow pathway from the hippocampus.
4. Tumors involving the posterior or anterior portion of the corpus callosum have been associated with defects in recent memory. In some of the anterior corpus callosum tumors, infiltration of both prefrontal areas has occurred.
5. Lesions of the basal forebrain: rupture of anterior communicating aneurysms.

D. The Amnestic Confabulatory Syndrome Following Lesions of the Hippocampus and Related Structures

In 1957, Scoville and Milner reported the significant effects on memory that followed bilateral resection of the anterior two-thirds of the hippocampus and parahippocampal gyri, including the uncus and amygdala. The surgical procedure had been performed in a patient with intractable seizures with bilateral temporal lobe epileptic spike foci. The syndrome that resulted was similar to the amnestic confabulatory syndrome described above. Subsequently, Penfield and Milner reported similar results following unilateral removal of the temporal lobe where preexistent disease was present in the un-operated contralateral temporal lobe.

Studies in the monkey subsequently indicated that selective bilateral lesions of the hippocampus or of the parahippocampal and entorhinal cortex produced a significant deficit for novel learning. Combining the lesions produced an even greater effect. Bilateral lesions of the amygdala produced a marked alteration of emotional response but no impairment of memory. Additional discussions of selective anterograde versus retrograde amnesia will be found in Marcus and Jacobson (2003). Damage to the hippocampus and related structures might occur in relation to trauma and vascular disease. In the case of vascular disease, the cause is usually embolic occlusion of the posterior cerebral arteries, although similar effects might occur in patients with severe upper basilar artery stenosis with decreased perfusion of the posterior cerebral arteries. In some patients, unilateral disease of the posterior cerebral artery, usually embolic, produces infarction of the hippocampus and or the thalamus. In 85% of the infarcts, the lesion is on the left side. Transient ischemia involving the posterior cerebral artery is probably the cause of a relatively common but misnamed syndrome of transient global amnesia (misnamed because the amnesia is selective although both anterograde and shrinking retrograde amnesias are present). The hippocampus, as discussed earlier, also has a selective vulnerability to the effects of anoxia, hypoglycemia, and herpes simplex infections of the brain.

E. Progressive Dementing Processes

Dementia refers to a progressive impairment of previously intact mental facilities. Overall, the incidence of dementia increases from 0.4% in the 60–64-year age group to 3.6% in the 75–79-year group and to 23.8% in the 85–93-year age group. In general, the most common cause of progressive impairment in the older population is the degenerative disease knows as Alzheimer's disease. Alzheimer's disease alone accounts for at least 55% of all cases of dementia. Alzheimer's disease combined with vascular disease accounts for an additional 12% of cases, with an overall figure of 67% of cases. When the process began before the age of 65 years, the designation presenile dementia has been used in the past; when the process began after the age of 65 years, the designation senile was employed. The basic disease however is the same. Grossly at end stage there is usually atrophy of the cerebral cortex involving the association cortex of the prefrontal, parietal, and temporal limbic regions but sparing the motor, sensory, and visual cortex. In early cases, the neuroimaging studies might not reflect the atrophy in the neocortical areas. Moreover, there are elderly patients without dementia who do demonstrate some degree of neocortical atrophy on imaging studies and also in the gross brain There is a significant correlation in Alzheimer's disease between the degree of atrophy of the hippocampus and the presence of dementia. This is best seen on MRI studies that utilize measurements of hippocampal volume (Figs. 14.9A and 14.9B); it is evident then that the earliest changes occur in the hippocampus and entorhinal cortex, both from the gross and microscopic standpoint.

The microscopic changes in *Alzheimer's* can be outlined as follows: (1) loss of large pyramidal neurons, (2) loss of neurons in certain subcortical neurons that

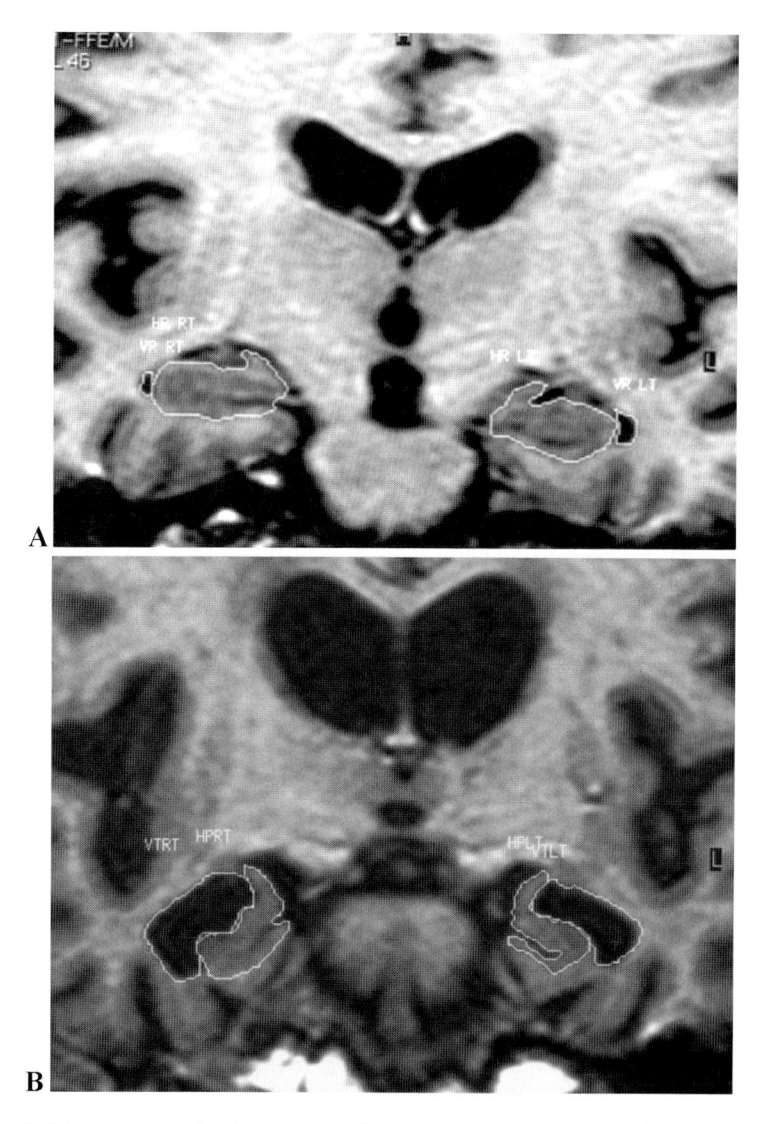

Fig. 14.9 Hippocampus. - **A** Hippocampus of a normal male patient, age 91; **B** hippocampus of a patient with Alzheimer's disease, age 72. Note the smaller hippocampus and larger ventricles in the patient with Alzheimer MRI, T2 weighted. (From EM Marcus and S Jacobson, Integrated neuroscience, Kluwer, 2003) Courtesy of Dr. Daniel Sax

project to the cerebral cortex, including the basal forebrain nucleus of Meynert (cholinergic), the locus ceruleus (noradrenergic), and the amygdala, (3) loss of dendritic spines of pyramidal neurons in the involved cortex, (4) neurofibrillary tangles composed of hyperphosphorylated and highly insoluble tau proteins that have formed aggregates, (5) dystrophic neurites: altered neuronal processes of axons and dendrites found free in the neuropil and surrounding senile plaques, and (6) extracellular plaques containing insoluble fibrils of amyloid beta protein.

From a clinical standpoint, the earliest changes are found in those memory functions mediated by the hippocampus and entorhinal cortex. In contrast to other common types of degenerative dementia (Lewy body and frontal-temporal) in the early stages, personality and behavior are generally well preserved. As the disease progresses, these functions and, eventually, motor functions are also affected. The magnitude of the problem from a societal and fiscal standing are staggering. The over-75 and over-85 age groups are the fastest growing sector of the population.

Alzheimer's disease has been called the silent epidemic; millions of patients will require care, straining the resources of families and institutions, as is demonstrated in the Case History 14.4.

Case History 14.4: Alzheimer's Disease

A 64-year-old, right-handed, white male was initially evaluated for impairment of recent memory on 05/01/97. Initially, the wife indicated this had been progressive over 4 years, but she was subsequently able to date this back to age 54 years, 10 years previously. The patient was seen 4 years previously by another neurologist in the department; at that time, there were no focal features, but his mental status examination indicated an inability to recall any of four objects. He was oriented for place and person but made errors on time orientation; his date was off by 1 month and the day of the week was incorrectly stated. The memory problems had progressed. In addition, there were now personality changes. He no longer participated in those activities that had previously been of great interest. He was also having some problems in finding words. There were no problems in the activities of daily living and he was not getting lost in his familiar environments.

Family history: There was no history of neurological disease in his siblings.

Neurological examination: Mental status: He had particular problems in time orientation. He was able to do the immediate recitation of three objects and of a test phrase but could remember none of these on a delayed recall test. He also had difficulty coping with a test figure. The patient was often tangential in his answers and demonstrated inappropriate joking.

Cranial nerves: Intact.

Motor system: Intact; however, premotor/frontal lobe functions were abnormal: (a) He had a release of the instinctive grasp reflex; (b) he had difficulties performing the Luria three-stage motor sequences demonstrated by the examiner (slap thigh with palmar surface, then with ulnar surface of hand, and then with ulnar surface of closed fist).

Clinical diagnosis: Alzheimer's disease.

Laboratory data: CT scan demonstrated significant dilatation of the temporal horns with atrophy of the hippocampus in addition to a general increase in lateral ventricular size. There was also blunting of the angles of the frontal horns. SPECT scan was within normal limits except for a slight decrease in perfusion in the left parietal region.

Subsequent Course: The patient was begun on treatment with 5 mg per day of donepezil (Aricept), a centrally acting acetylcholinesterase inhibitor, and a high dosage of vitamin E (a possible antioxidant). When reevaluated at 1 month, his family reported improvement in memory. There was particular improvement in the delayed-recall section of the exam. There was, however, no change in personality. He still showed inappropriate jocularity when he could not answer a question and at times was tangential in his answers. In January 1999, the family stopped the donepezil again because he was becoming more agitated, with the rationale that this might be due to this medication. Although the medication was restarted, behavioral disturbances became a major problem. He became aggressive, telling his wife to get out of the house, and began to wander at night. In December of 1999, he was placed in a nursing home, at which point he was still able to walk. Telephone follow-up with his wife in July of 2001 indicated that he was no longer agitated. He talked very little. He recognized his own name but shows no other response to questions. He occasionally recognized close relatives. He is usually in a wheelchair and requires assistance to walk. He was "stiff and afraid" when requested to walk. He had been incontinent for the last 2 years and requires diapers because of Alzheimer's disease.

Comments: Memory problems had begun insidiously at approximately age 54. Initially, these problems in the older adult might be attributed to a benign age-related process in which there is minor difficulty in accessing information such as names, although the ability to incorporate new information is still intact ("wait a minute, the name or word will come to me eventually"). As is often the case, this patient was not evaluated neurologically until several years later at age 60. Administration of the acetylcholinesterase inhibitor temporarily improved memory function for approximately 18 months However, by age 67, behavioral disturbance and urinary incontinence were becoming major problems and he could no longer be managed in his home and day care setting. This resulted in his placement in a nursing home. At age 69, he was now described as relatively nonfluent and very confused. He was usually restricted to a wheelchair but could walk with assistance. The "fear of walking" is often seen in patients with premotor/frontal gait apraxia.

Comment: This patient presents a typical example of this slowly progressive, dementing disease.

V. Prefrontal Granular Areas and Emotions

A. Anatomy and Functional Localization

The term *prefrontal* refers to those portions of the frontal lobe anterior to the agranular motor and premotor areas. Lateral–dorsal convexity, mesial, and orbital surfaces are usually delineated. Most studies of stimulation or of ablation lesions involving the prefrontal areas have included, within the general meaning of the term prefrontal, not only those sectors of areas 9, 10, 11, and 12 found on the medial and lateral aspects of the hemisphere but also several adjacent areas. These areas all share a common projection from the dorsal medial nucleus of the thalamus. Another area, the anterior cingulate gyrus (area 24), is also included in the prefrontal area, although it might be considered part of the limbic system and relates to the anterior thalamic nuclei.

In the prefrontal cortex three regions are usually delineated: 1) lateral dorsal convexity, 2) medial or mesial and 3) orbital/ventromedial. All three are concerned with executive functions and project to the dorsomedial thalamic nucleus. In general, although there is some overlap, the lateral dorsal cortex relates to motor association executive function; the planning of movement sequences and action. The medial and orbital relate to the limbic emotional control executive system. All or some of these functions may be altered by lesions of the prefrontal areas. It must be noted that lesions (e.g., meningiomas) which were initially parasagittal or subfrontal in relation to the prefrontal areas as they progress to involve the promotor areas may include not only the changes in personality but also the release of instincitve tactile grasp and of a sucking reflexes. an incontinence of urine and feces may be present in addition to an apraxia of gait, an unsteadiness of gait which is apparent as the patient attempts to stand and as he begins to walk but clears up once the act has been initiated.

B. Connections of the Prefrontal Cortex

The multiple connections of the prefrontal areas are discussed by Damasio (1985), Goldman-Rakic (1987), Jacobson and Trojanowski (1977), Nauta (1964), Pandya et al. (1971), and Stuss and Benson (1986). Essentially, all sensory association and polysensory areas and the olfactory cortex project to the prefrontal areas. The prefrontal areas, in turn, project to the premotor, temporal, inferior parietal, and limbic cortex. Subcortical bidirectional connections are prominent in relationship to the dorsomedial nucleus of the thalamus, amygdala, and hippocampus (uncinate fasciculus). The resultant clinical picture might then include not only the changes in personality but also the release of an instinctive tactile grasp and of a sucking reflex. An incontinence of urine and feces might be present in addition to an apraxia of gait—an unsteadiness of gait that is apparent as the patient attempts to stand and as he begins to walk but clears up once the act has been initiated. The following case history illustrates many of these features of focal disease involving the prefrontal areas.

C. The Case of Phineas P. Gage

There can be no better illustration of the effects of prefrontal damage than the first well-documented case of the frontal lobe syndrome the crowbar case of Mr. Phineas P. Gage reported by Harlow in 1868. While building a street railway in

Boston in 1848, a pointed tamping iron shot through the skull of the patient, an efficient, well-balanced, shrewd, and energetic railroad foreman. The bar, 3.5 feet in length and 1.25 inches in greatest diameter, entered below the left orbit and emerged in the midline vertex, anterior to the coronal suture, lacerating the superior sagittal sinus in the process (This skull can be seen in the Warren Museum at Harvard University School of Medicine.) Following the injury, a marked personality change was noted. The balance "between his intellectual faculties and animal propensities" had been destroyed. He was "impatient of restraint or advice which it conflicts with his desires; at times, obstinate yet capricious and vacillating, devising many plans of future operations which are no sooner arranged than they are abandoned in turn for others appearing more feasible." In brief, he was no longer the old Phineas P. Gage.

D. Studies of Jacobsen and Nissen

The modern era of experimental studies on the effects of prefrontal lesions in primates began with the work of Jacobsen (Jacobsen, 1935; Fulton and Jacobsen, 1935; Jacobsen and Nissen, 1937) on chimpanzees. They demonstrated two major dysfunctions following a lobotomy in delayed response and emotional response:

1. The **delayed response** test. The hungry test subject observes food placed under one of two containers. A solid opaque screen is then interposed and after a delay of at least 5 seconds, the screen is raised. Impairment in this test of working memory correlates with a lesion in the dorsolateral prefrontal region.
2. The **emotional response**. The second alteration was noted in behavior. In humans, there was a correlation between location of pathology (trauma and mass lesions) with major behavioral syndromes (Stuss and Benson, 1986). A syndrome of "frontal retardation" or "pseudodepression" (abulia) manifested by apathy, nonconcern, lack of motivational drive, a lack of motivational drive, and a lack of emotional reactivity has been associated with lesions involving the frontal poles and/or the medial aspects of both hemispheres. In its most severe form, a syndrome of akinetic mutism occurs.

   In contrast, the *pseudo-psychopathology* syndrome is distinguished by a lack of inhibition facetiousness, sexual and personal hedonism, delusions of grandeur, and a lack of concern for others. The lesion location in these patients has been associated with ventromedical/orbital lesions.

   Patients with lateral orbital and lateral convexity pathology have been described as restless, hypekekinetic, explosive, and impulsive.

   In contrast patients with dorsolateral lesions manifest a rigidity and concreteness in their cognitive functions with impairment of the ability to abstract.

E. Functional Neurosurgery

1. Prefrontal Lobotomy and Prefrontal Leucotomy (Fig. 14.10)

Based on the initial reports of Jacobsen regarding the effects of prefrontal lobotomy on reducing emotional responses of the chimpanzee, Moniz, a neuropsychiatrist and, Lima, a neurosurgeon, introduced in 1936 the surgical procedures of prefrontal lobotomy to modify the behavior and affect of psychotic patients. Subsequently, Moniz introduced the procedure of prefrontal leucotomy, a bilateral disconnection of prefrontal areas from subcortical regions (thalamus and basal ganglia). The procedure did in many cases reduce severe anxiety and manic activity, but often had un-desired side effects on on emotional and cognitive capacities. Following prefrontal lobotomy or leukotomy, these patients often were impulsive and distractable. Their emotional response often were uninhibited with an apparent lack of concern over the consequences. A related finding was an inability to plan aheadfor future goal: at items the patients were unable to postpone gratification, responding to their emotional content. (In a sense a loss of the reality principle.) Prefrontal lobotomy has also been performed on patients with intractable pain from carcinoma; the patient still perceives the pain but can ignore it and is no longer anxious or fearful about it.

2. More Selective Procedures

a. **Interruption of anterior thalamic radiation or destruction of dorsomedial nucleus**. A somewhat similar but less drastic personality change can be produced by bilateral severance of the anterior thalamic radiation from the dorsomedial

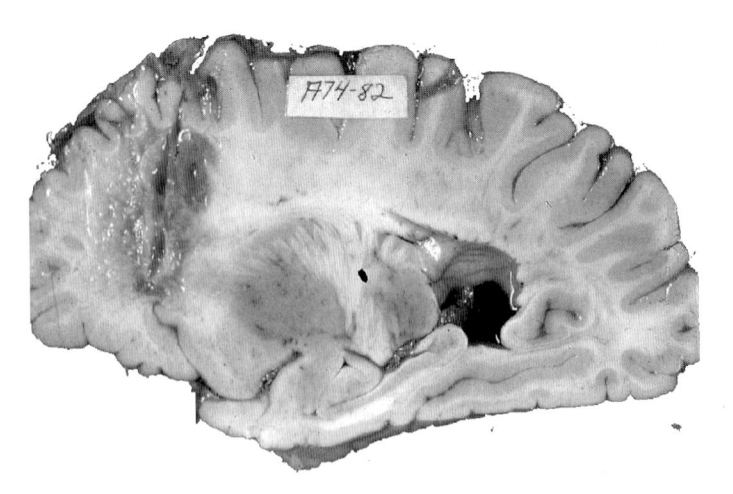

Fig. 14.10 Prefrontal lobotomy. A surgical section has separated the prefrontal connections with the thalamus. (Courtesy of Dr. Thomas Sabin and Dr. Thomas Kemper) (From EM Marcus and S Jacobson, Integrated neuroscience, Kluwer, 2003)

nucleus or by direct destruction of the dorsomedial nuclei or orbital cortex. This less massive resection of cortex lessens anxiety with fewer personality changes.
b. **Cingulotomy**. A more recent and more specific approach to problems of severe anxiety, manic behavior, and chronic pain was the approach of stereotaxic anterior cingulotomy introduced by Ballantine et al. (1967).

3. Decline of Functional Neurosurgery

Surgery to produce personality change (prefrontal lobotomy or leucotomy) has since fallen into disuse (Valenstein, 1986; Diering and Bell, 1991). A number of lobotomies were performed in the 1940s and early 1950s for psychiatric reasons or to modify the emotional response of patients with chronic pain problems previously requiring large doses of narcotics. Such studies must be interpreted with a certain degree of caution. The lesions are produced in individuals with preoperative abnormalities of personality function. The effects are not necessarily those that the same lesion would produce in otherwise normal individuals.

There is, however, a considerable resemblance to the effects produced by trauma (to the prefrontal areas) in relatively normal individuals. The early results did suggest that these procedures did produce an alteration in the emotional response with a reduction of anxiety generated in conflict and painful situations. Emotional response was often detached from the pain and conflicts. However, the effects on emotion produced by the procedure can now be produced by the use of the tranquilizing drugs that were developed beginning in the middle 1950s. The development of these drugs, in addition to the prominence of frequent postoperative complications, including seizures and personality alterations, led to the discontinuation of the procedure. Following prefrontal lobotomy or leucotomy, these patients often were impulsive and distractible. Their emotional responses often were uninhibited with an apparent lack of concern over the consequences of their actions. A related finding was an inability to plan ahead for future goals; at times, the patients were unable to postpone gratification, responding to their motivations of the moment. (In a sense, a loss of the reality principle had occurred.) Although distractible, a certain perseveration of response was noted with an inability to shift responses to meet a change in environmental stimuli or cues. A rigidity and concreteness of response was apparent with deficits in abstract reasoning. [Additional aspects of emotion and personality have been discussed earlier in relation to the temporal lobe.]

F. Role of the Limbic System in Psychiatric Disorders

The involvement of the limbic system either from a structural or functional standpoint is central to the psychiatric disorders that affect large numbers of patients. These disorders might affect perception, cognition, and affect. The neurological substrate of many psychiatric disorders is discussed in by Marcus and Jacobson (2003).

VI. The Limbic Brain as a Functional System

A. Hierarchy of Function

The emotional brain is organized into a hierarchy of function proceeding from the spinal cord and reticular formation, including mesencephalic midbrain nuclei to the hypothalamus and thalamus and onto the limbic neocortical regions, and, finally, the prefrontal cortical region.

B. Reticular Formation

The reticular formation is the site where information is received from the peripheral nerves. This system is so organized that only certain stimuli trigger the system: a gate, to alert the brain. If it were possible for any response to trigger this system, then the individual's survival would be threatened. The response, however, is selective because throughout our lives we have evolved a set of emotional responses that determine whether we will respond to situations calmly or with rage or fear. What happens is that the reticular system, based on the sensory information with probably some subconscious cortical assistance, focuses the attention by sorting out the relevant information, thus enabling the central nervous system to continue functioning efficiently throughout a crisis.

C. Hypothalamus

By the time the data reaches the hypothalamus, there are already distinct, well-organized emotional responses. The hypothalamus, with some assistance from the thalamus, sets the level of arousal needed for the emotional state and organizes and mobilizes the cortical and subcortical centers (especially the autonomic nervous system). If the situation is nonthreatening, the normal operations of the viscera continue. In a mildly stressful situation, such as one involving anger or heavy work, some of the digestive processes slow down and the heart rate and blood flow increase. If the situation is threatening—triggering the reactions of fear, pain, intense hunger or thirst, or sexual arousal—most of the digestive processes slow down and heart rate increases.

D. Pleasure/Punishment Areas

Throughout the limbic brain are found pleasure or punishment centers. These were located by implanting electrodes at various subcortical sites in an animal and training it to press a bar that connects the electrode to an electrical current (Olds, 1958).

If the electrode is in certain pleasure centers, the animal will self-stimulate until it is exhausted. In fact, the animal would rather press the lever than eat. The pleasure centers are located throughout the limbic system, but especially in the septum and preoptic region of the hypothalamus. Other brain centers when stimulated produce fear responses: pupil dilation, piloerection, and sweating. In these "punishment regions," located in the amygdala, hypothalamus, thalamus, and midbrain tegmentum, the animal quickly stops pressing the bar. If the situation is nonthreatening, the normal operations of the viscera continue. In a mildly stressful situation, such as one involving anger or heavy work, some of the digestive processes slow down and the heart rate and blood flow increase. If the situation is threatening—triggering the reactions of fear, pain, intense hunger or thirst, or sexual arousal—most of the digestive processes slow down and heart rate.

E. *Limbic Cortical Regions*

The limbic cortical regions are strongly influenced by the emotional patterns set by the hypothalamus and amygdale, which are then transmitted with only a very few synaptic interruptions through the dorsomedial thalamus and anterior thalamus into the limbic cortex, where the emotional pattern is elaborated and efficiently organized. The neocortex (prefrontal cortex and, to a lesser degree, the temporal lobe), based on past experience, examines the situation, sorts out the emotional responses from the intellectual, and inhibits or controls the situation based on what past experience has proven to be expedient for individual survival. Consider the following examples: a mother responds to her baby's crying, while the father sleeps on; in the middle of the night, a jet airplane thundering overhead causes no response, but a whiff of smoke or breaking glass quickly arouses the central nervous system and keeps it focused; blood flow and respiration increase. Once the threatening situation passes, conditions quickly return to a normal balance between the sympathetic and parasympathetic nervous systems.

Chapter 15
Higher Cortical Functions

What is a higher cortical function? As one examines the abilities of a human, one is struck by our ability to use tools and create wonderful buildings or works of art. However, our ability to communicate by speaking and writing and reading, we believe, is the best example of a higher cortical function. These centers that are responsible for language are primarily in the dominant hemisphere. The motor type of aphasia (Broca's area) originates from damage to the inferior frontal gyrus whereas the sensory type of aphasia originates from damage to the superior temporal gyrus (Wernicke's area).

It is well to warn the student beginning the study of language function that prior to the development of modern neuroimaging this had been an area of much confusion, with much disagreement and multiple hypotheses. This discussion will be limited to the more practical problems of anatomical localization.

Dysarthria refers to a difficulty in articulation of speech from weakness or paralysis or from mechanical difficulties and it is not related to a problem in the cerebral cortex but to disease in the lower motor neurons or the muscles that control the production of speech or to a lesion in the corticobulbar pathway. In this chapter we will discuss language dysfunction due to lesions in the cerebral cortex and introduce the following terms: aphasia, apraxia, and dyslexia.

I. Cerebral Cortex and Disturbances of Verbal Expression

There are complex disturbances of verbal expression that occur at a time when the basic motor and sensory systems for articulation are intact. Similarly, there are complex disturbances in language function with regard to comprehension of written and spoken symbolic forms that occur at a time when the basic auditory and visual receptor apparatus is intact. We refer to these more complex acquired disturbances of language function as *aphasias*. In general, it is possible to relate these language disturbances to disease of the dominant cerebral hemisphere, involving the cerebral cortex or cortical association fiber systems. The term *dysphasia*, which is used in England interchangeably with aphasias, is used in the United States in reference to developmental language disorders as opposed to the acquired aphasias.

A. Cerebral Dominance

It is perhaps appropriate at this point to consider the question of cerebral dominance Most individuals are right-handed (93% of the adult population), and such individuals almost always (>99%) are left-hemisphere dominant for language functions. A minority of individuals (some on a hereditary basis) are left-hand dominant. Baseball appears to have collected a high percentage of these individuals as left-handed pitchers. Some left-handers are right-hemisphere dominant (50%), but a certain proportion are left-hemisphere dominant. It has been estimated that 96% of the adult population are left-hemisphere dominant for speech.

One might inquire as to why the majority of humans are right-handed. Although hand preference does not become apparent until 1 year of age, the studies of Yakovlev and Rakic (1966) would suggest that even before this age, there is an underlying anatomical basis for the dominance of the right hand. In the study of the medullae and spinal cords of a large number of human fetuses and neonates, the fibers of the left pyramid were found to cross to the right side in the medullary pyramidal decussation at a higher level in the decussation than fibers of the right pyramid. Moreover, more fibers of the left pyramid crossed to the right side than vice versa. Although the majority of pyramidal fibers decussated, a minority of fibers remained uncrossed. It was more common for fibers from the left pyramid to decussate completely to the right side of the lower medulla (and eventually the spinal cord). The minority of fibers that remained uncrossed was more often those descending from the right pyramid into the right side of the lower medulla. The end result, in the cervical region at least, was for the right side of the spinal cord and presumably the anterior horn cells of the right side of the cervical cord to receive the greater corticospinal innervation. There is then an anatomical basis for the preference or dominance of the right hand. Because the majority of fibers supplying the right cervical area have originated in the left hemisphere, on this basis alone one could refer to the dominance of the left hemisphere.

A study of the adult brain results in similar conclusions (Kertesz and Geschwind, 1971). The studies of Geschwind and Levitsky (1968) established that an actual anatomical asymmetry was present in the adult brain between the two hemispheres in an area significant for language function. That area of the auditory association cortex, posterior to Heschl's gyrus on the superior-lateral surface of the temporal lobe (areas 22 and 42) bordering the sylvian fissure and including the speech reception area of Wernicke, was found to be larger in the left hemisphere. Similar differences have been found in the brain of the fetus and newborn infant (Wada et al., 1975). The studies of LeMay and her associates (1972, 1978a, and 1978b), Galaburda et al., (1978), and Chui and Damasio (1980) elaborated on the asymmetries of the sylvian fissure and occipital lobe demonstrated on arteriography and computed tomography (CT) scan during life. Similar asymmetries were demonstrated in the brains of the great apes(e.g., chimpanzee), but not in the Rhesus monkey (LeMay and Geschwind, 1978). LeMay and Culebras (1972) studied the endocranial casts of human fossil skulls in which imprints of the major fissure could be noted and demonstrated similar differences in Neanderthal man.

Table 15.1 Milestones of normal development of speech in the child there is considerable variability

At 9–12 months, babbles, "mama, dada"
At 12 months, single words utilized and echoes sounds
At 15 months, has several words
At 18 months, has vocabulary of six words and possible hand dominance.
At 24 months puts two to three words together into a sentence

B. Development Aspects (Table 15.1)

It is clear that from a functional standpoint that a considerable degree of flexibility exists in the child with regard to dominance for language function. Thus, in a child under the age of 5 years, destruction of the speech areas of the dominant left hemisphere during prenatal or postnatal life produces only a minor long-term language disturbance (Varqha-Khademe et al., 1985). If the damage to the speech areas in the dominant hemisphere occurs between the ages of 5 and 12 years, a more severe aphasia occurs; however, some limited recovery of language might occur within a year. These cases suggest an equipotentiality of the two hemispheres for language function during the early years of development. In these cases under the age of 5 years, the right hemisphere must assume a dominant role in mediating those learned associations important in language and in the use of symbols either through the process of compensation and/or reorganization. If total destruction of the speech area of the dominant hemisphere occurs in adult life, only a minor recovery of language functions will occur.

Patients with left-hemisphere damage sustained under the age of 5 develop a strong left-hand preference. Patients with damage sustained after the age of 5 have only weak left-hand preferences or are ambidextrous. Additional discussion of the role of genetics in dominance and of the effects of early damage to the left hemisphere can be found in Dellatolas et al. (1993).

II. Aphasia: Dominant Hemispheric Functions

In approaching the patient with aphasia, it is well to keep in mind that textbook discussions of this problem are often artificial, in the sense that such discussions tend to deal with relatively isolated pure types of language disturbance, which have relatively specific localization. The actual patient with aphasia more often presents a mixed disturbance with damage to several areas; some have damage to all major areas in a global aphasia. This is not unexpected when one realizes that all of the major speech areas of the dominant hemisphere are within the cortical vascular territory of the middle cerebral artery. Other areas involved in related disorders are supplied by the anterior cerebral artery (the superior speech area) or by the posterior cerebral artery (those areas involved in pure dyslexia and agnosias). Willmes and Poeck (1993) and Kirshner (2000) provide additional reviews.

A. Cortical Areas of the Dominant Hemisphere of Major Importance in Language Disturbances (Fig. 15.1)

The following cortical areas are of major importane in language disturbances:

1. **Broca's motor aphasia or expressive speech center.** This is localized to the opercular and triangular portions of the inferior frontal convolution (areas 44 and 45).
2. **Wernicke's receptive aphasia area.** The auditory association area is found in the superior and lateral surface of the posterior–superior temporal gyrus (area 22 and adjacent parts of area 42).
3. **Inferior parietal lobule.** The angular and supramarginal gyri (areas 39 and 40) have at times been associated with Gerstmann's syndrome of dysgraphia, dyscalculia, left-right confusion, and finger agnosia.
4. **Supplemental motor cortex (superior speech area).** This might also play a role in language function.
5. **Basal temporal speech area.** Stimulation studies by Luders et al. (1988) have also identified a basal temporal speech area.
6. **Association fiber systems and speech.** In addition to these cortical areas, the association fiber systems relating these areas to each other and to other cortical areas are of considerable importance for certain types of language disturbance, as we will indicate later.

From a practical localization standpoint, we can, in a general sense, speak of patients as presenting a nonfluent/anterior type of aphasia (anterior to the central sulcus) or fluent type (posterior to the central sulcus; see Fig. 15.1 and Table 15.2).

B. Types of Aphasia

1. Nonfluent Aphasia: Anterior Aphasia

The anterior type of aphasia relates to Broca's motor aphasia area. The patient's speech can be generally described as nonfluent with little spontaneous verbalization. Patients with other conditions, including global aphasia, might also be nonfluent.

2. Fluent Aphasia: Posterior Aphasia

The patients with the posterior types of aphasia are fluent and do speak spontaneously. The lesions in such cases involve Wernicke's area in the posterior portion of the superior temporal gyrus or the inferior parietal lobule. However, because the posterior areas are relatively close to one another, it is not unusual for a single disease process to involve both areas. The fluent aphasias include (a) Wernicke's receptive aphasia, (b) conduction aphasia, (c) Gerstmann's syndrome, (d) dyslexia

II Aphasia: Dominant Hemispheric Functions

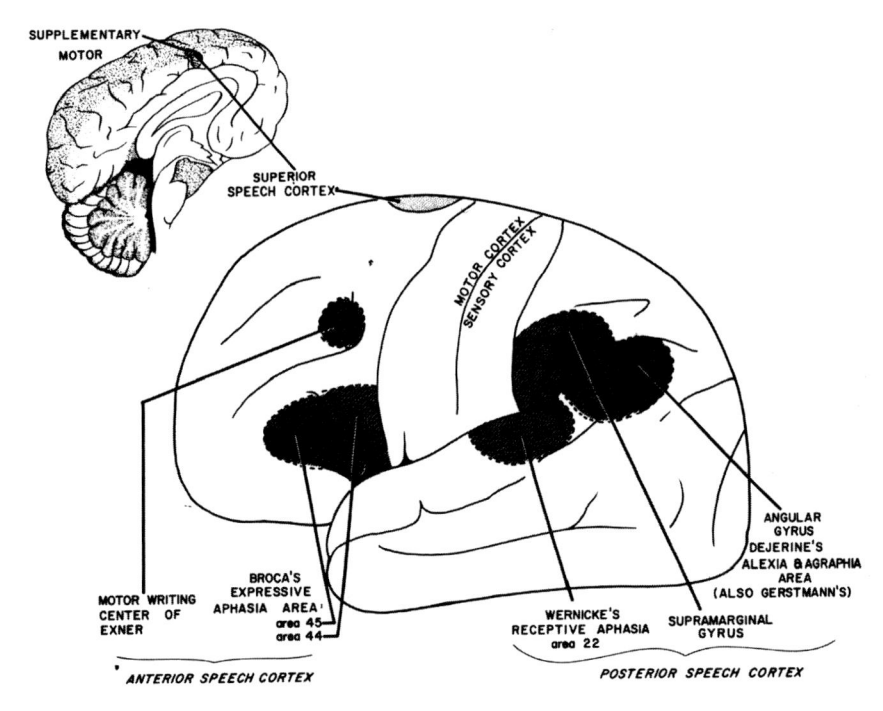

Fig. 15.1 Speech areas of the dominant hemisphere, summary diagram combining the data of pathological lesions and stimulation studies. Precise sharp borders are not implied. The designations of Penfield and Roberts (1959) the terms "anterior speech area", "posterior speech area", and "superior speech cortex. The basal temporal area is not labeled. Stimulation of these regions produces arrest of speech"

Table 15.2 Evaluation of language functions

1. Conversational speech:
   a. Amount: Nonfluent versus fluent. Mute refers to total absence.
   b. Articulation, rhythm, and melody, defective in nonfluent, normal in fluent. The nonverbal aspects of language prosody (the affective, intonation, and melodic aspects of speech) and gesture often involve the right hemisphere; refer to Ross (1993).
   c. Content:
      (1) Grammar (syntax): Nonfluent speech (e.g., Broca's aphasia) is often telegraphic (lacking the words or word parts and endings needed to express grammatical relations), whereas patients with fluent aphasia can produce long sentences or phrases with a normal grammatical structure.
      (2) Meaning and substance (semantics): Patients with fluent aphasia often demonstrate a number of abnormalities in terms of the choice of words.
   a. The precise word might be replaced by a nonspecific work or phrase.
   b. The correct word or phrase answer might be replaced by a phrase that describes the use of an object (circumlocution). Because such patients are fluent, casual examination might fail to detect such glib features.
   c. The correct word may be replaced by an inappropriate word or phrase substitution (verbal or semantic paraphasias) (e.g., "clocks for wristwatch").
   d. The correct word might contain replacement sounds or syllables, which are incorrect (literal or phonemic paraphasias) (e.g., "note for nose").
   e. The correct word might be replaced by a new word (neologism).

(continued)

Table 15.2 (continued)

2. **Repetition:** the ability to repeat words or phrases.
   a. Absent in a mute individual, in a severe nonfluent patient, or in a patient with a severe type of Wernicke's aphasia.
   b. Conduction aphasia, defective due to interruption of arcuate bundle.
   c. This might be preserved in an individual with otherwise intact language or preserved in isolation (so-called sylvian isolation syndrome).
3. **Comprehension of spoken language** is tested by providing spoken or written commands that do not require a verbal response, such as "hold up your left hand, close your eyes, and stick out your tongue."
4. **Word finding and selection:** Naming of objects or selection of the proper name of an object. Defect is defined as nominal aphasia (anomia).
5. **Reading:** Defect is defined as *alexia* or *dyslexia*.
6. **Writing:** Defect is defined as *agraphia* or *dysgraphia*
7. **Related functions:**
   a. Calculations: Defect defined as *acalculia* or *dyscalculia*.
   b. Perception in visual, auditory, or tactile sphere: Defects in recognition when the sensory modality and naming are otherwise intact are defined as agnosia. The main sensory modality involved is vision and this merges with optic aphasia, dyslexia, and nominal aphasia. The inability to recognize faces (prosopagnosia) and color agnosia are special examples (De Renzi, 2000).
   c. Apraxias: Defects in use of objects when motor and sensory functions are otherwise intact.

plus, (e) dyslexia without agraphia, and (f) the agnosias. Now, let us consider in greater detail language functions and the postulated anatomical correlations of these various language centers. In these correlations, it is important to make a distinction between acute and chronic effects. It is necessary first to review those components of language function that are evaluated in the neurological examination and reviewed in Table 15.2.

3. Effects of Stimulation of the Speech Areas

As demonstrated for the areas in Figure 15.1, stimulation of the dominant hemisphere (speech areas) produces arrest of speech.

Speech arrest or interference was elicited in five areas:

1. Primary motor cortex when stimulating the representation of muscles involved in speech of the dominant/non-dominant hemisphere
2. Negative motor areas (suppressor areas); that is, the supplementary motor areas and inferior frontal gyrus of the dominant hemisphere
3. Broca's area, dominant hemisphere only
4. Wernicke's area, dominant hemisphere only
5. Basal temporal language area, dominant hemisphere only.

Dinner and Luders concluded that any effect of stimulating the superior speech area was related to inhibition of motor activity. This area did not appear to have speech function per se. (Lesion-type studies do suggest possible speech functions.)

In stimulation of Broca's, Wernicke's, and the basal temporal areas at high stimulus intensities, complete speech arrest occurred with a global receptive and expressive aphasia; however, the patient could still perform nonverbal tasks. At lower intensities of

Table 15.3 Fluent aphasias

Location of aphasia/deficit	Dysfunction in language
1. Wernicke's. posterior temporal–parietal	Fluent, spontaneous speech, with poor comprehension and poor repetition
2. Pure word deafness disconnection from area 41	Fluent but inability to recognize new words
3. Conduction or repetition aphasia: disconnection of Wernicke's area from Broca's area; interruption of Arcuate bundle.	Relatively fluent spontaneous speech but with marked deficits in repetition
4. Inferior parietal lobe; angular and marginal gyri, disconnects parietal from visual association areas "region which turns written language into spoken and vice versa" (Geschwind, 1964)	Difficulty in writing (dysgraphia) and spelling, reading (dyslexia), and calculations (dyscalculia), and deficits in drawing. If patients also have finger agnosia and left-right confusion = Gerstmann's syndrome
5. Anomic or nominal aphasia. Lesion in posterior temporal–parietal area.	Fluent, normal comprehension and repeats. Recognizes objects when using visual or tactile cues but unable to name
6. Mixed Transcortical aphasia/isolation of speech area. Isolation of lateral temporal areas by interruption of border zone blood supply of MCA and ACA.	Repeats but with little comprehension and spontaneous speech
7. Transcortical aphasia: Sensory aphasia: rare. Isolation of Wernicke's area from posterior temporal–occipital areas with defect in supply from PCA	Speech pattern similar to Wernicke's except repetition is intact
8. Visual agnosia and dyslexia (alexia without agraphia); bBilateral destruction of visual associational areas 18 and 19 or destruction of area 17 of dominant hemisphere and areas 18 and 19 of no dominant hemisphere.	Lacks recognition of visual objects (word blindness)
9. Dyslexia; destruction of dominant visual cortex and selenium of corpus callous.	Inability to read

Source: Adapted from Benson (1985)

stimulation, the effects were incomplete: Speech was slow and simple but not complex tasks could be performed (both motor sequences and calculations). Simple but not complex material could be repeated. A severe naming defect (anomia) still occurred.

C. Nonfluent Aphasia: Anatomical Correlation of Specific Syndromes Involving Broca's area

In 1861, Broca described patients with nonfluent aphasia: little spontaneous speech with poor repetition but with considerable preservation of auditory comprehension The spontaneous speech that remained was agrammatic and telegraphic. Based on his examination of the lateral surface of the cerebral hemispheres at autopsy, he emphasized the damage to the third left frontal gyrus, the frontal operculum of the inferior frontal gyrus. Although considerable controversy was generated at the time, Broca's name has continued to be associated both with the disorder, nonfluent aphasia, and with the postulated anatomical area. As discussed by Damasio and Geschwind (1984), subsequent CT scans of the museum specimen of Broca's case indicated much more extensive damage than that described by Broca.

Broca's area is essentially a continuation of premotor cortex and can be considered a specialized motor association area with regard to the tongue, lips, pharynx, and larynx. This area is adjacent to the motor cortex representation of the face, lips, tongue, and pharyngeal muscles. It is not, therefore, unusual for these patients to also have a supranuclear-type weakness of the right side of the face and a right hemiparesis. This, by itself, is not sufficient to explain the problems in speech, which these patients manifest, because lesions involving the face area of the nondominant motor cortex do not produce the speech disturbance. Moreover, the tongue and pharynx receive a bilateral corticobulbar supply, so that a unilateral cortical lesion involving the motor cortex in this area is not a sufficient explanation. These patients appear to have lost the motor memories for the sequencing of skilled coordinated movements of the tongue, lips, and pharynx that are required for the vocalization of understandable single words, phrases, and sentences. They are usually able to vocalize sounds. At times, they might be able to vocalize single words into a proper grammatical sequence for telegraphic sentences. In the patient with a relatively pure form of Broca's motor aphasia, the capacity to formulate language and to select words in a mental sense is intact. The student might logically inquire as to whether the patient has also lost his motor memories or motor associations for making a nonverbal sequence of movements of the tongue and lips. Frequently, this is the case, as will be demonstrated in the illustrative case history. We can refer to this defect in skilled sequential motor function (at a time when motor, sensory, and cerebellar function are intact, and when the patient is alert and understands what movements are to be performed) as a motor apraxia. There are several types of apraxia with different localization. In general, we mght relate the more purely motor forms of apraxia to the motor association areas of the premotor cortex. One could then refer to patients with Broca's motor aphasia as patients with a motor apraxia of speech. The following presents an example of this type of aphasia. An embolus to the superior division of the left middle cerebral artery produced marked weakness of the face, tongue, and distal right upper extremity and a severe selective nonfluent aphasia. The hand weakness rapidly disappeared; the right central facial weakness, an apraxia of tongue movements and some expressive difficulties persisted. Most likely, the embolus fragmented and passed into the cortical branches supplying Broca's area. The Case History 15.1 is an example of Broca's motor aphasia.

Case History 15.1: Broca's Aphasia

A 55-year-old, right-handed, white housewife while working in her garden at 10:00 AM on the day of admission suddenly developed a weakness of the right side of her face and the right arm and was unable to speak. Nine years previously, the patient had been hospitalized with a 3-year history of progressive congestive heart failure secondary to rheumatic heart disease (mitral stenosis with atrial fibrillation). An open cardiotomy was performed with a mitral valvuloplasty (the valve opening was enlarged) and removal of a thrombus found in the left atrial appendage. Postoperatively, the patient was given an anticoagulant, bishydroxycoumarin (Dicumarol), to prevent additional emboli because atrial fibrillation continued until the present admission. She had done well in the interim. However, prothrombin time at the time of admission was close to normal; that is, the patient was not in a therapeutic range for anticoagulation.

General physical examination: Blood pressure was moderately elevated to 170/80. Pulse was 110 and irregular (atrial fibrillation). Examination of the heart revealed the findings of mitral stenosis: a loud first sound, a diastolic rumble, and an opening snap at the apex, as well as the atrial fibrillation.

Neurological examination: Mental status and language function: She had no spontaneous speech, could not use speech to answer questions and could not repeat words. She was able to indicate answers to questions by nodding (yes) or shaking her head (no), if questions were presented in a multiple-choice format. In this manner, it was possible to determine that she was grossly oriented for time, place, and person. She was able to carry out spoken commands and simple written commands, such as "hold up your hand, close your eyes." However, she had significant difficulty in performing voluntary tongue moments (such as "wiggle your tongue," "stick out your tongue") on command.

Cranial nerves: She tended to neglect stimuli in the right visual field and was unable to look to the right on command although the head and eyes were not grossly deviated to the left at rest. A marked right supranuclear (central) type facial weakness was present. The patient had difficulty in tongue protrusion and was unable to wiggle the tongue. When the tongue was protruded, it deviated to the right.

Motor system: There was a marked weakness without spasticity in the right upper extremity, most prominent distally and a minor degree of weakness involving the right lower extremity.

Reflexes: Deep tendon stretch reflexes were increased on the right in the arm and leg. Plantar responses were both extensor and more prominent on the right.

Sensory system: Intact within limits of testing.

Clinical diagnosis: Broca's: anterior aphasia due to embolus from heart to left middle cerebral artery, superior division.

Laboratory data: EEG was normal (48 hours after admission).

Electrocardiogram: Atrial fibrillation.

CSF was normal.

Subsequent course: Within 24 hours, a significant return of strength in the right hand had occurred; independent finger movements could be made. Within 48 hours after admission, the patient was able to repeat single words but still had almost no spontaneous speech. She appeared aware of her speech disability and was frustrated. She was able to carry out two- or three-stage commands, although tongue movements and perseveration remained a problem.

Within 6 days of admission, the patient used words, phrases, and occasional short sentences spontaneously and was better able to repeat short sentences. At this time, strength in the right arm had returned to normal, but a right central facial weakness was still present.

Two weeks after admission, she still had expressive disabilities consisting of word-finding and apraxic components in tongue placement and alternating tongue movements. Although complete sentences were used, sentence formulation in spontaneous speech was slow and labored with word-finding difficulties. Repetition was better performed. In reading sentences aloud, substitutions or word omissions were made. The patient did well in naming common pictures and in matching printed words to spoken words or printed words to pictures. She could write from dictation and would often respond preferentially in writing when difficulty in speaking was encountered. A right central facial weakness was still present.

With the passage of time, this type of patient would continue to show some degree of improvement (Mohr, 1973). Improvement could continue to occur over a 2-year period. Although lesions of Broca's area might produce the acute onset of a nonfluent aphasia, rapid amelioration of the deficit occurs even when the acute lesion involves underlying white matter, as well as the superficial cortex. In order to produce a more persistent severe nonfluent aphasia with good comprehension but poor repetition, more extensive lesions are required. In the CT scan correlation study of Naesar (1983), this larger lesion extended from Broca's area to the anterior parietal lobe, usually including the deep structures such as caudate, putamen and/or internal capsule. Ludlow et al. (1986) compared persistent versus nonpersistent nonfluent aphasia 15 years after penetrating head injuries of the left hemisphere in the Vietnam War. Both groups had nonfluent aphasia still present at 6 months after injury, with equal involvement of Broca's area. The group that failed to recover at 15 years had a more extensive left hemisphere lesion with posterior extension of the CT scan lesion into Wernicke's area and some involvement of the underlying white matter and basal ganglia.

Those patients with severe nonfluency had more extensive lesions of the subcortical white matter involving (1) the subcallosal fasciculus deep to Broca's area containing projections from the cingulate and supplementary motor areas to the caudate nucleus plus (2) the periventricular white matter near the body of the left ventricle deep to the lower motorsensory cortical area for the mouth. The MRI of Case History 15.2 with chronic residuals of a more persistent nonfluent aphasia is presented in Fig. 15.2.

Case History 15.2: Anterior motor aphasia. Figure 15.2

On May 30, at midday, this 37-year-old, right-handed woman with a past history of rheumatic fever, rheumatic heart disease, and "an irregular pulse" (atrial fibrillation) had the sudden onset of loss of speech, central weakness of right face, an inability

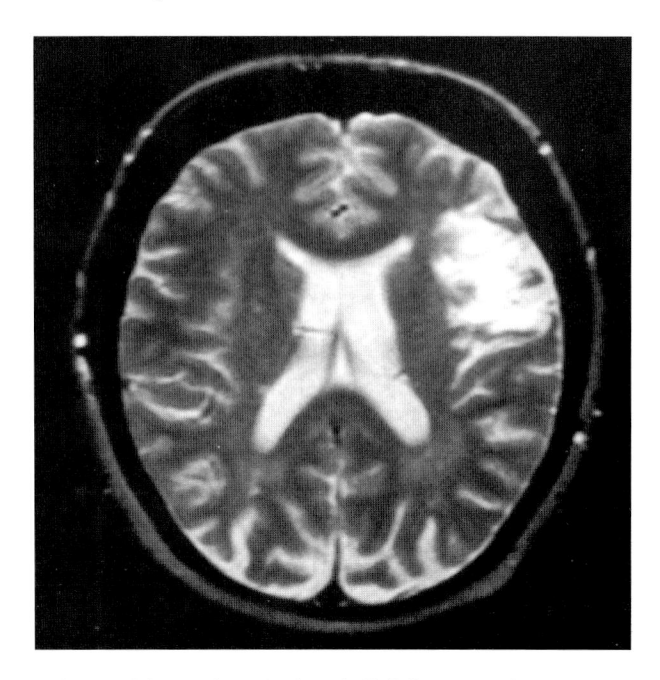

Fig. 15.2 Case History 15.2. Anterior aphasia embolic infarct secondary to rheumatic atrial fibrillation. MRI T2-weighted, horizontal section (From EM Marcus and S Jacobson, Integrated neuroscience, Kluwer, 2003)

to protrude the tongue, and a right hemiparesis Although hemiparesis, face and tongue problems disappeared within a week; recovery of speech was slower and limited. At 1 week, she could produce some two-syllable words.

Neurological examination 17 months later: Language function: Spontaneous speech was slow and relatively scanty. She could name objects slowly without difficulty, could do some simple repetitions, and could write from dictation. She could carry out two- and three-stage commands. She could read slowly aloud but had little comprehension of what she read.

Reflexes: A right Babinski sign and right-sided hyperreflexia were present.

Clinical diagnosis: Anterior aphasia: embolic infarct of superior division of middle cerebral artery.

Laboratory data: Although initial CT scans in May had been reported as negative, the MRIs in November and 4 years later demonstrated the infarct, which included Broca's area as well as adjacent frontal areas with predominant involvement of frontal operculum (inferior frontal gyrus) and adjacent middle frontal gyrus. In addition, a minor independent infarct was present in the right occipital area (presumably embolic to the calcarine artery).

Subsequent course: Similar findings were present at 5 years, June 1990.

D. Fluent Aphasias: Anatomical Correlation of Specific Syndromes (Wernicke's Aphasia and Wernicke's Area) (See Table 15.3)

In 1874, Wernicke described a type of aphasia that differed significantly from the nonfluent aphasia described by Broca and that had a more posterior localization. This type of patient has fluent spontaneous speech with poor comprehension and poor repetition. The patient has difficulty in understanding symbolic sounds i.e., words that have been heard) and is unable to carry out verbal commands. The patient is unable to use those same auditory associations in the formulation of speech. Moreover, the patient is essentially unable to monitor his own spoken words as he talks, because he lacks the ability to interpret and to compare the sounds, which he himself is producing to previous auditory associations. The end result is a patient who is fluent but who uses words combined into sentences that lack any meaning to the listeners. Word substitution (paraphasia) is frequent. In severe cases, not only are phrases and words combined into meaningless sentences, but also the patient combines syllables into words that have no meaning (jargon aphasia). The patient, however, is usually unaware of his errors. The patient usually fails to show the "rationale" frustration, which is characteristic of patients who have Broca's motor aphasia and who are aware of their errors. At times, the patient shows some awareness that his verbal responses are failing to deal with the environment. At times, the problems in communication are so severe that the patient becomes agitated and is mistaken for psychotic, as demonstrated in Case History 15.3. The anatomical localization for Wernicke's aphasia based on early autopsy studies has been confirmed by subsequent CT scan and MRI studies: the posterior temporal–parietal area. In the studies of Naeser (1983), the essential lesions involved Brodmann's area 22 (Wernicke area) of the posterior–superior temporal gyrus and the adjacent supramarginal gyrus of the inferior parietal area. In general, the responsible lesion is usually an infarct within the territory of the inferior division of the left middle cerebral artery, often embolic. There is usually very little evidence of a hemiparesis. In contrast, Broca's aphasia usually reflects infarction within the territory of the superior division of the left middle cerebral artery, and a right facial weakness and right hemiparesis are often present. (See Chapter 17 for additional discussion of vascular anatomy.). Case History 15.3 provides an example of a patient with a Wernicke's-type aphasia.

Case History 15.3: Wernicke's Aphasia

A 40-year-old, right-handed, white male had experienced a series of myocardial infarctions beginning 10 months prior to consultation. Angiographic studies indicated occlusion of the right coronary artery and the left anterior descending coronary artery, with stenosis of the circumflex artery, and a coronary artery bypass procedure was performed using extracorporeal circulation. The anesthetist noted that the patient was nonreactive 15 minutes after the last administration of halothane anesthesia. When the patient did wake up, he was quite agitated and combative and, in retrospect, he had transient difficulty using the right arm; and he continued to behave, despite medications throughout the postoperative period, in a disoriented and combative manner. Transfer to the psychiatry service was actively planned, but a neurology consultation was obtained.

Neurological examination: Mental status: He was alert but agitated. Speech was fluent and usually grammatical. He made use of many paraphasias and neologisms (nonsense words). He often appeared unaware of his errors. He was unable to name single objects such as pencil, cup, and spoon although he could recognize and demonstrate their use. He could read occasional individual words (e.g., cat, house). He could not comprehend or carry out either spoken or written commands. His attempts at spontaneous writing produced only a few letters. He was unable to write words or phrases from dictation. He did significantly better when asked to copy a printed sentence or phrase. Repetition of spoken phrases or sentences was poor. Calculations of even a simple nature were poorly performed.

Reflexes: A minor increase in deep tendon stretch reflexes was present on the right side. The plantar response on the right was equivocal. The left was flexor.

Clinical diagnosis: Wernicke's type of fluent aphasia, probably secondary to an embolus to the inferior division of the middle cerebral artery.

Pure Word Deafness

When seen in pure form, this is a relatively rare problem. More often, infracts tend to destroy adjacent areas as well. In cases of pure word deafness, the hypothetical center of auditory word association and comprehension (Wernicke's area) appears to be intact, but auditory information is unable to reach this center from the auditory projection cortex; in a sense, a disconnection has occurred. Both Wernicke's receptive aphasia and word deafness are often grouped together under the general term *auditory agnosia.*

Conduction or Repetition-Type Fluent Aphasia

This type of aphasia is characterized by a relatively fluent spontaneous speech but with marked deficits in repetition. Literal paraphasias are present and word-finding problems might be present. The patient is usually aware of the errors made. In contrast to the Wernicke's type of aphasia, auditory comprehension is usually relatively well preserved. The anatomical location of the lesion is somewhat variable (Geschwind, 1965; Benson et al., 1973; Naeser, 1983; Damasio and Geschwind, 1984). The lesion essentially serves to disconnect Wernicke's area from Broca's area (the posterior from the anterior speech areas). The essential fiber system is the arcuate fasciculus (Fig. 15.3). This fiber system arches around the posterior end of the sylvian fissure to join the superior longitudinal fasciculus. In the study of

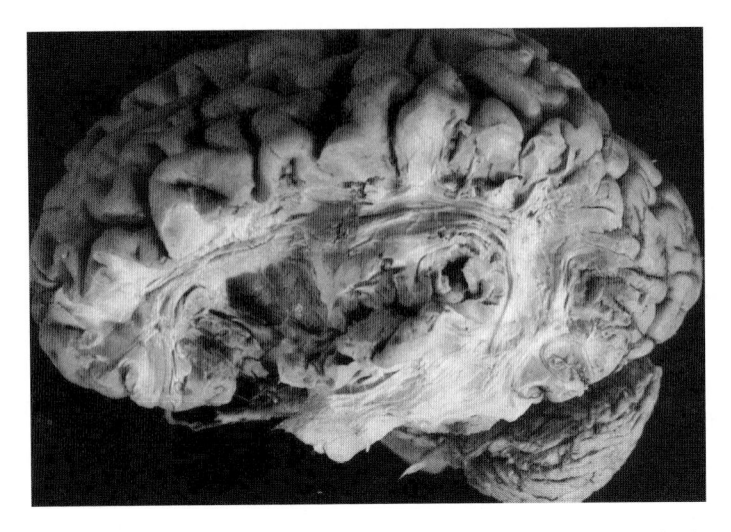

Fig. 15.3 The arcuate fasciculus: dissection of the long-fiber system, the superior longitudinal fasciculus and the arcuate fasciculus, passing beneath the cortex of inferior parietal, and posterior temporal areas. A small bundle of U fibers are also seen interconnecting adjacent gyri in the occipital and temporal lobes (From EM Marcus and S Jacobson, Integrated neuroscience, Kluwer, 2003)

Naeser (1983), this system was involved deep to the supramarginal gyrus and/or deep to Wernicke's area. In other cases, the white matter deep to the insular cortex has been involved (Damasio and Geschwind, 1984). In the resultant conduction aphasia, the patient understands spoken commands and is usually able to carry out complex instructions that do not require a repetition of language. Although able to speak spontaneously with little or no evidence of a Broca's aphasia, the patient is unable to repeat test phrases and sentences (e.g., "The rain in Spain falls mainly on the plain" or "no ifs, ands or buts")and is unable to write from dictation. Often, the syndrome is seen in a less than pure form, as in Case History 15.4. This is not surprising because the involved territory is that of the middle cerebral artery, predominantly the inferior division. The essential lesion must spare Broca's and Wernicke's areas to a considerable degree but significantly damage the interconnection of these areas. If there is significant damage to either Wernicke's or Broca's area, repetition will be defective on either the input or output side of the loop. A case demonstrating conduction aphasia will be presented as part of the next section.

Fluent Aphasias Associated with Lesions of the dominant inferior parietal Areas: Angular And Supramarginal Gyri

Geschwind (1965) indicated the critical location of the inferior parietal lobule in the human situated among the visual, auditory, and tactile association areas. As such, it might act as a higher association area between these adjacent sensory association areas. In earlier chapters, we have already identified this area as the posterior parietal multimodal sensory association area. Geschwind has suggested that correlated with the development of this area in the human (the angular and supramarginal gyri cannot be recognized as such in the monkey and are present only in rudimentary form in the higher apes), there has developed the capacity for cross-modality sensory–sensory associations without reference to the limbic system. He contrasts this situation with that in subhuman forms, where the only readily established sensory–sensory associations are those between a nonlimbic (visual, tactile, or auditory) stimulus and a limbic stimulus: those stimuli related to primary motivations such as hunger, thirst, and sex. (Such considerations, of course, do not rule out the establishment of limbic–nonlimbic stimuli associations in the human, particularly during infancy and childhood.) These theoretical concepts would suggest the underlying basis for human language functions. It is not surprising then that those aspects of language function that are most dependent on the auditory and visual stimuli and auditory–visual–tactile stimuli (reading, writing, and calculations) are most disturbed by lesions of the angular and supramarginal gyri in the dominant hemisphere. In Gerstmann's syndrome, the following features are present to a variable degree: dysgraphia, dyscalculia, finger agnosia, left–right confusion, and deficits in drawing plus or minus dyslexia.

Dysgraphia is a difficulty in writing (Table 15.3). This deficit usually due to a lesion in the posterior speech area might also result from disease involving the premotor cortex anterior to the motor cortex representation of the hand. It is likely that information is conveyed from the inferior parietal area to the premotor motor association cortex and then to the precentral motor cortex. This area of premotor cortex (sometimes termed the *writing center of Exner*) must function in a manner analogous to Broca's area. It would be reasonable to consider dysgraphia from a lesion in this premotor location as essentially a motor apraxia due to destruction of an area concerned in the motor association memories for writing. Dyslexia is defined as difficulty in reading and may occur with lesions of the supramarginal and angular gyri of the inferior parietal areas. However pure dyslexias might occur without actual involvement of the angular supramarginal areas. The basic lesion, as we will indicate, is a lesion that deprives the inferior parietal cortex of information from the visual association areas. Many lesions in the inferior parietal area extend into the subcortical white matter involving the long-association fiber system, the arcuate fasciculus, as discussed earler. Patients with dominant inferior parietal lobule lesions often demonstrate problems in drawing (constructional drawing). Whether this reflects damage to the cortex of the angular and supramarginal gyri or a disconnection of these areas from the more anterior motor areas because of involvement of subcortical association fibers is usually uncertain. As we have already discussed, many patients have involvement of all of the components of the posterior speech areas, the inferior parietal, the arcuate fasciculus, and Wernicke's area.

Case History 15.4 demonstrates the effects of a tumor infiltrating the posterior language area resulting in great difficulty in repetition as the most prominent initial feature.

Case History 15.4: Fluent Posterior Aphasia

A 42-year-old, right-handed truck driver noted fatigue and irritability 6–8 weeks prior to admission. Three to 4 weeks prior to admission, he noted that he was using words that he did not mean to use. Certain words would not come to him. During the week prior to admission, he had two generalized convulsive seizures, preceded by a ringing in the ears.

Neurological examination: Mental status: The patient was alert and oriented to time, place, and person cooperative and able to carry out the four-step command: "stick out your tongue, close your eyes, hold up your hand and touch the thumb to your ear." There was, however, evidence of significant left-right confusion when laterality was introduced (e.g., "left hand to right ear"). The ability to do even simple calculations was markedly impaired with or without paper (e.g. $100-9 = 99$).

Language functions. There was little evidence of an expressive aphasia. Flow of speech was slow, with only minor mispronunciations. Reading was slow but with few errors. The patient's did have minor difficulty in naming objects (a mild nominal aphasia). The patient's greatest difficulty was in repetition of simple test phrases. There were moderate defects in drawings of a house and a clock but few errors in copying simple figures. There was a significant dysgraphia with marked difficulty in writing a simple sentence spontaneously or in writing from dictation. On the other hand, the patient was much better able to copy a simple sentence. Significant errors were made in spelling. Memory was impaired; with immediate recall, object recall limited to 2/5 objects and delayed recall in 5 minutes limited to 0/5. Digit span was limited to 2 forward and 0 in reverse.
Cranial nerves: A mild right central facial weakness was present.

Motor system and reflexes were intact

Sensory system: Tactile localization was impaired on the right side.

Clinical diagnosis: Fluent posterior aphasia, disconnection syndrome plus elements of Gerstmann's syndrome (left–right confusion, dyscalculia, dysgraphia) plus seizures of focal origin (Heschl's gyrus posterior temporal lobe) with secondary generalization. All findings suggested a lesion in the left posterior temporal–posterior parietal area.

Laboratory data: Left carotid arteriogram: A vascular mass was present in the region of the left angular gyrus with tumor stain in the area.

Subsequent course: Papilledema soon developed. Dr. Robert Yuan performed a craniotomy, which revealed a palpable firm mass slightly above and posterior to the angular gyrus. At 2 cm below the surface, firm yellow tumor tissue was encountered. All visible tumor and adjacent cortex of the angular and supramarginal gyri were removed and a partial left temporal lobectomy was performed. Histological examination indicated a highly malignant glial tumor with active mitosis and necrosis (a glioblastoma). Postoperatively, the patient had a marked expressive and receptive aphasia. By 30 days after surgery, the patient had a significant improvement in fluency; he continued to have a severe posterior type of aphasia. He was unable to carry out simple commands. His speech was incoherent, with nonsense words and neologisms. He appeared unaware of his errors and did not appear frustrated by his failure to follow commands. In addition to this Wernicke's-type aphasia, the patient was unable to repeat words of two syllables. He could do no calculations, had severe difficulties in copying drawings, and was unable to even write his name. The patient received radiotherapy but his condition continued to worsen with the development of a progressive expressive and receptive aphasia and hemiparesis, all suggesting the spread of a rapidly growing glioblastoma. Coma, pupillary changes, and respiratory changes then intervened, suggesting the effects of herniation of the residual temporal lobe.

Comment: Death occurred 7 months after onset of symptoms, and 5 months following surgery. Unfortunately, this patient had a very malignant and aggressive tumor. The median survival of glioblastomas with surgery alone is 26 weeks; with the addition of radiotherapy, it is 52–60 weeks. In earlier chapters we have already identified this area as the posterior parietal multimodal sensory association areas. Geschwin has suggested that correlated with the development of this area in the human(there is no equivalent area in the monkey or great apes) there has developed the capacity for cross modalities sensory-sensory associations without reference to the limbic system, and that this could be the basis of human language.

The following case is an example of Wernicke's left posterior temporal aphasia.

Case History 15.5: Wernicke's Posterior Aphasia (Fig. 15.4)

A 57-year-old female, with a long history of hypertension in the summer, had a brief episode of difficulty in word-finding and a more prolonged episode in December with confusion, "visual problems," and "inappropriate speech and difficulty finding the appropriate words." Following an additional episode in March of the next year, she had persistent problems in reading and apraxia of the right hand.

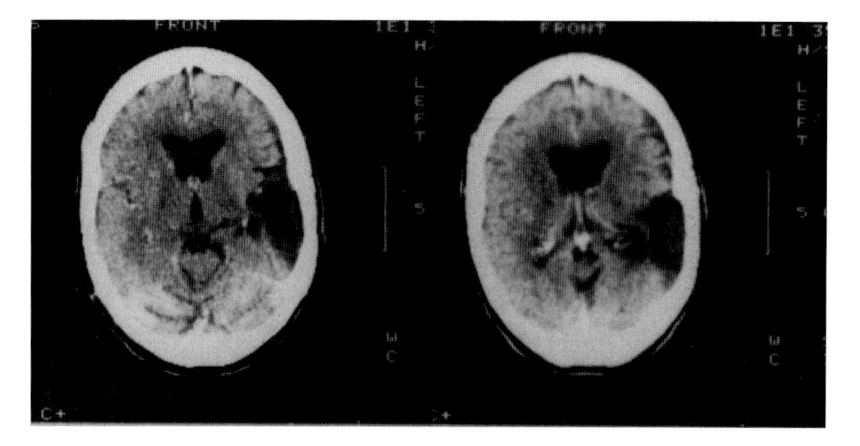

Fig. 15.4 Case History 15.5. Posterior aphasia, infarction left posterior-temporal and parietal areas, and inferior division of left MCA (probably embolic following left carotid occlusion). CT scan

Neurological examination: Language function: Spontaneous speech was usually quite fluent. She described the room, the environment, and her friend. In contrast, capacity for repetition was very poor both with regard to speech and writing. She had little comprehension of a story read to her aloud. Spontaneous writing was very poor except for name and address. She was able, however, to follow single one-stage spoken or written commands but unable to do a sequence of commands. Naming of objects was well performed.

Cranial nerves: Extinction occurred in the right visual field for bilateral simultaneous stimuli.

Motor system: No hemiparesis was present.

Reflexes: Right-sided hyperreflexia and a right Babinski sign were present.

Clinical diagnosis: Posterior aphasia, infarction left posterior temporal and parietal areas, due to embolic occlusion of inferior division of the left MCA territory.

Laboratory data: CT scan: Infarct left posterior temporal and parietal areas.

Angiograms demonstrated occlusion of the left internal carotid artery. The left anterior and middle cerebral artery and its branches filled from the right carotid. An artery-to-artery embolus was presumed.

Selective Dyslexia: This lesion has destroyed the visual cortex of the dominant left hemisphere and at the same time damaged the posterior portion or, at least, the splenium of the corpus callosum (or the fibers radiating from the splenium). The fibers conveying visual data from the right hemisphere must presumably reach the speech areas of the dominant hemisphere by passing through the posterior segment of the corpus callosum.

The following case 15.6 is a unilateral lesion producing a transient and partial but selective type of dyslexia.

Case History 15.6: Selective Dyslexia

Seven days prior to admission, following a period of athletic activity, this 15-year-old, right-handed, white, male, high school student had the onset of a 48-hour period of sharp pain in the left jaw and supraorbital area. The next day, the patient noted blurring of vision, and the following day, he had the onset of vomiting. Four days prior to admission, the patient noted that he was unable to read and to translate his Latin lessons. No definite aphasia was apparent. Two days prior to admission, the patient had the sudden onset of severe left facial pain, accompanied by tingling paraesthesias of the right leg. The patient became stuporous and then unresponsive for approximately 90 minutes. An examination of spinal fluid at that time revealed pressure elevated to 290 mm of water with 250 fresh blood cells.

Neurological examination: Mental status and language function: Intact.

Cranial nerves: Small flame-shaped hemorrhages were present on funduscopic examination. A slight, right, supranuclear-central facial weakness was present.

Neck: A slight degree of resistance was present on flexion of the neck (nuchal rigidity) consistent with the presence of blood in the subarachnoid space.

Clinical diagnosis: SAH of uncertain etiology. In view of age, the transient dyslexia, and the transient paresthesia of the right leg, an arteriovenous malformation in the left parasagittal parietal–occipital area might be suspected.

Laboratory data: Arteriogram: An arteriovenous malformation was present in the left occipital region. The main arterial supply was the posterior temporal branch of the left posterior cerebral artery. The malformation drained into the lateral sinus. The posterior portion of the anterior cerebral artery was shifted across the midline, indicating a hematoma as well as the malformation.

Subsequent course: On the evening after admission, the patient had the sudden onset of a bifrontal headache and was unconscious for a 2–3-minute period with deviation (repetitive driving) of both eyes to the right and with sweating and slowing of the pulse rate. Over the next 2 days, an increase in blurring of the optic disk margins was noted (early papilledema). Four days after admission, the patient suddenly complained of being unable to see from his right eye. He became restless and agitated. Additional headache and neck pain with numbness of right arm and leg were reported. Examination now disclosed a dense right visual field defect (homonymous hemianopsia). The deep tendon reflexes were now slightly more active on the right. Funduscopic examination suggested that recent additional subarachnoid bleeding had occurred, as new retinal hemorrhages were present. The next day, the patient was noted to have a significant reading disability (dyslexia), although other language functions were intact. Because of the progressive evolution of neurological findings due to an expanding intracerebral hematoma, Doctor Robert Yuan performed a craniotomy, which revealed blood present in the subarachnoid space. The left lateral occipital cortical surface had numerous areas of bluish discoloration, indicating an intracerebral clot. The main arterial vessel was a lateral and inferior branch of the posterior cerebral artery, bearing no relationship to the calcarine region. The draining vein was located in the occipital–temporal area, close to the tentorium. At a depth of 1 cm, a hematoma of 50 cc of clot was found. The clot extended down to the occipital horn of the lateral ventricle but did not enter the ventricle. The malformation, hematoma, and related cerebral tissue of the lateral occipital area were removed. The calcarine area remained intact at the end of the procedure. Over a period of several days, the visual field deficits disappeared. Detailed language and psychological testing approximately 2 weeks after surgery indicated that language function was intact except for reading. Oral reading was very slow (four times slower than normal for a test paragraph), halting and stumbling. Comprehension of the material read was good. Minor errors were made in the visual recognition of letters. This was reflected in occasional errors in spelling. Writing and drawing were intact but slow. Visual, auditory, and tactile recognition and naming of objects was intact. Calculations were intact. No right–left confusion or finger agnosia was present. Repetition of speech and spontaneous speech were intact. Follow-up evaluation at 20 and 30 months indicated that, although the patient was doing well in school, receiving A's and B's with excellent grades in mathematics, he still was described as slow in reading compared to his level prior to illness. No actual errors were made in reading when this was tested.

Geschwind (1965) has reviewed many of the previously reported cases of selective dyslexia and he concluded (1) that information can pass from area 18 of the nondominant hemisphere to the nondominant angular gyrus and then cross to the dominant temporal parietal areas or (2) these fibers can pass directly to area 18 of the dominant hemisphere with information then conveyed to the dominant temporal parietal areas. The relatively minor nature of the final residual deficit in the present case would be consistent with such multiple pathways.

Apraxia can be defined as impairment in motor performance in the absence of a paralysis or sensory receptive deficit and at a time when cerebellar and cognitive functions are otherwise intact. In the carrying out of a skilled movement on command, there are several stages that are similar to those considered under language function. The command, if auditory, must be received at the cortical sensory projection area of the dominant hemisphere. The information must then be relayed to the auditory association areas for the words of the command to be comprehended (i.e., to evoke the appropriate auditory associations). From this area, information must be relayed to the multimodal association area of the dominant inferior parietal association area. Information is then conveyed to the premotor and supplementary motor association areas of the dominant hemisphere and then on to the dominant motor cortex and, via the corpus callosum, to the premotor and motor areas of the nondominant hand so that hand can also perform skilled movements on command. Apraxia then could result from damage at any point in this series of association centers and their interconnections. A lesion of the anterior half of the corpus callosum will result in an apraxia limited to the nondominant hand. In contrast, a lesion of the dominant premotor association areas would be more likely to produce a bilateral impairment in certain tongue and hand movements. The concept that many varieties of apraxia represent a disconnection between various cortical areas, as reintroduced by Geschwind (1965), provides a more anatomically based approach than the use of terms such as motor or limb kinetic (motor association areas) or ideomotor apraxia (parietal lobe apraxia). Heilman et al. (2000) and Ochipa and Rothi (2000) have provided a more recent analysis of this topic. The student will also encounter the term *ideational apraxia*. This term is now employed to indicate a defect in performing a series of acts in their proper sequence. This problem might occur in diffuse disorders such as Alzheimer's disease, frontal lobe disorders, or left parietal disorders.

III. Language Functions in the Nondominant Parietal Hemisphere

As we have previously indicated in Chapter 12, certain symptoms appear to follow damage to the nondominant parietal areas such as the following:

1. Lack of awareness of hemiplegia
2. Neglect syndrome
3. Defects in spatial construction and perception of three-dimensional space

Some have questioned whether a true dominance of these functions actually exists in the hemisphere that is nondominant for speech. Certainly, neglect syndromes can also follow lesions of the dominant parietal area or of either premotor area. With regard to concepts of visual space, there is some evidence from the studies of Gazzaniga et al, (1965) on section of the corpus callosum that the dominant left hemisphere is dependent on information that must be conveyed from the right hemisphere. Perception for visual spatial material might be localized to the right parietal–occipital and to right frontal operculum. A defect in the recognition of faces (prosopagnosia) has been related to lesions of the right parietal–occipital or inferior temporal–occipital junction (see Luders et al., 1988).

IV. Role of Corpus Callosum in Transfer of Information

The corpus callosum allows the nondominant hemisphere access to the special language centers of the dominant hemisphere. Thus, the intact right-handed individual can name an object, such as a key, independent of whether the object is placed in the left or right hand. Following section of the corpus callosum, the object can be named when placed in the right hand but not when placed in the left hand. In a similar manner, the corpus callosum allows the dominant hemisphere access to those specialized areas of the nondominant hemisphere concerned with concepts of visual space. There is also evidence that the corpus callosum is involved in the transfer of information concerned with learned sensory discriminations. In an intact human who has been trained to press a key with the right hand when a particular visual pattern appears in the right visual field has no difficulty in performing the same discrimination without additional learning when the left visual field and left hand are employed. Section of the corpus callosum interferes with such transfer of learned information.

Part III
The Non-Nervous Elements within the Central Nervous System

Chapter 16
Meninges, Ventricular System and Vascular System

The central nervous system(CNS) is enwrapped by protective coverings, the meninges. In addition in the center of the CNS is a fluid-filled space (the ventricular system) that shock mounts the CNS. The cerebrospinal fluid (CSF) is formed in the ventricular system and fills the ventricular space and meninges. The CSF also forms the extracellular fluid within the CNS. The CNS is also provided with a remarkable blood supply. Finally, the CNS is surrounded by the bony vertebrae and skull, which can be reviewed in one of the many excellent gross anatomy texts.

I. Meninges: Coverings of the Brain (Fig. 16.1)

The brain is enclosed by three membranes: the dura mater, arachnoid, and pia mater. These protective fluid-filled membranes are formed by connective tissue with embedded nerves, especially in the dura. The dura mater is the most external membrane, followed by the arachnoid, and, finally, the pia mater, which adheres to the central nervous system (CNS).

A. Dura Mater

The externally located dura mater (Fig. 16.1) consists of a tough fibrous connective tissue. The two layers of the dura are fused together except where they seperate for the venus sinuses. In the cranial vault, the dura adhere to the periosteum on many of the adjacent bones and also forms an inner, or meningeal, portion that forms partition within regions of the brain:

Falx cerebri: between the cerebral hemispheres
Tentorium cerebelli: between the cerebellum and cerebrum
Falx cerebelli: between the cerebellar hemispheres
Diaphragma sellae: covers over the pituitary gland in the sellae turcica of the body of the sphenoid.

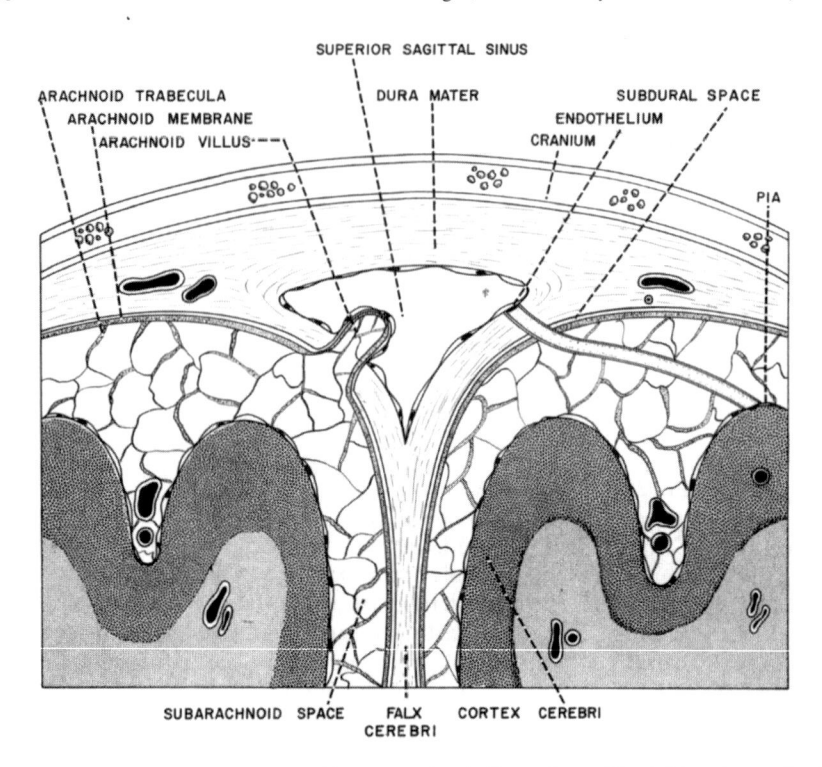

Fig. 16.1 Diagram showing relationship among the skull, meninges (dural falx, pia, arachnoid) subarchnoid space, arachnoid granulations, and venous sinus. (From BA Curtis, S Jacobson, and EM Marcus, Introduction to the neurosciences, Saunders, 1972)

Meckel's cave: located in the medial region of middle cranial fossa and contains the primary ganglion of the trigeminal nerve, cranial nerve V.

The cranial dura is securely attached to the bone at the actives and at the base of the skull.

The spinal dura forms a sac surrounding the spinal cord. The vertebrae have their own periosteum. The dural sac attaches at the margin of the foramen magnum to the occipital bone and to the inner surface of the second and third cervical vertebrae. It is also continuous with the perineurium on the spinal nerves and covers the filum terminale and becomes continuous with the periosteum on the coccyx as the coccygeal ligaments.

B. Arachnoid (Fig. 16.1)

This thin membrane is located between the dura and pia. The space between the pia and arachnoid, the subarachnoid, is filled with the cerebrospinal fluid (CSF).

II Ventricular System

The arachnoid bridges the cerebral sulci and extends from the posterior surface of the medulla to the cerebellum (cisterna magna) and below the neural ending of the spinal cord (lumbar cistern). All of these spaces contain CSF. The arachnoid is usually separated from the dura by the subdural space. The arachnoid membranes form arachnoid granulations (Pacchionian bodies) that permit passage of the CSF from the higher pressure in the ventricular system into the lower pressure of the venous sinuses.

C. Pia Mater (Fig. 16.1)

The pia adheres to the entire CNS and is continuous with the perineurium of cranial and spinal nerves. It attaches to the blood vessels entering and leaving the CNS and fuses with the dura. The cranial pia actually invests the cerebrum and cerebellum and extends into the sulci and fissures. It also forms the non-neural roof of the third ventricle, the lateral ventricle, and fourth ventricle. The pia forms the denticulate ligaments that anchor the spinal cord to the dura between the exiting spinal nerve rootlets.

II. Ventricular System (Fig. 16.2)

The ventricular system consists of the lateral ventricles, third ventricle, cerebral aqueduct, fourth ventricle, and spinal canal. The CNS is surrounded by CSF. Externally, the meninges, especially the subarachnoid space, contain CSF. Internally, the ventricular system and cisterns contain CSF, and a specialized structure, the choroid plexus, excretes CSF. CSF also is found in the extracellular space in the brain.

The *lateral ventricles* are found within the cerebral hemispheres and consist of the body and the anterior, posterior, and inferior horns. A horizontal section of the brain shows the relationship between the ventricles and the CNS.

The *third ventricle* lies in the midline between the left and right diencephalons and is continuous superiorly with the frontal horn of the lateral ventricles and inferiorly with the cerebral aqueduct. Several recesses associated with the third ventricle are important radiological landmarks: the optic recess, hypophyseal recess, and pineal recess.

The *cerebral aqueduct* is the narrow ventricular space in the center of the midbrain connecting the third and fourth ventricles.

The *fourth ventricle* forms that portion of the ventricle seen in pontine and medullary levels. The fourth ventricle is continuous laterally (foramina of Luschka) and medially (foramen Magendie) with the subarachnoid space. The ventricle is continuous inferiorly with the narrow central canal of the spinal cord. The narrow *spinal canal* is seen in the center of the gray commissure of the spinal cord.

The ventricular system is lined by ependymal cells (see Fig. 2.13 of Chapter 2), Primarily, the choroid plexus in the lateral ventricles, third ventricle, and fourth ventricle, which are supplied by the choroidal arteries, forms CSF. It is a clear, colorless, basic fluid that resembles protein-free plasma but with many significant

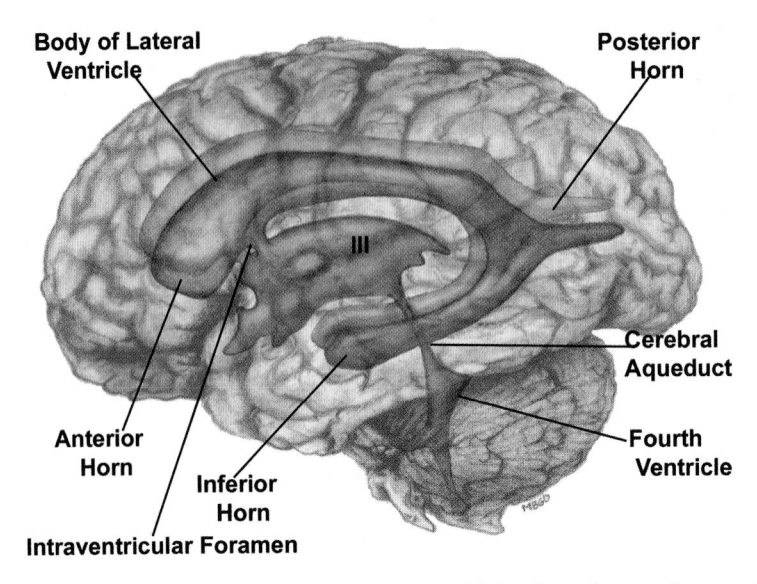

Fig. 16.2 The ventricular system (From EM Marcus and S Jacobson, Integrated neuroscience, Kluwer, 2003)

Table 16.1 Components of plasma and CSF

Plasma (mg/dl)	CSF (mg/dl)
Na^+ 330	310
HCO_3^- 1200	1310
CA^{2+} 10	11
K^+ 17	12
HPO_4^- 3	1.8
SO_4^- 1.9	0.6
Glucose 100	70
Protein 8000	45

differences (Table 16.1). The total volume of CSF is usually between 100 and 150 ml. The rate of formation is 20 ml/hour, or about 500 ml/day. Excess fluid is readily absorbed into the venous sinuses through the arachnoid granulations because the pressure in the CSF is higher.

Cerebrospinal fluid flows from the lateral ventricles through the foramen of Monro into the third ventricle and then throughout the cerebral aqueduct into the fourth ventricle. In the fourth ventricle, it is continuous with the subarachnoid space at the foramen of Luschka (lateral) and Magendie (medial). In the subarachnoid space, the fluid passes into the venous sinuses through the arachnoid granulations.

Hydrocephalus results from a blockage in the system either within the ventricle or within the subarachnoid space. When the block is in the ventricles or at the foramen of Monro, of at the foramer of Luschka or Magandie this is called noncommunicating; a frequent site is at the aqueduct of Sylvius, usually on a congenital basis. When the bloc is in the subarachnoid space, it is frequently a consequence of meningitis or hemorrhage; this is a communicating hydrocephalus.

III. Blood Supply to the Brain

A. Arterial Supply to the Brain (Fig. 16.3)

The CNS is dependent on a continuous supply of enriched, oxygenated arterial blood. This has been accomplished by prioritizing blood supply to the brain with regard to the initial branches of major artery that leaves the heart, the aorta. The arteries that supply the brain originate from the arch of the aorta with the first branch the brachiocephalic, the next branch is the left common carotid, and the final branch is the left subclavian. The blood supply to the brain from the arch of the aorta has the following organization:

1. **The brachiocephalic** for innominate artery divides behind the right sternoclavicular joint into the right subclavian and the right common carotid. The major branches of the right common carotid are the right external carotid to the head and

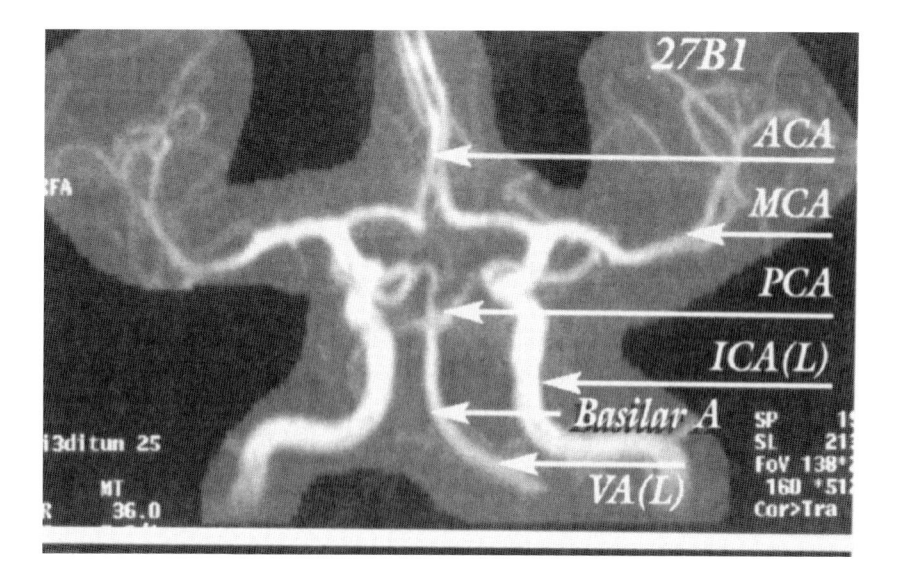

Fig. 16.3 The blood supply to the brain. MRA. ACA = anterior cerebral artery; MCA = middle cerebral artery; PCA = posterior cerebral artery, ICA = Internal carotid artery; VA = vertebral artery. (From EM Marcus and S Jacobson, Integrated neuroscience, Kluwer, 2003)

neck and the right internal carotid that supplies the brain. The right subclavian gives rise to the vertebral artery, which supplies the head, neck, and posterior regions in the brain.

2. **The left common carotid** arises to the left of the brachiocephalic trunk and divides into the left external carotid, which supplies the head and neck and the left internal carotid that supplies the brain.

3. **The left subclavian**, the final branch off the arch of the aorta, arises behind and to the left of the left common carotid and gives rise to the left vertebral that supplies the head and neck and posterior regions in the brain. Blood flow to the brain is divided into the anterior and posterior circulation.

4. **Blood supply** to the brain:

   Internal Carotid. The anterior circulation is provided by the internal carotids and provides 75% of the blood to the brain supplying the basal ganglia, anterior diencephalon, the lateral surface of the cerebral hemispheres, anterior two-thirds of the medial surfaces. The internal carotid has the following divisions:

   a. **Cervical:** in the neck.
   b. **Petrous:** in the carotid canal (petrous portion of temporal bone). The caroticotympanic and artery of the pterygoid canal arise here.
   c. **Cavernous:** in the medial part of middle cranial fossa found between layers of the dura. Ascends toward posterior clinoids, passes lateral to the body of the sphenoid bones, perforates dura near the anterior clinoids and enters the cranial vault. It gives arterial branches in the cavernous sinus, hypophyseal, semilunar ganglia (of cranial nerve V), anterior meningeal, ophthalmic and terminal intracranial branches.
   d. **Cerebral/Supraclinoid/Intradural** branches (**Fig. 16.4**): Cerebral portion forms the anterior cirlculation of the brain and divides into anterior cerebral (ACA), middle cerebral (MCA), posterior comnunicating and choroidal. It connects to the posterior circulation (from vertebrals, basilar, posterior cerebrals) in the arterial circle of Willis.

   Above the clinoid process of sphenoid bone, one finds its most important branches: two terminal branches (middle cerebral and anterior cerebral) and. the posterior communicating and anterior choroidal.
   e. Terminal branches of internal carolid-anterior circulation.

   Anterior Circulation: The anterior circulation (internal carotid, middle cerebral, anterior communicating, and anterior cerebral arteries) supplies the basal ganglia, anterior diencephalon, and the lateral surface of the cerebral hemispheres, anterior two-thirds of the medial surfaces of the cerebral hemisphere including the corpus callosum and the orbital frontal surfaces of the cerebral hemispheres.

   1. **Anterior Cerebral Artery (ACA).** The anterior cerebral is one of the terminal branches of the internal carotid that irrigates the medial surface of the hemisphere, including the corpus callosum, frontal, parietal, and cingulate cortex. The short anterior communicating artery found under the optic

III Blood Supply to the Brain

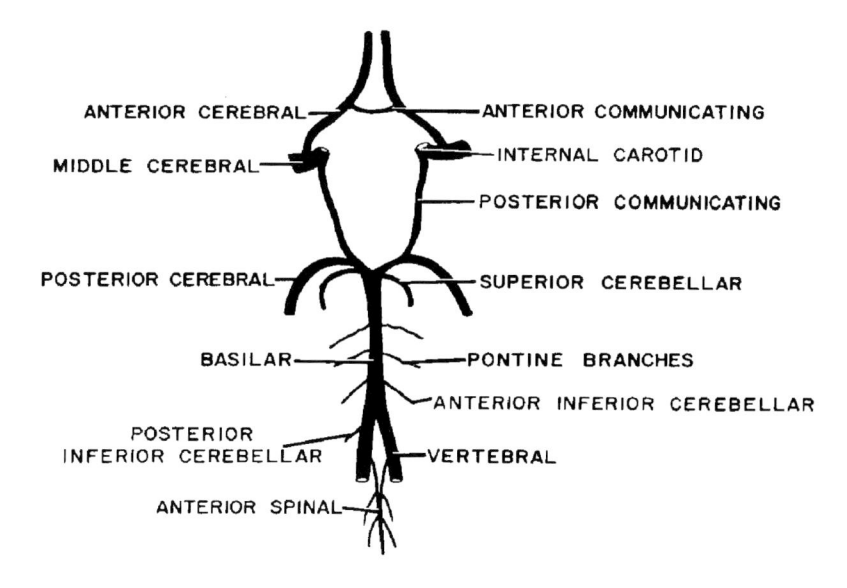

Fig. 16.4 The arterial circle of Willis (From BA Curtis, S Jacobson, and EM Marcus, Introduction to the neurosciences, Saunders, 1972)

chiasm interconnects the left and right anterior cerebral arteries, forming the anterior part of the arterial circle of Willis.

2. **Middle Cerebral Artery (MCA).** The middle cerebral is the largest terminal branch of the internal carotid and passes from medial to lateral below the anterior perforated substance into the lateral sulcus and supplies much of the lateral surface of the brain.
3. **Anterior Choroidal Artery.** This artery supplies the base of the brain, and in its course, it supplies the optic tract, cerebral peduncle, lateral geniculate, and posterior limb of the internal capsule tail of the caudate and choroidal plexus of the lateral ventricle.
4. **Ophthalmic Artery.** This vessel arises as the internal carotid emerges from the cavernous sinus and has orbital and ocular divisions.
5. **Anterior Communicating Artery (ACA).** This artery interconnects the anterior cerebral arteries.

b) **Posterior Circulation:** Vertebral arteries, basilar artery, posterior cerebral Arteries, and posterior communicating arteries. At the base of the brain is a set of arteries called the circle of Willis (Fig. 16.4) that interconnects the main portions of the anterior and posterior circulation by the formation of anterior and posterior communicating vessels. The posterior circulation provides 25% of the arterial blood to the brain and supplies the upper cervical spinal cord, cerebellum, brain stem, most of the diencephalon, and the inferior and posterior surfaces of the temporal and occipital lobes of the cerebral hemispheres. It is organized as follows: Vertebrals → Basilar → Posterior Cerebrals → Posterior Communicating.

The vertebral and basilar vessels supply the medulla, pons, and cerebellum and are divided into the paramedian, short circumferential, and long circumferential vessels.

1) **Vertebrals.** The vertebral arteries arise from the subclavian artery and enter the cranial cavity via the foramen magnum. Each vertebral artery terminates by fusing with the other vertebral arteries to form the basilar artery at the inferior border of the pons. The anterior spinal artery descends as a paramedian branch of the vertebral arteries.
2) **Basilar.** The basilar artery is a midline, single vessel that supplies the pons, midbrain, and the anterior surface of the cerebellum; it divides at the, upper end of the pons into the left and right posterior cerebral arteries.
3) **Posterior cerebral arteries (PCA).** The posterior cerebrals arteries are the continuation of the basilar and supply the rostral midbrain, posterior thalamus, and the medial surface of the temporal lobe, including the hippocampus and amygdala, and it supplies the medial surface of the occipital lobe, including the posterior visual cortex.
4) **Circle of Willis (Fig. 16.4).:** This is an anastomotic, pentagonal circle at the base of the brain that interconnects the anterior and posterior circulation via the posterior communicating arteries.
5) **Posterior communicating arteries.** These connect the posterior cerebrals and internal carotid or middle cerebral arteries, closing the circle of Willis. The circle is closed in the front with the anterior cerebrals interconnected via the short anterior communicating artery found behind the optic chiasm.

B. Venous Circulation of the Brain

The cerebral veins (Fig. 16.5) drain blood from the brain into cranial venous sinuses, which are the channels for draining the great volume of blood from the brain. The vessels have very thin walls, with no muscular layer and no valves. They are found on the surface of the brain or within the substance of the central nervous system. The follows is a list of the cerebral veins:

a. **Superficial cerebral veins** lie in the sulci and are thinner and usually more external than the arteries. They drain the superior surface of the cerebral cortex and empty into the superior sagittal sinus.
b. **Middle cerebral veins** lie in the lateral fissure and empty into the cavernous sinus. They drain the bulk of the lateral and inferior surface.
c. **Inferior cerebral veins** open directly into the transverse sinus. They drain the lateral occipital gyrus.
d. **Basal vein** (of Rosenthal) drains the base of brain and empties into the cerebral vein.

Internal cerebral veins (of Galen) drains the deep portion of the cerebral hemispheres unite and form the single Great cerebral vein, which ends in the straight sinus.

The venous sinuses are either single or paired (see Table 16.2), are located in the dura, and ultimately drain the blood into the internal jugular or subclavian veins.

III Blood Supply to the Brain

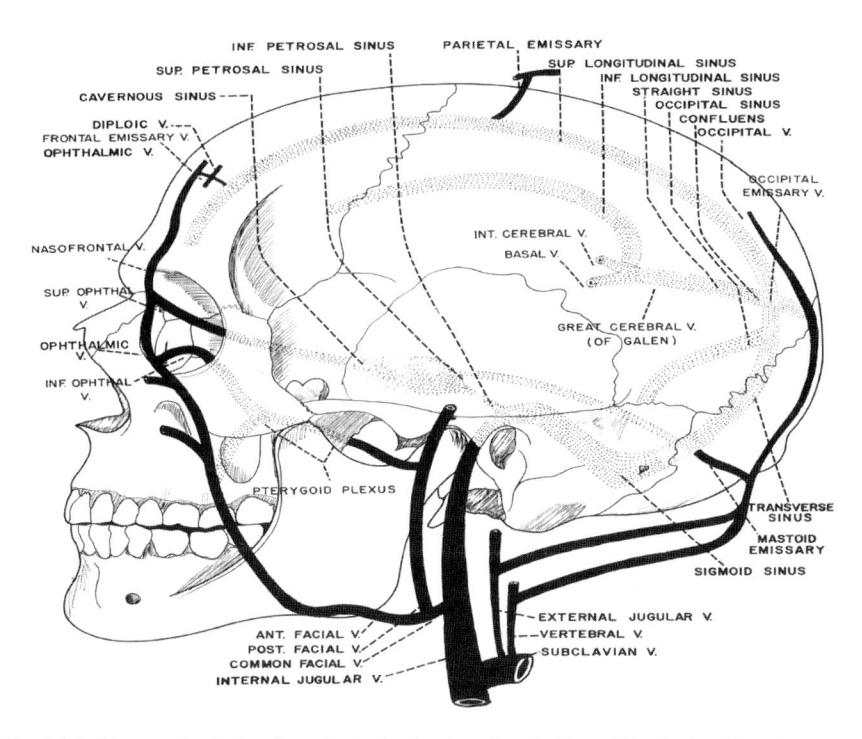

Fig. 16.5 Venous circulation from the brain, head, and neck (From BA Curtis, S Jacobson, and EM Marcus, Introduction to the neurosciences, Saunders, 1972)

Table 16.2 Venous sinus

Single midline venous sinus	Paired lateral venous sinuses
Superior sagittal sinus	Transverse
Inferior sagittal sinus	Sigmoid
Straight sinus	Cavernous
Confluens of sinuses	Intercavernous
Occipital	Superior petrosal
basilar plexus	Inferior petrosal

Chapter 17
Cerebral Vascular Disease

Cerebral vascular disease is the primary cause of damage in the brain. At least 60% of all neurological admissions to a large general hospital will consist of patients with cerebral vascular disease. Therefore, we have included an overview of cerebral vascular disease and the major syndromes one will encounter.

I. Overview

A. Definitions

Cerebral vascular disease refers to disorders of the blood vessels supplying the brain and spinal cord, which result in neurological symptoms and signs. The term refers primarily to disease of arteries. At present, disease of cerebral veins and sinuses is not common.

The term *stroke* or *cerebral vascular accident* refers to an event characterized by the acute onset of these neurological signs and symptoms. The term *ictus* is used synonymously.

B. Demographics

Cerebral vascular disease is the third leading cause of death in the United States accounting for 10% of all deaths. Approximately 300,000 new strokes and 300,000 recurrent strokes occur per year in the United States. Age adjusted incidence varies between 100 and 300 per 100,000 population depending on ethnic background and socioeconomic class. Stroke mortality in the United States began to decline in the early 1900's and this decline continued throughout the 1900's. This decrease pre-dated the development of anti-hypertensive drugs, antibiotics, anticoagulants, and anti-arrhythmia drugs. The explanation for the earlier years of decline is unclear; this may relate to improved hospital care of stroke patients, resulting in improved survival. The prevalence of stroke survivors in the United States is approximately 4,000,000 constituting a significant percentage of the disabled population.

There are two major categories of cerebral vascular disease: *ischemic–occlusive* and *nontraumatic intracranial hemorrhage*. Within the second category are two subtypes: intracerebral hemorrhage and subarachnoid hemorrhage, both with a high mortality. Because the ischemic–occlusive constitutes 77–82% of all cases of cerebral vascular disease in the United States, Canada, and Western Europe, this category of disease will receive major attention. Intracerebral hemorrhage accounts for 5–17% of all strokes depending on the era and ethnic group surveyed. At the present time in the United States, the figure is closer to 5%. Subarachnoid hemorrhage accounts for approximately 9% of all stroke cases. In Japan and other countries of the Orient, intracerebral hemorrhage is more common than the ischemic–occlusive category of disease.

II. Ischemic–Occlusive Cerebrovascular Disease

A. *Definitions*

Ischemia refers to a condition produced by decreased perfusion whereby tissue receives less than the necessary amount of blood. As a result, the tissue or organ, in this case *the brain* or an *area of the brain*, receives less than the required amount of oxygen, glucose, and other essential nutrients. When perfusion is maintained at least greater than 40% of normal value, normal spontaneous and evoked activities of nerve cells are still present. At between 30% and 40% of normal flow, the neurons are unable to produce sufficient energy to continue transmission of impulses. At less than 30% of perfusion, transmission of impulse by the neuron fails completely, although the neuron is still viable. As perfusion falls below 15% of normal, membrane failure begins to occur and transmembrane ion gradients are no longer maintained. Unless perfusion and energy production are restored, irreversible damage to the neuron occurs.

Transient ischemic attacks (TIAs) are characterized by brief episodes of neurological symptoms and signs that clear completely within 24 hours. A more modern definition recognizes that these attacks are usually minutes in duration and rarely rarely more than 1 hour. Cerebral hemispheres or the retina might be involved when the carotid circulation is involved. Less often, symptoms referable to the brain stem, diencephalons, and occipital/mesial temporal areas occur when the vertebral basilar circulation is involved. These attacks can be related to defects in perfusion or to cardioembolic or to artery-to-artery embolic phenomena.

Infarction or *encephalomalacia* refers to the irreversible tissue damage that occurs as a result of failure to maintain perfusion and energy production as described earlier. The extent of tissue damage might be limited or might involve the entire territory supplied by the artery involved.

Reversible ischemic neurological deficit (RIND) refers to neurological deficits that are present for a period greater than 24 hours but resolve within 3 weeks. An actual area of infarction has occurred, but the area is small, as in a lacunar infarct.

Note that alterations in metabolic state, temperature, or hypotension might again result in the emergence of the neurological deficit.

Penumbra refers to the ischemic zone adjacent to the area of infarction where total membrane failure has not yet occurred and where function could still be restored by therapeutic measures.

Stenosis and *occlusion of arteries* are the result of narrowing of the arterial lumen by a process such as atherosclerosis. This generally occurs at points of bifurcation or angulation of the large and medium-sized extracranial and intracranial portions of the cerebral arteries. In the Caucasian population, the process primarily involves the extracranial large arteries; in the non-Caucasian population, there is increased involvement of the intracranial arteries, but considerable involvement of the extra cranial vessels might also occur.

B. Role of Anastomoses

It is important to recognize that there is not a one-to-one correlation between stenosis (occlusion of an artery) and the development of ischemia and infarction. Thus, perfusion of tissue can be maintained by collateral (anastomotic) blood flow from several sources:

1. The circle of Willis at the base of the brain involves the anterior communicating artery between the two anterior cerebral arteries and the posterior communicating arteries between the carotid arteries and the posterior cerebral arteries. Note that variations in the circle of Willis also account for variation in patterns of infarction
2. Leptomeningeal anastomoses are present over the surface of the cerebral hemispheres among the anterior, middle, and posterior cerebral arteries. Consequently, for each cerebral artery there is an area of central supply and one of peripheral supply. The area of peripheral supply constitutes a border zone or watershed overlapping with the adjacent arterial territory. Similar leptomeningeal anastomoses also occur over the surface of the cerebellum among the circumferential branches of the posterior inferior, anterior inferior, and superior cerebellar arteries.
3. Anastomoses are present between the external and internal carotid arteries (e.g., between external carotid branches and the ophthalmic artery). Anastomotic flow can also occur from the external carotid artery to the muscular branches of the vertebral artery in the neck.

C. Major Types of Ischemic–Occlusive Disease

Three major categories may be identified:

1. *Atherosclerosis* produces infarction secondary to stenosis/occlusion.

2. *Embolism* of cardiac or arterial origin constitutes 60% of all cases. Embolic occlusion is an acute phenomenon. Emboli tend to occur during the active waking hours. In addition, the anatomical pattern of involvement often suggests embolic disease (e.g., occlusion of the cortical branches of the middle or posterior cerebral arteries or of the stem of the middle cerebral artery). In terms of cardioembolic infarcts, approximately 40% of patients have atrial fibrillation.
3. *Lacunar infarction* secondary to occlusion of small penetrating arteries is a small infarct that is usually within the distribution of a penetrating artery, this has been occluded by a process of fibrinoid degeneration or lipohyalinosis occurring on a background of long-standing hypertension or diabetes mellitus.

III. Clinical Correlates of Vascular Territories: Syndromes

A. Internal Carotid Artery

This is the most common artery involved in patients with infarction secondary to atherosclerosis. The majority (77%), but not all, of patients with occlusion of the carotid artery will have an actual infarct. When infarction occurs, the distributions of the infarcts is as follows: combined middle cerebral artery (MCA) and anterior cerebral artery (ACA) territory = 37.5%, total MCA territory =13%, proximal core cortical and deep territory of MCA = 21%, watershed cortical territory = 29%, and terminal or deep border zone = 5%.

1. The Carotid Border Zone Syndrome (Fig. 17.1)

The carotid border zone corresponds to that area of the motor and sensory cortex representing the upper extremity. The patient will experience episodes of weakness and paresthesias in the upper extremity. When an infarct occurs, a persistent weakness and sensory deficit will involve the upper extremity. This area of cortex is dependent on an overlapping blood supply from both the MCA and ACA. With increasing stenosis of the carotid, there is decreased perfusion within this border zone/watershed region The adjacent MCA proximal cortical territory might be involved if the ischemia involves a larger area of cortex with the presence of a central facial (upper motor neuron) paralysis plus aphasia, if the dominant hemisphere is involved. The major differential diagnosis is transient ischemia in the brachial artery distribution due to the benign thoracic outlet syndrome resulting in transient usually positional tingling paresthesias and weakness of the hand.

III Clinical Correlates of Vascular Territories: Syndromes 413

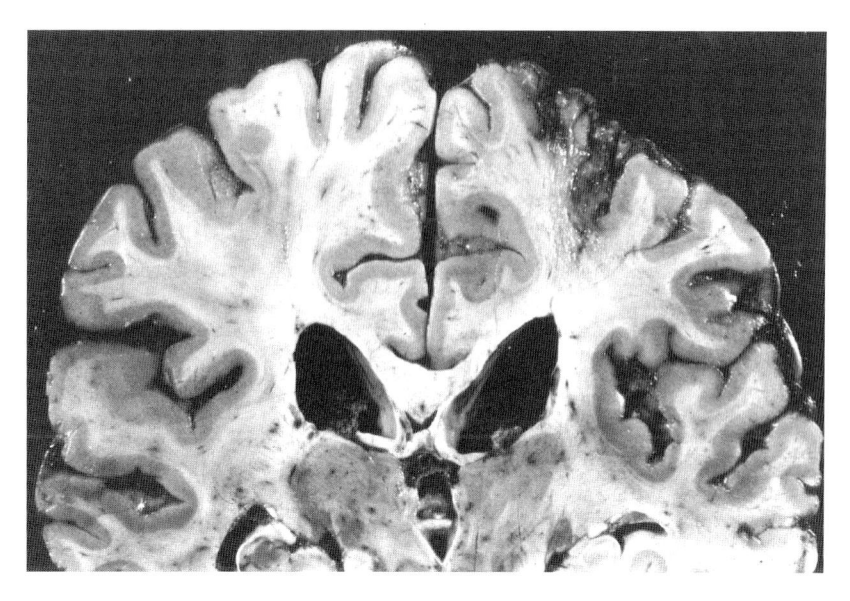

Fig. 17.1 Left carotid stenosis with old infarction predominantly within the border zone between anterior and middle cerebral arteries. This 72-year-old male had a persistent paralysis of the right upper extremity with defective position sense at fingers. (Courtesy Dr. John Hills and Dr. Jose Segarra) (From EM Marcus and S Jacobson, Integrated neuroscience, Kluwer, 2003)

2. The Ophthalmic Artery

Transient retinal ischemia produces transient monocular blindness (amaurosis fugax). This is often described as a shade or curtain descending over the vision of the involved eye.

Case History 17.1. Carotid Occlusion with Retinal and Hemispheric TIAs

A 54-year-old, right-handed, white male, for 7 years, had experienced twice per month intermittent 30–60-second episodes of blurring and blacking out of vision in the left eye. Ten days prior to admission, the patient had a 45-minute episode of minor weakness of the right face, arm, and leg plus numbness of the right side of the face accompanied by a transient difficulty in speech (possibly dysarthria, possibly difficulty in word-finding). The patient's mother died of a "stroke" at age 66 years, and his father died of heart disease at age 57.

General physical examination: Blood pressure was elevated to 160/100 in both arms. Bruits (murmurs) were present over each carotid artery. The retinal artery pulsation was easily obliterated by pressure on the globe of the left eye. Peripheral pulses in the lower extremities were poor.

Neurological examination: Intact except for a minor right central facial weakness.

Clinical diagnosis: Left carotid transient ischemic attacks.

Laboratory data: Cholesterol was elevated to 284 mg%. *Arteriography* (aortic arch study) demonstrated a complete occlusion at the origin of the left internal carotid artery. A *right brachial arteriogram* demonstrated excellent collateral circulation with filling of the left anterior and middle cerebral arteries by cross-flow through the anterior communicating artery from the right side. The intracranial portion of the left internal carotid artery filled from the left posterior communicating artery.

Subsequent course: A left carotid endarterectomy performed by Dr. Allan Callow restored blood flow following removal of the occlusive lesion (atherosclerosis with recent thrombosis) at the carotid bifurcation.

Comment: This case demonstrates the significant protective effects of collateral circulation. The patient had sustained multiple retinal TIAs related to stenosis. No hemispheric TIA occurred until the final occlusion by a thrombus.

B. Middle Cerebral Artery Syndromes

Three types of syndrome can occur: (1) those relevant to the penetrating branches, (2) those relevant to the cortical branches and (3) those relevant to the combined involvement of penetrating and cortical branches when the stem of the MCA has been involved. A similar scheme of analysis will be followed for each of the subsequent vessels.

1. Syndromes of the Penetrating Lenticulostriate Branches

Occlusion of these branches produces a small deep infract (a lacunar syndrome) involving the internal capsule. The most common syndrome is the pure motor syndrome involving weakness of contralateral face, arm, and leg that resolves over days or weeks. Involvement of several branches will produce a larger lesion with a paralysis that is more persistent, as demonstrated in Figure 17.2. Note that a less common cause of the pure motor syndrome is an infarct of the upper basilar pons due to occlusion of basilar artery paramedian branches, which is illustrated below in Fig. 17.7.

2. Syndromes of the Cortical Branches (Fig. 17.3)

In general, these syndromes are usually embolic, particularly in the Caucasian population. Either the main cortical stem or the superior division supplying the precentral and postcentral gyri or the inferior division supplying the posterior parietal lobules and superior temporal gyri can be involved.

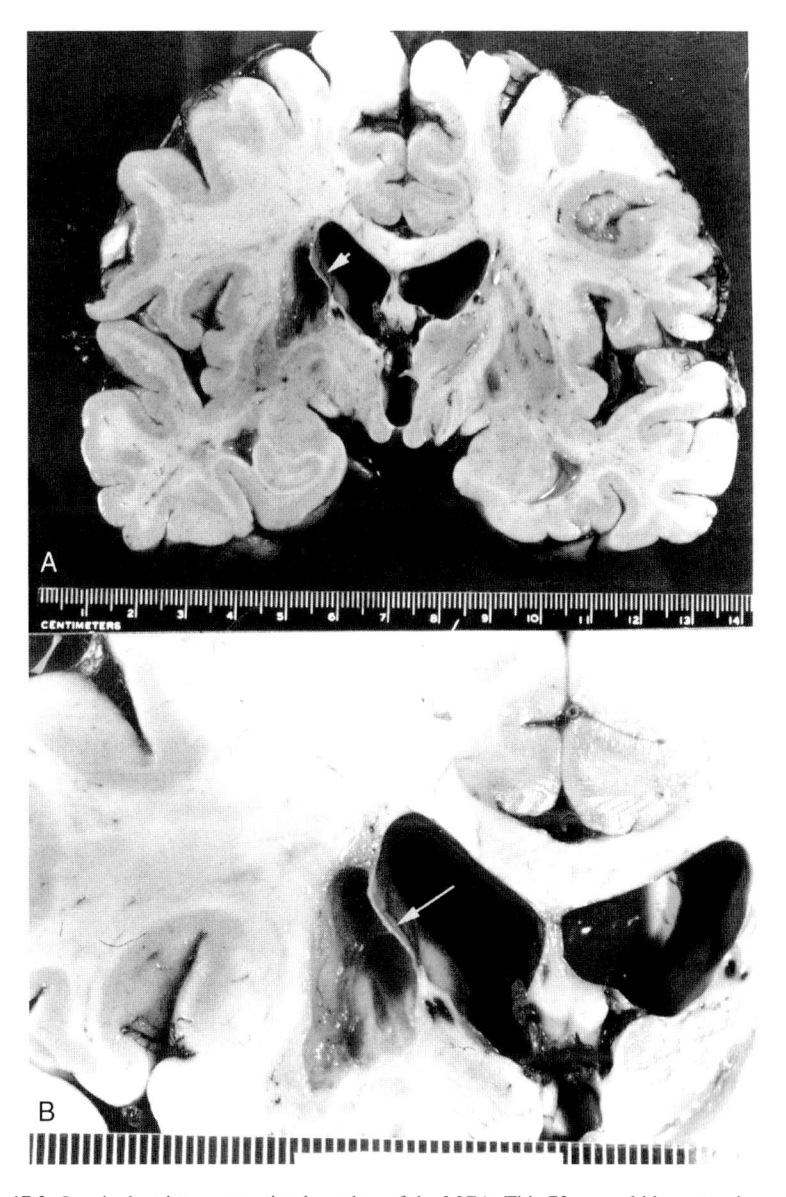

Fig. 17.2 Lenticulostriate penetrating branches of the MCA. This 73-year-old hypertensive, at age 53, had sustained a right hemiplegia with persistent and dense motor deficits involving the right central face, arm, and leg. **A** Coronal section; **B** close-up of the cavity. (Courtesy of Dr. John Hills and Dr. Jose Segarra) (From EM Marcus and S Jacobson, Integrated Neuroscience, Kluwer, 2003)

Occlusion of the *main cortical stem* produces a contralateral mixed cortical motor–sensory syndrome involving the face and arm. If the dominant hemisphere were involved, a mixed or global type of aphasia would be present. The patient would be mute, with little understanding of speech. The degree of involvement of

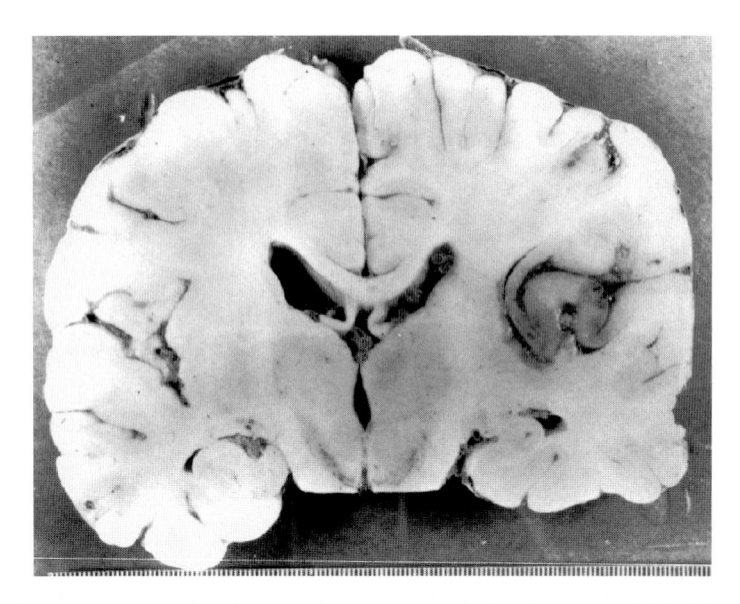

Fig. 17.3 Acute hemorrhagic infarction of the central cortical territory of the LT-MCA: calcified embolus occluded MCA following atheromatous occlusion of the carotid artery at the bifurcation. This 74-year-old, right-handed, white male, 5 days prior to death, had the sudden onset of pain in the left supraorbital region and lost consciousness. Upon regaining consciousness, the patient had a right central facial weakness and a right hemiparesis and was mute but was transiently able to follow commands. (Courtesy of Drs. John Hills and Dr. Jose Segarra) (From EM Marcus and S Jacobson, Integrated neuroscience, Kluwer, 2003)

the hand would be less than the involvement of the face (and of language in the case of the dominant hemisphere). Involvement of this main cortical stem in the nondominant hemisphere would produce a contra lateral face–arm upper motor neuron syndrome plus neglect and denial syndrome and other components of the syndrome of the nondominant hemisphere.

Occlusion of the *superior division* in the *dominant* hemisphere would produce a contralateral mixed cortical motor–sensory syndrome involving the face and arm plus an anterior-type nonfluent aphasia. In the *nondominant* hemisphere, a contralateral face–arm syndrome would occur, again with greater involvement of face than hand.

Occlusion of the *inferior division* in the *dominant* hemisphere would produce a fluent aphasia syndrome with a Wernicke's-type aphasia or a Gerstmann-type syndrome with involvement of reading, writing, and calculations. In the *nondominant* hemisphere, a neglect–denial syndrome would occur. In both the dominant and nondominant hemisphere syndromes, a hemiparesis would be absent.

3. Combined Syndrome: Total Occlusion of the Initial Segment of the MCA

As with the cortical branch syndromes, this syndrome is usually embolic, particularly in the Caucasian population. The patient will present the acute onset of a dense contralateral paralysis and hemi-anesthesia involving the face, arm and leg, plus a contralateral homonymous hemianopia, plus deviation of the eyes to the ipsilateral side of the infarct. When the dominant hemisphere is involved, the patient will be mute, with no understanding of speech (a mixed anterior–posterior global aphasia will be present). Although initially alert, within 24–72 hours the patient will often become drowsy due to swelling of the hemisphere secondary to significant edema plus or minus a hemorrhagic infarction component. In 30% of cases, transtentorial herniation will occur, with dilatation and paralysis of the ipsilateral pupil. Progressive brain stem compression will occur, resulting in death within a week of the onset of symptoms. In those surviving into the second week, pneumonia secondary to aspiration often occurs.

Case History 17.2. Total Middle Cerebral Artery Occlusion

This 61-year-old, right-handed, hypertensive black housewife 1 week prior to admission suddenly fell to the floor while taking a bath and lost consciousness. She was found by her son, who noted that she was unable to move her left arm and leg. Her speech had been thick, but no aphasia had been present. No headache had been noted. Both parents had died of heart disease and hypertension.

Physical examination: Blood pressure was 150/100 and pulse was regular at 84.

Neurological examination: Mental status: The patient was obtunded and slow to respond but grossly oriented to time, place, and person.

Cranial nerves: Papilledema was present bilaterally, particularly in the right fundus, where a recent hemorrhage was present. The head and eyes were deviated to the right. The eyes did not move to the left on command but did move to the left on vestibular stimulation. The patient neglected stimuli in the left visual field. Pain sensation was decreased or neglected on the left side of the face. A marked left central facial weakness was present.

Motor system: A complete flaccid paralysis of the left arm and leg was present.

Reflexes: Deep tendon reflexes were increased on the left compared to the right. The left plantar response was extensor.

Sensory system: All modalities of sensation were decreased on the left side.

Clinical diagnosis: Acute occlusion of the stem of the middle cerebral artery, possibly embolic, although primary occlusions of intracranial arteries can occur in the non-Caucasian population.

Laboratory data: The EEG indicated severe focal damage with a relative absence of electrical activity in the right temporal area and focal 3–5-Hz slow waves most prominent in the right frontal area. *Radioactive brain scan* demonstrated a marked right hemisphere uptake of isotope (Hg 197) extending from the surface to the deep midline outlining the total territory of the middle cerebral artery *Right brachial arteriogram* revealed a complete occlusion of the right middle cerebral artery at its origin. Cerebrospinal fluid contained no significant cells.

Hospital course: The patient showed no significant improvement during a 4-week hospital course.

Comment: Prognosis for survival during the acute state of the first week depends on the degree of edema. Fifty-five percent of patients with massive middle cerebral artery territory infarcts due to MCA occlusion who already had early CT scan evidence of brain swelling within 24 hours of the ictus expired, overwhelmingly of tentorial herniation effects during the first week. Patients older than 45 years had a poorer prognosis than younger patients. Infection of the lungs is a serious complication of large infarcts and a major cause of death during the second week. If the patient survives, prognosis for recovery is related to the size of infarct. Infarcts less than 3 cm are associated with a good recovery; infarcts greater than 3 cm are associated with severe disability.

C. Anterior Cerebral Artery Syndrome

Because of the extensive collateral supply via the anterior communicating artery and leptomeningeal anastomoses, infarcts involving the territory of this artery are uncommon, accounting for 1.8% of all infarcts. Penetrating and cortical branch syndromes might be recognized, but most patients will have involvement of both territories.

1. Penetrating Branches

These supply the anterior head of the caudate nucleus and the anterior limb of the internal capsule. A slow apathetic abulic state or a hyperactive restless state may result.

2. Cortical Branches (Fig. 17.4)

These supply the orbital and medial prefrontal cortex, the genu and body of the corpus callosum, the medial aspects of the premotor, motor, and sensory cortex (supplementary area and paracentral lobule), and parietal cortex. The resulting infarct will produce a weakness and cortical sensory deficit of the contralateral lower extremity, impairment of bladder control, release of grasp and other premotor frontal reflexes, and possible alteration of personality and level of consciousness. The patient might have an abulic or hyperactive restless state.

D. Posterior Cerebral Artery Syndromes

The penetrating branches or the cortical branches might be involved. In some cases with occlusion of the proximal segment, a combined syndrome might be present.

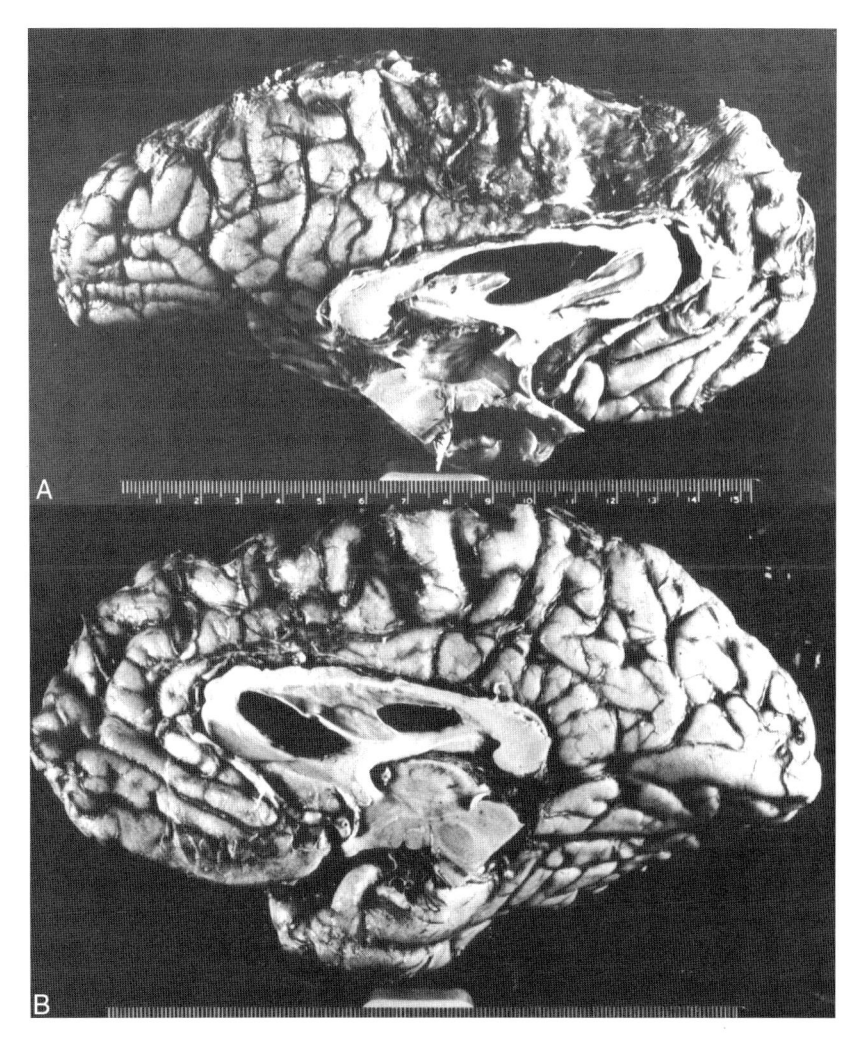

III Clinical Correlates of Vascular Territories: Syndromes 419

Fig. 17.4 Anterior cerebral arteries: old bilateral infarcts due to severe bilateral stenosis ACA (complete occlusion on right). (**A**) Left hemispheric involvement was marked with (**B**) a lesser involvement of the right hemisphere. This 61-year-old, right-handed female 16 months prior to death developed difficulty in speech, right-sided weakness, apraxia of hand movements, and progressed 2 days later to an akinetic mute state with urinary incontinence. (Courtesy of Dr. John Hills and Dr. Jose Segarra) (From EM Marcus and S Jacobson, Integrated neuroscience, Kluwer, 2003)

1. Penetrating Branch Infarcts

The penetrating branches supply the paramedian sectors of the midbrain, plus the thalamus, subthalamus and hypothalamus. The infarcts are usually small but involve critical areas to produce several distinctive syndromes. These syndromes have been included within the syndromic term "top of the basilar."

Paramedian Rostral Midbrain Syndrome: Weber's Syndrome (Fig. 17.5)

This classical syndrome combines ipsilateral involvement of the third (oculomotor) cranial nerve with a lesion of the cerebral peduncle, producing contralateral upper motor neuron paralysis of the face, arm, and leg. The adjacent substantia nigra also sustains asymptomatic damage.

Paramedian Rostral Midbrain Syndrome with Associated Tegmental Involvement (Benedikt's Syndrome)

Involvement of the adjacent midbrain tegmentum (red nucleus and dentate–rubral–thalamic fibers) will produce a contralateral tremor and movement disorder in association with an ipsilateral third-nerve syndrome.

Paramedian Periventricular Mesencephalic–Diencephalic Junction

This infarct usually bilateral involves the periaqueductal, pretectal, posterior medial thalamic, and subthalamic areas. The lesion interrupts the ascending reticular system within the upper mesencephalic tegmentum and its intralaminar thalamic extension. The patient presents a drowsy, relatively immobile state (apathetic akinetic mutism). Third cranial nerve oculomotor findings are often present. The

Fig. 17.5 Weber's syndrome. An area of infarction is noted in the right cerebral peduncle (arrow), so located as to involve the fibers of the right third cranial nerve. This 61-year-old patient with rheumatic heart disease, endocarditis, and auricular fibrillation had the sudden onset of paralysis of the left face, arm, and leg, plus partial right third nerve palsy. (Courtesy of Dr. John Hills and Dr. Jose Segarra) (From EM Marcus and S Jacobson, Integrated neuroscience, Kluwer, 2003)

major differential is the frontal akinetic mute state described earlier in relation to the anterior cerebral artery in which extraocular movements are intact and in which the behavioral state can be either abullic or restless (hyperkinetic).

Subthalamic Syndrome

Infarcts involving the subthalamic nucleus of Luys produce a contralateral hemichorea and or hemiballismus (see Chapter 10).

Thalamic Syndrome

Infarcts involving the thalamo-geniculate supply to the ventral posterior lateral and ventral posterior medial nuclei might produce a contralateral loss of sensation and a "thalamic pain syndrome."

Bilateral Paramedian Thalamic Infarcts

The patient might present the acute onset of excessive daytime drowsiness associated with changes in personality and problems in recognition of familiar individuals and photographs. New learning might be defective. The syndrome is usually transient, clearing over several weeks. The memory problems might raise the differential of hippocampal ischemia. Note that both diencephalic and hippocampal areas are supplied by branches of the posterior cerebral artery.

2. Cortical branch syndromes of the Posterior Cerebral Artery (Fig. 17.6)

Overall, infarcts involving the cortical territory of this artery account for 5% of all infarcts. The entire cortical territory or the branches might be involved. In general, occlusions of the cortical branches are a result of emboli, particularly in the Caucasian population. At times, emboli to the top of the basilar artery might produce involvement of both posterior cerebral arteries.

Anterior and Posterior Temporal Branches of the Posterior Cerebral Artery

These branches supply the medial and inferior surfaces of the temporal lobe. Ischemia and infarction of the *dominant mesial temporal area including the hippocampus* can produce a confusional state. Bilateral involvement of the *mesial temporal areas* will produce a persistent defect in the ability to record new memories (a persistent anterograde amnesia). Infarction of the *inferior temporal areas* can disconnect the visual cortex from the memory recording areas and from the speech areas. There might be problems in visual recognition.

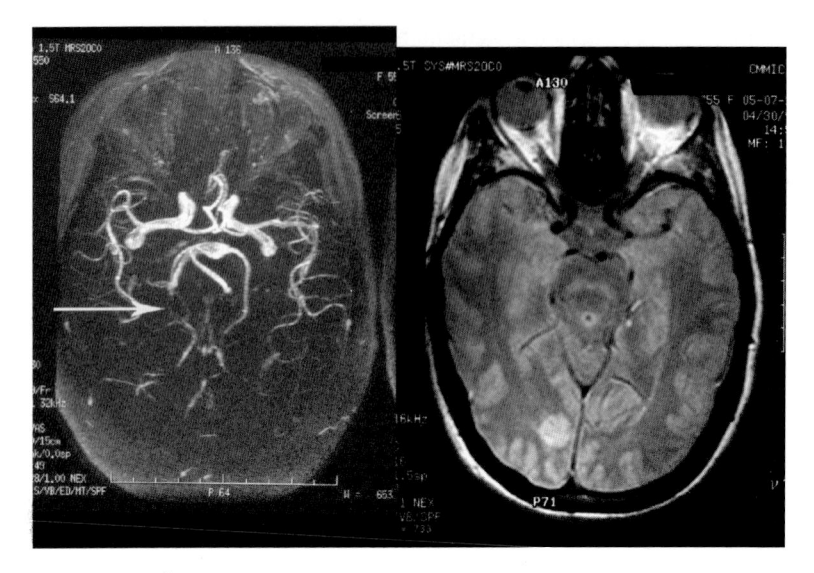

Fig. 17.6 Posterior cerebral artery syndrome. Left: MRA. Embolic occlusion of distal right posterior cerebral artery (arrow) with filling of more distal branches by anastamotic flow. Right: MRI. Infarct right occipital area, calcarine artery branch territory. This 55-year-old white female had the sudden onset of a left visual field defect followed by tingling paresthesias of the left face and arm. On exam, she was hypertensive and obese, with a small ecchymosis under the left second digit toenail. A noncongruous left homonymous hemianopsia was present greater in the left temporal field than the right nasal field. The left plantar response was possibly extensor. Pain and graphesthesia were decreased over the left foot. Reevaluation of the patient 22 hours after the onset of symptoms demonstrated clearing of all neurological findings. Holter monitoring demonstrated multiple brief episodes of paroxysmal atrial fibrillation, a cause of emboli. Echoencephalogram demonstrated a patent foramen ovale, so that shunting of blood occurred from the right to left atrium, which is also a cause of emboli. She received long-term anticoagulation and treatment for hypertension. (From EM Marcus and S Jacobson, Integrated neuroscience, Kluwer, 2003)

Calcarine Artery Branches of the Posterior Cerebral Artery

These branches supply the visual cortex. Infarction of this territory produces a contralateral homonymous hemianopia. The macular area is often spared because this central area of vision is located near the occipital pole and receives collateral circulation from the leptomeningeal branches of the middle cerebral artery. Figure 17.6 demonstrates a case with embolic occlusions of the posterior cerebral artery (PCA).

Occlusion of the Stem of the Cortical Branches

This syndrome is usually embolic and is characterized by sudden onset of headache confusion and visual impairment. If unilateral, a visual field defect is seen; if bilateral, cortical blindness will be seen with preservation of papillary responses.

As the embolus fragments and collateral circulation reduce the ischemic area, the confusion might clear and vision return to some degree.

Combined Syndrome

The combined syndrome involves cortical plus penetrating branches. This syndrome is also due to an embolus to the top of the basilar artery and combines the midbrain syndromes plus cortical syndromes. The sympotms and findings maybe blialeral or unilaaeral.

E. Vertebral and Basilar Artery Syndromes of the Brain Stem

The two vertebral arteries join to form the basilar artery at the junction of the pons and medulla. The top of the basilar artery gives rise to the posterior cerebral arteries at the junction of the pons and midbrain. The various ischemic vascular syndromes can be divided into the paramedian (penetrating) and circumferential branch syndromes. In many cases however, multiple levels and areas of involvement are combined as in a basilar artery thrombosis.

1. Paramedian Branch Syndromes

These syndromes always involve the pyramidal tract producing a contralateral hemiparesis. Often the contralateral upper motor neuron hemiparesis is associated with a lower motor neuron ipsilateral cranial nerve syndrome (e.g. Weber's syndrome of ipsilateral third nerve and contralateral hemiparesis).

Medullary Syndrome

Occlusion of the penetrating branches of the vertebral artery will produce a syndrome characterized by an ipsilateral paralysis and atrophy of the tongue combined with a contralateral upper motor neuron hemiparesis of the arm and leg. At times, there might be a contralateral deficit in position sense and vibratory sensation and other higher modalities of sensation due to involvement of the medial lemniscuses.

Lower Pontine Syndrome

If unilateral, the syndromes will combine an ipsilateral lower motor neuron involvement of cranial nerves 7 (facial) and 6 (abducens/lateral rectus) with a contralateral upper motor neuron hemiparesis. If the area of infarction extends into the tegmentum to involve the area of reticular formation around the abducens nucleus, an ipsilateral gaze paralysis might be present as well as an internuclear ophthalmoplegia due to involvement of the medial longitudinal fasciculus. Bilateral infarcts that might result from a basilar artery thrombosis are more common, producing a locked-in syndrome. Vertical eye movement and eyelid movements are preserved because the midbrain has been spared.

Upper pontine syndrome

If unilateral, the resultant syndrome reflects the unilateral involvement of the corticospinal and corticobulbar (corticonuclear) tracts producing a contralateral pure upper motor neuron syndrome of face, arm, and leg (Fig. 17.7) This pure motor syndrome is more often due to an infarct in the lenticulostriate MCA branches supplying the posterior limb of the internal capsule. Basilar artery thrombosis more often produces a bilateral infarct in which attacks of left or right hemiparesis might

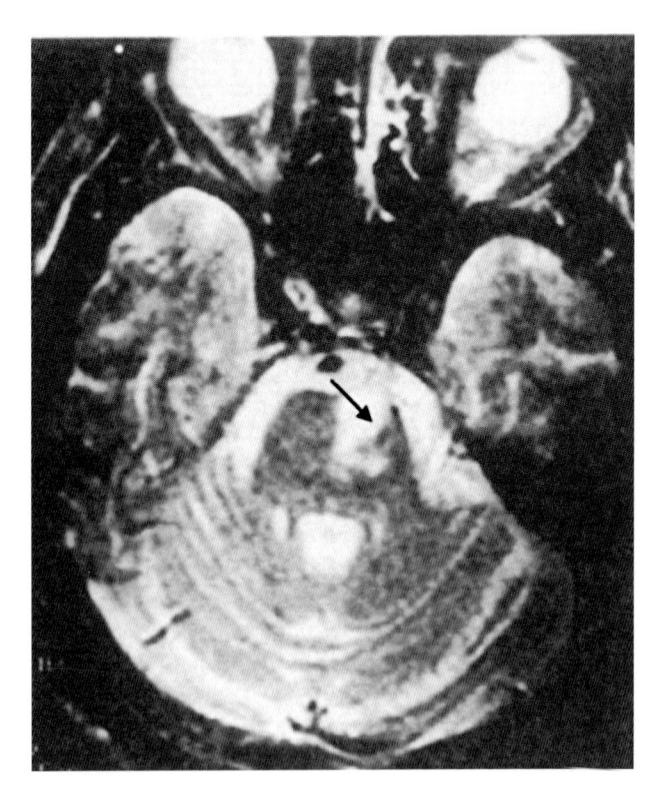

Fig. 17.7 Paramedian infarct upper pons. MRI T2. This 70-year-old, right-handed male had the acute onset of weakness of the right face (central), arm, and leg plus slurring of speech. A similar episode had occurred 2 months previously but had cleared over 2 weeks. His findings indicated a relatively pure motor syndrome. (From EM Marcus and S Jacobson, Integrated neuroscience, Kluwer, 2003)

be followed by a quadriparesis. If the upper midbrain is spared, the patient might present a "locked-in syndrome" in which the patient is able to answer questions by a system in which opening or closure of the eyes (i.e., blinking provides a yes or no system of response). These responses might not be recognized immediately by the medical observers, so there might be a failure to recognize that consciousness is preserved. When the area of thrombosis extends to the top of the basilar artery, the midbrain will also be involved; the patient will no longer have any eye movements and could be potentially totally "locked in."

2. Circumferential Branch Syndromes

These syndromes reflect the involvement of the lateral tegmental territories of the three cerebellar arteries. These arteries are named from that surface of the cerebellum that each supplies. Occlusion of these arteries might produce an infarct of the lateral tegmentum of the brain stem and/or the cerebellar territory. Prior to the development of modern imaging (CT and MRI), the actual involvement of the cerebellum in patients with the lateral tegmental syndrome was often missed because the cerebellar pathways also were involved within the brain stem.

Lateral (Dorsolateral) Medullary Syndrome/Wallenberg's syndrome (Fig. 17.8)

This syndrome reflects the infarction of the territory of the posterior inferior cerebellar artery (PICA) supplying the rostral dorsolateral medulla. The actual point of occlusion is more often located at the intracranial segment of the vertebral artery than at the PICA. In the pure or modified form, it is one of the more common of the classical brain stem syndromes.

Included within this territory are the following structures: the inferior cerebellar peduncle (ipsilateral finger-to-nose and heel-to-shin dysmetria and intention tremor), the vestibular nuclei (vertigo and nausea and vomiting plus nystagmus), the descending spinal tract and nucleus of the trigeminal nerve (ipsilateral selective loss of pain over the face), the spinothalamic tract (contra lateral loss of pain and temperature over the body and limbs),the descending sympathetic pathway (ipsilateral Horner's syndrome), and the nucleus ambiguus (dysarthria, hoarseness, ipsilateral vocal cord paralysis difficult in swallowing with ipsilateral failure in palatal elevation).

Case History 17.3. Lateral Medullary Syndrome

A 53-year-old male on the morning of admission, while seated, had the sudden onset of vertigo accompanied by nausea and vomiting, incoordination of his left arm and leg, and numbness of the left side of his face. On standing, he leaned and staggered to the left. Several hours later, he noted increasing difficulty in swallowing and his family described a minor dysarthria. He had a two-pack per day cigarette smoking history for more than 10 years.

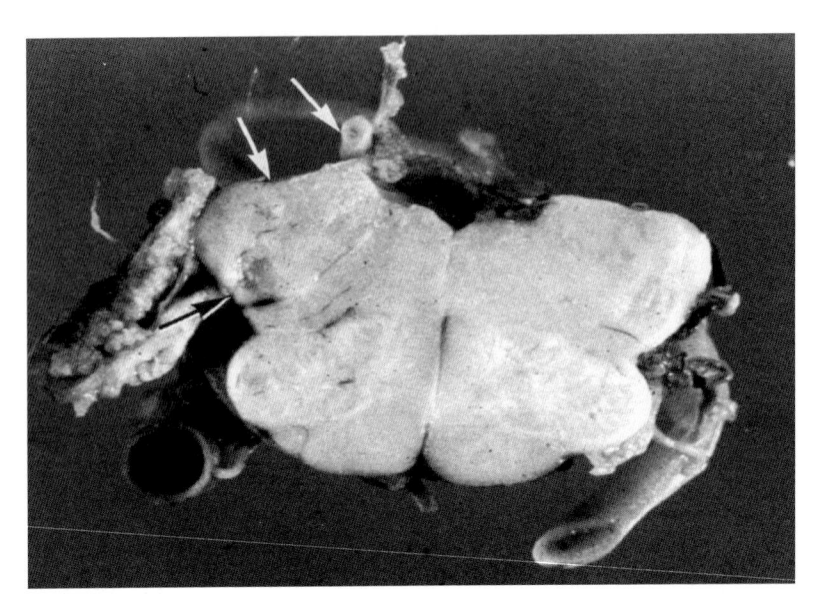

Fig. 17.8 Dorsal–lateral medullary infarct (arrows) due to occlusion of the posterior inferior cerebellar artery (arrow) in a 72-year-old diabetic with the correlated clinical syndrome of acute onset nystagmus on left lateral gaze, decreased pain sensation on the left side of face and right side of body, with difficulty swallowing, slurring of speech, and deviation of the uvula to the right. (Courtesy of Dr. Jose Segarra) (From EM Marcus and S Jacobson, Integrated neuroscience, Kluwer, 2003)

Physical Examination: Elevated blood pressure of 166/90.
Neurologic examination: Mental status: Intact.

Cranial nerves: The right pupil was 3 mm in diameter; the left was 2 mm in diameter. Both were reactive to light. There was partial ptosis of the left eyelid. Both of these findings were consistent with Horner's syndrome. Horizontal nystagmus was present on left lateral gaze, but extraocular movements were full. Pain and temperature sensation was decreased over the left side of the face, but touch was intact. Speech was mildly dysarthric for lingual and pharyngeal consonants. Secretions pooled in the pharynx. Although the palate and uvula elevated well, the gag reflex was absent.

Motor system: A dysmetria was present on left finger-to-nose test and heel-to-shin tests. He tended to drift to the left on sitting and was unsteady on attempting to stand, falling to the left.

Reflexes: Deep tendon reflexes were physiologic and symmetrical except for ankle jerks, which were absent. Plantar responses were flexor.

Sensory system: Pain and temperature sensations were decreased on the right over the right arm and leg, but all other modalities were normal.

Clinical diagnosis: (1) Dorsolateral medullary infarct. (2) Decreased ankle jerks due to possible diabetic peripheral neuropathy.

Laboratory data: The MRI indicated a significant acute or subacute infarct involving the posterior inferior cerebellar artery territory of the left dorsolateral medulla and left posterior inferior cerebellum. The MRA suggested decreased flow in the left vertebral artery. Note that prior to the era of modern imaging, the additional involvement of the posterior inferior cerebellum would not have been detected based only on the clinical findings.

Subsequent course: A percutaneous endoscopic gastrostomy (PEG) tube was placed because he was unable to swallow for the first 6 days. By the time of transfer to a rehabilitation facility, 12 days after admission, finger-to-nose tests and gait had improved; he could walk with a walker. Speech and sensory examination had returned to normal.

Inferior Pontine Lateral Tegmental Syndrome

This syndrome reflects infarction of the territory supplied by the anterior inferior cerebellar artery that usually arises from the basilar artery and supplies the cochlear nucleus and the auditory portion of the entering eighth cranial nerve (sudden ipsilateral deafness) and the facial nerve (ipsilateral peripheral facial paralysis). The following structures overlap with the lateral medullary syndrome: the vestibular nuclei and the entering vestibular portion of the eighth cranial nerve (sudden vertigo, nausea and vomiting, plus nystagmus), cerebellar peduncle (ipsilateral cerebellar symptoms), descending spinal tract and nucleus of the trigeminal nerve (ipsilateral selective loss of pain and temperature over the face), spinothalamic tract (contralateral loss of pain and temperature over the body and limbs), and descending sympathetic pathway (ipsilateral Horner's syndrome). This syndrome is sometimes seen as one component of a wider basilar artery syndrome.

Superior Pontine Lateral Tegmental Syndrome

The superior cerebellar artery supplies this territory. At this level, the following structures are involved: the medial lemniscuses and the lateral spinothalamic pathway (contralateral hemianesthesia and hemianalgesia of body and limbs), secondary trigeminothalamic fibers (contralateral deficit in pain and touch sensation over the face), and superior cerebellar peduncle (brachium conjunctivum) below or above its decussation (ipsilateral or contralateral or bilateral cerebellar symptoms of intention tremor, etc.).

F. Ischemic Occlusive Disease Involving the Cerebellum

Autopsy studies localize 1.5–4.5% of all infarcts to the cerebellum. However, CT scan studies indicate that cerebellar infarcts account for 15% of all intracranial infarcts; the majority (80–90%) of infarcts are not large and are not fatal. The classical syndrome seen in large infarcts with edema presents the acute onset of headache, vertigo, and ataxia. A similar syndrome might occur with a cerebellar hemorrhage.

G. Ischemic Occlusive Disease of the Spinal Cord (the Anterior Spinal Artery Syndrome)

The spinal cord is supplied by a major single anterior spinal artery and two posterior spinal arteries. The major vessel, the anterior spinal artery, has a bilateral origin from the intracranial vertebral arteries. It is dependent on additional supply from the radicular arteries that originate from intercostal arteries. The inputs from the middle thoracic artery at T7 and the artery of the lumbar enlargement (of Adamkiewicz), which usually arises between T10 and L2, are of critical importance. There are border zones of arterial circulation at segments T4 and L1. The lesser arterial supply from the paired posterior spinal arteries has a similar origin from the vertebral arteries.

When infarcts occur, the territory of the anterior spinal artery is usually involved. This artery supplies the anterior two-thirds of the spinal cord: essentially all of the spinal cord, except the posterior columns. The usual area of infarction is the middle–lower thoracic spinal cord. The usual syndrome is characterized by the acute onset of a flaccid paraplegia with an absence of deep tendon reflexes in the lower extremities due to spinal shock and a bilateral pain sensory level at T7–T8. Position and vibratory sensation will be preserved. If the area of infarction also involves the lumbar sacral spinal cord, with destruction of the anterior horn cells at that level, the legs will remain flaccid with muscle atrophy and a loss of deep tendon reflexes. If only the middle–lower thoracic cord is infarcted, deep tendon reflexes will recover and spasticity could develop.

Primary vascular disease of the spinal cord is not common. Actual occlusion of the anterior spinal artery is rare. The usual cause of the anterior spinal artery syndrome relates to disease of the aorta or to surgical procedures involving the heart, aorta, or related vessels: (1) clamping of the aorta for treatment of an aortic aneurysm, (2) clamping of intercostal branches, particularly during surgical procedures on the kidney and sympathetic ganglia, (3) dissecting aneurysms of the aorta might also occlude the critical intercostal vessels.

IV. Primary Intracerebral Hemorrhage

There are two major categories within the primary group:

1. Hypertensive intracerebral hemorrhage related to degenerative changes in small penetrating arteries and arterioles supplying the basal ganglia, thalamus, cerebral white matter, pons, and cerebellum.
2. Lobar hemorrhage related to amyloid angiopathy.

Note that secondary intracerebral hemorrhage may occur in trauma and in subarachnoid hemorrhage.

IV Primary Intracerebral Hemorrhage

A. Demographics and Risk Factors

This category refers to primary hemorrhages into the parenchyma of the brain. The most common site for primary intracerebral hemorrhage is the basal ganglia or cerebral white matter due to the involvement of penetrating arteries by hypertension. However, in the elderly, lobar hemorrhages might also occur secondary to amyloid degeneration of arteries. The incidence among African Americans in the United States is 50 per 100,000 population, approximately double the incidence in the Caucasian population. These differences are correlated with differences in prevalence of hypertension, level of attained education, lack of awareness of primary prevention, and lack of access to health care. In the United States, hypertension detection and treatment programs have been able to achieve a 33–46% reduction in the overall risk of stroke, including intracerebral hemorrhage. In the younger adult population, the use of amphetamines and of illicit street drugs, such as "crack" cocaine, has been implicated as a risk factor. Note that cocaine might also induce vasoconstriction, producing cerebral infarction. The use of anticoagulants and of fibrinolytic agents is also complicated by intracerebral hemorrhage. In Japan, where stroke is the leading cause of death, intracerebral hemorrhage is more common than ischemic occlusive disease.

B. Location

Of all CT-scan-diagnosed cases of nontraumatic primary intracerebral hemorrhage in the NIH stroke databank, 85% are supratentorial and are 15% infratentorial. Table 17.1 indicates the location, frequency, and clinical correlation of primary intracerebral hemorrhages.

C. Diagnostic Studies in Intracerebral Hemorrhage

The following imaging studies can be performed:

CT scan (nonenhanced). The evidence of an intracerebral hematoma should be evident on acute presentation at the emergency room.

MRI scan will demonstrate changes in time related to the type of hemoglobin breakdown product present.

MRA or CT angiography or conventional angiography. These studies should be performed in selected patients primarily based on age, presence or absence of hypertension, and location of the hematoma.

Lumbar puncture. There is no indication for this procedure. As a matter of fact, this procedure carries a risk of herniation, particularly in the case of lobar hematomas in the temporal lobe (tentorial herniation) and cerebellar hematomas (tonsillar herniation).

Table 17.1 Intracerebral hemorrhage

Location	Vessel	Frequency	Clinical manifestations
Putamen, lateral ganglionic mass, internal capsule	Lenticulostriate penetrators of the middle cerebral artery	33–40%, with overall mortality of 37%.	Progressive headache, hemiparesis, hemianesthesia, confusion, herniation and coma
Temporal lobe stem and lobar (lobar often secondary to amyloid angiopathy)	Superficial small and medium arteries	23–32%; lower mortality than putamen unless temporal herniation	Progressive temporal lobe or other cortical mass with focal seizures; recurrent or multiple.
Thalamus	Penetrators of posterior cerebral artery	20%; lower mortality than putamen	Unilateral numbness, hemianesthesia, and vertical gaze impairment
Caudate	Penetrators of middle or anterior cerebral arteries	5%; low mortality	Sudden headache and confusion, without focal findings
Pons	Paramedian penetrators of the basilar artery	4–7%; high mortality if large	Sudden quadriplegia, coma, basilar artery respiratory impairment, death; small lateral tegmental lesions might survive
Cerebellum	Branches posterior inferior or superior cerebellar arteries	8%; lower mortality, if mass effects respond to surgery	Sudden headache, vomiting, vertigo, ataxia, then extraocular arteries findings and coma

Based on series of Hier et al. (1977), Kase et al. (1982), Kase and Caplan (1986) (After EM Marcus and S Jacobson, Integrated neuroscience, Kluwer, 2003)

D. *Clinical Correlates of Intracerebral Hemorrhage*

Refer to Table 17.1

Case history examples of intracerebral hemorrhage are provided in Marcus and Jacobson (2003).

V. Subarachnoid Hemorrhage (Table 17.2)

A. *Demographics*

This category refers to primary bleeding into the subarachnoid space. As outlined below, the major cause in 85% of cases is the rupture of intracranial aneurysms. These aneurysms have a defect in the media and internal elastic membrane of the

V Subarachnoid Hemorrhage

Table 17.2 Summary of the causes of SAH, aneurysm location, and manifestations

Location or type	Frequency as cause of Sah	Frequency of all aneurysms	Early manifestations or symptoms of compression prior to rupture
I. Aneurysms	85%	—	Sudden onset of worst ever headache plus stiff neck, in some cases, sudden coma
Junction of post communicating and internal carotid		30%	Third-nerve paralysis, as third cranial nerve runs close to and parallel to the posterior communicating artery
Bifurcation MCA		20%	Focal symptoms in MCA territory: focal weakness or seizure face, speech, etc
Junction anterior communicating- anterior cerebral		30%	Compression optic chiasm, or bilateral prefrontal or bilateral lower extremity or coma or mute state or if giant dementia a mass effect may be present. Often nonlocalized SAH
Basilar vertebral system		5–10%	Variable
Multiple aneurysms		15%	Variable, usually only one is the site of SAH
II. Other SAH causes	15% Total		
Perimesencephalic	10%		Nonlocalized: worst headache, stiff neck drowsiness
Other: arteriovenous malformation (AVM) and mycotic aneurysms	5%		Variable depending on location (often distal cortical branches in mycotic); seizures frequent in both

arterial wall. They are located at critical junction or bifurcation points. Unless promptly recognized and treated, a significant percentage of patients will die of the initial subarachnoid hemorrhage (SAH) or its complications. Mortality is high, particularly if rebleeding occurs. A significant percentage (12%) of patients will die before they reach the hospital or before they reach a neurosurgical center (30%). In a significant proportion of cases, the bleeding extends into the parenchyma or the ventricles of the brain and these cases have a poor prognosis. The incidence of incidental unruptured aneurysms found at autopsy or on MRA is high.

Autopsy studies suggest an overall frequency of intracranial aneurysms of 5% with a population-based incidence of subarachnoid hemorrhage secondary to aneurysms of 10 per 100,000 per year (see Phillips et al. 1980). Patients with autosomal-dominant polycystic kidney disease or a family history are at increased risk for aneurysms and require noninvasive screening with MRA (Wiebers and Torres, 1992). Whether an aneurysm ruptures depends on the size of the aneurysm. Those smaller than 10 mm are unlikely to rupture and do not require clipping.

B. Major Clinical Features

The classical essential feature is the sudden onset of the worst possible headache ever experienced by the patient. The headache has been described as explosive or as a "thunderclap." Such patients should be assumed to have a SAH until proven otherwise.

Approximately 11% of patients also report in retrospect a severe headache, sometimes referred to as a warning or sentinel headache, in the day or weeks prior to the emergency room presentation. Meningeal irritation with nuchal rigidity usually takes several hours to develop, as the blood must migrate to the infratentorial and cervical area. An immediate CT scan should be obtained and will be positive for subarachnoid blood in 90–95% of patients if obtained in the first 24 hours after the SAH. In the other 5–10% of suspected cases where the CT scan is normal, a lumbar puncture should be obtained to determine whether blood is present in the subarachnoid space. A diagnosis of subarachnoid hemorrhage is dependent on the demonstration of blood in the cerebrospinal fluid.

C. Complications

The major complication is rebleeding, which might occur in at least 30% of patients. This carries a mortality of 50%. Additional complications include vasospasm with infarction of tissue, hematoma formation with herniation effects, early or late hydrocephalus, and seizures.

D. Management and Treatment

Immediate four-vessel substraction angiography is required to demonstrate the source of the SAH. This is the gold standard. The neurosurgeon must have information concerning the details of the neck of the aneurysm and of any critical small vessels emerging from the aneurysm. At some point in the near future, developments in neuroimaging might allow the use of MRA or CT angiography. To prevent rebleeding, immediate clipping of the neck of the aneurysm must be undertaken. If clipping of the neck of the aneurysm is not feasible, then endovascular procedures must be performed in an attempt to occlude the aneurysm

For additional discussion, see Marcus and Jacobson (2003).

Case History 17.4. Subarachnoid Hemorrhage Secondary to an Aneurysm of the Posterior Communicating Artery

This 50-year-old right-handed, white male factory foreman was referred for evaluation of ptosis and diplopia involving the left eye. Approximately 2 weeks prior to admission, the patient developed a bifrontal headache. During the week prior to his evaluation, the headache had become a left-sided aching pain. It increased in intensity during the 2 days prior to admission and was present as a constant pain interfering with sleep. If the patient were to cough, he had additional pain in the left eye. On the day of admission, the headache became much more severe; it was now the worst headache he had ever experienced. In retrospect, the patient reported that lights had been brighter in the left eye for approximately 1 week. At 3:00 PM on the day of admission, the patient noted the sudden onset of diplopia, which was more marked on horizontal gaze to the right and much less marked on horizontal gaze to the left. At approximately the same time, he noted the rapid onset of ptosis involving the left lid. His past history was not relevant to his present neurological problem.

General physical examination: Moderate resistance to flexion of the neck.

Neurological examination: The complete neurological examination was within normal limits except for findings relevant to the left third cranial nerve:

1. The left pupil was fully dilated to approximately 7 mm.

There was no response to light or accommodation.

2. Total ptosis of the left eye lid was present.
3. No medial movement of the left eye was present, there was no upward movement of the left eye possible. The patient had minimal downward gaze of the left eye. He had full lateral movement of the left eye.

Clinical diagnosis: Subarachnoid hemorrhage secondary to an aneurysm at the junction of the posterior communicating and internal carotid arteries.

Laboratory data: Lumbar puncture: Demonstrated bloody spinal fluid with no definite clearing from Tube 1 to Tube 3. There were approximately 40,000 fresh red blood cells in each tube, indicating that this was not a traumatic lumbar puncture.

Four-vessel angiography: The left posterior cerebral artery arose from the left internal carotid artery as a continuation of the posterior communicating artery. At the junction of the posterior communicating artery and the internal carotid artery, a 13-mm aneurysm was present.

Subsequent course: The neck of the aneurysm was clipped by the neurosurgeon, Dr. Bernard Stone. Adhesions and clots were present between the aneurysm and the third cranial nerve. The patient did well except that the third nerve paralysis was still present at the time of discharge 3 weeks after surgery.

Chapter 18
Movies on the Brain

This chapter is a series of films that depict the effects of various disorders on the brain. The arrangement of films corresponds to the chapter sequence of Marcus and Jacobson, *Integrated Neurosciences* published in 2003. An asterisk indicates a more significant films from a medical and cinematic standpoint.

I. Developmental Disorders

Neurological Disorder: Mental Retardation. Film: *Charly*
   In this film, released in 1968, a mentally retarded man Charley (Cliff Robertson, who won an Academy Award) develops remarkable intelligence after experimental brain surgery, but then begins to slip back to his former state. Claire Bloom plays his case worker with whom the intelligent Charly has a romance. The movie is based on the novel by Daniel Keyes: *Flowers for Algernon*.

*__Neurological Disorder: A Blind Deaf and Mute Infant after Fever. Film:__ *The Miracle Worker*
   This film, released in 1962, depicts the story of the young 7-year-old Helen Keller, played by Patty Duke, and of her teacher Annie Sullivan, played by Anne Bancroft, who taught the child to communicate initially with touch. Both received Academy Awards. Although Helen Keller never attained hearing or sight, she did graduate from Radcliffe College with honors in 1904. The director was Arthur Penn; movie adapted by William Gibson from his play and based on the autobiography of Helen Keller *The Story of My Life*.

*__Neurological Disorder: Autism. Movie:__ *Rain Man*
   In this film, released in 1988, an ambitious young man (played by Tom Cruise) discovers, when his father dies, that he has an autistic older brother (played by Dustin Hoffman) who has been institutionalized for years. He needs him to claim the inheritance. He liberates him from the institution and they set off cross-country. As they travel, the relationship between the brothers evolves. The film and screenplay won Oscars, as did Dustin Hoffman and the director Barry Levinson.

II. Spinal Cord/Brain Stem Disorders

*Neurological Disorder: Traumatic High Cervical Spinal Cord Transection Producing Quadriplegia: Ethical Issues: The Rights of the Patient to Make Decisions About How Long to Continue Life Support. Film: *Million Dollar Baby*.

This film, released in 2004, concerns a woman of 32 years (played by Hilary Swank) who trains to be a boxing champion. In the championship fight she is fouled by the other boxer and sustains a cervical 1-2 transection. The last third of the film deals with the ethical questions involved in the decision to discontinue life support in a patient who is conscious and can clearly express her wishes. The film, Swank (Best Actress), Morgan Freeman (Best Supporting Actor), and Clint Eastwood (director and producer) won Oscars. Eastwood, the co-star, was also nominated but did not receive an Oscar as Best Actor.

Neurological Disorder: Thoracic Spinal Cord Injury. Film: *Passion Fish*

In this film, released in 1992, a soap opera actress (played by Mary McDonnell) sustains a thoracic spinal cord injury in a New York taxi accident, resulting in paraplegia. She is reduced to an embittered wheelchair existence. With the help of a strong nurse (played by Alfre Woodard), she must face up to her disability.

Neurological Disorder: Poliomyelitis (FDR Subsequently Died of a Hypertensive Intracerebral Hemorrhage). Film: *Sunrise at Campobello*

This film, released in 1960, traces the career of a wartime president Franklin Roosevelt (portrayed by Ralph Bellamy). Who. after serving as Assistant Secretary of the Navy and running for Vice President, developed poliomyelitis. He overcame this severe disability to become governor of the state of New York. Eleanor Roosevelt was portrayed by Greer Garson. The original play from which the film was adapted received a Tony Award. A later dramatization of the personal lives of the two and of the impact of polio was presented on television in 1976. The film entitled *Eleanor and Franklin* was based on the book by Joseph Lash and starred Edward Hermann and Jane Alexander.

*Neurological Disorder: Amyotrophic Lateral Sclerosis. Film: *Pride of the Yankees*

This famous 1942 movie provided a biography of the famous baseball player Lou Gehrig, who died with amyotrophic lateral sclerosis. His record for career grand slam home runs and of single-season runs batted in for the American League remain to this day. The film starred Gary Cooper as Lou Gehrig and Theresa Wright as his wife. Former teammates Babe Ruth, Bill Dickey, and Bob Meusel also appeared. A later retelling of the story (*Love Affair: The Eleanor and Lou Gehrig Story*) appeared on television in 1977 and starred Edward Hermann and Blythe Danner.

Neurological Disorder: Amyotrophic Lateral Sclerosis. Film: *Tuesdays with Morrie*

This made-for-TV movie presented in 1999 was based on a real-life drama. A sports writer, Mitch Albom (played by Hank Azira), in Detroit finds out that his old college professor in Boston, Morrie Schwartz (played by Jack Lemmon). has developed ALS. He visits the professor and they undertake a series of weekly discussions regarding life, purpose, and death.

III. Disorders of Motor Systems and Motor Control

*Neurological Disorder: Cerebral Palsy: Spastic Diplegia and Choreo-Athetosis. Film: *My Left Foot*

This film, released in 1989, was based on the autobiography of a young Irish writer/artist Christy Brown, who had with severe cerebral palsy. Using his left foot, the only extremity over which he has voluntary control, he teaches himself to write and goes on to become an acclaimed writer and painter. Daniel Day Lewis played Christy Brown and won an Academy Award for Best Actor. Brenda Flicker (Academy Award for Best Supporting Actress) played his very supportive mother, who recognizes that he is not an imbecile and insists on integrating him into her large family.

*Neurological Disorder: Parkinson's Disease. Film: *Awakenings*

This 1990 film was based on a book by the neurologist Oliver Sacks (played by Robin Williams) concerning his experiences at a chronic disease hospital in the initial treatment of postencephalitic Parkinsonism with an apparent miracle drug L-dopa. Leonard, the youngest patient treated, was played by Robert Di Niro. Julie Kravner played the experienced nurse who knows more about the patients than the neophyte physician. Both the effects and the subsequent complications and loss of effect of the drug are shown.

*Neurological Disorder: Foreign Arm/Hand Syndrome Film: *Dr. Strangelove or: How I Learned to stop Worrying and Love the Bomb*

This classic film, released in 1964 and directed by Stanley Kubrick, dealt with one of the great fears of the Cold War: a nuclear confrontation between the great powers, the United States and the Soviet Union. The National Security Adviser or Dr. Strangeglove has been injured during World War II. As a result, he is confined to awheel chair and has a prosthetic right hand. At times of stress he cannot control the right arm, which performs in a foreign manner, taking on postures that reflect his previous life as a Nazi leader and adviser. Peter Sellers in a remarkable performance portrayed three characters: Dr. Strangeglove as well as the sensible Canadian Group Captain Mandrake and the American President Maffley. George Scott played the American Air Force Commander General Buck Turgedson, a character based on General Curtis LeMay, who masterminded the fire bombing of the Japanese cities in World War II. He was one of the hawks of the Cold War. Sterling Hayden played General Ripper, the squadron commander. The nuclear confrontation is triggered when General Jack Ripper sends off a squadron of nuclear armed B52 bombers. He is concerned with a plot by the Soviets to rob Americans of their vital fluids and he aims to preserve "the purity of essence."

*Neurological Disorder: Obsessive–Compulsive Disorder, Psychosis and Posttraumatic Disorder with Complex Partial Seizures: Film: *The Aviator*

This film, released in 2004, is a biographic study of the early adult life of Howard Hughes (Leonard De Caprio), a brilliant filmmaker, airplane designer, test pilo,t and aviation record holder who, unfortunately, was increasingly incapacitated by a severe obsessive–compulsive disorder, periods of psychosis, and,in the film, episodes characterized by speech automatisms and confusion. He produced and/or directed *Hell's Angels, Scarface*, and *The Outlaw*. Hughes was involved in romances with several Hollywood actresses: Katherine Hepburn (Cate Blanchet), and Ava Gardner (Kate Beckinsale), Jean Harlow (Gwen Stefani), possibly but probably only from a professional standpoint), and Jane Russell, only professionally. The film was directed by Martin Scorsese and also featured Alan Alda as Senator Brewster and Alex Baldwin as Juan Trippe, the founder and CEO of Pan American Airways.

Neurological Disorder: Obsessive–Compulsive Disorder Plus Personality Disorder. Film: *As Good as It Gets*

This film, released in 1997 and directed by James L Brooks, concerns the romance of an unpleasant obsessive–compulsive novelist and the only person who can tolerate his behavior, a waitress at his favorite local restaurant. He can only sit at his table and be waited on by his waitress. During the course of the film, his behavior improves and his attitudes change. Jack Nicholson played Melvin Udall, the writer with an obsessive–compulsive disorder. He received an Academy Award as Best Actor. Helen Hunt played Carol, the waitress, and received an Academy Award as Best Actress.

Neurological Disorder: Huntington's Disease. Films: *Bound for Glory* **and** *Alice's Restaurant*

The first film, released in 1976, presented an excellent biography of the famous American song writer and folk singer Woody Guthrie (portrayed by David Carridine) for the period 1936–1940 a time of protest and union-organizing activities. He subsequently manifested the progressive neurological disease Huntington's disease characterized by a movement disorder and cognitive changes. The film does demonstrate the personality changes already beginning to occur.

The second film, released in 1969 and directed by Arthur Penn, is centered on another period of protest, the hippie era of the late 1960s, when the Vietnam War, the draft,, drugs and free love were prominent issues. Arlo Guthrie, Woody's son and also a folk singer, and Pete Seeger, the folk singer, played themselves. The terminal bed-ridden stages of Woody's progressive Huntington's disease are demonstrated.

IV. Limbic System

*Neurological Disorder: Prefrontal Lobotomy/Psychosurgery. Film: *Suddenly Last Summer*

This film, released in 1960 and directed by Joseph Mankiewicz, was adapted from the play by Tennessee Williams "Suddenly Last Summer." A wealthy woman, played by Katherine Hepburn, has arranged for her niece, played by Elizabeth Taylor, to be committed to a state mental hospital in the deep South because of her supposed delusions and hallucinations regarding the death of the woman's son at the hands of cannibals. She attempts to have the young neurosurgeon at the hospital, played by Montgamery Cliff, perform a psychosurgical procedure to eliminate the "delusion and hallucinations." She is willing to endow the psychosurgery program at the institution. The neurosurgeon discovers that the delusions are in fact true. He faces the ethical dilemma of having to say "no" to the wealthy donor.

***Neurological Disorder: The Effects of Neuroleptics, Electroshock Therapy and the Complications of Prefrontal Lobotomy. Film:** *One Flew over the Cuckoo's Nest*

This film, released in 1975, was the first film since the 1934 film *It Happened One Night* to win all five major Academy Awards: Best Picture, Best Actor (Jack Nicholson), Best Actress (Louise Fletcher), Best Director (Milos Forman) and Best Screenplay (Ken Kesey). The film concerns a patient who enters an insane asylum to escape prosecution for his criminal activities. He inspires the other inmates of his ward to assert themselves, upsetting the ward routine dictated by a rigid strong-willed nurse. Unfortunately, the system deals with his behavior by altering his neurological function by means of medications, electroshock, and, eventually, a prefrontal lobotomy. After that surgery, he is left in an abulic, poorly responsive bed-ridden state.

See also *Fifty First Dates* under Brain Trauma for Short-Term Memory
See also *Memento* under Memory for Short-Term Memory
See also *Dead Ringers* under Toxic Metabolic
See also *Arsenic and Old Lace* under Toxic Metabolic
See also *Dr. Strangelove* under Disorders of Motor Control
See also *Regarding Henry* under Brain Trauma
See also *Rain Man* under Developmental Disorders

V. Cerebrovascular Disease

Neurological Disorder: Massive Stroke. Film: *Wilson*

This film, released in 1944, which was a critical success but a box office failure, traces the political career of an actual American President, Woodrow Wilson (Alexander Knox). After a career as a Professor of History at Princeton University, he became president of that institution, then governor of New Jersey, and then President of the United States in 1912. As president, he "kept us out of war" but then entered a war (World War I) "to make the world safe for democracy." At the conclusion of that war, while campaigning for acceptance of the peace treaty, he sustained a massive stroke resulting in a severe left hemiparesis and cognitive changes. His wife (Geraldine Fitzgerald) played a major role in limiting access to the president and helping him to make decisions during this period of actual but undeclared presidential disability.

Neurological Disorder: Massive Stroke Producing Irreversible Coma. Film: *Dave*

This 1993 film concerned a presidential look-alike who is asked to impersonate the actual president at a function so that the president can keep an assignation with his mistress. During the tryst, the actual president suffers a massive stoke, resulting in a state of irreversible coma. The fill-in is manipulated to continue the impersonation by a presidential adviser who hopes to thus control the power of the position. However, the impersonator takes on a life of his own and begins to make decisions in the public interest. Five actual US senators had cameo roles in this film, as did the McLaughlin group, Larry King, and Oliver Stone. Kevin Kline played Dave, the impersonator, and the President and Sigourney Weaver played the President's wife.

VI. Brain Trauma

*****Neurological Disorder: Traumatic Frontal Lobe Injury. Film:** *Regarding Henry*

This film, released in 1991 and directed by Mike Nichols, dealt with the disturbances in memory and personality in a high-powered lawyer following a bullet wound to the brain. "I remember a woman in a blue dress standing on grass. I think it's my mother but I'm not certain it's my mother" Whereas previously he manipulated the truth, cheated on his wife, and had a terrible relationship with his daughter, he now became an honest man with solid relationships with his wife and daughter. He concludes that he can no longer be a lawyer. Harrison Ford played Henry Turner, the lawyer, and Annette Benning played his wife.

*****Neurological Disorder: Posttraumatic Short-Term Memory Loss due to Right Temporal Lobe (Hippocampal) Damage Which Spares her Amygdala. Film:** *50 First Dates (A.K.A 50 First Kisses)*

In this recent film, a romantic comedy released in 2004, a veterinarian in Hawaii, Henry Roth (Adam Sandler) falls in love with a charming young lady, Lucy (Drew Barrymore). She has had a head injury in an auto accident and has a short-term memory loss. So, each day she awakens with no memory of the previous day. Every day her family re-creates the last day she remembers. The veterinarian has to woo her again each day starting from scratch. According to the psychologist, "scar tissue" in the temporal lobe prevents her from converting short-term memories to long-term memories during sleep. Much more realistic is the portrayal of "10 second Tom," who had a hunting accident (shot in the head) after which he has only working memory and can remember no longer than 10 seconds. The cast included Dan Aykroyd as Dr Keats, the psychologist who is the memory expert at the brain trauma institute. Watch him interpret MRIs. Sean Astin and Rob Schneider star in other roles.

Neurological Disorder: Brain Trauma Producing Coma. Film *Fantastic Voyage*

This science fiction film released in 1966 concerns a defecting Russian scientist who sustains a significant head injury in a vital but relatively inaccessible brain region. In order to correct the lesion with the least possible damage, a crew, including a neurosurgeon and his nurse, are placed in a remarkable submarine that is then reduced to microscopic size and injected into the vascular system. Stars on the voyage included Raquel Welch, Stephen Boyd, Arthur Kennedy and Donald Pleasance, with Edmond O'Brien as the commander of the facility.

See also *Cleopatra* under Seizures
See also *The Aviator* under Motor System Control

VII. Brain Tumors and Increased Intracranial Pressure

*Neurological Disorder: Brain Tumor. Film: *Dark Victory*

In this classic movie of 1939, from Warner Brothers, the hero, Dr. Frederick Steele (George Brent), the most successful neurosurgeon of New York, decides, after 9 years of sending flowers to the funerals of his patients with gliomas and other brain tumors, to close his practice and to return to medical research. He poses the question "Why do healthy cells go berserk producing gliomas." He will work in the laboratory that he has established at his farm in Vermont, where he will culture cells from the brain. "Some day some one will discover a serum like insulin." The concept, while true in terms of gliomas of the brain, was extraordinary for the time. The film provides a notable depiction of the symptoms of increased intracranial pressure The film also reflects the prevailing concept of the time regarding what a patient with incurable disease was to be told about their prognosis. Bette Davis played the heroine Judith Terherne the young beautiful and wealthy socialite patient with a glioma in love with her neurosurgeon. The performance by Bette Davis, who had previously won two Academy Awards, is considered one of her finest. Although nominated for an Academy Award, she did not win; 1939 was the year of *Gone with the Wind*. She was also nominated for her other remarkable performance in *All About Eve* in 1950 but did not win. The actors and actresses reflected almost the full acting company of the studio system at Warner Brothers, including Humphrey Bogart, Ronald Reagan, Henry Travers as the caring sympathetic general practitioner Dr. Parsons, and Geraldine Fitzgerald as the close friend and secretary. This film provided Ronald Reagan with his worst role portraying a drunken playboy, followed by Bogart as her Irish horse trainer.

Other films concerned with brain tumors include *Crisis*. In this film, released in 1950, a famous somewhat cynical American neurosurgeon (played by Cary Grant) vacationing with his wife in a Latin American country is forced to operate on a paranoid and oppressive dictator (played by Jose Ferrer) with a brain tumor (frontal) while his wife is held hostage. The patient's recovery is complicated by a successful revolution led by Gilbert Roland.

Neurological Disorder: Glioma in an Adolescent. Film: *Death Be Not Proud*

In this 1975 film made for television, an adolescent (played by Robby Benson) is treated for a malignant glioma. The film, which starred Arthur Hill and Jane Alexander as the parents, is based on the true story by John Gunther concerning his son's illness and death at age 17.

VIII. Infections

*Neurological Disorder: Severe Neurological Disability and Death Following Partially Treated Bacterial Meningitis. Film: *The Third Man*

In this classic film, released in 1949, Holly Martins, an American writer of Western pulp fiction (played by Joseph Cotton). arrives in Vienna at the invitation of his best old friend Harry Lime (played by Orson Welles), to find that the old friend might have been murdered. He also eventually discovers that the old friend has been the author of a complex black market scheme in which penicillin has been stolen from the Children's Hospital and the solutions diluted. As a result, children with bacterial meningitis have died or have been left in a state of severe neurological disability. The film is notable for the haunting musical score provided by Anton Karas on an unaccompanied zither. The cast included Trevor Howard as Major Calloway, the British military police director who knows the real evil posed by Lime, Bernard Lee as the British military police sergeant who is an admirer of Westerns and is shot by Lime in the sewers of Vienna during the climatic pursuit, (he later played M in the James Bond films) and Wilfred Hyde-White, the British cultural attaché who manages to enjoy all that Vienna has to offer. Alida Valli played Anna, Harry's girl friend, an central/eastern European refugee with false documents whom the Russians wish to deport to her native country now under Communist control. The director was Carol Reed and the screen writer was Graham Greene.

*Neurological Disorder: Neurosyphilis (General Paresis) Producing Dementia and A Change in Personality. Film: *Young Winston*

This film, released in 1972 and directed by Richard Attenborough, depicts the childhood and early adult life of the famous British politician Winston Churchill, played by Simon Ward, who trained as a military officer at Sandhurst but later pursued a career as a combat reporter and author. He subsequently rose to political prominence just prior to and during World War I as the First Lord of the Admiralty. Our interest here is in his father, Lord Randolph Churchill, played by Robert Shaw, who at the peak of a brilliant political career as Lord of the Exchequer began to demonstrate, in his late thirties/early forties, erratic behavior and a progressive impairment of his mental and physical capacities, all of which is extensively demonstrated in the film. His wife, portrayed by Anne Bancroft (remember Mrs. Robinson in *The Graduate* and Annie Sullivan in *The Miracle Worker*), was told that the disorder was an inflammation of the brain and that intimate contact with the patient was to be avoided. The film received the 1973 Golden Globe Award as Best Foreign Film. Ward bore a remarkable resemblance to the young Churchill. The cast included a veritable gallery of fine English actors portraying various politicians, journalists, and military leaders: John Mills, Jack Hawkins, Ian Holm, Patrick Magee, andEdward Woodward. The movie was actually based on Winston Churchill's autobiography *My Roving Life: A Roving Commission*.

Neurological Disorder: Syphilis. Film: *Dr. Ehrlich's Magic Bullet*

This film, released in 1940, presented a biography of Dr. Ehrlich (played by Edward G Robinson), the physician who helped to develop a treatment for diphtheria and then went on to develop a treatment for syphilis, which was utilized until penicillin came into use. There were questions at the time as to whether the film should be released, but the US Public Health Service supported the release of the film. Robinson was more famous for his role of Rico in the 1930 drama *Little Caesar*.

See also *Sunrise at Campobello* under Spinal Cord: Polio Myelitis

IX. Toxic and Metabolic Disorders

*****Neurological Disorder: Metabolic Induced Psychosis. Film:** *The Hospital*

This film, released in 1971, concerned the sequence of mis-adventures that befell a healthy retired physician, Dr. Drummond (played by Bernard Hughes), admitted to a major teaching hospital, the Manhattan Medical Center, for a routine medical checkup. He undergoes an unnecessary renal biopsy, has a botched nephrectomy, following which he has an allergic reaction and develops renal failure, resulting in coma. He recovers from coma but continues to present a picture of pseudo-coma. He now manifests a highly developed paranoid psychosis with hallucinations. With the assistance of the hospital's own inefficient systems and the incompetence of multiple house officers, attending physicians, and nurses, he begins to secretly eliminate the physicians and nurses who have caused the various mishaps. The chief of medicine of the hospital Dr. Bock (George Scott), who has his own problems of excessive alcohol intake, impotence, and depression, attempts to sort out the series of deaths, assisted by Dr. Drummond's daughter (played by Diana Rigg). The film also featured Richard Dyshart as Dr. Welbeck, the incompetent urologic surgeon more interested in business than in quality of care, and Nancy Marchand as Ms Christie, the director of the nursing service who has to deal with multiple incompetents. She was later cast as the matriarch of the *Sopranos*.

Filmed at Metropolitan Hospital in New York, the movie exposed the problems of operating a large city teaching hospital at a time of fiscal restraints and inner-city conflicts. The screenplay was written by Paddy Chayefsky (Academy Award).

Neurological Disorder: Drug Addiction: Barbituates Plus. Film: *Dead Ringers*

This disturbing film, directed by David Cronenberg and released in 1988, was based on a real series of events and concerns the fictional named identical twins Elliot and Beverly Mantle, both played by Jeremy Irons). They are brilliant students from Toronto, educated and trained at Harvard, and now back in Toronto as academic gynecologists working together. They share their medical patients, research, and women. One, Elliot, is confident and ruthless; the other, Beverly, is modest and sensitive. A failed love affair with an actress, who is a patient with a congenitally malformed uterus, triggers a descent into mental collapse and barbiturate addiction. Both die, possibly as a result of drug overdose or barbiturate withdrawal or murder and suicide. The real-life twins were Stewart and Cyril Marcus, who were recruited from Harvard to the Cornell New York Hospital. Jeremy Irons won the New York Film Critics Award for Best Actor and Genevieve Bujold won the award as Best Supporting Actress.

*Neurological Disorder: Cocaine Addiction. Film: *The Seven-Percent Solution*

This film, released in 1987, combines the activities of the most famous fictional detective, Sherlock Holmes (portrayed by Nicol Williamson), with the real-life detective of the mind, Sigmund Freud, portrayed by Alan Arkin. *The Seven-Percent Solution* refers to the usual solution of cocaine. The plot concerns the addiction of Sherlock Holmes and of the damsel in distress (played by Vanessa Redgrave), as well as the background of Holmes's obsession with Professor Moriarty (Lawrence Olivier). Freud's early methods of analysis are portrayed, as well as his work with cocaine and cocaine addiction. Dr. Watson is played by Robert Duvall. Joel Grey played one of several villains. The author of the original book and screenplay, Nicholas Meyer, has resurrected the characters originally created by Dr. Arthur Conan Doyle.

Neurological Disorder: Acute Arsenic Poisoning and Criminal Psychosis. Film: *Arsenic and Old Lace*

This film, directed by Frank Capra, was produced in 1941 but not released until 1944. Gary Grant portrays a drama critic, Mortimer Brewster, who finds that his two dear old aunts are poisoning lonely old men with arsenic in a solution of elderberry wine. Among his not-so-normal relatives are the escaped criminally insane cousin Jonathan Brewster (Raymond Massey) and Teddy (John Alexander), a Brewster cousin who believes that he is Teddy Roosevelt building canals in the basement. The two aunts are portrayed by childlike Josephine Hull and the more intelligent Jean Adair. Peter Lorre portrays the plastic surgeon of the criminal madman Jonathan Brewster and Edward Everett Horton plays the psychiatrist who locks up almost all of these characters by the end of the movie. The screenwriters for this romp were Julius and Philip Epstein. (They also did the screen play for *Casablanca*.)

Neurological Disorder: A Metabolic Disorder (Pophyria). Film: *The Madness of King George*

This film, released in 1994, concerns the effects of porphyria on the brain of a government leader, King George III (portrayed by Nigel Harthorne), "who had lost the American colonies." The effects on his family interactions and on the affairs of government are chronicled. The movie also provides information on the treatment of medical and mental disorders in 1788. Helen Mirren portrays his wife, Queen Charlotte. She is more famous for her portrayal of a police inspector in a series of television dramas, and in 2006, she has provided a remarkable performance as Queen Elizabeth II. She had been previously knighted by the Queen for her contributions to the cinema.

Neurological Disorder: Mercury Poisoning. Film: *Alice in Wonderland*

Various films, both with real and cartoon actors, have been made of this story by Lewis Carroll, which features, among other characters, an individual with mercury poisoning resulting in psychosis acquired in the hatting industry: the Mad Hatter. Actually, cerebellar effects producing the hatters shakes were more common than the cerebral/psychotic involvement.

Neurological Disorder: Effects of Hallucinogens and Sleep Deprivation. Film: *Altered States*

In this film, released in 1980, a research professor (played by William Hurt) hopes to discover the inner self. He takes hallucinogenic drugs and subjects himself to sensory deprivation with severe alterations in his behavior.

X. Disorders of Myelin

*Neurological Disorder: Progressive Multiple Sclerosis. Film:** *Jackie and Hilary*

This film, released in 1998, was based on the real-life story of an outstanding cellist, Jacqueline Du Pre (Emily Watson), her relationship with her sister and their husbands, and her decline and death after developing progressive multiple sclerosis. The early symptoms of change in behavior, cerebellar ataxia and urinary incontinence, as well as her progressive decline are well demonstrated in the second half of the film. Rachel Griffiths played Hilary, her sister, who was also an excellent musician (flutist). The husband was the famous conductor Daniel Barenbom.

Neurological Disorder: Adrenoleukodysdrophy: A Disorder That Affects Brain, Spinal Cord, and Peripheral Nerves as Well as the Adrenal Gland. There Is Defective Formation of Myelin. Film: *Lorenzo's Oil*

This film, released in 1992, was based on a true story of a 5-year-old boy with adrenoleukodystrophy and the determined efforts of his mother and father, portrayed by Susan Sarandon and Nick Nolte, to find a cure.

XI. Memory

*Neurological Disorder: Alzheimer's Disease. Film:** *Iris*

This film, released in 2001, was based on real lives and chronicled the love story of two English writers from their early days at Oxford in the 1950s to Iris' long decline and death from a progressive neurological disorder in 1999. The wife, Iris Murdock, was both a novelist and philosopher. The film won an Academy Award for Best Supporting Actor and several British Academy Awards and Golden Globe awards. This film has also been recognized by the Alzheimer's Association as providing the most realistic portrayal of the problems of this disease from the standpoint of the patient and family. Iris Murdock was portrayed as the young scholar by Kate

Winslet and as the mature scholar and writer by Judi Dench. Jim Broadbent portrayed John Bayley, her husband and biographer (Oscar for Best Supporting actor).

Neurological Disorder: Short-Term Memory Loss. Film: *Memento*
In this thriller, released in 2000, a man searches for the killer of his wife. The plot is rather twisted since the protagonist has a problem: Since the attack in which he also sustained severe head trauma, he forgets current events after 15 minutes. As a remedy, he has tattooed the essential information on his body and leaves Post-its and Polaroids about of the usual suspects.

Other Film in this Category:

****Gone With the Wind***. 1939. This blockbuster Academy Award winner does include a sequence of scenes in which Scarlett O'Hara (Vivian Leigh) returns home to the devastated plantation Tara to find that her father (Thomas Mitchell), age 65, is confused, delusional, and demented. She must take responsibility for the management of the estate if the entire family and their dependents are to survive. The cast of thousands also included Clark Gable as Rhett Butler, Leslie Howard as Ashley Wilkes, Olivia De Haviland as Melanie Wilkes, and Hattie McDaniel as her wise nanny (Academy Award for Best supporting Actress). The picture, Vivian Leigh, Director and Producer all won Academy Awards in 1939.

See also *50 First Dates* under Trauma
See also *Regarding Henry* under Trauma

XII. Seizures and Epilepsy

Neurological Disorder: Temporal Lobe Seizures with Secondary Generalization. Film: *The Terminal Man*
In this 1974 film, a patient, a computer engineer with frequent seizures played by George Segal, has a computer-controlled stimulator implanted in his brain, designed to interfere with seizure discharges. It malfunctions and actually triggers seizures during which he inadvertently kills several people. Richard Dysart plays the neurosurgeon who conceives of computer control of seizures. The screen writer was Dr. Michael Critchton. He was a Harvard medical student and based his book on the actual work of Mark and Ervin at Harvard Medical School, which was published as *Violence and the Brain*.

Neurological Disorder: Posttraumatic Focal Motor Seizures with Secondary Generalization. Film: *Cleopatra*
In this film, released in 1963, the manipulative, well-endowed, and well-educated young woman ruler, Cleopatra (Elizabeth Taylor), secretly witnesses the posttraumatic seizures of a visiting military leader, Julius Caesar (Rex Harrison).

*Neurological Disorder: Generalized Convulsions and& Learning Disabilities. Film: *The Lost Prince*

This premiere presentation of Masterpiece Theater's 2004 season was based on true events and recalls a long-ago Europe when most of the royal houses were related as cousins. The youngest child of the ruling King and Queen (Miranda Richardson) of England born in 1905 suffers from epileptic seizures and learning disabilities. He is sent off to live in the country away from public gaze, supervised by a dedicated Nanny (Gina McKee). As an adolescent, he dies after a prolonged seizure. A series of generalized seizures, post-ictal states, and the effects of the poorly controlled seizures on his cognitive functions are demonstrated. The attitudes of the medical experts, the family, and the public regarding epilepsy are demonstrated. The film includes remarkable performances by two boys as the prince: Daniel Williams and Matthew Thomas.

*Neurological Disorder: Seizures Induced by Photic Stimulation and Sleep Deprivation. Film: *The Andromeda Strain*

In this film, released in 1971 and based on the Michael Critchon novel, concerns a group of scientists who are working together in a secret government facility attempting to deal with a deadly biologic agent. The woman scientist has a generalized convulsive seizure triggered by intermittent photic stimulation and possibly related to sleep deprivation. Among those fighting the biologic agent are Arthur Hill, David Wayne, James Olson, and Kate Reid.

Other Film in this Category

Lust For Life. 1956. The life of Vincent Van Gogh, who certainly had a bipolar disorder and might possibly have had complex partial seizures. The film is remarkable not only for the performances of Kirk Douglas and Anthony Quinn (as his friend Gauguin) but also for the number of paintings presented in color. A hyperreligiosity at an early stage of his career is suggested.

See also *The Aviator* Under Motor System Disorders

XIII. Coma

*Neurological Disorder: Anoxic Encephalopathy. Film: *Coma*

In this movie, released in 1978, written by Dr Robin Cook and directed by Michael Critchon, MD, young healthy patients are experiencing hypoxia during the anesthesia for elective surgery at Boston Memorial Medical Center. A young and very competent surgical house officer, Dr. Susan Wheeler (Genevieve Bujold) discovers the series of patients reduced to a vegetative state of irreversible coma after an "anesthetic accident" involving her closest friend (Lois Chiles). As she pursues the solution to this mystery, her own life is threatened and she also comes close to ending up at the Jefferson Institute for patients in irreversible coma, where body parts are salvaged for transplantation. The cast includes Michael Douglas as her boyfriend/senior resident, who is anxious to move ahead in the medical hierarchy to become the chief surgical resident and is not interested in making waves, and Richard Widmark as Dr Harris, the chief of the surgical service, who has built a great institute and transplantation program that requires a source of body parts.

***Neurological Disorder: Irreversible Coma Following Insulin Overdose. Film:** *Reversal of Fortune*

In this film from 1990, based on an actual case, a wealthy woman Sunny von Bulow (Glenn Close) lies in irreversible coma in a New York hospital after a near-lethal overdose of insulin, which might have been administered by her husband Klaus von Bulow (Jeremy Irons, Academy Award as Best Actor) or, then again, she might have self-administered the insulin and he is being framed for her murder. The movie centers on the appeal and second trial of her husband, who is defended by a famous Harvard law professor, Alan Dershowitz (Ron Silver). Barbet Schroeder directed based on the book by Alan Dershowitz.

***Neurological Disorder: Anesthetic Accident Producing Anoxic Encephalopathy with Irreversible Coma. Film:** *The Verdict*

This film, released in 1982, is based on an actual case in Boston, although, for dramatic effects, the institutional location has been transposed. A down-on-his-luck alcoholic lawyer Frank Galvin (Paul Newman), attempts to find justice for his client, a previously healthy woman who entered a prominent hospital of the archdiocese for a routine delivery. She receives the wrong anesthetic agent and remains in a state of irreversible coma. In the search for justice, he must battle the establishment forces of organized religion, medicine, and the law. He is assisted or opposed by the following actors Jack Warden, James Mason, and Milo O'Shea.

Other Films in this Category:

Critical Care. 1997, Ethical issues of end-of-life decisions
Flatliners. 1990, Medical students experiment with transient hypoxic encephalopathy

Bibliography

Chapter 1 Introduction to the Central Nervous System

General References

Denny-Brown, D. 1957. Handbook of Neurological Examination and Case Recording, 2nd ed. Cambridge, MA, Harvard University Press.
Edelman, E.R., and S. Warach. 1993. Medical progress: Magnetic resonance imaging. N. Engl. J. Med. 328:708, 716, 785–791.
Greenberg, J.O. (ed.). 1999. Neuroimaging: A Companion to Adams and Victor's Principles of Neurology. 2nd ed. New York, McGraw-Hill.
Kandel, E.R., J.H. Schwartz, and T.M. Jessell. 2000. Principles of Neurosciences. New York, McGraw Hill.
Marcus, E.M., and S. Jacobson. 2003. Intregrated Neuroscience. Boston, Kluwer Academic.
Martin, J.H., J.C.M. Brust, and S. Hilal. 1991. Imaging the living brain. In. E. R. Kandel, J.H. Schwartz, E. Niedermeyer, Nolte, J. 1990. The Human Brain. St. Louis, MO, Mosby.

Chapter 2 Neurocytology

Specific References

Barr, M., and R. Bertram. 1949. A morphological distinction between neurons of the male and female, and the behavior of the nucleolar satellite during accelerated nucleoprotein synthesis. Nature 163:676–678.
Bentivoglio, M.H., G.J.M. Kuypers, C.E. Catsman-Berrevoets, H. Loewe, and O. Dann. 1980. Two new fluorescent retrogradeneuronal tracers which are transported over long distances. Neurosci. Lett. 18:25–30.
Bodian, D. 1970. An electron microscopic characterization of classes of synaptic vesicles by means of controlled aldehyde fixation. J. Cell Biol. 44:115.
Brightman, J. 1989. The anatomic basis of the blood-brain barrier. In E.A. Neuwelt (ed.). Implications of the Blood-Brain Barrier and Its Manipulation, Vol. 1. New York, Plenum, p. 125.
Brady, S.T. 1985. A novel brain ATPase with properties expected for the fast axonal transport motor. Nature (London) 317:73–75.
Colonnier, M. 1969. Synaptic patterns on different cell types in the different laminae of the cat visual cortex. An electron microscopic study. Brain Res. 33:268–281.

Cowan, W.M., D.I. Gottlieb, A.E. Hendrickson, J.L. Price, and T.A. Woolsey. 1972. The autoradiographic demonstration of axonal connections in the central nervous system. Brain Res. 37:21–51.

Darnell, J., H. Lodish, and D. Baltimore. 1990. Molecular Cell Biology, 2nd ed. New York,: Scientific American Books.

Davis, E.J., T.D.Foster, and W.E. Thomas. 1994. Cellular forms and functions of brain microglia. Brain Res. Bull. 34:73–78.

De Duve, C., and R. Wattiaux. 1966. Functions of lysosomes. Annu. Rev. Physiol. 28:435.

Esposito, P., D. Gheorghe, K. Kandere, X. Pang, R. Connolly, S. Jacobson, and T.C. Theoharides. 2001. Acute stress increases permeability of the blood-brain barrier through activation of brain mast cells. Brain Res. 888:117–127.

Finger, S. 1994. Origins of Neuroscience: A History of Exploration into Brain Function. New York: Oxford University Press.

Gage, F.H. 2002. Neurogenesis in the adult brain. J. Neurosci. 612–613.

Geren, B.B. 1954. The formation from the Schwann cell surface of myelin in the peripheral nerves of chick embryos. Exp. Cell Res. 7, 558–562

Goodman, C.S. 1994. The likeness of being; phylogenetically conserved molecular mechanisms of growth cone guidance. Cell 78:353–356.

Graftstein, B., and D.S. Forman. 1980. Intracellular transport in neurons. Physiol. Rev. 60:1167–1183.

Gray, E.G. 1959. Axosomatic and axodendritic synapses of the cerebral cortex; an electron microscopic study. J. Anat. 93:420.

Guth, L., and S. Jacobson. 1966. The rate of regeneration of the cat vagus nerve. Exp. Neurol. 14:439.

Haase, A. 1986. Pathogenesis of lentivirus infections. Nature 322:130–136.

Hall, Z.W., and J.R. Sanes. 1993. Synaptic structure and development: The neuromuscular junction. Cell 71/Neuron 10: 99121.

Herman, I., and S. Jacobson. 1988. In situ analysis of microvascular periocytes in hypertensive rate brain, Tissue Cell 20:1–12.

Horner, P.J., and F.H. Gage, 2000. Regenerating the damaged central nervous system, Nature 407:963–970.

Hutton, M. 2000. Ann. NY Acad. Sci. 920 63.

Jacobson, S. 1963. Sequence of myelinization in the brain of the albino rat. A. Cerebral cortex, thalamus and related structures. J. Comp. Neurol. 121:5–29

Jacobson, S., and L. Guth. 1965. An electrophysiological study of the early stages of peripheral nerve regeneration. Exp. Neurol. 11:48.

Kandel, E., J.H. Schwartz, and T.M. Jessell. Principals of Neuroscience 2000 4th Ed. McGraw Hill.

Kernakc, D.R., and P. Rakic. 1976. Continuations of neurogenesis in the hippocampus of the adult macaque monkey. Proc. Natl. Acad. Sci. USA 96:5768–5773

Lasek, R.J., and B.S. Joseph. 1967. Radioautography as a neuroanatomic tracing method. Anat. Rec. 157:275–276.

LaVail, J.H., and M.M. LaVail. 1972. Retrograde axonal transport in the central nervous system. Science 176:1416–1417.

Levi-Montalcini, R., and P.U. Angeletti. 1968. Biological aspects of the nerve growth factor. In E.E. Woolstenholme and, M.O. Connor (eds.): Growth of the Nervous System. Boston: Little, Brown.

Ling, E.A., and W Wong Glai. 1993. The origin and nature of ramified and amoeboid microglia: A historical review and current concepts. Glia 7:9–18.

McQuarrie, I.G. 1988. Cytoskeleton of the regenerating nerve. In P.J. Reier, R.D. Bunge, and F.J. Seil (eds.). Current Issues in Neural Regeneration Research. New York, A.R. Leiss, pp. 23–32.

Nauta, W.J.H. 1957. Silver impregnation of degeneration axons. In W.F. Windle (ed.). New Research Techniques of Neuroanatomy. Springfield, IL, Charles C Thomas.

Neuwelt, E.A., and S.A. Dahlborg. 1989. Blood-brain barrier disruption in the treatment of brain tumors: Clinicalimplications. In E.A. Neuwelt (ed.). Implications of the Blood-Brain Barrier and Its Manipulation, Vol. 2, pp. 195–262.

Nogo A. myelin-associate mark granites factor Nature 403: 434–439

Oppenheim, R.W. 1991. Cell death during development of the nervous system. Annu. Rev. Neurosci. 14:453–501.
Palay, S.L. 1967. Principles of cellular organization in the nervous system. In G.C. Quarton, T. Melnechuk, and F.O..Schmitt (eds.). The Neurosciences: A Study Program. New York, Rockefeller University Press.
Price, S.D., R.B. Brew, J. Sidtis, M. Rosenblum, A. Scheck, and P. Cleary. 1988. The brain in AIDS: Central nervous system, HIV-1 infection and AIDS dementia complex. Science 231:586–592.
Ramón y Cajal, S. 1909. Histologie du système nerveux de l'homme et des vertèbres. Paris, J.A. Maloine.
Ramón y Cajal, S. 1928. Degeneration and Regeneration of the Nervous System. London, Oxford.
Rassmussen, G.T. 1957. Selective silver impregnation of synaptic endings. In W.F. Windle (ed.). New Research Techniques of Neuroanatomy. Springfield, IL, Charles C Thomas.
Reese, T.S., and M..J. Karnovsky. 1968. Fine structural localization of the blood-brain barrier to exogenous peroxidase. J. Cell Biol. 34:207.
Rio-Hortega, P. del. 1919. El Atercer elemento@ de los centros nerviosos. Bol. Soc. Española Delbiol. 9:69–120.
Robertson, J. D. 1955. The ultrastructure of adult vertebrate peripheral myelinated nerve fibers in relation to myelinogenesis. J. Biophys. Biochem. Cytol.1, 271–278.
Scharrer, E. 1966. Endocrines and the Central Nervous System. Baltimore, Williams & Wilkins.
Schwartz, J. 1980. The transport of substances in nerve cells. Sci. Am. 242:152–171.
Sedgwick, J.D., and R. Dorries. 1991. The immune system response to viral infection. Neurosciences 3:93–100.
Serafini, T., T.E. Kennedy, M.J. Galko, C. Mizrayan, T.M. Jessel, and M. Tessler-Lavigne. 1994. The netrins define a familyof axon outgrowth-promoting proteins homologous to *C. elegans* UNC-6. Cell 78:409–424.
Sladek, J.R., Jr., and D.M. Gash. 1984. Neural Transplants: Development and Function. New York, Plenum Press.
Vale, R.D. 1987. Intracellular transport using microtubule based molecules. Annu. Rev. Cell Biol. 3:347–378.
Vale, R.D., T.S. Reese, and M.P. Sheetz. 1985. Identification of a novel force-generating protein, kinesin, involved in microtubule based motility. Cell 42:39–50.
Vaughn, J.E., and A.E. Peters. 1968. A third neuroglial cell type. J. Comp. Neurology 133:269–288.
Weiss, P.A., and M.B. Hiscoe. 1948. Experiments on the mechanism of nerve growth. J. Exp. Zool. 197:315–396.
Wislocki, G.B., and E.H. Leduc. 1952. Vital staining of the hematoencephalic barrier by silver nitrate and trypan blue and cytological comparisons of neurohypophysis, pineal body, area postrema, intercolumnar tubercle and supraoptic crest. J. Comp. Neurol. 96:371.
Wujek, J.R., and R.K. Lasek. 1983. Correlation of axonal regeneration and slow component B in two branches of a single axon. J. Neurosci. 3:243–251.
Young, JZ. 1942. Functional repair of nervous tissue. Physiol. Rev. 22:318.

Chapter 3   Spinal Cord

General References

Abel Majuid, T. and D. Bowsher. 1985. The gray matter of the dorsal horn of the adult human spinal cord. J. Anat. 142:33–58.
Asanuma, H. 1981.The pyramidal tract. In V.B. Brooks (ed). Handbook of Physiology. Section I. The Nervous System. Vol II. Motor Control, Bethesda, MD, American Physiological Society, pp. 703–733.

Becksted, R. J. Morse, and R. Norgeren. 1980. The nucleus of the solitary tract in the monkey. J. Comp Neurol 198:259–282.
Burton, H., and A. Craig. 1979. Distribution of trigeminothalamic neurons in cat and monkey. Brain Res. 161:515–521.
Carpenter, M.B. 1991. Core Text of Neuroanatomy, 4th ed. Baltimore, Williams and Wilkins.
Everts, E. 1955. Role of motor cortex in voluntary movement in primates. In V.B. Brooks (ed). Handbook of Physiology. Section I. The Nervous System. Vol. II. Motor Control. Bethesda, MD, American Physiological Society, pp. 1083–1120.
Finger. T. and W. Silver. 1987. Neurobiology of Taste and Smell. New York, Wiley.
Foote, S., F. Bloom, and G. Aston-Jones. 1983. Nucleus locus cereuleus. New evidence of anatomical and physiologicalspecificity. Physiol. Rev. 63:844–914.
Hobson, J., and M. Brazier. 1983. The reticular formation revisited. Brain Res. 6:1–564.
Moore, B., and F. Bloom. 1978. Central catecholamine neuron systems; anatomy and physiology of the norephinephrine and epinephrine systems. Annu. Rev. Neurosci. 2:113–168.
Norgren, R. 1984. Central mechansims of taste. In Darian-Smith (ed.). Handbook of Physiology. Section I. The Nervous System. Vol III. Sensory Processes. Bethesda, MD, American Physiological Socienty, pp. 1087–1128
Riley, H.A. 1960. Brain Stem and Spinal Cord. New York, Hafner.
Sinclair, D. 1982. Touch in primates. Ann. Rev. Neurosci. 5:155–194.
Wall, P., and R. Melzak. 1985. Textbook of Pain. Edinburgh, Churchill Livingston.
Willis, W. 1985. Nocioceptive pathways. Anatomy and physiology of nociceptive ascending pathways. Phil, Trans, Roy, Soc. London (Biol) 308:252–268.
Victor, M., and A.H. Ropper. 2001 Principles of Neurology, 7th ed. New York, McGrawHill.
Brodal, A. 1965. The Cranial Nerves. Anatomy & Anatomical Clinical Correlation. London, Oxford Blackwell.
Clemente, C. 1988. Gray's Anatomy, 30th ed. Philadelphia, Lea & Febiger.
DeJong, R.N. 1978. The Neurologic Examination. New York, Hoeber Medical Division, Harper & Row.
Herrick, C.J. 1984. The Brain of the Tiger Salamander. Chicago, University of Chicago Press.

Chapter 4   Brain Stem

General References

Marcus, E.M., and S. Jacobson. 2003. Intregrated Neuroscience. Boston, Kluwer Academic..
Riley, H.A. 1960. Brain Stem and Spinal Cord. New York, Hafner.
Sinclair, D. 1982. Touch in primates. Annu. Rev. Neurosci. 5:155–194.
Victor, M., and A.H. Ropper. 2001 Principles of Neurology, 7th ed. New York, McGrawHill.

Chapter 5   Cranial Nerves

Brodal, A. 1965. The Cranial Nerves. Anatomy and Anatomical Clinical Correlation. London, Oxford Blackwell.
Brodal, A. 1957. The Reticular Formation of the Brain Stem. Springfield, IL, Charles C Thomas.
Marcus, E.M., and S. Jacobson. 2003. Intregrated Neuroscience. Boston, Kluwer Academic.
Riley, H.A. 1960. Brain Stem and Spinal Cord. New York, Hafner.
Victor, M., and A.H. Ropper. 2001 Principles of Neurology, 7th ed. New York, McGraw Hill.

Chapter 6   Diencephalon

General References

Carpenter, MB 1978. Core text of neuroanatomy. Baltimore, Williams and Wilkins.
De Wulf, A. 1971. Anatomy of the Normal Human Thalamus, Topometry and Standardized Nomenclature. Amsterdam, Elsevier.
Goldman-Rakic, P.S., and L.J. Porrino. 1989. The primate mediodorsal (MD) and its projection to the frontal lobe. J. Comp. Neurol. 242:535–560.
Guillery, R.W. 1995. Anatomical evidence concerning the role of the thalamus in corticocortical communications: A brief review. J. Anat. 187:583–592.
Jones, E.G. 1985. 1985. The Thalamus. New York, Plenum.
Jones. E.G. 1989. A new parcellation of the human thalamus on the basis of histochemical staining. Brain Res. Rev. 14:1–34.
Purpura, D.P., and M.D. Yahr. 1966. The Thalamus. New York, Columbia University Press.
Trojanowski, J.Q., and S. Jacobson. 1976. Areal and laminar distribution of some pulvinar cortical efferents inrhesus monkey. J. Comp. Neurol. 76:371–391.
Van Buren, J.M., and R.C. Borke. 1972. Variations and Connections of the Human Thalamus. 2. Variations of the Human Diencephalon. New York, NY: Springer- Verlag, p. 81
Walker, A E. 1938. The Primate Thalamus. Chicage, University of Chicago Press.

Chapter 7   Hypothalamus

General References

Albers, W.B., and P.B. Molinoff. Basic Neurochemistry. Molecular, Cellular and Medical Aspects. New York, Raven Press.
Bodian, D. 1963. Cytological aspects of neurosecretion in opossum neurohypophysis. Bull. Johns Hopkins Hosp. 113:57.
Bodian, D. 1966. Herring bodies and neuroapocrine secretion in the monkey: An electron microscopic study of the fate of the neurosecretory product. Bull. Johns Hopkins Hosp. 118:282.
Becker, K.L. 2001. Principles and Practice of Endocrinology and Metabolism, 3rd edn. Philadelphia, Lippincott/Williams and Wilkins.
Brownstein, M.J. 1989. Neuropeptides. In G.J. Siegel, B.W. Agranoff, R.W. Albers, and P.B. Molinoff (eds.). Basic Neurochemistry. Molecular, Cellular and Medical Aspects. New York, Raven Press, pp. 287–309.
Gershon, M.D. 1981. The enteric nervous system. Annu Rev. Neurosciences 4:227–272.
Gershon, M.D. 2000. Second Brain: A groundbreaking new understanding of nervous disorders of the stomach and intestine. New York Harper collins Publishers.
Green, J.D. 1966. The comparative anatomy of the portal vascular system and of the innervation of the hypophysis. In G.W. Harris and B.T. Donovan (eds.). The Pituitary Gland. Berkeley, University of California Press, Vol. I, p. 127.
Haymaker, W., E. Anderson, and W.J.H. Nauta.1969. The Hypothalamus. Springfield, IL, Charles C Thomas.
Karczmar A.G. K. Koketsu, and S. Nishi (eds.). 1986. Autonomic and Enteric Ganglia. New York, Plenum Press.
Martin, J.B., S. Reichlin, and G.M. Brown. 1987. Clinical Neuroendocrinology. 2nd ed. Philadelphia, F.A. Davis.

McGeer, P.L., and E.G. McGeer. 1989. Amino acid neurotransmitters. In G.J. Siegal, B.W. Agranoff, R.W. Muller,. and G. Nistico. Brain Messengers and the Pituitary. New York, Academic Press.

Palay, S.L. 1957. The fine structure of the neurohypophysis. In H. Waelsch (ed.). Progress in Neurobiology: II. Ultrastructure and Cellular Chemistry of Neural Tissue. New York, Paul B. Hoeber, p. 31.

Siegal, G.J., B.W. Agranoff, R.W. Albers, and P.B. Molinoff. 1989. Basic Neurochemistry. Molecular, Cellular and Medical Aspects. New York, Raven Press.

Scharrer, E. 1966. Endocrines and the Central Nervous System. Baltimore, Williams and Wilkins.

Szentagothai, J., B. Flerko, B. Mess, and B. Halasy. 1968. Hypothalamic Control of the Anterior Pituitary. Budapest, Akademiai Kiado.

Wislocki, G.B.1938. The vascular supply of the hypophysis cerebri of the rhesus monkey and man. Proc. Assoc. Res. Nerv. Dis. 17:48.

Chapter 8 Cerebral Cortex Functional Localization

General References

Bailey, P., and G. von Bonin. 1951. The Isocortex of Man. Urbana, IL. University of Illinois Press.

Brodal, A. 1981. Neurological Anatomy in Relation to Clinical Medicine, 3rd ed. New York,, Oxford University Press.

Brodmann, K. 1909. Localization in the Cerebral Cortex.Vergleichende Lokaizations der Grosshirninde. JA Basth,. Leipzig.

Brazier, M.A.B. 1977. The Electrical Activity of the Nervous System, 4th ed. Baltimore, Williams and Wilkins, pp. 66–74, 175–208, 217–236.

Colonnier, M.L. 1966. The structural design of the neocortex. In J,C. Eccles (ed.). Brain and Conscious Experience. New York, Springer-Verlag, pp. 1–23.

Commission on Classification and terminology of the International League against epilepsy. 1981. Proposal fpr revised clinical and EEG classification of epileptic seizures. Epilepsia. 22:489–510.

Conel, J. 1939–1968. The Postnatal Development of the Human Cerebral Cortex. Volumes I to VI. Cambridge, MA, Harvard University Press.

Curtis, B.A., S. Jacobson., and E.M., Marcus. 1972. An Introduction to the Neurosciences. Philadelphia, WB Saunders.

Eccles, J.C. (ed.). 1966. Brain and Conscious Experience. New York, Springer-Verlag.

Engel, J., Jr. 1989. Seizures and Epilepsy. Philadelphia: F.A. Davis, pp. 41–70.

Geschwind, N., and W. Levitsky. 1968. Human brain; left-right asymmetries in temporal speech region. Science 161:186–187.

Lorente de No. R. 1949. Cerebral cortex: Architecture, intracortical connections, motor projectionsIn J.F. Fulton (ed.). Physiology of the Nervous System, 3rd ed. New York, Oxford University Press, pp. 288–330.

Peters, A., and E.G. Jones (eds.). 1984. Cerebral Cortex I: Cellular Components of the Cerebral Cortex. New York, Plenum.

Penfield, W., and H. Jasper, 1954. Epilepsy and the Functional Anatomy of the Human Brain. Little Brown Boston

Rakic, P., and W. Singer (ed.). 1988. Neurobiology of Neocortex. New York, Wiley.

Ransom, S., and S. Clark. 1959. The Anatomy of the Nervous System. Philadelphia, WB Saunders.

Reeves, A.G. (ed.). 1985. Epilepsy and the Corpus Callosum. New York, Plenum.

Specific References

Angevine, J., and M.S. Smith. 1982. Recent advances in forebrain anatomy and their clinical correlates.In R.A. Thompson and J.R. Green (eds.). New Perspectives in Cerebral Localization. New York, Raven Press.
Clemente, C.D., and M.B. Sterman. 1967. Basal forebrain mechanisms for internal inhibition and sleep. Res. Publ. Assoc. Nerv. Mental Dis. 45:127–147.
Curtis, H.H. 1940. Inter cortical connections of corpus callosum as indicated by evoked potentials. J. Neurophysiol. 3:407–413.
Dehay, C., H. Kennedy, J. Bullier, et al. 1988. Absence of interhemi-spheric connections of area 17 during development in the monkey. Nature 331:348–350.
Duchowny, M., P. Jayakar, and B. Levin. 2000. Aberrant neural circuits in malformation of cortical development and focal epilepsy. Neurology 55:423–428.
French, J.D., B.E. Gernandt, and R.B. Livingston. 1956. Regional differences in seizure susceptibility in monkey cortex. Arch.bNeurol. Psychiatry 72:260–274.
Geschwind, N., and W. Levitsky. 1968. Human brain left right asymmetries in temporal speech region. Science 161:186–187.
Jacobson, S., and E. M. Marcus. 1970. The laminar distribution of fibers of the corpus callosum: A comparative study in rat, cat, monkey and chimpanzee. Brain Res. 24:517–528.
Jacobson, S., and J. Q. Trojanowski. 1974. The cells of origin of the corpus callosum in rat, cat and rhesus monkey. Brain Res.74:149–155.
Jasper, H. 1949. Diffuse projection system: the integrative action of the thalamic reticular system. Electroenceph. Clin. Neuro-physiol. 11:121–128.
Jasper, H. 1958. The international ten-twenty electrode placement system. EEG, Clin. Neurophysio. 10:374
Jones, E.G. 1985. Anatomy, development and physiology of the corpus callosum. In A.G. Reeves (ed.). New York, Plenum.
Killackey, H.P. 1985. The organization of somato-sensory callosal projections: a new interpretation. In A.G. Reeves (ed.). Epilepsy and the Corpus Callosum.. New York, Plenum, pp. 41–53.
Killackey, H.P., and L.M. Chalupa. 1986. Ontogenetic changes in the distribution of callosal projection neurons in the post central gyrus of the fetal rhesus monkey. J. Comp. Neurol. 244:331–348.
Marcus, E.M. 1985. Generalized seizure models and the corpus callosum. In A.G. Reeves (ed.). Epilepsy and the Corpus Callosum. New York, Plenum, pp. 131–206.
Marcus, EM., W. Watson, and S. Jacobson. 1969. Role of the corpus callosum in bilateral synchronous discharges induced by intravenous pentylenetetrazol. Neurology 19:309.
Martin, J.H. 1985. Cortical neurons, the EEG and the mechanisms of epilepsy. In E.R. Kandel and J.H. Schwartz (eds.). Principles of Neural Science, 2nd ed. New York, Elsevier. pp. 636–647.
McCulloch, W.S. 1944. Cortico-cortical connections. In P. C. Bucy (ed.). The Precentral Motor Cortex. Urbana, IL, University of Illinois Press, pp. 211–242.
McNamara, J.C. 1993. Excitatory amino acid receptors and epilepsy. Curr. Opin. Neurol. Neurosurg. 6:583–587.
Meyer, B.U., S. Roricht, H. Grafin von Einsiedel, et al. 1995. Inhibitory and excitatory interhemispheric transfers between motor cortical areas in normal humans and patients with abnormalities of the corpus callosum. Brain 118:429–440.
Morrison, R.S., and E.W. Dempsey. 1942. A study of thalamocortical relations. Am. J. Physiol. 135:281–292.
Myers, R.E. 1965. General discussion: phylogenetic studies of commissural connections. In E.G. Ettlinger (ed.). Functions of the Corpus Callosum. Boston, Little Brown & Company, pp. 138–142.
Pandya, D.N., and D.L. Rosene. 1985. Some observations on trajectories and topography of commissural fibers. In A.G. Reeves (ed.). Epilepsy and the Corpus Callosum. New York, Plenum Press, pp. 21–39.

Purpura, D.P. 1959. Nature of electrocortical potentials and synaptic organizations in cerebral and cerebellar cortex. Int. Rev. Neurobiol. 1:47–163.

Purpura, D.P. 1964. Relationship of seizure susceptibility to morphologic and physiologic properties of normal and abnormal immature cortex. In F. Kellaway and I. Petersen (eds.). Neurological and electroencephalographic correlative studies in infancy. New York, Grune and Stratton, pp. 117–157.

Scheibel, M.E., and A.B. Scheibel. 1967. Structural organization of non-specific thalamic nuclei and their projection toward the cortex. Brain Res. 6:60–94.

Skinner, J.E., and D.B. Lindsley. 1967. Electrophysiological and behavioral effects of blockade of the non-specific thalamocortical system. Brain Res. 6:95–118.

Spencer, W.A., and E.R. Kandel. 1961. Electrophysiology of hippocampal neurons: IV. Fast prepotentials. J. Neurophysiol. 24:272–85.

Spencer, W.A., and E.R. Kandel. 1968. Cellular and integrative properties of hippocampal pyramidal cells and the comparative electrophysiology of cortical neurons. Int. J. Neurol. 6:266–296.

Starzl, T.E., and D.G. Whitlock. 1952. Diffuse thalamic projection system in the monkey. J. Neurophysiol. 15:449–468.

Steriade, M., and R. Llinas. 1988. The functional states of the thalamus and the associated neuronal interplay. Physiol. Rev. 68: 649–742.

Velasco, M., and D.B. Lindsley. 1965. Role of orbital cortex in regulation of the thalamocortical electrical activity. Science 149:1375–1377.

Von Economo, C. 1929. The Cytoarchitectonics of the Human Cerebral Cortex. London, Oxford University Press.

Walker, A.E. 1964. The patterns of propagation of epileptic discharge. In G. Schaltenbrand and C.W. Woolsey (eds.).. Cerebral Localization and Organization. Madison, University of Wisconsin Press, pp. 95–111.

Chapter 9   Motor Systems I: Movement and Motor Pathways

General References

Alexander, G.E., and M.R. Delong.1992. Central mechanisms of initiation and control of movements. In A.K. Asbury et al. (eds.). Diseases of the Nervous System. Philadelphia, W.B. Saunders. pp. 285–308.

Carew, T.J., and D.B Kelley. 1989. Perspectives in Neural Systems and Behavior. MBL Lectures in Biology 10. New York,. Alan Liss.

Denny-Brown, D. 1960. Motor Mechanisms: Introduction, the general principles of motor integration. In Handbook of Physiology. Section 1. Neurophysiology. Washington, DC, American Physiologic Society, Vol. 2, pp. 781–796.

Denny-Brown, D. 1966. The Cerebral Control of Movement. Springfield, IL, Charles C. Thomas.

Desmedt, J.E. (ed). 1983. Motor Control Mechanisms in Health and Disease. New York, Raven Press.

Foerster, O. 1936. The motor cortex in the light of Hughlings Jackson's doctrines. Brain 59: 135–159.

Freund, H.J. and H. Hummelshein. 1985. Lesions of the premotor cortex in man. Brain. 108 697–733.

Fulton, J.F. 1951. Physiology of the Nervous System, 3rd ed. New York, Oxford University Press.

Ghez, C. 1991. The Control of Movement. In E.R. Kandel, J.H. Schwartz, and T.M. Jessell (eds.). Principles of Neural Science, 3rd ed. New York, Elsevier, pp. 553–547.

Hitzig, E. 1900. Hughlings Jackson and the cortical motor centres in light of physiological research. Brain. 23:545–581.

Humphrey, D.R., and Freund, H.J.1991. Motor Control Concepts and Issues. New York, Wiley.

Klemm, W.R., and Vertes, R.P. (eds.). 1990. Brain Stem Mechanisms of Behavior. New York, Wiley Interscience.

Jacobson, S., N. Butters, and N.J. Tovsky. 1978. Afferent and efferent subcortical projections of behaviorally defined sectors of prefrontal granular cortex. Brain Res. 159:279–296.

Kertesz, A. (ed.). 1983. Localization in Neuropsychology. New York, Academic Press.

Kuypers, H.G.J.M. 1987. Some aspects of the organization and output of the motor cortex. Ciba foundation Symposium. 132:63–82.

Lawrence, D.G., and H.G.J.M. Kuypers. 1968. The functional organization of the motor system in the monkey. I: The effect of bilateral pyramidal tract lesions. Brain 91:1–14.

Luders, H.O., and S. Noachtar (eds.). 2000. Epileptic Seizures: Pathophysiology and Clinical Semiology. New York, Churchill Livingstone.

Magoun, H.W., and R. Rhines. 1946. An inhibitory mechanism in the bulbar reticular formation. J Neurophysiology 9:161–171.

Martin, J.H., J.C.M. Brust, and S. Hilal. 1991. Imaging the living brain. In J.W. lance and J.G. McLeod (eds.). Principles of Neural Science. 1981. A Physiological Approach to Clinical Neurology - 3rd Ed., London, Butterworths. Chapter 4: Spinal Reflexes, pp.73-100, Chapter 5: Muscle Tone and Movement, pp.101-127, Chapter 6 - Disordered Muscle Tone, pp.28-152.

Mesulam. Neuroanatomical aspects of cerebral localization. pp. 21-61.

Pandya, D.N. and H.J.M. Kuypers. 1969. Cortico-cortical connections in the rhesus monkey. Brain Research. 13:13–16.

Penfield, W., and H. Jasper. 1954. Epilepsy and the Functional Anatomy of the Human Brain. Boston, Little, Brown, pp. 41–154, 350–539.

Penfield, W., and T. Rasmussen (eds.). 1957. The Cerebral Cortex of Man: A Clinical Study of Localization. New York, MacMillan.

Phillips, C.G., and R. Porter. 1977. Corticospinal Neurons. London. Academic Press.

Reynolds, E.H. 1988. Hughlings Jackson: A Yorkshireman's Contributions to Epilepsy. Arch. Neurol. 45:675–678

Sanes, J.N., and J.P. Donoghue, V. Thangoraj et al. 1995. Shared neural substrate controlling hand movements in the human motor cortex. Science 268:1775–1777.

Sanes, J.N., and J.P. Donoghue, 2000. Plasticity and Primary motor cortex. Ann Rev Neurosci. 23:393–416.

Schomer, D.L. 1983. Partial epilepsy. N.Engl. J.Med. 309:536–559.

Taylor. J. (ed.). 1931. Selected Writings of John Hughlings Jackson. Reprint New York: Basic Books, Inc. 1959, pp.1–36, 385–405, 406–411, 458–463.

Twitchell, T.E. 1965. The automatic grasping response in infants. Neuropsychologica. 3:247–259.

Wolff, P.H. and Ferber. 1979. The development of behavior in human infants: premature and newborn. Ann Rev Neurosci 2:291–307.

Wilkins, R.H., and I.A. Brody. 1970. Jacksonian epilepsy. Arch. Neurol. 22:183–188. (See also J. Taylor).

Willis, W.D., and R. G. Grossman. 1981. Medical Neurobioloqy; Neuroanatomical and Neurophvsiological Principles of Basic Clinical Neuroscience3rd ed., St.Louis, MO, Mosby, pp. 347–402.

Wise, S.P. 1985. The primate premotor cortex: Past, present and preparatory. Annu. Rev. Neurosci. 8:1–19.

Yakolev, P.I., and P. Radic. 1966. Patterns of decussation of bulbar pyramids and distribution of pyramidal tracts on two sides of the spinal cord. Trans. Am. Neurol. Assoc. 91:366–367.

Specific References: Central Pattern Generators

Bizzi, E., and E.V. Evarts. 1971. Translational mechanisms between input and output. Neurosci. Res. Program Bull. 9:31–59.

Burke, R.E. 1971. Control systems operating on spinal reflex mechanisms. Neurosci. Res. Program Bull. 9:60–85.

Delong, M. 1971. Central patterning of movement. Neurosci. Res. Program Bull. 9:10–30.

Grillner, S., and P. Wallen. 1985. Control pattern generators for locomotion with special reference to vertebrates. Annu. Rev. Neurosci. 8:233–261.

Harris-Warrick, P.M., and B.R. Johnson. 1989. Motor pattern networks: flexible foundations for rhythmic pattern production. In T.J. Carew and D.B Kelley (eds.). Perspectives in Neural Systems and Behavior. NewYork, Alan R. Liss, pp. 51–71.

Stein, P.S.G. 1978. Motor systems with specific reference to the control of locomotion. Annu. Rev. Neurosci. 1:61–81.

Gait Disorders of the Elderly

Adams, R.D., C.M. Fisher, S. Hakim. et al. 1965. Symptomatic occult hydrocephalus with "normal" cerebrospinal fluid pressure: A treatable syndrome. N. Engl. J. Med. 273:117–126.

Fisher, C.M. 1982. Hydrocephalus as a cause of disturbances of gait in the elderly. Neurology 32:1358–1363.

Fishman, R.A. 1985. Normal pressure hydrocephalus and arthritis (editorial). N. Engl. J. Med. 312:1255–1256.

Hachinski, V.C., P. Potter, and H Merskey. 1987. Leuko-araiosis.Arch. Neurol. 44:21–23.

Inzitar, D., F. Diaz, A. Fox, et al. 1987. Vascular risk factors and leuko-araiosis. Arch.Neurol. 44:42–47.

Jacobs, L., D. Conti., W.R Kinkel,., E.J. Manning. 1976. "Normal pressure" hydrocephalus: relationship of clinical and radiographic findings to improvement following shunt surgery. JAMA. 235:510-12.

Masdeu, J.C., L. Wolfson, and G., Lantos, et al. 1989. Brain white matter changes in the elderly prone to falling. Arch. Neurol. 46:1292–1296.

Rasker, J.J., E.N.H. Jansen, J. Haan, and J. Oostrom. 1985. Normal pressure hydrocephalus in rheumatic patients: A diagnostic pitfall. N. Engl. J. Med., 312:1239–1241.

Steingart, A., V.C. Hackinski, C., Lau, et al. 1987. Cognitive and neurological findings in subjects with diffuse white matter changes on computed topographic scan (leukoariosis). Arch. Neurol. 44:32–34.

Sudarsky, L. 1990. Current Concepts—Geriatrics: Gait disorders in the elderly. N. Engl. J. Med. 322:1441–1446.

Sudarsky, L., and M. Ronthal. 1983. Gait disorders among elderly patients: A survey study of 50 patients. Arch. Neurol. 40:740–743.

Thompson, P.D., and C.D. Marsden. 1987. Gait disorder of subcorttical arteriosclerotic encephalopathy: Binswanger's disease. Movement Disord. 2:1–8.

Tinetti, M.D., and M. Speechly. 1989. Current Concepts: Geriatrics: Prevention of falls among the elderly. N. Engl. J. Med.. 320:1055–1059.

Tinetti, M. D., M. Speechly, and S. F. Ginter. 1988. Risk factors for falls among elderly persons living in the community. N. Engl. J. Med. 319:1701–1707.

Chapter 10 Motor Systems II: Basal Ganglia

General References

Albin, D., A.B. Young, and J.B. Penney. 1989. The functional anatomy of basal ganglia disorders. Trends Neurosci. 12:366–375.

Cooper, I.S. 1969. Involuntary Movement Disorders. New York, Hoeber Medical Division.

Cooper, J.R., F.E.Bloom, and R.H. RothH. 1991.The Biochemical Basis of Neuropharmacology. New York, Oxford University Press. (See, in particular, Chapter 10: Dopamine pp. 285–337.)
Denny-Brown, D. 1962. The Basal Ganglia and Their Relation to Disorder of Movement. London, Oxford University Press.
Galvez- Jimenez, N. 2001. Movement disorders. Semin. Neurol. 21:1–123.
Hallett, M. 1991. Classification and treatment of tremor. JAMA 266:1115–1117.
Haymaker, W., W.F. Mehler, and F. Schiller. 1969. Extrapyramidal motor disorders. In W. Haymaker (ed.). Bing's Local Diagnosis in Neurological Disease. St.Louis, MO,.Mosby, pp. 404–440.
Hurtig, H.I., and M.B. Stern 2001. Movement disorders. Neurol. Clin.19:523–788.
Jankovic, J., and E. Tolosa, E. (eds.). 1988. Parkinson's Disease and Movement Disorders. Baltimore, Urban & Schwarzenberg.
Lowe, J., G. Lennox, and P.N. Leigh.1997. Disorders of movement and system degenerations. In D.I. Graham and P.L. Lantos (eds.). Greenfield's Neuropathology. Vol, 2, pp. 281–366.
Marcus, E.M., and S. Jacobson. 2003. Integrated Neuroscience: A clinical problem solving approach. Boston Kluwer Academic.
Marsden. C.D., and S. Fahn (eds.).1987. Movement Disorders II. London, Butterworths.
Noback, C.R., N.L. Strominger, and R.J. Demarest. 1991.The Human Nervous System: Introduction and Review, 4th ed.. Philadelphia, Lea & Febiger, pp. 375–395.
Oppenheimer, D.R. 1984. Diseases of the basal ganglia, cerebellum and motor neurons. In J.H. Adams, J.A.N. Corsellis, and L.W. Dicker (eds.). Greenfield's Neuropathology. London, Edward Arnold, pp. 700–747.
Quinn, N. 1995. Fortnightly review: Parkinson's disease: Recognition and differential diagnosis. Br. Med. J. 310:447–452.
Rascol, O., and A.J. Lees. 1999. Dyskinesias. Movement Disord. 14(Suppl. 1):1–80.
Riley, D.E., and A.E. Lang. 2000. Movement disorders. In W.G. Bradley, RB. Daroff, G.M. Fenichel, and C.D. Marsden. Neurology in Clinical Practice, Boston, Butterworth Heinemann, pp. 1889–1930.

Specific References and Citations

Agid, Y., D. Cervera, E. Hirsh, et al. 1989. Biochemistry of Parkinson's disease 28 years later: A critical review. Movement Disord. 4(Suppl. 1):126–144.
Agid, Y., F. Javoy-Agid, and M. Ruberg. 1987. Biochemistry of neurotransmitters in Parkinson's disease. In C.D. Marsden and S. Fahn. Movement Disorders 2. London. Butterworths, pp. 166–230.
Albin, R.I., A.B. Young, J.B. Penney, et al. 1990. Abnormalities of striatal projection neurons and N-methyl D-aspartate receptors in presymptomatic Huntington's disease. N. Engl. J. Med. 322:1293–1298.
Alexander, G.E., M.R. De Long, and P.L. StrickL. 1986. Parallel organization of functionally segregated circuits linking basal ganglia and cortex. Annu.Rev.Neurosci. 9:375–381.
Alyward, E.H., A.M. Codori, A. Rosenblatt, et al. 2000. Rate of caudate atrophy in pre-symptomatic and symptomatic stages of Huntington's disease. Movement Disord. 15:552–560.
Ascherio, A.H., H. Chen, M. Weisskopf et al. 2006. Pesticide exposure and risk for Parkinson's disease. Ann Neruology. 60:197–203.
Aziz.T.Z., D. Peggs, M.A. Sambrook, and A.R. Crossman. 1991. Lesion of the subthalamic nucleus for the alleviation of l-methyl-4 phenyl 1,2,3,6, tetrahydropyridine (MPTP) INduced Parkinsonism in the primate. Movement Disord. 6:288–292.
Ballard, P.H., J.W. Tetrud, and J.W. Langston. 1985. Permanent human parkinsonism due to l-methyl-4 phenyl, 1,2,3,6, tetra-hydropyrine (MPTP): Seven cases. Neurology 35:949–956.
Bandmann, O., M.G. Sweeney, S.E. Daniel, et al. 1997. Multiple system atrophy is genetically distinct from identified inherited causes of spinocerebellar degeneration. Brain 49:1598–1604.

Berciano, J. 1988. Olivopontocerebellar Atrophy (in) Jankovic, J., and Tolosa, E.: Parkinson's Disease and Movement Disorders. Baltimore/Munich, Urban & Schwartzenberg. pp. 131–151.

Bergman, H., T. Wichmann, and M.R. DeLong. 1990. Reversal of experimental Parkinsonism by lesions of the subthalamic nucleus. Science 249:1436–1438.

Bhatia, K.P. 2001. Familial (idiopathic) paroxysmal dyskinesias.: An update. Semin. Neurol. 21: 69–74.

Bhatia, K.P., R.C. Griggs, and L.J. Ptacek. 2000. Episodic movement disorders as channelopathies. Movement Disord. 15:429–423.

Brashear, A. 2001. The botulinum toxins in the treatment of cervical dystonia. Semin. Neurol. 21: 85–90.

Brooks, D.J. 1991. Detection of preclinical Parkinson's disease with PET. Neurology 41 (Suppl. 2): 24–27.

Cooper, I.S. 1969. Involuntary Movement Disorders. New York, Paul B. Hoeber.

Cote, L., and Crutcher, M.D. 1991.The basal ganglia. In E.R. Kandel, J.H. Schwartz, and T.M. Jessell (eds.). Principles of Neural Science, 3rd ed.. New York, Elsevier, pp. 647–659.

Cotzias, G.C., P.S. Papavasiliou, and R. Gellen. 1969. Modification of Parkinsonism: Chronic treatment with L-Dopa. N. Engl. J. Med. 280:337–345.

Cotzias, G.C., M.H. Van Woert, and L.M. Schiffer. 1967. Aromatic amino acids and modification of Parkinsonism. N. Engl. J. Med. 276:374–379.

Crossman, A.R. 1990. A hypothesis on the pathophysiological mechanisms that underlie levodopa or dopamine agonist induced dyskinesia in Parkinson's disease: Implications for the future strategies in treatment. Movement Disord. 5:100–108.

Crossman, A.R., I.J. Mitchell, M.A. Sambrook, and A. Jackson. 1988. Chorea and myoclonus in the monkey induced by gamma amino butyric acid antagonism in the lentiform complex. The site of drug action and a hypothesis for the neural mechanism of chorea. Brain 121:1–33.

Daniel, S.E., V.M. de Bruin, and A.J. Lees. 1995. The clinical and pathological spectrum of Steele-Richardson Olszewski syndrome (progressive supranuclear palsy): A reappraisal. Brain 118:759–770.

De Bie, R.M.A., P.R. Schuurman, P.S. de Haan, et al. 1999. Unilateral pallidotomy in advanced Parkinson's disease: A retrospective study of 26 patients. Movement Diosord. 14:951–957

Delwaide, P.J., and M. Gance. 1988. Pathophysiology of Parkinson's signs. In J. Jankovic and E. Tolosa (eds.). Parkinson's Disease and Movement Disorders. Baltimore, Urban and Schwarzenberg, pp. 59–73.

Deuschl, G. 2000. Treatment options for tremor. N. Engl. J. Med. 342:505–507.

Dunnett, S.B., and A. Bjorklund. 1999. Prospects for new restorative and neuroprotective treatments in Parkinson's disease. Nature 399(Suppl.):A32–A39.

Duvoisin, R.C. 1998. Familial Parkinson's disease. NeuroSci. News 1:43–46.

Ebersbach, G.M., F. Sojer, Valladeoriopla, et al. 1999. Comparative analysis of gait in Parkinson's disease.cerebellar ataxia and subcortical arteriosclerotic encephalopathy. Brain 122:1349–1355.

Eidelberg, D., A. Sortel, C. Joachim, et al. 1987. Adult onset Lallervorden–Spatz disease with neurofibrillary pathology. A discrete clinicalpathological entity. Brain 110:993–1013.

Fearnley, J.M., and A.J. Lees. 1990. Striatatonigral degeneration. A clinicopathological study. Brain 113:1823–1824.

Forno, L.S., and E.C. Alford. 1987. Pathology of Parkinson's disease. In W.C. Roller (ed.). Handbook of Parkinson's Disease. New York, Marcel Dekker, pp. 209–236.

Freed, C.R., P.E. Greene, R.E. Breeze, et al. 2001. Transplantation of embryonic dopamine neurons for severe Parkinson's disease. N. Engl. J. Med. 344:710–719. See also editorial by G.D.

Fischbach and G.M. Mc Khann. 2001. Cell therapy for Parkinson's disease. N. Engl. J. Med. 344:763–765.

Furukawa, Y., and S.J. Kish. 1999. Dopa responsive dystonia: Recent advances and remaining issues to be addressed. Movement Disord. 14:709–715.

Gash, D., K. Rutland, N. Hudson et al. 2008. Trichlorethylene: Parkinsonism and complex mitrochondrial neurotoxicity. Ann Neurology DOI:10:10021:ANA.21288. (online)

Gerfan, C.R., and T.M. Engbar. 1992. Molecular neuroanatomic mechanisms of Parkinson's disease. Neurol. Clin. 10:435–449.
Gibb, W.R.G. 1988.The neuropathology of Parkinsonian disorders. In J. Jankovic and E. Tolosa. Parkinson's Disease and Movement Disorders. Baltimore, Urban & Schwarzenberg, pp. 205–223.
Gibb, W.R., P.J. Luthert, and C.D. Marsden. 1989. Corticobasal degeneration. Brain 112:1171–1192.
Goerdert, M., R. Jakes, and M.G. Spillantini. 1998. Alpha-Synuclein and the Lewy body. NeuroSci. News 1:47–51.
Goetz, C.G. 1986. Charcot on Parkinson's disease. Movement Disord. 1:27–32.
Goldblatt. D., W. Markesbery, and A.G. Reeves. 1974. Recurrent hemichorea following striatal lesions. Arch. Neurol. 31:51–54.
Graybiel, A.M., andvRagsdale, C.W. 1978. Histochemically distinct compartments in stratum of human, monkey and cat demonstrated by acetylthiocholinesterase staining. Proc. Natl. Acad. Sci. USA 75:5723–5726.
Growdon. J.H., M.J. Hirsh, R.J. Wurtman, and W. Wiener. 1977. Oral choline administration to patients with tardive dyskinesia. N. Engl. J. Med. 297:524–527.
Guridi, J., and J.A. Obesco. 2001. The subthalamic nucleus, hemiballismus and Parkinson's disease: Reappraisal of a neurosurgical dogma. Brain 124:5-19.
Gusella, J.F., N.S. Wexler, D.M. Conneally, et al. 1983. A polymorphic DNA marker genetically linked To Huntington's disease. Nature 306:234–238.
Guttman, M. 1992. Dopamine receptors in parkinson's disease. Neurol. Clin. 10:377–386.
Hallett, M., I. Litvan, et al. 2000. Scientific position paper of the Movement Disorder Society Evaluation of Surgery for Parkinson's Disease. 15:436–438.
Hornykiewicz, O.D. 1970. Physiological, biochemical and pathological backgrounds of Levo Dopa and possibilities for the future. Neurology 20(Pt. 2):1–13.
Hornykiewicz, O., S. Kish, L.E. Beckler, et al. 1986. Brain neurotransmitters in dystonia musculorum deformans. N. Engl. J. Med. 315:347–531.
Huntington, G. 1872. On chorea. Med. Surg. Rep. 26:317–321.
Huw, R.M., A.J. Lees, and N.W. Wood. 1999. Neurofibrillary tangles Parkinsonian disorders: Tau pathology and taugenetics. Movement Disord. 14:731–736.
Jankovic, J. 1989. Parkinsonism plus syndromes. Movement Disord. 4(Suppl. 1) 95–119.
Jancovic, J. 2001. Tourette's syndrome. N. Engl. J. Med. 345:1184–1192.
Jankovic, J., and S. Fahn. 1986. The phenomenology of tics. Movement Disord. 1:17–26.
Jankovic, J., and S. Fahn. 1988. Dystonic syndromes. In J. Jankovic and E. Tolosa (eds.). Parkinson's Disease and Movement Disorders. Baltimore, Urban & Schwarzenberg, pp. 283–314.
Jankovic, J. 1989. Parkinsonism plus syndrome. Movement Disord. 4(Suppl.1):S95–S119.
Jarman, P.R., M.B. Davis, S.V. Hodgson, et al. 1997. Paroxysmal dystonic choreoathetosis: Genetic linkage studies in a British family. Brain 120:2125–2130.
Jarman, P.R., K.P. Bhatia, C. Davie, et al. 2000. Paroxysmal dystonic choreoathetosis: Clinical features and investigation of pathophysiology ina large family. Movement Disord. 15:648–657.
Kandel, E. 1991. Disorders of thought: Schizophrenia. In E.R. Kandel, J.H..Schwartz, and T.M. Jessell (eds.).. Principles of Neural Science, 3rd ed. New York, Elsevier, pp. 853–868.
Keenen, R.E. 1970.The Eaton Collaborative Study of Levo Dopa Therapy in Parkinsonism. Neurology 20(Pt. 2):46–59.
Kertesz, A. 2000. Corticobasal degeneration. J. Neurol. Neurosurg. Psyciatry 68:275–276.
Kertesz, A., P. Martinez-Lage, W. Davidson, and D.G. Munoz. 2000. The corticobasal degeneration syndrome overlaps progressive aphasia and frontotrmporal dementia. Neurology 55:1368–1375.
Klawans, H. L., H. Moses, D.A. Nausieda, et al. 1976. Treatment and prognosis of hemiballismus. N. Engl. J. Med. 295:1348–1350.
Klawans, H.L., G.W. Paulson, S.P. Ringel, and A. BarbeauA. 1972. Use of L-Dopa in the detection of presymptomatic Huntington's chorea. N. Engl. J. Med. 286:1332–1334.
Klawans. H.L. 1988. The Pathophysiology of drug-induced movement disorders. In J. Jankovic and E. Tolosa (eds.). Parkinson's Disease and Movement Disorders. Baltimore, Urban and Schwarzenberg, pp.315–326.

Koller, W.C. 1992. Role of the controlled release formulation of carbidopa-levodopa in the treatment of Parkinson's disease. Neurology 42(Suppl. 1):S-l-S-60.

Koller, W.C. 1992. Initiating TREATMENT of Parkinson's disease. Neurology 42(Suppl. 1):29–32.

Kopin, I.J. 1988. MPTP toxicity: Implications for research in Parkinson's disease. Annu. Rev. Neurosci. 11:81–96.

Lamousin, P., P. Krack, P. Pollak, et al. 1998. Electrical stimulation of the subthalamic nucleus in advanced Parkinson's disease. N. Engl. J. Med. 339:1105–1111.

Lang, A.E., and A.M. Lozano.1998. Medical progress: Parkinson's disease. N. Eng.J. Med. 339:1044–1053.

Lang, A.E., A.M. Lozano, M.E. Duff., et al. 1997. Posteroventral medial pallidotomy in advanced Parkinson's disease. N. Engl. J. Med. 337:1036–1042.

Langston, J. W. and P. Ballard. 1984. Parkinsonism induced by 1-methyl,4-phenyl-1,2,3,6, tetrahydropyridine (MPTP): Implications for treatment and pathogenesis of Parkinson's disease. Can. J. Neurol. Sci.11:160–165.

Langston, J.W., P. Ballard, J.W. Tetrud, and I. Irwin. 1983. Chronic Parkinsonism in humans due to a product of meperidine: Analog synthesis. Science 219:979–980.

Langston, J.W., H. Widner, C.G. Goetz, et al. 1992. Case Assessment Program for Intracerebral Transplantation (CAPIT). Movement Disord. 7:2–13.

Laplane, D., M. Levasseur, B. Pillone, et al. 1989. Obsessive compulsive and other behavioral changes with bilateral basal ganglia lesions. Brain 112:699–725.

Lees, A.J., and E. Tolosa. 1988.Tics. In J. Jankovic and E. Tolosa (eds.). Parkinson's Disease and Movement Disorders. Baltimore, Urban & Schwarzenberg, pp.275–281.

Lees, A.J. 1987.The Steele-Richardson, Olszewski Syndrome (Progressive Supranuclear Palsy) (in) Marsden, C. D., and Fahn. S. (Ed.).Movement Disorders 2 London/Boston, Butterworths. Pp 272–287,.

Lewin, R. 1983.Trail of ironies to Parkinson's disease. Science 224:1083–1085.

Lewin. R. 1984. Brain enzyme is the target of drug toxin. Science 225:1460–1462.

Le Witt, D.A. 1992. Clinical studies with and pharmacokinetic considerations of sustained release levodopa. Neurology 42(Suppl. 1):29–31.

Lieberman. A. 1992. Emerging perspectives in Parkinson's disease. Neurology 42(Suppl. 4):5–7.

Lin T.S., Okumura and I.S. Cooper. 1961. Symposium supseay for tremor: Electroeneeph. Clin Neurophysiol 13:633.

Lindvall, O. 1999. Cerebral implantation in movement disorders: state of the art. Movement Disord. 14:201–205.

Lindvall. O., P. Brundin, H. Widner, H., et al. 1990. Grafts of fetal dopamine neurons survive and improve motor function in Parkinson's disease. Science 247:514–577.

Litwin, I. 2001. Diagnosis and management of progressive supranuclear palsy. Semin. Neurol. 21:41–48.

Litwin, I., G. Campbell, C.A. Mangone, et al. 1997. Which clinical features differentiate progressive supranuclear palsy (Steele–Richardson–Olszewski syndrome) from related disorders? A clinicopathological study. Brain 120

Martin, J.R. 1951. Hemichorea (hemiballismus) without lesions in the corpus Luysii. Brain 80:1–10.

Martin, J.B., and J.F. Gusella. 1986.Huntington's disease: Pathogenesis and management. N. Engl. J. Med 315:1267–1276.

Mathuranath. P.S., J.H. Xuereb, T. Bak, and J.R. Hodges. 2000. Corticobasal degeneration and/or frontal temporaldementia? A report of two overlap cases and review of the literature. J. Neurol. Neurosurg. Psychiatry 68:304–312.

Mayeux, R., Y Stern, and S. Spanton. 1985. Heterogeneity in dementia of the Alzheimer type: Evidence of subgroups. Neurology 35:453–461.

Testing for Huntington's disease with use of a Linked DNA marker. N. Engl. J. Med. 318:535-42.

Mena, I., J. Court, S. Fuenzalida, et al. 1970. Chronic manganese poisoning: Treatment with L-Dopa or 5-OH tryptophane. N. Engl. J. Med. 282:5–10.

Meyers, R., D.B. Sweeney, and J.T. Swidde. 1947. Hemiballismus: Etiology and surgical treatment J. Neurol. Neurosurg. Psychiatry. 58: 672–692.

Mitchell, I. J., A. Jackson, M.A. Sambrook, and A.R. Crossman. 1989. The role of the subthalamic nucleus in experimental chorea. Evidence from 2-desoxyglucose metabolic mapping and horse radish peroxidase tracing studies. Brain112:1533–1548.

Montastruc, J.L., O. Rascol, and J.M. Senard. 1999. Treatment of Parkinson's disease should begin with a dopamine agonist. Movement Disord. 14:725–730.

Myers, R.H., D.S. Sax, W.J. Koroshetz, et al. 1991. Factors associated with slow progression in Huntington's disease. Arch. Neurol. 48:800–804.

Myers, R.H., J.P. Vonsattel, T.J. Stevens, et al. 1988. Pathologic assessment of severity in Huntington's disease. Neurology 38:341–347.

Olanow, C.W., and W.G. Tatton. 1999. Etiology and pathogenesis of Parkinson's disease. Annu. Rev. Neurosc. 22:123–144.

Papp, M.I., and P.L. Lantos. 1994. The distribution of oligodendroglial inclusions in multiple system atrophy and its relevance to clinical symptomatology. Brain 117:235–243.

Parkinson, J. 1817. An Essay on the Shaking Palsy. London, Sherwood, Neely & Jones. Reprinted: The Classics of Neurology Neurosurgery Library, Birmingham, Gryphen Editions, 1986.

Patten, B.M. 1988. Wilson's disease. In J. Jankovic and F. Tolosa (eds.). Parkinson's Disease and Movement Disorders. Baltimore, Urban & Schwarzenberg, pp. 179–190.

Pauls, D.L., and J.F. Leckman. 1986.The inheritance of Gilles de La Tourette's syndrome and associated behaviors: Evidence for autosomal dominant transmission. N. EngI. J. Med.. 315:993–997.

Pearce, R.K.B.1999. L- Dopa and dyskinesias in normal monkeys. Movement Disord.14(Suppl. 1):9–12.

Penney, J.B., Jr., and A.B.Young. 1988. Huntington 's disease. In J. Jankovic and E. Tolosa (eds.). Parkinson's Disease. Baltimore, Urban & Schwarzenberg, pp. 167–178.

Rinne, J.O., M.S. Lee, P.D. Thompson, and C.D. Marsden. 1994. Corticobasal degeneration. A clinical study of 36 cases. Brain 117:1183–1196.

Roberts, L. 1990. Huntington's gene: So near and so far. Science 247:624–617.

Sarma, K., J.M. Waltz, M. Riklan, M. Koslow, and I.S. CooperS. 1970. Relief of intention tremor by thalamic surgery.vJ.Neurol.vNeurosurg..Psychiatry 33:7–15.

Sax. D.S., and J.P. Vonsattel. 1992. Chorea and progressive dementia in an 88-year-old woman. N. EngI. J. Med. 326:117–124.

Schmidt, W.R., and L.W. Jarcho. 1966. Persistent dyskinesias following phenothiazide therapy. Arch. Neurol. 14:369–377.

Schneider. J.S., A. Pope, K. Simpson, et al. 1992. Recovery from experimental Parkinsonism in primates with GM1 ganglioside treatment. Science 256:843–846.

Scheinberg, I.H, and I. Steinlieb. 1984. Wilson's Disease. Philadelphia, W. B. Saunders.

Schuurman, P.R., D.A. Bosch, P.M.M. Bossuyt, et al. 2000. A comparison of continuous thalamic stimulation and thalamotomy for suppression of severe tremor. N. Engl. J. Med. 342:461–468

Schwarz, G.A., and L.J. Barrows. 1960. Hemiballism without involvement of Luy's body. Arch. Neurol. 2:420–434.

Sethi, K.P. 2001. Movement disorders induced by dopamine blocking agents. Semin. Neurol. 21:59–68.

Siemers, E., and V. Reddy. 1991. Profile of patients enrolled in a new movement disorders clinic. 6:336–341.

Singer, H.S. 2000. Current issues in Tourette's syndrome. Movement Disord. 15:1051–1063.

Sladek, J.R., B.E. Redmond, Jr., T.J. Collier, et al. 1989. Transplantation advances in Parkinson's disease. Movement Disord. 4 (Suppl.1):S120–S125. 248.

Wexler, N.S., E.A. Rose, and D.E. Houseman. 1992. Molecular approaches to hereditary disease of the nervous system: Huntington's disease as a paradigm. Ann Rev Neuroscience 15:402–442.

Young, A.B., T. Grenamyre, Z. Hollingsworth et al. 1988. NMDA receptor loss in putamen from patients with Huntington's disease. Science 241:981–983.

Chapter 11 Motor Systems III: Cerebellum and Movement

General Anatomy, Physiology, and Functional Localization:

Allin, M., H. Matsumoto, A.M. Santhouse, et al. 2001. Cognitive and motor function and the size of the cerebellum in adolescents born very preterm. Brain 124:60–66.
Amarenco, P., C. Chevrie-Muller, E. Roullet, and M.G. Bousser. 1991. Paravennal infarct and isolated cerebellar dysarthria. Ann. Neurol. 30:211–213.
Carrera, R.M.E., and F.A. Mettler. 1947. Physiologic consequences following extensive removals of cerebellar cortex and deep cerebellar nuclei and effect of secondary cerebral ablations in the primate. J. Comp. Neurol. 87:167–288.
Courchesne, E., J. Townsend and O. Saitoh. 1994. The brain in infantile autism: posterior fossa structures are abnormal. Neurology. 44: 214–223.
Diener, H.C., and J. Dichgans. 1992. Pathophysiology of cerebellar ataxias. Movement Disord. 7:95–109.
Dow, R.S., and G. Moruzzi. 1958. The Physiology and Pathology of the Cerebellum. Minneapolis, University of Minnesota Press.
Drepper, J., D. Timmann, F.P. Kolb, and H.C. Diener. 1999. Non-motor associative learning inpatients with isolated degenerative cerebellar disease. Brain 122:87–97.
Fiez, J.A., S.E. Petersen, M.K. Cheney, and M.E. Raichle. 1992. Impaired non motor learning and error detection associated with cerebellar damage. A single case study. Brain 115:155–178.
Gilman, S. 1986. Cerebellum and motor function. In A.K. Asbury, G.M. McKhann, and W.I. Mc Donald (eds.). Diseases of the Nervous System: Clinical Neurobiology. Philadelphia, W.B. Saunders, Vol.1, pp. 401–422.
Holmes, G. 1922. Clinical symptoms of cerebellar disease and their interpretation. The Croonian Lectures. Lancet 1:1177–1182,1231–1237; 2:59–65, 111–115.
Holmes, G. 1939. The Cerebellum of Man. Brain 62:1–30.
Ito, M. 1984. The Cerebellum and Motor Control. New York, Raven Press.
Lechtenberg, R. 1988. Ataxia and other cerebellar syndromes. In J. Jankovic and E. Tolosa (eds.). Parkinson's Disease and Movement Disorders. Baltimore, Urban & Schwarzenberg, pp. 365–376.
Lechtenberg, R., and S. GilmanS. 1978. Speech disorders in cerebellar disease. Ann. Neurol. 3:285–290.
Levinsohn, L., A. Cronin-Golomb, and J.D. Schmahmann. 2000. Neuropsychological consequences of cerebellar tumor resection in children. Brain 123:1041–1050.
Mettler, F.A., and F. OrioliF. 1958. Studies on abnormal movement: Cerebellar ataxia. Neurology 8:953–961.
Noback, C.R. and R.J. Demarest. 1991. The Human Nervous System. 4th edition, Philadelphia, Lea and Febiger.
Orioli, F.L., and F.A. Mettler. 1958. Consequences of section of the simian restiform body. J. Comp. Neurol. 109:195–204.
Raymond, J., L.S.G. Lisberger, and M.D. Mauk. 1996. The cerebellum: A neurononal learning machine? Science 272:1126–1131.
Riva, D., and C. Giogi. 2000. The cerebellum contributes to higher functions during development. Brain.123:1041–1061.
Sanes, J.N., B. Dimitrov, and M. Hallett. 1990. Motor learning in patients with cerebellar dysfunction. Brain 113:103–120.
Schmahmann, J.D., and J.C. Sherman.1998. The cerebellar cognitive affective syndrome. Brain 121:561–579.
Thach, W.T. 1987. Cerebellar inputs to motor cortex. Ciba Foundation Symposium #132: Motor Areas of the Cerebral Cortex. Chichester, Wiley, pp. 201–220
Thach, W.T., M. Goodkin, and J.G. Keating. 1992. The cerebellum: The adaptive coordination of movement. Annu. Rev. Neurosci. 15:402–442.

Topka, H., J. Valls-Sole, S.G. Massaquoi, and M. Hallett. 1993. Deficit in classical conditioning in patients with cerebellar degeneration. Brain 116:961–969.
Wood, N.W., and A.E. Harding. 2000. Ataxic disorders. In W.G. Bradley, R.B. Daroff, G.M. Fenichel, and C. D. Marsden. Neurology in Clinical Practice. Boston, Butterworth Heinemann, Vol. I, pp. 309–317.
Wood, N.W., and A.E. Harding. 2000. Cerebellar and spinocerebellar disorders. In W.G. Bradley, R.B. Daroff, G.M. Fenichel, and C.D. Marsden. Neurology in Clinical Practice. Boston, Butterworth Heinemann. Vol. II, pp. 1931–1951.

Vascular Syndromes of the Cerebellum

Amarenco, P. 1991. The Spectrum of cerebellar infarctions. Neurology 41:973–979.
Amarenco, P., and J.J. Hauw. 1990a. Cerebellar infarction in the territory of the superior cerebellar artery: A clinicopathologic study of 33 cases. Neurology 40:1383–1390.
Amarenco, P., and J.J. Hauw. 1990b. Cerebellar infarction in the territory of the anterior and inferior cerebellar artery. Brain 113:139–155.
Amarenco, P., C.S. Kase, A. Rosengart, et al. 1993. Very small (border zone) cerebellar infarcts: Distribution, causes, mechanisms and clinical features. Brain 116:161–186.
Amarenco, P., E. Roulellet, C. Goujon, et al. 1991. Infarction in the anterior rostral cerebellum (the territory of the lateral branch of the superior cerebellar artery). Neurology 41:253–258.
Caplan, L.R. 1986. Vertebrobasilar occlusive disease. In H.J.M. Barnett, et al. (eds.). Stroke: Pathophysiology. Diagn. Manag. 1:549–619.
Chaves, C.J., L.R. Caplan, C.S. Chung, and P. Amarenco. 1994. Cerebellar infarcts. Curr. Neurol. 14:143–177.
Chaves, C., M.S. Pessin L.R. Caplan, et al. 1996. Cerebellar hemorrhagic infarction. Neurology 46:346–349.
Greenberg, J., D. Skubick, and H. Shenkin. 1979. Acute hydrocephalus in cerebellar infarct and hemorrhage. Neurology 29:409–413.
Heros, R. 1982. Cerebellar infarction and hemorrhage. Stroke 13:106.
Kase, C.S., and L.R. Caplan. 1986. Hemorrhage affecting the brain stem and cerebellum. In H.J.M. Barnett.., et al. (eds.). Stroke: Pathophysiology. Diagnosis and Management. 1:621–641.
Kase, C.S., B.O. Norrving, S.R. Levine, et al. 1993. Cerebellar infarction. Clinical and anatomical observations in 66 patients. Stroke 24:76–83.
Skenkin, H.A., and M. Zavala. 1982. Cerebellar strokes: Mortality, surgical indications and results of ventricular drainage. Lancet 11:429–431.
St. Louis, E.K., E.F. Wijdicks, and H. Li. 1998. Predicting neurologic deterioration in patients with cerebellar hematomas. Neurology 51:1364–1369.
Sypert, G.W., and E.C. Alvord. 1975. Cerebellar infarction. A clinicopathological study. Arch. Neurol. 32:357–363.

Cerebellum Degenerations and Systemic Disorders

Baloh, R.W., R.D. Yee, and V. Honrubia. 1986. Late cortical cerebellar atrophy: Clinical and oculographic features. Brain109:159–180.
Berciano, J. 1988. Olivopontocerebellar atrophy. In J. Jankovic and E. Tolosa (eds.). Parkinson's Disease and Movement Disorders. Baltimore, Urban & Schwarzenberg, pp.131–151.
Bhatia, K.P., R.C. Griggs, and L.J. Ptacek. 2000. Episodic movement disorders as channelopathies. Movement Disord. 15:429–433.

Burk, K., M. Abele, M. Fetter, et al. 1996. Autosomal dominant cerebellar ataxia type I clinical features and MRI in families with SCA 1, SCA 2 and SCA 3. Brain 119:1497–1505.
Fujigasaki, H., I.C. Verma A. Camuzat, et al. 2001. SCA 12 is a rare locus for autosomal dominant cerebellar ataxia: A study of an Indian family.Ann. Neurol. 49:117–121.
Giunti, P., G. Sabbadini, M.G. Sweeney, et al. 1998. The role of the SCA 2 trinucleotide repeat expansion in 89 autosomal dominant cerebellar ataxia. Frequency, clinical and genetic correlates. Brain 121:459–467
Giunti, P., M.G. Sweeney, and A.E. Harding. 1995. Detection of the Machado–Joseph disease: Spinocerebellar ataxia three trinucleotide repeat expansion in families with autosomal dominant motor disorders including the Drew family of Walworth. Brain 118:1077–1085.
Greenfield. J.G. 1954. The Spinocerebellar Degenerations. Springfield, IL, Charles C Thomas.
Harding, A.E. 1981. Friedreich's ataxia: A clinical and genetic study of 90 families with an analysis of early diagnostic criteria and intra-familial clustering of clinical features. Brain 104:589–620.
Klockgether, T., U. Wullner, A. Spauschus, and B. Evert. 2000. The molecular biology of the autosomal-dominant cerebellar ataxias. Movement Disord. 15:604–612.
Mason, W.P., F. Graus, B. Lang, et al. 1997. Small cell lung cancer, paraneoplastic cerebellar degeneration and the Lambert–Eaton myasthenic syndrome. Brain 120:1279–1300.
Rosen, F.S., and N.L. Harris. 1987. A 30-year-old man with ataxia telangiectasia and dysarthria. Case records of the Massachusetts General Hospital: Case #2-1987. N. Engl. J. Med. 316:91–100.
Rosenberg, R.N. 1992. Machado–Joseph disease: An autosomal dominant motor system degeneration. Movement Disord. 7:193–203.
Rosenberg, R.N. 1996. DNA: Triplet repeats and neurologic disease. N. Engl. J. Med.335: 1222–1224.
Smitt, P.S., A. Kinoshita, B. De Leeuw, et al. 2000. Paraneoplastic cerebellar ataxia due to autoantibodies against a glutamate receptor. N. Engl. J. Med. 342:21–27.
Sudarsky, L., L. Corwin, and D.M. Dawson. 1992. Machado–Joseph disease in New England: Clinical description and distinction from olivopontocerebellar atrophy. Movement Disord. 7:204–208.
Swift, M., D. Morrell, R.B. Massey, and C.L. Chase. 1991. Incidence of cancer in 161 families affected by ataxia-telangiectasia. N. Engl. J. Med. 325:1831–1836.
Tolosa, E., and J. Berciano. 1993. Choreas, hereditary and other ataxias and other movement disorders. Curr. Opin. Neurol. Neurosurg. 6:358–368.
Truman, J.T., E.P. Richardson, Jr., and H.F. Dvorak. 1975. Case records of The Massachusetts General Hospital, Case 22-1975 (ataxia-telangiectasia). N. Engl. J. Med. 292:1231–1237.
Victor, M., R.D. Adams, and G.H. Collins. 1989. The Wernicke Korsakoff Syndrome and Related Neurological Disorders due to Alcoholism and Malnutrition, 2nd ed. Philadelphia, F.A. Davis.
Victor, M., R.D. Adams, and E.L. Mancall. 1959. A Restricted form of cerebellar cortical degeneration occurring in alcoholic patients. Arch. Neurol. 1:579–688.
Yount, W.J. 1981. IgG2 deficiency and ataxia telangiectas (editorial). N. Engl. J. Med. 306:541–543.

Cerebellum and Tremor

Colebatch. J.G., T. Britton, L.J. Findley, et al. 1990. The cerebellum is activated in essential tremor. Lancet 2:1028–1030.
Deuschl, G. 1998. Tremor: Basic mechanisms and clinical aspects. Movement Disords.!3(Suppl. 3): 1–149.
Deuschl, G., R. Wenzelburger, K. Loffler, et al. 2000. Essential tremor and cerebellar dysfunction. Clinical and kinematic analysis of intention tremor. Brain 123:1568–1580.

Dupuis. M.J.M., P.J. Delwaide,., D. Boucguey, and R.E. Gonette. 1989. Homolateral disappearance of essential tremor after cerebellar stroke. Movement Disord. 4:183–187.
Elble, R.J. 1998. Animal models of action tremor. Movement Disord. 13(Suppl. 3):35–39.
Findley, L.J. 1988. Tremors: Differential diagnosis and pharmacology. In J. Jankovic and E. Tolosa (eds.). Parkinson's Disease and Movement Disorders. Baltimore, Urban & Schwarzenberg, pp. 243–261.
Hallett, M. 1991. Classification and treatment of tremor. JAMA 266:1115–1117.
Hallett, M. 1998. Overview of human tremor physiology. Movement Disord. 13 (Suppl. 3):43–48.
Hua, S., S.G. Reich, A.T. Zirh, et al. 1998. The role of the thalamus and basal ganglia in Parkinsonian tremor. Movement Disord. 13 (Suppl. 3):40–42.

Chapter 12 Somatosensory Function and the Parietal Lobe

Corkin S., Milner, B., Rasmussen T. 1964. Effects of different cortical exisions on sensory thresholds in man. Transactions of the American Neurological Association 89: 112–116.
Darian-Smith I., Johnson K.A., LaMotte C., Kenins P., Shigenaga Y., Ming VC., 1979. Coding of incremental changes in skin temperature by single warm fibers in the monkey. J Neurophysiol. 42(5):1316–1331.
Foerster 1936. Supplementary-motor wrtex in the cerebral corton of man. Handbook of Neurology, Berlin: Springer
Galaburda, A., and M.M. Mountcastle. 1957. Modality and topographic properties of single neurons of cats somatic sensory cortex. J. Neuro-physiol. 20:408–434.
Jones, E.G., and T.P.S. Powell. 1970. An anatomic study of converging sensory pathways within the cerebral cortex of the monkey. Brain 93:793–820.
Kasdon, D.L., and S. Jacobson. 1978. The thalamic afferents to the inferior parietal lobule. J. Comp. Neurol. 177:685–706.
Kass, J.H., M.M. Merzenich, and H.P. Killackey. 1983. The reorganization of somatosensory cortex following peripheral nerve damage in adult and developing mammals. Annu. Rev. Neurosci. 6:325–356.
Kass, J.H., R.J. Nelson, M. Sur, et al. 1979. Multiple representations of the body within somatosensory cortex of primates. Science 204:521–523.
Kass, J.H., R.J. Nelson, M. Sur, et al. 1981. Organization of somatosensory cortex in primates. In F.O. Schmitt (eds.). The Organization of the Cerebral CortexCambridge, MA, MIT Press, pp. 237–261.
Meyer, A. 1978. The concept of a sensory motor cortex: Its early history with special emphasis on two early experimental contributions by W. Bechterew. Brain 101:673–685.
Nagel-Leiby, S., Butchel, H.A., Welch K.M.A. 1990. Cerebral control of directed visual attention and orienting saccades. Brain. 113: 237–276.
Palmini, A., and P. Glood. 1992. The localizing value of auras in partial seizures: A prospective and retrospective analysis. Neurology 42:801–808. [In particular: Kertesz, A. Issues in localization. pp.1–20;
Pause, M., Kunesch, E., Binkofski, F., Freund, HJ. (1989). Sensorimotor disturbances in patients with lesions of the parietal cortex. Brain 112, 1599–1625.
Randolph, M. and Semmes, J. 1974. Behavioral consequences of selective subtotal ablations in the postcentral gyrus of Macaco mulatta. Brain Res 70: 55–70.

Chapter 13 Visual System and Occipital Lobe

General References

Aldrich, M.S., C.W. Vanderzant, A.C. Alessi, et al. 1989. Ictal cortical blindness with permanent visual loss. Epilepsia 30:116–120.
Anand, I., and E.B. Geller. 2000. Visual auras. In H.O.Luders and S. Nocachtar (eds.). Epileptic Seizures: Pathophysiology and Clinical Semiology. New York, Churchill Livingstone, pp. 298–303
Barbur, J.L., J.D.G. Watson, R.S.J. Frackowiak, and S. Zeki.1993. Conscious visual perception without V 1. Brain 116:1293–1302.
Benton, S., I. Levy, and M. Swash.. 1980. Vision in the temporal crescent in occipital lobe infarction. Brain 103:83–97.
Brindley, G.S., P.E.K. Donaldson, M.A. Falconer, and D.N. Rushton. 1972. The extent of the occipital cortex that when stimulated gives phosphenes fixed in the visual field. J. Physiol. (London) 225:57–58.
Brindley, G.S., and W.S. Lewin. 1968. The Sensations produced by electrical stimulation of the visual cortex. J. Physiol. (London) 196:474–493.
Brookhart, J.M. (ed.). 1984. The Nervous System (Handbook of Physiology, Section 1, Vol. III, Part I, Vision), Bethesda, MD, American Physiological Society.
Geller, E.B., H.O. Luders, J.C. Cheek, and Y.G. Comair. 2000. Electrical stimulation of the visual cortex. In H.O. Luders and S. Noachtar (eds.). Epileptic Seizures: Pathophysiology and Clinical Semiology. New York, Churchill Livingstone, pp. 219–227.
Horton, J.C., and W.F. Hoyt. 1991. Quadrontic visual field defects: A hallmark of lesions in extrastriate V2/V3 cortex. Brain114:1703–1718.
Hubel.D 1988. Segregation of form, color, movement and depth: Anatomy,Physiology and Perception, Science 240:740-749.
Hubel, D.H., and T.N. Wiesel. 1977. Ferrier Lecture: Functional architecture of Macague monkey visual cortex. Proc. Roy. Soc. London (Biol.) 198:1–59.
Humphrey, N.K., and L. Weiskrantz. 1967. Vision in monkeys after removal of the striate cortex. Nature 215:595–597.
Kennard, C., and F.C. Rose. 1988. Physiological Aspects of Clinical Neuro-ophthalmology. Chicago,Year Book Medical Publishers.
Livingstone, M., and D. Hubel. 1988. Segregation of form, color, movement and depth: Anatomy, physiology and perception, Science 240:740–749.
Leeson and Leeson. 1970. Histology. Philadelphia. WB Saunders.
Lolley, R.N. and R.H. Lee. 1990. Cyclic GMP and photoreceptor function. FASEB J. 4:3001–3008.
Mishkin, M. 1972. Cortical visual areas and their interactions. In A.G. Karzmar and J.C. Eccles (eds.). Brain and Human Behavior. Berlin, Springer-Verlag, pp.187–208.
Moses,. W.M. 1987. Adler's Physiology of the Eye: Clinical Application. St. Louis, MO, Mosby.
Pandya, D.N. and H. Kuypers. 1969. Cortico-cortical connections in the rhesus monkey brain. BrainRes. 13:13–16.
Penfield, W., and H. Jasper. 1954. Epilepsy and the Functional Anatomy of the Human Brain. Boston, Little, Brown, pp. 116–126, 168–173, 401–408.
Plant, G.T., K.D. Laxer, and N.M. Barbaro. 1993. Impaired visual motion perception in the contralateral hemifield following unilateral posterior cerebral lesions in humans. Brain 116:1303–1335.
Pollen, D.A. 1975. Some perceptual effects of electrical stimulation of the visual cortex in man. In T.N. Chase (ed.). The Nervous System, Volume 2: The Clinical Neurosciences. New York, Raven Press, pp. 519–528.

Poppel, E., R. Held, and D. Frost. 1973. Residual visual function after brain wounds involving the central visual pathways in man. Nature 243:295–296.
Salanova, V., F. Andermann, and F.B. Rasmussen. 1993. Occipital lobe epilepsy. In E. Wyllie (ed.). The Treatment of Epilepsy: Principles and Practices, Philadelphia, Lea & Febiger, pp. 533–540.
Stryer, L. 1986. Cyclic GMP cascade in vision. Annu. Rev. Neurosci. 9:87–119.
VanEssen, D.C. 1979. Visual areas of the mammalian cerebral cortex. Annu. Rev. Neurosci. 2:227–263.
Weiskrantz, L., E.K. Warrington, M.D. Sanders, and J. Marshall. 1974. Visual capacity in the hemianopic field following a restricted occipital ablation. Brain 97:709–728.
Wiesel, T.N., and D.H. Hubel. 1963. Efforts of visual deprivation on morphology and physiology of cells in the cat's lateral geniculate lobes. J. Neurophysiol. 26:978–993.
Wiesel, T.N., and D.H. Hubel. 1963. Single cell responses in striate cortex of kittens deprived of vision in one eye. J. Neurophysiol. 26:1003–1007.
Williamson, P.D., P.A. Boon, V.M. Thadani, R.M. Darcey, D.D. Spencer, S.S. Spencer, R.A. Novelly, and R.H. Mattson. 1992. Occipital lobe epilepsy. Clinical characteristics, seizure spread patterns and results of surgery. Ann. Neurol. 31:3–13.
Zeki, S. 1992. The visual image in mind and brain. Sci. Am. 267:69–76.

Chapter 14 Limbic System, Temporal Lobe, and Prefrontal Cortex

Temporal Lobe References

Altman, J. 1962. Are neurons formed in the brains of adult mammals? Science 135:1127–1128.
Anderson, P. 1975.Organization of hippocampal neurons and their interconnections. In R.L. Isaccson and K.H. Probram (eds.). The Hippocampus, Vol. I. Structure and Development. New York, Plenum.
Albert, D. J., M.L. Walsh, and R.H. Jonik. 1993. Aggression in humans: What is its biological foundation? Neurosci. Biobehav. Rev. 17:405–425.
Adrian, E.D. 1950. Sensory discrimination with some recent evidence from the olfactory organ. Br. Med. Bull. 6:330.
Bancaud J. 1987. Semiologie clinique des crises epileptiques d'origine temporale. Rev. Neurol. (Paris) 143:392–400.
Bancaud, J. F., R. Brunet-Adolphs, D. Tranel, H. Damasio, and A.R. Damasio. 1995. Fear and the human amygdala. J. Neurosci. 15:5879–5891.
Blaettner, V., M. Scherg, and D. VonCramon. 1989. Diagnosis of unilateral telencephalic hearing disorders: Evaluation of a simple psychoaucoustic pattern discrimination test. Brain 112:177–196.
Bourgin, P. Chauvel, et al. 1994. Anatomical origin of déjà vu and vivid memories inhuman temporal lobe epilepsy. Brain 117:71–90.
Blume, W.T., J.P. Girvin, and P. Stenerson. 1993. Temporal neocortical role in ictal experiential phenomena. Ann. Neural. 33:105–107.
Breuer, J., and S. Freud (Translated and edited by J. Stachey). 1957. New York. BasicBooks.
Breuer, J., and S. Freud (Translated and edited by J. Stachey). Studies in Hysteria. 1957. New York. BasicBooks.
Burchell, B. 1991. Turning on and turning off the sense of smell. Nature 350:16–18.
Burgerman, R.S., M.S. Sperling, J.A. French, et al. 1995. Comparison of mesial versus neocortical onset of temporal lobe seizures. Epilepsia. 36:662–672.
Cendes, F., F. Andermann, P. Gloor, et al. 1994. Relationship between atrophy of the amygdala and ictal fear intemporal lobe epilepsy. Brain 117:739–746.

Crichton, M. 1982. The Terminal Man. New York, Avon Press.
Davidson, R.J, K.M. Putnam, and CL Larson. Dysfunction in the neural circuitry of emotion regulation: A possible prelude to violence. Science 289:591–594.
Devinsky, O.M.J. Morrell, and B.A. Vogt. 1995. Contribution of anterior cingulate cortex to behavior. Brain 118:279–306.
Dionne, V. E. How do you smell? The principle in question. Trends In Neurosciences 11:188-189, 1988.
Engel, J., Jr. 1989. Seizures and Epilepsy. Philadelphia, F.A. Davis, p. 536.
Ferguson, S.M., M. Rayport, and W.S. Carrie, W.S. 1986. Brain correlates of aggressive behavior in temporal epilepsy. In B.K. Doane and K.E. Livingstone. The Limbic System. New York, Raven Press, pp. 183–193.
Fish, D.R., P. Gloor, F.L. Quesney, and A. Olivier. 1993. Clinical responses to electrical brain stimulation of the temporal and frontal lobes in patients with epilepsy. Pathophysiological implications. Brain 116:397–414.
Fletcher, P.C., and R.N.A. Henson. 2001. Frontal lobes and human memory insight from functional neuroimaging. Brain 124:849–881.
Foong, J., M.R. Symms, G.J. Barker, et al. 2001. Neuropathological abnormalities in schizophrenia. Brain 124:882–892.
Fulton, J.F. 1953. The limbic system: A study of the visceral brain. In primates and man. Yale J. Biol. 26:107.
Gage, F.H. 1994. Challenging an old dogma; neuronogenesis in the adult hippocampus. J. NIH Res. 6:53–56.
Geschwind, N. 1965. Disconnection syndromes in animals and man. Brain 88:237–294, 585-644.
Geschwind, N. 1983. Interictal behavior changes in epilepsy. Epilepsia 24(Suppl. l):523–530.
Gloor, P. 1990. Experiential phenomena of temporal lobe epilepsy. Facts and hypotheses. Brain 113:1673–1694.
Gloor, P. 1997. The Temporal Lobe and Limbic System. New York, Oxford University Press.
Gloor, P., A. Olivier, L.F. Quesney, F. Andermann, and S. Horowitz. 1982. The role of the limbic system inexperiential phenomena of temporal lobe epilepsy. Ann. Neurol. 12:129–144.
Green, J.D.1958. The rhinencephalon. Aspects of its relation to behavior and the reticular activating system. In H.H. Jasper (ed.). Reticular Formation of the Brain. Boston, Little, Brown and Company, p. 607.
Halgren, E., R.D. Walter, D.G. Cherlow, and P.H. Crandall. 1978. Mental phenomena evoked by electrical stimulation of the human hippocampal formation and amygdala. Brain 101:83–117.
Honavar, M., and B.S. Meldrum.1997. Epilepsy. In D.I. Graham and P.L. Lantos (eds.). Greenfield's Neuropathology, 6th ed. New York, Oxford University Press, pp. 931–971.
Horel, J.A., and L.G. Misantone. 1975. Partial Kluver–Bucy syndrome produced by destroying temporal neocortex or amygdala. Brain Res. (Amsterdam) 94:347-359.
Isaacson, R. 1982. The Limbic System. 2nd ed. New York, Plenum,.
Ishibashi, T., H. Hori, K. Endo, and T. Sato. 1964. Hallucinations produced by electrical stimulation of temporal lobe seizures in schizophrenic patients. Tohoku J. Med. 82:124–139.
Jackson, J.H., and C.E. Beevor. 1889. Case of tumor of the right temporo-sphenoidal lobe, bearing on the localization of the sense of smell and on the interpretation of a particular variety of epilepsy. Brain 12:346–357. [Reprinted in Taylor, J. (ed.), pp. 406–411.]
Jackson, J.H., and W.S. Colman. 1898. Case of epilepsy with tasting movements and "dreamy state" –very small patch of softening in the left uncinate gyrus. Brain 21:580–590. [Reprinted in Taylor, J. (ed.), pp. 458–463.]
Jackson, J.H., and P. Stewart. 1899. Epileptic attacks with a warning of crude sensation of smell and with the intellectual aura (dreamy state) in a patient who had symptoms pointing to gross organic disease of the righttemporo-sphenoidal lobe. Brain 22:534–549.,
Jasper, H.H., and L.O. Proctor (eds.). 1958. Reticular Formation of the Brain. Henry Ford Hospital Symposium. Boston, Little, Brown and Company.

Johnston, D., and D.G. Amaral. 1998. Hippocampus. In G.M. Shepherd (ed.). The Synaptic Organization of the Brain, 4th ed. NewYork,, Oxford University Press, pp. 417–458.

Kauer, J. 1988. Real-time imaging of evoked activity in local circuits of the salamander olfactory bulb. Nature 331:166–168.

Kauer, J. 1991. Contributions of topography and parallel processing to odor coding in the vertebrates olfactorypathway. Trends Neurosci. 14:79–85.

Kluver, H. 1952. Brain mechanisms and behavior with special reference to the rhinencephalon. J. Lancet (Minneapolis) 72:567.

Kluver, H.1958. The "temporal lobe syndrome" produced by bilateral ablations. In E.E. Wolstenhoime and C.M. O'Connor (eds.). Neurological Bases of Behavior. Ciba Foundation Symposium. London, J. A. Churchill, p. 175.

Kluver, H., and P.C. Bucy. 1937. Psychic blindness and other symptoms following bilateral temporal lobectomy in Rhesus monkeys. Am. J. Physiol. 119:352.

Kluver, H., and P.C. Bucy. 1939. Preliminary analysis of functions of the temporal lobes of monkeys. Arch. Neurol. Psychiatry 42:979–1000.

Lazard, D., Y. Barak, and D. Lancet. 1989. Bovine olfactory cilia preparation: Thiol-modulated odorant-sensitive adenylylcyclase. Biochem. Biophys. Acta 1013:68–72.

Liegeois-Chauvel. C., A. Musolino, and P. Chauvel. 1991. Localization of primary auditory area in man. Brain 114:139–153.

MacLean, P.D. 1955. The limbic system (visceral brain) and emotional behavior. Arch. Neurol. Psychiatry (Chicago) 73:130.

Margerison, J.H., and J.A.N. Corsellis. 1966. Epilepsy and the temporal lobes. Brain 89:499–530.

Mark, V.P., and Ervin, F.R. 1970. Violence and the Brain. New York, Harper and Row, pp.170.

Meyer, A.1963. Intoxications. In J. Greenfield, W. Blackwood, W.H. McMenemy, A. Meyer, and R.M. Norman (eds.). Neuropathology. Baltimore, Williams and Wilkins, pp. 235–287.

Milner, B. 1972. Disorders of learning and memory after temporal lobe lesions in man. Clin. Neurosurg. 19:421–446.

Mishkin, M. 1972. Cortical visual areas and their interactions. In A.G. Karzmar and J.C. Eccles (eds.). Brain and Human Behavior. Berlin, Springer-Verlag, pp. 187–208.

Moniz, E. 1936. Tentatives operatoires dans Ie traitement de certaines psychoses. Paris, Masson et Cie.

Mullan, S., and W. Penfield. 1959. Illusions of comparative interpretation and emotion. Arch. Neurol. Psychiatry 81:269–284.

Nauta, W.J.H. 1963. Central nervous organization and the endocrine motor system. In A.V. Nalbandov (ed.). Advances in Neuroendocrinology. Urbana, IL, University of Illinois Press, p. 5.

Nauta, W.J.H., and V.B. Domesick. 1981. Ramifications of the limbic system. In S. Matthysse (ed,). Psychiatry and the Biology of the Human Brain. Amsterdam, Elseveier/North-Holland Biomedical Press.

Nauta, W.J. H, and H.G.J.M. Kuypers. 1958. Some ascending pathways in brainstem reticular formation. In H.H. Jasper and L.O. Proctor (eds.). Reticular Formation of the Brain. Henry Ford Hospital Symposium, Boston, Little, and Company.

Olds, J. 1958. Self-stimulation experiments and differentiated reward systems. In H.H. Jasper and L.O. Proctor (eds.). Reticular Formation of the Brain. Boston, Little, Brown and Company, p. 671.

Papez, J.W. 1958. The visceral brain, its components and connections. In H.H. Jasper (ed.): Reticular Formation of the Brain. Boston, Little Brown and Company, p. 591.

Papez, J.W. 1937. A proposed mechanism of emotion. Arch. Neurol. Psychiatry (Chicago) 38:725–743.

Penfield, W., and H. Jasper. 1954. Epilepsy and the Functional Anatomy of the Human Brain. Boston, Little, Brown and Company.

Penfield, W., and J.P. Evans. 1934. Functional defects produced by cerebral lobotomies. Res. Publ. Assoc. Nerv. Ment. Dis.13:352–377.

Penfield, W., and P. Perot. 1963. The Brain's record of auditory and visual experience. Brain 86:595–696.

Penfield, W., and T. Rasmussen. 1950. The Cerebral Cortex of Man: A Clinical Study of Localization of Function. New York, MacMillan Company.

Reese, T.S., and G. M. Shepherd. 1972. Dendrodenritic synapses in the central nervous system. In G.D. Papas and D.P.Purpura (eds.). Structure and Function of Synapses. New York, Raven Press, pp. 121–136.

Sachdev, P, J.S. Smith, J. Matheson, et al. 1992. Amygdalo-hippocampectomy for pathological aggression. Aust NZ. J Psychiatry 26:671–674

Seifert, W. (ed). 1983. Neurobiology of the Hippocampus. New York, Academic Press.

Sem-Jacobsen, C.W., and A. Torkildsen. 1960. Depth stimulation and electrical stimulation in the human brain. In E.R. Ramey and D.S. O'Doherty (eds.). Electrical Studies of the Unanesthetized Brain. New York, Hoeber, pp. 275–290.

Sheer, D.E. (ed.). 1961. Electrical Stimulation of the Brain. Austin, University of Texas Press.

So, N.K., G. Savard, F. Andermann, et al. 1990. Acute postictal psychosis: A stereo EEG study. Epilepsia 31:188–193.

Steinberg, J., R. Taylor, and K. Haglund. 1994. Pheromones: A new term for a class of biologically active substances. J. NIH Res. 6:63–66.

Stevens, J.R. 1975. Interictal clinical manifestations of complex partial seizures. Adv. Neurol. 11:85–112.

Swanson, S.J., S.M. Rao, J.Grafman, et al. 1995. The relationship between seizure subtypes and interictal personality. Results from the Vietnam head injury study. Brain 118:91–103.

Tatnaka, Y., T. Kamo, M. Yoshide, et al. 1991. So-called cortical deafness: Clinical, neuropsychological and radiologic observations. Brain 114:238–240.

Taylor, J. (ed.). 1931. Selected Writing of John Hughlings Jackson. Vol. 1. On Epilepsy and Epileptiform Convulsions. Reprinted New York: BasicBooks, 1958, pp. 385–405, 406–411, 458–463.

Valenstein, E.S. 1986. Great and Desperate Cures: The Rise and Decline of Psychosurgery and Other Radical Treatments for Mental Illness. New York, Basic Books.

Vassar, R., S.K. Chao, R. Sitcheran, et al. 1994. Topographic organization of sensory projections to the olfactory bulb. Cell, 79:981–991.

Walczak, T. S. 1995. Neocortical temporal lobe epilepsy. Epilepsia. 36:623–635.

Waxman, S.G., and N. Geschwind. 1975. The interictal behavior syndrome of temporal lobe epilepsy. Arch. Gen. Psychiatry 32:1580–1586.

Wieser, H.G. 1988. Selective amygdalo-hippocampectomy for temporal lobe epilepsy. Epilepsia 29(Suppl. 2):100.

Amygdala, Emotion, Autism, and Psychiatric Disorders: Specific References

Adams, R.D., M. Victor, and A.H. Ropper. 1997. Psychiatric disorders. In Principles of Neurology. New York, McGraw Hill, pp.1501–1562.

Adolph, R., L. Sears, and J. Piven. 2001. Abnormal processing of social information from faces in autism. J. Cogn. Neurosci. 13:232–240.

Adolph, R., D. Tranel, and A.R. Damasio. 1998. The human amygdala in social judgment. Nature 393:417–418.

Amaral, D., and J. LeDoux. 2001. Fear and social conditioning in the amygdala. Lectures at the Boston Society of Neurology and Psychiatry. 924th meeting, June 14 2001. Boston, MA.

Anderson, A.K., and Phelps. 2000. Expression without recognition: contributions of the human amygdala to motional communication. Psychol. Sci. 11:106–111

Baron-Cohen, S., H.A. Ring, E.T. Bullmore, et al. 2000. The amygdala theory of autism. Neurosci. Biobehav. Rev. 24:355–364.

Cendes, F., F. Anderman, P., Gloor. et al. 1994. Relationship between atrophy of the amygdala and ictal fear in temporal lobe epilepsy. Brain 117:739–746.
Critchley, H.D., E.M. Daly, E.T. Bullmore. 2000. The functional neuroanatomy of social behaviour. Brain 123:2203–2212.
Foong, J., M.R. Symms, G.J. Barker, et al. 2001. Neuropathological abnormalities in schizophrenia. Brain 124:882–892
Friston, K.J., P.F. Liddle, C.D. Firth, et al. 1992. The left medial temporal region and schizophrenia. A PET study. Brain 115:367.
Howard, M.A., P.E. Cowell, J. Boucher, et al. 2000. Convergent neuroanatomical and behavioural evidence of an amygdalahypothesis of autism. Neuroreport 11:2931-2935.
Kallmann, F.J. 1946. The genetic theory of schizophrenia. An analysis of 691 twin index families. Am. J Psychiatry 103:309.
Levine, D.N., and A. Grek. 1984. The anatomic basis of delusions after right cerebral infarction. Neurology 34:577

Learning Memory and the Temporal Lobe: General References

Brown, P., D.C. Gajdusek, C.J. Gibbs, Jr., et al. 1985. Potential epidemic of Creutzfeldt–Jakob disease from human growth hormone therapy. N. Engl. J. Med. 313:728–731.
Brown, P., L.G. Goldfarb, J. Kovanen, et al. 1992. Phenotypic characteristics of familial Creutzfeldt–Jakob disease associated with the Codon 178 Asn PRNP mutation. Ann. Neurol. 31:282–285.
Brown, P., C.J. Gibbs, H.L. Amyx, et al. 1982. Chemical disinfection of Creutzfeldt–Jakob disease virus. N. Engl. J. Med. 306:1279–1282.
Brown, P., L. Ceruera'kova', L.G. Goldfarb, et al. 1994. Latrogenic Creutzfeldt–Jakob disease: An example of the interplay between ancient genes and modern medicine. Neurology 44:291–293.
Burkhardt, C.R., C.M. Filiey, B.K. Meinschmidt-DeMasters, et al. 1988. Diffuse Lewy body disease and progressive dementia. Neurology 38:1520–1528.
Calford, M.B., and R. Tweedale. 1990. Interhemispheric transfer of plasticity in the cerebral cortex. Science 249:805–807.
Caplan, L., F. Chedru, F. Lhermitte, et al. 1981. Transient global amnesia and migraine. Neurology 31:1167–1170.
Caplan, L.R., and W.C. Schoene. 1978. Clinical features of subcortical arteriosclerotic encephalopathy (Bunswanger disease). Neurology 28:1206–1215.
Carpenter, W.T., Jr., and R.W. Buchanan. 1994. Schizophrenia. N. Engl. J. Med.. 330:681–690.
Corder, E.H., A.M. Saunders, W.J. Strittmatter, et al 1993. Gene dose of apolipoprotein E type 4 allele and the risk of Alzheimer's disease in late onset families. Science 261:921–923.
Desimone, R. 1992. The physiology of memory: recordings of the past. Science 258:245–246.
Dollard, J., and N.E. Miller. 1950. Personality and Psychotherapy. New York, McGraw Hill.
Drachman, D.A., and J. Leavitt. 1974. Human memory and the cholinergic system: A relationship to aging? Arch. Neurol. 30:113–121.
Drachman, D.A. 1993. New criteria for the diagnosis of vascular dementia: Do we know enough yet? Neurology 43:243–245.
Evans, D.A., H. H. Funkenstein, M.S. Albert et al. 1989. Prevalence of Alzheimerˆs disease in a community population of Older adults. JAMA. 262:2551–2556.
Friedland, R.P. 1993. Epidemiology, education and the ecology of Alzheimerˆs disease. Neurology. 43:246–249.

Fristoen, K.J., P.F. Liddle, C.D. Firth, et al. 1992. The left medial temporal region and schizophrenia. A PET study. Brain 115:367–382.
Gajdusek, D.C., C.J. Gibbs, D.M. Asher, et al. 1977. Precautions in medical care of, and in handling materials from patients with transmissible virus dementia (Creutzfeldt–Jakob disease). N. Engl. J. Med. 297:1253–1258.
Gallassie, R., A. Morreale, S. Lorusso, et al. 1977. Epilepsy presenting as memory disturbances. Epilepsia 29:624–629.
Gilden, D.H. 1983. Slow virus diseases of the CNA: Part II. Scrapie, Kuru, and Jakob–Creutzfeldt disease. Postgrad. Med. 73:113–118.
Goldenberg, G., I. Podreka, N. Phaffel-Meyer, et al. 1991. Thalamic ischemia in transient global amnesia: A SPECTstudy. Neurology 41:1748–1752.
Graff-Radford, N.R., D. Tranel, G.W. VanHoesen, et al. 1990. Diencephalic amnesia. Brain 113:1–25.
Graff-Radford, N.R. Dementia: Neurologic Clinics 25:577–865.
Greenlee, J.E. 1982. Infection control: Containment precautions in hospitals for cases of Creutzfeldt–Jakob disease (editorial). Infect. Control. 3:222–223.
Growdon, J.H. 1992. Treatment for Alzheimer's disease? (editorial). N. Engl. J. Med. 327:1306–1308; see also David et al. 1992. N. Engl. J. Med.. 327:1253–1259.
Grundman, M., and L. J. Thal. 2000. Treatment of Alxheimer's disease. Neurol. Clin.18:807–828.
Hansen, L., D. Salmon, and D. Galasko. 1990. The Llewy body variant of Alzheimer's disease: A clinical and pathologicentity. Neurology 40:1–8.
Harlow, H. 1974. Learning to Love. New York, James Aronson.
Heiligenberg, W. 1991. The neural basis of behavior: A neuroethological view. Annu. Rev. Neurosci. 14:247–267.
Hellmuth, J. (ed.). 1963. Disadvantaged Child. Vol. 2: Headstart and Early Intervention. New York, Brunner-Mazel.Inc.
Hsiao, K., and S.B. Prusiner. 1990. Inherited human prion diseases. Neurology 40:1820–1827.
Hubel, D., and H. Eve. 1988. Brain and Vision. New York, Scientific American Library.
Irle, E., B. Wowra, Kunert, et al. 1992. Memory disturbances following anterior communicating artery rupture. Ann. Neurol. 31:473–480.
Jarvis, W.R. 1982. Precautions for Creutzfeldt–Jakob disease. Infect. Control. 3:238–239.
Jenkins, W.M., M.M. Merzenich, M. Ochs, et al. 1990. Functional reorganization of primary somatosensory motor cortex in adult owl monkeys after behaviorally control led tactile stimulation. J.Neurophysiol. 63:82–104.
Kaas, J.H., L.A. Krubitzer, Y.M. Chino, et al. 1990. Reorganization of retinotopic cortical maps in adult mammalsafter lesions of the retina. Science 248:229–231.
Kass, J. 1991. Plasticity of sensory and motor maps in adult mammals. Annu. Rev.Neurosci. 14:137–167.
Kandel, E.R., and T. Jessell. 1991. Early experience and the fine tuning of synaptic connections. In E.R. Kandel, J.H. Schwartz, and T.M. Jessel (eds.). Principles of Neural Science. New York, Elsevier, pp. 945-958.
Kandel, E.R., and T.J. O'Dell. 1992. Are adult learning mechanisms also used for development? Science 258:243–295.
Kapur, N., D. Ellison, M.P. Smith, et al. 1992. Focal retrograde amnesia following bilateral temporal lobe pathology: A neuropsychological and magnetic resonance study. Brain 115:73–85.
Katzman, R. 1993. Education and the prevalence of Alzheime's disease. Neurology. 43:13–20.
Kemper.T.L., and M.L. Bauman. 1993. The contribution of neuro-pathological studies to the understanding of autism: Neurol. Clin. 11:175–187.
Lin, K.N., R.S. Liu, T.P. Yeh, et al. 1993. Posterior ischemia during an attack of transient global amnesia. Stroke 24:1093–1095.
Madison, D.V., R.C. Malenka, and R.A. Nicoll. 1991. Mechanisms underlying long term potentiation of synaptictransmission. Rev. Neurosci. 14:379–397.
Markesbery, W.R. 1992. Alzheimer's disease. In C.M. McKhann and W. McDonald (eds.). Diseases of the Nervous System: Clinical Neurobiology, 2nd ed. Philadelphia, W.B. Saunders, pp.755–803.

Marx, J. 1990. Human brain disease recreated in mice. Science 250:1509–1510.
Marx, J. 1992a. A new link in brain's defenses: Research news. Science 256:1278–1280.
Marx, J. 1992b. Alzheime's debate boils over: News & comments. Science 257:1336–1338.
Marx, J. 1992c. Familial Alzheimer's linked to chromosome 14 gene: Neurobiology. Science 258:550.
Masters, C.L., D.C. Gajdusek, and C.J. Gibbs. 1981. The familial occurrence of Creutzfeldt–Jakob disease and Alzheimer's disease. Brain 104:535–558.
Masters, C.L., J.O. Harris, D.C. Gajdusek, et al. 1979. Creutzfeldt–Jakob disease: Patterns of world-wide occurrence and the significance of familial and sporadic clustering. Ann. Neurol. 5:177–187.
Mathew, N.T., and J.S. Meyer. 1974. Pathogenesis and natural history of transient global amnesia. Stroke 5:303–311.
Mayeaux, R., and S. Sternspanton. 1985. Heterogenicity in dementia of the Alzheimer type: Evidence of subgroups. Neurology 35:453–461.
Mazzucchi, A., G. Moretti, P. Caffarra, et al. 1980. Neuropsy: Chological functions in the follow-up of transient global amnesia. Brain 103:161–178.
McClelland, D.C., and J.W. Atkinson. 1948. The projective expression of needs. J. Psychol. 25:206.
McKee, A.C., K.S. Kosik, and N.W. Kowall. 1991. Neuritic pathology and dementia in Alzheime's Disease. Ann. Neurol.30:156–165.
Mc Keith, I.G. 2000. Spectrum of Parkinson's disease, Parkinson's dementia and Lewy body dementia. Neurol. Clin. 18:865–884.
McKhann, G., D. Drachman, M. Folstein, et al. 1984. Clinical diagnosis of Alzheimer's Disease. Neurology 34:939–944.
Melo, T.P., J.M. Ferro, and H. Ferro. 1992. Transient global amnesia: a case control study. Brain 115:261–270.
Merzenich, M.M. 1985. Sources of intraspecies and interspecies cortical map variability in mammals: conclusions and hypothesis. In J.J. Cohen and Strumwasser (eds.). Comparative Neurobiology Modes of Communication in the Nervous System. New York, Wiley, pp. 105–116.
Mesulam, M.M. 1982. Slowly progressive aphasia without generalized dementia. Ann. Neurol. 11:592–598.
Mesulam, M.M. 1990. Schizophrenia and the brain. N. Engl. J. Med. 322:842–844.
Nestor, P.N., and J Hodges. 2000. Non Alzheimer's dementias.. Semin. Neurol. 20:439–446.
Ojemann, G.A., O. Creutzfeldt, E. Lettich, et al. 1988. Neuronal activity in human lateral temporal cortex related to short-term verbal memory, naming and reading. Brain. Ill:1383–1403.
Ott, B.R. and J.L. Saver. 1993. Unilateral amnesic stroke, six new cases and review of the literature. Stroke 24:1033–1042.
Penfield, W., and B. Milner. 1958. Memory deficit produced by bilateral lesions in the hippocampal zone. Arch. Neurol. Psychiatry 79:475–497.
Perl, DP.2000. Neuropathology of Alzheimer's disease and related disorders. Neurol. Clin.18:847– 864.
Perry, R.H., D. Irving, G. Blessed, et al. 1989. Clinically and neuropathologically distinct form of dementia in the elderly. Lancet 1:166.
Pollen, D.A. 1993. Hannah's heirs: the guest for the genetic origin of Alzheimer's disease. New York,. Oxford University Press.
Pons, T.P., P.E. Garraghty, A.K. Ommaya, et al. 1991. Massive cortical reorganization after sensory deafferentation in adult macaques. Science 252:1857–1860.
Powell-Jackson, J., P. Kennedy, E.M. Whitcomb, et al. 1985. Creutzfeldt–Jakob disease after administration of human growth hormone. Lancet 2:244–246.
Prusiner, S.B. 1991. The molecular biology of prion diseases. Science. 252:1515–1522.
Riesen, A.H. 1966. Sensory deprivation in progress. In E. Stellar and J.M. Sprague (eds.). Psyiologic Psychology. New York, Academic Press, Vol. 1, pp.117–147.
Rocca, W.A., A. Hofman, C. Brayno, et al. 1991. Frequency and distribution of Alzheimer's disease in Europe: A collaborative study of 1980–1990 prevalence findings. Ann. Neurol. 30:381–390.

Rocca, W.A., A. Hofman, C. Brayne, et al. 1991. The prevalence of vascular dementia in Europe: Facts and fragments from 1980–1990 studies. Ann. Neurol. 30:817–824.

Rogers, J.D., D. Brogan, and S.S. Miran. 1995. The nucleus basalis of Meynert in neurological disease: A quantitative morphological study. Ann. Neurol. 17:163–170.

Roman, G.C., T.R. Tatemich, T. Erkinjuntti, et al. 1993. Vascular dementia: Diagnostic criteria for research studies: Report of the NINDS-AIREN International Workshop. Neurology 43:250–260.

Rosen, HJ, J. Lengenfelder, and B. Miller. 2000. Frontotemporal dementia. Neurol. Clin. 18:979–992.

Rosenzweig, M.B., and A.L. Leiman. 1986. Brain functions. Annu. Revi. Psychol. 19:55–98.

Russell, R.W. 1966. Biochemical substrates of behavior. In R.W. Russell (ed.).Frontiers in Physiological Psychology. New York, Academic Press.

Saunders, Strittmatter, D. Schmechel, et al. 1993. Association of apolipoprotein E. allele-4 with late onset and sporadic Alzheimer's disease. Neurology 43:1467–1472.

Schatz, C.J. 1992. The developing brain. Sci. Am. 267:61–76.

Scoville, W.B., and B. Milner. 1957. Loss of recent memory after bilateral hippocampal lesions. J. Neurol. Neurosurg. Psyvchiatry 20:11–21.

Selkoe, D.J. 2000. The genetics and molecular pathology of Alzheimer's disease: Roles of amyloid and the presenilins. Neurol. Clin. 18:903–922.

Selkoe, D.J. 2000. Presenilins, B amyloid precursors and the molecular basis of Alzheimer's disease. Clin. Neurosci. Res. 1:91–103.

Shenton, M.E., R. Kikinis, F.A. Jolesz, et al. 1992. Abnormalities of the left temporal lobe and thought disorder in schizophrenia: A quantitative magnetic resonance imaging study. N. Engl. J. Med. 327:604–612.

Silva, A.J., C.F. Stevens, S. Tonegawa, et al. 1992. Deficient hippocampal long term potentiation in B-calcium calmodulin kinase II mutant mice. Science 257:201–206.

Silva, A.J., R. Paylor, J. M. Wehner, et al. 1992. Impaired spatial learning in B-calcium-calmodulin kinase II mutantmice. Science. 257:206–211.

Skoog, I., L. Nilsson, B. Palmertz, et al. 1993. A population-based study on dementia in 85-year-olds. N. Engl. J. Med. 328:153–158.

Sparks, D.L., and W.R. Markesbery. 1991. Altered serotonergic and cholinergic synaptic markers in Pick^s disease. Arch. Neurol. 48:796–799.

Spillantini, M.G., and M. Goedert. 2000. Tau mutations in familial frontotemporal dementia. Brain123:857–859.

Spitz, R. 1945. Hospitalism. Psychoanal. Study Child. 1:53.

Squire, L.R., and S. Zola-Morgan. 1991. The medial temporal lobe memory system. Science 253:1380–1386.

Stillhard, G., T. Landis, R. Schiess, et al. 1990. Bitemporal hypoperfusion in transient global amnesia: 99mTCHM-PAO SPECT and neuropsychological findings during and after an attack. J. Neurol. Neurosurg. Psvchiatry 53:339–342.

Strittmatter, W.J., A.M. Saunders, D. Schmechal, et al. 1993. Apolipoprotein E. High avidity binding to B. amyloid and increased frequency of type 4 allele in late onset familial Alzheimer's disease. Proc. Natl. Acad. Sci. USA 90:1977–1981.

Suddath, R.L., G.W. Christison, E.F. Torrey, et al. 199 0. Anatomical abnormalities in the brains of monozygotic twins discordant for schizophrenia. N. Engl. J. Med. 322:789–794.

Tassinari, C.A., C. Ciarmatori, C. Alesi, et al. 1991. Transient global amnesia as a postictal state from recurrent partial seizures. Epilepsia 32:882–885.

Tatemichi, T.K., W. Steinke, C. Duncan, et al. 1992. Paramedian thalamopeduncular infarction: clinical syndromes and MRI . Ann. Neurol. 32:162–171.

Teitelbaum, J.S., R.J. Zatorre, S. Carpenter, et al. 1990. Neurologic sequelae of domoic acid intoxication due to the ingestion of contaminated mussels. N. Engl. J. Med. 322:1781–1787.

Tomlinson, B.E., G. Blessed, and M. Roth. 1970. Observations on the brains of demented old people. J. Neurol. Sci.11:205–242.

Tulving, E., and D.L. Schacter. 1990. Priming and human memory systems. Science 247:301–306.

Victor, M., J.B. Agevine, E.L. Mancall, et al. 1961. Memory loss with lesions of hippocampal formation. Arch. Neurol. 5:244–263.

von Cramon, D.Y., N. Hebel, and V. Schuri. 1988. Verbal memory and learning in unilateral posterior cerebral infarction. A report on 30 cases. Brain 111:1061–1078.
von Cramon, D.Y., N. Hebel, and V. Schri. 1985. A contribution to the anatomical basis of thalamic amnesia. Brain 108:993–1008.
Whitehouse, P.J., D.L. Price, A.W. Dark, et al. 1981. Alzheimer's disease: Evidence for selective loss of cholinergic neurons in the nucleus basalis. Ann. Neurol. 10:1226.
Will, R.G., and W.B. Matthews. 1982. Evidence for case-to-case transmission of Creutzfeldt–Jakob disease. Neurol. Neurosurg. Psychiatry 45:235–238.
Williams, M., and J. Pennybacker. 1954. Memory disturbances in third ventricle tumors. J. Neurol. Neurosurg. Psvchiatry 17:115–123.
Winker, M.A. 1994. Tacrine for Alzheimer's disease. Which patient, what dose? JAMA 271:1023–1024. (See also Knapp et al. 1994. JAMA. 271:985–991; Watkins et al. 1994. JAMA. 271:992–998.)
Woods, B.T., and A.C. McKee. 1992. Case records of the Massachusetts General Hospital: A 67-year-old man with aphasia and memory loss followed by progressive dementia. N. Engl. J. Med. 326:397–405.
Yankner, B.A., L.F. Dawes, S. Fisher, et al. 1989. Neuro-toxicity of a fragment of the amyloid precursor associated with Alzheimer's disease. Science 245:117–210.
Yankner, B.A., L.K. Duffy, and D.A. Kirschner. 1990. Neurotrophic and neurotoxic effects of amyloid B protein: Reversal by tachykinin neuropeptides. Science 250:279–282.
Yankner, B.A., and M.M. Mesulam. 1991. B-amyloid and the pathogenesis of Alzheimer's disease. N. Engl. J. Med. 325:1849–1857.
Zola-Morgan, S., L.R. Squire, and D.G. Amaral. 1986. Human amnesia and the medial temporal region: Enduring memory impairment following a bilateral lesion limited to field CA1 of the hippocampus. J. Neurosci. 6:2950–2967.
Zola-Morgan, S.M., and L.R. Squire. 1990. The primate hippocampal formation. Evidence for a time limited role in memory storage. Science 250:288–296.

Prefrontal Lobe: General References

Bookman, J.M., D.T. Kingsbury, M.P. McKinley, et al. 1985. Creutzfeldt–Jakob disease prion proteins in human brains. N. Engl. J. Med. 312:73–78.
Boothe, R.G., V. Dobson, and DY. Teller. 1985. Postnatal development of vision in human and nonhuman primates. Annu. Rev. Neurosci. 8:495-545.
Butter, N.M., K. Mishkin, and A.F. Mirsky. 1968. Emotional response in monkeys with selective frontal lesions. Physiol. Rev. 3:213–215.
Butter, C.M., D.R. Snyder, and J. McDonald. 1970. Effects of orbital frontal lesions on adversive and aggressive behavior in rhesus monkeys. J. Comp. Physiol. Psychol. 72:832–144.
Butters, N., and D. Pandya. 1969. Retention of delayed alternation: effect of selective lesions of sulcus principalis. Science 165:1271–1273.
Calleja, J., R. Carpizo, and J. Berciano. 1988. Organismic epilepsy. Epilepsia. 29:635–639.
Cauvel, P., A.V. Delgado-Escueta, E. Halgren, et al. (eds.). 1992. Frontal Lobe Seizures And Epilepsies. New York, Raven Press. (See also Delgado-Escueta et al. 1988. Epilepsia 29:204–221 for abstracts.) Damasio, A.R. 1985. The frontal lobes. In K.M. Heilman and E. Valensrein (eds.). Clinical Neuropsychology, 2nd ed. New York, Oxford University Press, pp. 339–375.
Damasio, A.R. 1992. Medical progress: Aphasia. N. Engl. J. Med. 326:531–539.
Damasio, A.R., and N. Geschwind. 1984. The neural basis of language. Ann. Rev. Neurosci. 7:127–143.
Fulton, J.F., and C.F. Jacobsen. 1935. The functions of the frontal lobes: A comparative study in monkeys, chimpanzees, and man. Adv. Med. Biol. (Moscow). 4:113–123.

Jacobsen, C.F. 1935. Functions of the frontal association area in primates. Arch. Neurol. Psychiatry 33:558–569.
Jacobsen, C.F., and H.W. Nissen. 1937. Studies of cerebral function in primates: IV. The effects of frontal lobe lesion on the delayed alteration habit in monkeys. J. Comp. Phys. Psychol. 23:101–112.
Jacobsen, S., and J. Trojanowski. 1977. Prefrontal granular cortex of the rhesus monkey: I. Intrahemispheric cortical afferents. II. Interhemispheric cortical afferents. Brain Res. 132:209–233, 235–246.
Jones, E.G., D.P. Friedman, and S.H.C. Hendra. 1982. Thalamic basis of place and modality specific columns in monkey cerebral cortex: A correlative anatomical and physiological study. J. Neurophysiol. 48:545–568.
Ledoux, J. 2000. Emotion circuits in the brain. Annu. Rev. Neurosci 23:155–184.
Valenstein. New York: Oxford University Press. pp.49-73.

Chapter 15   Higher Cortical Functions

General or Historical References

Ajamone-Marsan, C., and B.L. Ralston. 1957. The epileptic seizure: Its Functional Morphology and Diagnostic Significance. Springfield, IL. Charles C Thomas.
Alexander, G.E., and M.R. DeLong. 1992. Central mechanisms of initiation and control of movement. In A.K. Asbury, G.M. McKhann and W.I. McDonald (eds.). Diseases of the Nervous System: Clinical Neurobiology, 2nd ed. Philadelphia, W.B. Saunders, pp. 285–308.
Brumback, RA. (Ed). 1993. Behavioral neurology. Neurol. Clin.11:1–237.
Coslett, H.B. (ed.). 2000. Behavioral neurology/higher cortical function. Semin. Neurol. 20:405–515.
Commission on Classification and Terminology of the International League Against Epilepsy. 1981. Proposal for re-revised clinical and electroencephalographic classification of epileptic seizures. Epilepsia 22:489–501.
Commission on Classification and Terminology of the International League Against Epilepsy. 1989. Proposal for a revised classification of epilepsies and epileptic syndromes. Epilepsia 30:389–399.
Damasio, H., and A.R. Damasio. 1989. Lesion Analysis in Neuropsychology. New York, Oxford University Press.
Engel, J., Jr. 1989. Seizures and Epilepsy. Philadelphia, F.A. Davis.
Ferrier, D. 1873. Experimental researches in cerebral physiology and pathology. Rep.West Riding Lunatic Asylum. 3:30–96.
Foerster, O. 1936. The motor cortex in the light of Hughlings Jackson's doctrines. Brain 59:135–159.
Fulton, J.F. 1949. Physiology of the Nervous System, 3rd ed. New York, Oxford University Press.
Green, J.R. 1987. Sir Victor Horsley: A centennial recognition of his impact on neuroscience and neurological surgery. Barrow Neurological Institute Quarterly 3:2–16.
Heilman, K.M., and E.D. Valenstein (eds.). 1985. Clinical Neuropsychology, 2nd ed. New York, Oxford University Press.
Chui, H., and A. Damasio. 1980. Human cerebral asymmetries evaluated by computed tomography. J. Neurol. Neurosurg. Psychiatry 43:873–878.
Cole, J. 1957. Laterality in the use of eye, hand and foot in monkeys. J. Comp. Physiol. Psychol. 50:296–299.
Corkin, S., B. Milner, and T. Rasmussen. 1964. Effects of different cortical excisions on sensory thresholds in man. Trans. Am. Neurol. Assoc. 89:112–116.
Coslett, H.B., H.P. Brashear, and K.M. Heilman. 1984. Pure word deafness after bilateral primary auditory cortex infarcts. Neurology 34:347–352.
Coslett, H.B. 2000. Acquired dyslexia. Semin. Neurol. 20:419–426.
Coslett, H.B., and E.M. Saffran. 1989. Evidence for preserved reading in pure alexia. Brain 112:327–359.

Cracco, R.Q., V.E. Amassian, and P.J. Maccabee. 1989. Comparison of human transcallosal responses evoked by magnetic coil and electrical stimulation. Electroencephalogr. Clin. Neurophysiol. 74:417–424.

Damasio, A.R. 1985. The frontal lobes. In K.M. Heilman and E. Valenstein (eds.). Clinical Neuropsychology, 2nd ed. New York, Oxford University Press, pp. 339–375.

Damasio, A.R. 1992. Medical progress: Aphasia. N. Engl. J. Med. 326:531–539.

Damasio, A.R., and N. Geschwind. 1984. The neural basis of language. Annu. Rev. Neurosci. 7:127–147.

Damasio, H., and A. Damasio. 1980. The anatomical basis of conduction aphasia. Brain 103:337–350.

Damasio, H., T. Grabowski, R. Frank, et al. 1994. The return of Phineas Gage: Clues about the brain from the skullof a famous patient. Science 264:1102–1105.

Dellatolas, G., S. Luciani, A. Castresana, et al. 1993. Pathologic left-handed: Left-handedness correlatives in adultepileptics. Brain 116:1565–1574.

De Renzi, E. 2000. Disorders of visual recognition. Semin. Neurol. 20:479–486.

DeRenzi, E., and F. Luchelli. 1988. Ideational apraxia. Brain 111:1173–1185.

Duffner, K.R., L.L. Ahern, and S. Weintraub. 1990. Dissociated neglect behavior following sequential strokes in the right hemisphere. Ann. Neurol. 28:97–101.

Duffner, K.R., M.M. Mesulam, L.F.M Scinto, et al. 2000. The central role of prefrontal cortex in directing attentionto novel events. Brain 123:927–939.

Dusser de Barenne, J.G., and W.S. McCulloch. 1941. Suppression of motor response obtained from area 4 by stimulation of area 4S. J. Neurophysiol. 4:311–323.

Ettlinger, F.G. (ed.). 1965. Function of the Corpus Callosum. Boston, Little, Brown and Company.

Fangel, C., and B.R. Kaada. 1960. Behavior "attention" and fear induced by cortical stimulation in the cat. Electroenceph. Clin. Neurophysiol. 12:575–588.

Feinberg, T.E., and M.J. Farah. 2000. Agnosias. In W.G. Bradley, et al. (eds.). Neurology in Clinical Practice. Boston, Butterworth Heinemann, pp.131–139.

Fletcher, P.C., and R.N.A. Henson. 2001. Frontal lobes and human memory. Insights from functional neuroimaging. Brain 124:849–881.

Freund, H.J. 1987. Differential effects of cortical lesions in humans. In G. Bock, M. O'Connor, and J. Marsh (eds.). Motor Areas of the Cerebral Cortex. Ciba Foundation Symposium No.132. hichester, Wiley, pp. 269–281.

Freund, H.J., and H. Hummelshein. 1985. Lesions of the premotor cortex in man. Brain 108:697–733.

Fried, I., C. Materr, G. Ojemann, et al. 1982. Organization of visuospatial functions in human cortex: Evidence from electrical stimulation. Brain 105:349–371.

Friedman, R.B., and M.L. Albert. 1985. Alexia. In K.M. Heilman and E. Valestein (eds.). Clinical Neuropsychology, 2nd ed. New York, Oxford University Press.

Galaburda, A.M., M. LeMay, T.L. Kemper, et al. 1978. Right–left asymmetries in the brain. Science. 199:852–856.

Gandevia, S.C., D. Burke, and B. McKeon. 1984. The projection of muscle afferents from the hand to cerebral cortex in main. Brain 107:1–13.

Gazzaniga, M.S., J.E. Bogen, and R.W. Sperry. 1965. Observations on visual perception after disconnections of the cerebral hemispheres in man. Brain 88:221–236.

Geier, S., J. Bancaud, J. Talairach, et al. 1977. The seizures of frontal lobe epilepsy: A study of clinical manifestations. Neurology 27:951–958.

Geschwind, N. 1965. Disconnection syndromes in animals and man. Brain 88:237–294.

Geschwind, N. 1971. Aphasia. N. Engl. J. Med. 284:654–656.

Geschwind, N., and W. Levitsky. 1968. Human brain: Left–right asymmetries in temporal speech region. Science 161:186–187.

Geschwind, N., F.A. Quadfasel, and J. Segarra. 1968. Isolation of the speech area. Neuropsychologia 6:327–340.

Goldman-Rakic, P.S. 1987. Motor control function of the prefrontal cortex. In G. Bock, M. O'Connor, and J. Marsh (eds.). Motor Areas of the Cerebral Cortex. Ciba Foundation Symposium. No.132. Chichester, Wiley, pp.187–201.

Grafman, J., S.C. Vance, H. Weingartner, et al. 1986. The effects of lateralized frontal lesion on mood regulation. Brain 109:1127–1148.

Halsband, U. and H.J. Freund. 1990. Premotor cortex and conditional motor learning in man. Brain 113:207–222.

Halsband, U., N. Ito, J. Tanji, and H.J. Freund. 1993. The role of premotor cortex and the supplementary motor area in the temporal control of movement in man. Brain 116:243–266.

Harlow, J.M. 1868. Recovery from the passage of an iron bar through the head. Mass. Med. Soc. Publ. 2:327–346.

Hausser-Hauw, C., and J. Bancaud. 1987. Gustatory hallucinations in epileptic seizures: Electrophysiological, clinical and anatomical correlates. Brain 110:339–360.

Heilman, K.M., and L.J.G. Roth. 1985. Apraxia. In K.M. Heilman and E. Valenstein (eds.). Clinical Neuropsychology, 2nd ed. New York, Oxford University Press, pp.134–150.

Heilman, K.M., R.T. Watson, E. Valenstein, et al. 1983. Localization of lesions in neglect. In A. Kertesz (ed.). Localization in Neuropsychology. New York, Academic Press, pp. 471–492.

Heilman, K.M., E. Valenstein, and RT Watson. 2000. Neglect and related disorders. Semin. Neurology 20:463–470.

Helmstaedter, C., M. Kurher, D.B. Linke, and C.E. Elger. 1994. Right hemisphere restution of language and memory functions in right hemisphere language dominant patients with left temporal lobe epilepsy. Brain 117:729–737.

Kertesz, A., and N. Geschwind. 1971. Patterns of pyramidal decussation and their relationship to handedness. Arch. Neurol. 4:326–332.

Killackey, H., and F. Ebner. 1973. Convergent projection of three separate thalamic nuclei into a single cortical area. Science 179:283–285.

Kirsher, H.S. 2000. Aphasia. In W.G. Bradley. et al. (eds). Boston, Butterworth Heinemann, pp. 141–159.

Kennard, M.A., and L. Ectors. 1938. Forced circling in monkeys following lesions of the frontal lobes. J. Neurophysiol. 1:45–54.

Landis, T., M. Regard, A. Bliestel, et al. 1988. Prosopagnosia and agnosia for non-canonical views: An autopsied case. Brain. 111:1287-1297.

Legarda, S., P. Jayakar, M. Duchowny et al. 1994. Benign rolandic epilepsy: High central and low central subgroups. Epilepsia 35:1125–1129.

Lehman, R., F. Andermann, A. Olivier, et al. 1994. Seizures with onset in the sensorimotor face area: Clinical patterns and results of surgical treatment in 20 patients. Epilepsia 35:1117–1124.

Leiguarda, R.S., and C.D. Marsden. 2000. Limb apraxias. Higher order disorders of sensori motro integration. Brain 123:860–879.

Leiguarda, R., S. Starkstein, and M. Berthier. 1989. Anterior callosal hemorrhage: A partial interhemispheric disconnection syndrome. Brain 112:1019–1037.

LeMay, M., and A. Culebras. 1972. Human brain morphologic differences in the hemispheres demonstrable by carotid arteriography. N. Engl. J. Med. 287:168–170 (See also N. Geschwind. 1972. Editorial. N. Engl. J. Med. 187:194–195.)

LeMay, M., and D.K. Kido. 1978. Asymmetries of the cerebral hemispheres on computed tomograms. J. Computer Assist Tomogr. 2:471.

LeMay, M., and N. Geschwind. 1978. Asymmetries of the human cerebral hemispheres. In A. Caramazza and E.B. Zurif (eds.). Language Acquisition and Breakdown. Baltimore, Johns Hopkins University Press, pp. 311–328.

Luders, H., R.P. Lesser, D.S. Dinner, et al. 1988. Localization of cortical function: New information from extraoperative monitoring of patients with epilepsy. Epilepsia 29(Suppl. 2):56–65.

Luders, H., R.P. Lesser, D.S. Dinner, et al. 1985. The second sensory area in humans: Evoked potential and electrical stimulation studies. Ann. Neurol. 17:177–184.

Ludlow, C.L., J.R. Rosenberg, C. Fair, et al. 1986. Brain lesions associated with nonfluent aphasia, fifteen years following penetrating head injury. Brain 109:55–80.

Lynch, J.C., V.B. Mountcastle, W.H. Talbot, et al. 1977. Parietal lobe mechanisms for directed visual attention. J. Neurophysiol. 40:362–89.

Marcus, E.M., S. Jacobson, C.W. Watson, et al. 1970. An experimental model of "petit mal epilepsy" in the monkey: Additional studies of the anterior premotor area. Trans. Am. Neurol. Assoc. 95:279–281.

Marcus, E.M., C.W. Watson, and S.A. Simon. 1968. Behavioral correlates of acute bilateral symmetrical epileptogenic foci in monkey cerebral cortex. Brain Res. 9:370–373.

Meador, K.F., R.T. Watson, D. Bowers, et al. 1986. Hypometria with hemispatial and limb motor neglect. Brain 109:293–305.

Mesulam, M.M., G.W.V. Hoeser, D.N. Pandya, et al. 1977. Limbic and sensory connections of the inferior parietal lobule (area PG) in the rhesus monkey. Brain Res. 136:393–414.

Meyer, B.-U., S. Roricht, H. Gafin von Einsiedel, et al. 1995. Inhibitory and excitatory interhemispheric transfers between motor cortical areas in normal humans and patients with abnormalities of the corpus callosum. Brain 118:429–440.

Milner, B., and H.L. Teuber. 1968. Some cognitive effects of frontal lobe lesions in man: Reflections on methods. In L. Weiskrantz (ed.). Analysis of Behavioral Changes. New York, Harper and Row, pp. 268–375.

Mohr, J.P. 1973. Rapid amelioration of motor aphasia. Arch. Neurol. 28:77–82.

Moniz, E. 1936. Tentatives operatories dans le traitement de certaines psychoses. Paris, Masson et Cie.

Morris, H.H., III, D.S. Dinner, M.D. Luders, et al. 1988. Supplementary motor seizures: Clinical and electroencephalographic findings. Neurology 38:1075–1082.

Morris, H.H., III. 1993. Supplementary motor seizures. In E. Wyllie (ed.). The Treatment of Epilepsy: Principles and Practices. Philadelphia, Lea & Febiger, pp. 541–546.

Morrow, M.J., and J.A. Sharpe. 1995. Deficits of smooth pursuit eye movement after unilateral frontal lobe lesions. Ann. Neurol. 37:443–451.

Mountcastle, V.B. 1957. Modality and topographic properties of single neurons of cats somatic sensory cortex. J. Neuro-physiol. 20:408–434.

Naeser, M.A. 1983. CT scan lesion size and lesion locus in cortical and subcortical aphasias. In A. Keertesz (ed.). Localization in Neuropsychology. New York, Academic Press, pp. 63–119.

Naeser, M.A., C.L. Palumbo, N.H. Estabrook, et al. 1989. SAevere non- fluency in aphasia: Role of the medialsubcallosal fasciculus and other white matter pathways in recovery of spontaneous speech. Brain 112:1–38. Nagel-Leiby, S., H.A. Buchtel, and K.M.A. Welch. 1990. Cerebral control of directed visual attention and orienting saccades. Brain.113:237-276.

Nauta, W.J.H. 1964. Some efferent connections of the prefrontal cortex in the monkey. In J.M. Warren and K. Akert (eds.). The Frontal Granular Cortex and Behavior. New York, McGraw Hill, pp. 397–409.

Ochipa, C., and L.J. Gonzalez Rothi. 2000. Limb apraxia. Semin. Neurol. 20:471–478.

Ochs, R., P. Gloor, F. Quesney, et al. 1984. Does head turning during seizure have lateralizing or localizing significance? Neurology 34:884–890.

Pandya, D.N., P. Dye and N. Butters. 1971. Efferent corticocortical projections of the prefrontal cortex in the rhesus monkey. Brain Research. 31:35–46.

Pandya, D.N. and H. Kuypers. 1969. Cortico-cortical connections in the rhesus monkey. Brain Research. 13:13–16.

Pandya, D.N., and L.A. Vignolo. 1971. Intra and inter hemispheric projections of the precentral premotor and arcuate areas in the rhesus monkey. Brain Res. 26:217–233.

Passingham, R.E. 1987. Two cortical systems for directing movement. In G. Bock, M. O'Connor, and J. Marsh (eds.). Motor Areas of the Cerebral Cortex. Ciba Foundation Symposium No.132. Chichester, Wiley, pp.151–164.

Pause, M., E. Kunesch, F. Binkofsi, et al. 1989. Sensorimotor disturbances in patients with lesions of the parietal cortex. Brain 112:1599–1625.

Perenin, M.T. and A. Vighetto. 1988. Optic ataxia: A specific disruption in visuomotor mechanisms. I. Differentaspects of the deficit in reaching for objects. Brain. 111:643–674.

Peretz, I, R. Kolinsky, M. Tramo, et al. 1994. Functional dissociations following bilateral lesions of auditory cortex. Brain 117:1283–1301.

Phillips, C.G. 1966. Changing concepts of the precentral motor area. In J.C. Eccles (ed.). Brain and Conscious Experience. New York, Springer-Verlag, pp. 389–421.

Pierrott-Deseilligny, C.H., F. Cray, and P. Brunot. 1986. Infarcts of both inferior parietal lobules with impairment of visually guided eye movement, peripheral visual attention and optic ataxia. Brain 109:81–97.

Pierrott-Deseilligny, C.H., S. Rivaud, B. Gaymard, et al. 1995. Neurological progress: Cortical control of saccades. Ann. Neurol. 37:557–567.

Quesney, L.F., M. Constan, D.R. Fish, et al. 1990. The clinical differentiation of seizures arising in the parasagittal and anterolateral dorsal frontal convexities. Arch. Neurol. 47:677–679.

Randolph, M., and J. Seemes. 1974. Behavioral consequences of selective subtotal ablations in the post central gyrus of Macaca mulatta. Brain Res. 70:55–770.

Reeves, A.G. (ed.) 1985. Epilepsy and the Corpus Callosum. New York, Plenum.

Rizzolatti, G. 1987. Functional organization of inferior area 6. In G. Bock, M. O'Connor, and J. Marsh (eds.). Motor Areas of the Cerebral Cortex. Ciba Foundation Symposium No.132. Chichester, Wiley, pp. 171–186.

Ward, A.A., I.K. Pededen, and O. Sugar. 1946. Cortico-cortical connections in the monkey with special reference toarea 6. J. Neurophysiol. 9:453–461.

Warner, J.J. 1988. Ictal alexia, agraphia, anomia without speech arrest. Epilepsia 29:652.

Watson, R.T., B.D. Miller, and K.M.Heilman. 1978. Nonsensory neglect. Ann. Neurol. 3:505–508.

Weiller, C., C. Isensee, M. Rijntjes, et al. 1995. Recovery from Wernicke's aphasia: A positron emmission tomographic study. 37:723–732.

Weiskrantz, L., L.J. Liharlovic, and G.G. Gross. 1962. Effects of stimulation of frontal cortex and hippocampus on behavior in the monkey. Brain 85:487–504.

Welch, K., and P. Stuteville. 1958. Experimental production of unilateral neglect in monkeys. Brain 81:341–347.

Wiesendanger, M., H. Hummelshein, M. Bianchetti, et al. 1987. Input and output organization of the supplementary motor area. In G. Bock, M. O'Connor, and J. Marsh (eds.). Motor Areas of the Cerebral Cortex. Ciba Foundation Symposium No.132. Chichester, Wiley, pp. 40–62.

Wiesendanger, M., and S.P. Wise. 1992. Current issues concerning the functional organization of motor cortex in non-human primates. In P. Chavel et al. (eds.). Frontal Lobe Seizures and Epilepsies. New York, Raven Press,. pp.117–134.

Wilkins, R.H., and I.J.A. Brody. 1969. The thalamic syndrome. Arch. Neurol. 20:550–562.

Williamson, P.D., P.A. Boon, V.M. Thadani, et al. 1992. Parietal lobe epilepsy: Diagnostic considerations and results of surgery. Ann. Neurol. 31:193–201.

Williamson, P.D., D.D. Spencer, S.S. Spencer, et al. 1988. Complex partial seizures of frontal lobe origin. Ann. Neurol. 18:497–504.

Willmes, K., and K. Poeck. 1993. To what extent can aphasic syndromes be localized. Brain. 116:1527–1540.

Chapter 17   Cerebral Vascular Disease

Suggested Reading

I: Overview

Marcus, E.M., and S. Jacobson. 2002. Integrated Neuroscience: A Clinical Problem Solving Approach. Boston. Kluwer Academic, Chap. 26.

Mohr, J.P., L.R. Caplan, J,W, Melski, et al. 1978. The Harvard Cooperative Stroke Registry: A prospective registry of patients hospitalized with stroke. Neurology 28:754–762.

Ropper, A.H., and R. Brown. 2005. Adams and Victor's Principles of Neurology, 8th ed. New York, McGraw-Hill, pp. 660–746.

Sacco, R.L. 2005. Pathogenesis, classification and epidemiology of cerebrovascular disease. In Rowland, L.P. (ed.). Merritt's Neurology. 11th ed. Philadelphia, Lippincott/Williams and Wilkins, pp. 275–290.

Sacco, R.L. Boden-Albala, B., Gan, R., et al. 1998. Stroke incidence among white black and Hispanic residents of an urban community. Am. J. Epidemiol. 147:259–268.

Sacco, R.L., P.A. Wolf, N.F. Bhariche, et al. 1984. Subarachnoid and intracerebral hemorrhage. Neurology 34:847–854.

II: Ischemic–Occlusive

General

Biller, J., and B.B. Love. 2000. Vascular diseases of the nervous system. A. Ischemic cerebrovascular disease. In Bradley, W.G., Daroff, R.B., Fenichel, G.M., et al. (eds.). Neurology in Clinical Practice. Boston, Butterworth Heinemann, pp. 1125–1166.

Brust, J.C.M. 2005. Transient ischemic attack. In Rowland, L.P. (ed). Merritt's Neurology, 11th ed. Philadelphia, Lippincott/Williams and Wilkins, pp. 293–294.

Brust, J.C.M. 2005. Cerebral infarction. In Rowland, L.P. (ed). Merritt's Neurology, 11th ed. Philadelphia, Lippincott/Williams and Wilkins, pp. 295–303.

Marcus, E.M., and S. Jacobson. 2002. Integrated Neuroscience: A Clinical Problem Solving Approach. Boston, Kluwer Academic, pp. 2-9–2-12, 2-25–2-27, 9-10–9-13, 13-8–13-18, 26-1–26-22.

Sacco, R.L. 2005. Pathogenesis, classification, and epidemiology of cerebrovascular disease. In Rowland, L.P. (ed). Merritt's Neurology, 11th ed. Philadelphia, Lippincott/Williams and Wilkins, pp. 275–290.

Wolf, J. 1971. The Classical Brain Stem Syndromes. Sprigfield, IL, Charles C Thomas.

Specific

Benavente, O., M. Eliasziw, J.Y. Streifler, et al. 2001. Prognosis after transient monocular blindness associated with carotid artery stenosis. N. Engl. J. Med. 345:1084–1090.

Brott, T., and J. Bogousslavsky. 2000. Treatment of acute ischemic stroke. N. Engl. J. Med. 343:710–722.

Caplan, L.R. 1980. "Top of the basilar" syndrome. Neurology 30:72–79.

Caplan, L.R. 1998. Vertebrobasilar disease and thrombolytic treatment. Arch. Neurol. 55:450–451.

Caplan, L.R., V. Babikian, C. Helgason, et al. 1985. Occlusive disease of the middle cerebral artery. Neurology 35: 975–982.

Fisher, C.M. 1982. Lacunar strokes and infarcts. Neurology 32:871–876.

Fisher, C.M. 1986. Posterior cerebral artery occlusions. Can. J. Neurol. Sci. 13:232–235.

Hart, R.G. 2003. Atrial fibrillation and stroke prevention. N. Engl. J. Med. 349:1015–1016.

Inzitari, D., M. Eliasziw, P. Gates, et al. 2000. The causes and risk of stroke in patients with asymptomatic internal artery stenosis. N. Engl. J. Med. 342:1693–1701.

Kistler, J.P., and K.L. Furie. 2000. Carotid endarterectomy revisited N. Engl. J. Med. 342: 1743–1745.

Kubik, C.S., and R. Adams. 1946. Occlusion of the basilar artery: Clinical and pathological study. Brainn 69:6–121.

Lyden, P.D., M. Lu, S.P. Levine, et al. 2001. A modified National Institutes of Health Stroke Scale for use in stroke clinical trials. Stroke 32:1310–1317.

Mas, J.L., C. Arquizan, C. Lamy, et al. 2001. Recurrent cerebrovascular events associated with patentforamen ovale, atrial septal aneurysm or both. N. Engl. J. Med. 345:1740–1746.

Mehler, M.F. 1991. The rostral brain stem syndrome. Neurology 39:9–16.

Norrving, B. 1991. Lateral medullary infarcts. Prognosis in an unselected series. Neurology 41:244–248.
Pessin, M.S., P.B. Gorelick, E.S. Kwan, and L.R. Caplan. 1987. Basilar artery stenosis: Middle and distal segments. Neurology 37:1742–146.
Pessin, M.S., E.S. Lathi, M.B. Cohen, et al. 1987. Clinical features and mechanism of occipital infarction. Ann. Neurol. 21:85–89.
Torvik, A., and L. Jorgenson. 1966. Thrombotic and embolic occlusions of the carotid arteries. J. Neurol. Sci. 3:410–432.
Widdick, E.F., and M.N. Diringer. 1998. Middle cerebral artery territory infarction and early brain swelling. Progression and effect of age on outcome. Mayo Clin Proc. 73:829–836.

III. Intracerebral Hemorrhage

Kase, C.S. 2000. Vascular diseases of the nervous system. B. Intracerebral hemorrhage. In Bradley, W.G., et al (eds.). Neurology in Clinical Practice, 3rd ed,. Boston Butterworth Heinemann, pp. 1167–1183.
Marcus, E.M., and S. Jacobson. 2003. Integrated Neuroscience: A Clinical Problem Solving Approach. Boston, Kluwer Academic, pp. 13-18–13-19, 20-16–20-18, 26-22–26-26.
Massaro, A.R., R.L. Sacco, J.P. Mohr., et al:. 1991. Clinical discriminators between lobar and subcortical hemorrhage. Neurology 41:1881–1885.
Quereshi, A.I., S. Tuhrim, J. Broderick., et al. 2001. Medical progress: Spontaneous intracerebral hemorrhage. N. Engl. J. Med. 344:1450–1460.
Zhu, X.L., M.S.Y. Chan, and W.S Poon. 1997. Spontaneous intracranial hemorrhage: Which patients need diagnostic cerebral angiography? A prospective study of 206 cases and review of the literature. Stroke 28:1406–1409.

IV: Subarachnoid Hemorrhage and Cerebral Aneurysms

Broderick, J.P., T.G. Brott, J.E. Duldner, et al. 1994. Initial and recurrent bleeding are the major causes of death following subarachnoid hemorrhage. Stroke 25:1342–1347.
Brisman, J.L., J.K. Song, and D.W. Newell. 2006. Cerebral aneurysum. N. Eng. J. Med 355; 928–939.
Dodick, D.W. 2002. Thunderclap headache. J. Neurol. Neurosurg. Psychiatry 72:6–11.
International Study of Unruptured Intracranial Aneurysms Investigators. 1998. Unruptured intracranial aneurysms-risk of rupture and risks of surgical intervention. N. Engl. J. Med. 339:1725–1733.
Kirkpatrick, P.J. 2002. Subarchnoid haemorrhage and intracranial aneurysms: What neurologists need to know. J. Neurol, Neurosurg. Psychiatry 73:i28–i33.
Marcus, E.M., and S. Jacobson. 2003. Integrated Neuroscience: A Clinical Problem Solving Approach. Boston, Kluwer Academic, pp. 26-26–26-32.
Olafsson, E., A.H. Hauser, and G. Gudmundsson. 1997. A population based study of prognosis of ruptured cerebral aneurysm: mortality and recurrence of subarachnoids hemorrhage. Neurology 48:1191–1195.
Phililips, L.H., J.P. Whisnant, W.O. Fallon, and T.M. Sundt. 1980. The changing patterns of subarchnoid Hemorrhage in a Community. Neurology 30:1034–1040.
Sacco, R.L., P.A. Wolf, N.F. Bhariche, et al. 1984. Subarachnoid and intracerebral hemorrhage. Neurology 34:847- 854.
Schievink, W.I. 1997. Intracranial aneurysms. N. Engl. J. Med. 336:28–40.

Selman, W.R., R.W. Tarr, and R.A. Ratcheson. 2000. Vascular disease of the nervous system. C. Intracranial aneurysms and subarachnoid hemorrhage. D. Arteriovenous malformations. In Bradley, W.G., et al. (eds.). Neurology in Clinical Practice. Boston, Butterworth Heinemann, pp. 1185–1213.

Van Gijn, J., and G.J.E. Rinkel. 2001. Subarachnoid haemorrhage: Diagnosis, causes, and management. Brain 124:249–278.

Wardlaw, J M., and P.M. White. 2000. The detection and management of unruptured intracranial aneurysms. Brain 123:205–221.

Wiebers, D.O., and V.E. Torres. 1992. Screening for unruptured intracranial aneurysms in autosomal dominant polycystic kidney disease. N. Engl. J. Med 327:953–955.

Index

A

Accommodation, 315
Action tremors, 293
Acute transection, of spinal cord in human, 225–226
Adenohypophysis, 177–181
Afferent inputs and efferent projections of neocortex, 211–212
Afferent pathways, 170–173
Alzheimer's disease, 368–369, 447–448
Amnestic confabulatory syndrome following lesions of the hippocampus and related structures, 365–366
Amygdala, 344–347
An immunologically privileged site, 43
An inability to interpret drawings, 304–308
Anatomical correlates, 348
Anatomical correlation
   of specific language syndromes, 383–392, 393–396
Anatomical substrate of learning in humans, 361–362
Anatomy
   cerebellum
   eye, 313–318
Anosmia, 126
Anterior Cerebral Artery (ACA), 406–407, 420
Anterior choroidal artery, 407
Anterior circulation, 406
Anterior commissure, 210
Anterior cranial fossa (CN I and II), 121
Anterior group, 169
Anterior Inferior Cerebellar Artery (AICA), 291–292
Anterior limb of the internal capsule, 159
Anterior nuclei, 151, 153
Anterior radiations, 160
Anterolateral pathway and pain, 75

Aphasia –dominant hemispheric functions, 379–396
Apraxia, 390
Arachnoid, 402–403
Archicerebellum, 276
Area 17 corresponds to the striate cortex, 208–209
Area 22, 208
Area 4 lesions, 235
Area 4 stimulation, 233–235
Area 42 and 22, 356
Area 6, 201–204
Area 6 stimulation- stimulation of SMA and PMA, 239–240
Area 8-Premotor, 240
Area V3 and V3A (area 18)-selective for form, 325
Area V5 (area 19)-selective for motion and directionality, 326
Areas 18, 19 form surrounding stripes around area 17, 209
Areas 41, 42 the transverse gyri of Heschl, 207
Areas 44 and 45, 204, 237–238
Areas 5, 7, 39, 40, parietal lobules, 205
Areas 6, premotor cortex (Areas 6 and 8), 232, 237–240
Areas in occipital lobe-17, 18, 19 (V1-V 5), 322–323
Aristotle, 5
Arterial blood supply to the brain, 407
Arterial supply to the brain, 405–408
Arteriosclerotic or vascular (multi infarct), 263
Ascending tracts in the spinal cord, 74
Association fiber systems and speech, 380
Associational fibers, 211
Astrocytes, 40–42
Atrophic change, 48
Audition, 155

Auditory (Transverse Temporal Gyri of Heschel) 41, 206–207
Auditory and auditory association. area 41, the primary auditory cortex, 356
Auditory associational 42 and 22, 206–207
Autonomic effects
　of stimulation of anterior hypothalamus, 181
　of stimulation of posterior hypothalamus, 181–182
Autonomic nervous system, 185–189
Axon and axon origin, 33
Axon hillock, 33
Axoplasmic flow, 31–32

B

"Basal ganglia" originally included the deep telencephalic nuclei: the caudate, putamen, globus pallidus, the claustrum, and nucleus accumbens), 251
Basal nuclei, 17–18
Basal temporal Speech Area, 380
Basic
　design and functional organization of cerebral cortex, 196–199
　principal of sensory system, 308
　principals of voluntary motor system, 246
Basilar, 408
Bell's palsy, 145
Beta rhythm, 217–218
Bilateral
　lesions limited to areas 18 and 19, 328
　necrosis of the globus pallidus, 263
Bitemporal hemianopsia, lesion in optic chiasm, 329
Blind spot, 319
Blood brain barrier, 52–54
Blood supply to the brain, 405–409
Body temperature, 183
Brachiocephalic artery, 405–406
Brain stem, 9–11, 85–120
Brain stem levels
　level 1: spinomedullary junction with motor decussation, 91–92
　level 4: lower pons at level of facial nerve and facial colliculus ventricle equivalent, 98–100
　level 2: lower/narrow medulla at sensory decussation, 92–94
　level 3: wide medulla at level of inferior olive, 95–97
　level 5: upper pons at the motor and main sensory nuclei of nerve V, 100–103
　level 6: inferior colliculus and pontine basis, 103–106
　level 7: midbrain superior collicular level and pontine basis, 106–110
Brain, MRI sagittal plain, 10
Broca's motor aphasia or expressive speech center, 380

C

Callosal fibers, 323
Cardinal signs of parkinson disease, 258–259
Cardiovascular centers, 114–116
Carotid sinus nerve, 139
Case 12-1, 299
Case 12-2, 300–302
Case 12-3, 305–308
Case 3-1, 79–80
Case 4-1, lateral (dorsolateral) medullary syndrome/wallenberg's syndrome, 119
Case histories, from the visual system, 329–338
Case history 10-1, 265–267
Case history 10-2, 271–272
Case history 11-1, 283–284
Case history 13-2, 332–333
Case history 13-3, 334–336
Case history 13-4, 336–338
Case history 14-1, 359–361
Case history 14-3, wernicke korsakoff syndrome, 363–364
Case history 14-4, Alzheimer's disease, 368–369
Case history 15-1, Broca's aphasia, 394–395
Case history 15-3, wernicke's aphasia, 385–386
Case history 15-4, fluent posterior aphasia, 386–387
Case history 15-5, wernicke's posterior aphasia, 387–388
Case history 15-1, selective dyslexia, 388–390
Case history 16-1, 179–180
Case history 4-1, lateral medullary syndrome of wallenberg, 119
Case history 9-1, 235–237
Case history 15-2, anterior motor aphasia, 388, 396
Case of Phineas P. Gage, 370–371
Causes of cranial nerve III dysfunction, 131

Causes of hemorrhage into the cerebellum,
Central control of saccades, 244
Central nervous system, 7–21
  in situ, 8
  pathways, 18, 35
Central tegmental tract, 112–113
Central/cerebral innervation of VII, 136
Cerebellar dysarthria, 290
Cerebellum, 12
Cerebellar syndromes
  anterior lobe, 285–288
  floccular nodular lobe and other
    midline cerebellar tumors, 283–285
  lateral cerebellar hemispheres
    neocerebellar or middle-posterior
      lobe syndrome, 288–291
  cerebellar peduncles, 291
Cerebral dominance-development aspects, 378, 379
Crebral aqueduct, 403
Cerebral cortex, 13, 15
  and disturbances of verbal expression, 377–379
  functional localization, 191–219
Cerebral cortical gray matter, 191–193
Cerebral cortical motor functions, 223–231
Cerebral dominance, 378
Cerebral veins, 408, 409
Cerebral palsy, 241, 242
Cerebrospinal fluid, circulation, 402, 404
Cerebrum, 13–17
Cervical sympathetic ganglia, 188
Chapter One-, Case History, 1
Cholinergic nuclei, 113–114
Chorda tympani, 117
Chorea, hemichorea and hemiballismus, 268–273
Choroid, 314
Chromatolysis, 47–48
  of RNA, 47
Ciliary body, 314
Cingulate cortex, 351–352
Cingulotomy, 372
Cingulum, 211, 355
Circle of Willis, 407, 408
Circuits in emotional brain, 353
Classification of the various types of neocortex, 198–199
Climbing fibers, 279
Clinical lesions of posterior columns, 310
Clinical symptoms and signs of dysfunction, 257–273
Commissural fibers, 16, 209–211

Complete paralysis of nerve, 129–131
Complete unilateral ablation of area 17/V1, 327
Complex cell, 325
Comprehension of spoken language, 382
Concept of central pattern generators, 223–225
Conduction or repetition type fluent aphasia, 391–392
Cones - color vision cones, 318
Connections of amygdala, 345–346
Connections of the prefrontal cortex, 370
Control center
  for heat loss, 183
  for heat production and conservation, 183
Conus, 55
Conversational speech, 381
Correlation of neocortical cytoarchitecture and function, 199–209
Cortical areas of the dominant hemisphere of major importance in language disturbances, 380
Cortical control of the cranial nerves. the Corticobulbar Pathway, 146
Cortical control of eye movements, 243–246
Cortical neurons, 118
Cortical nucleus-the amygdala, 348
Cortical structures, in limbic system, 343–352
Cortical system, 246
Cortical white matter, 15–17
Corticomesencephalic system, 250
Corticonuclear/corticobulbar system–
  voluntary control of the muscles
    controlled by cranial nerves V, VII,
    and IX to XII, 248–250
Corticorubral spinal system, 237
Corticospinal tracts–voluntary control of the limbs, thorax, and abdomen, 246–248
Coughing, 116
Cranial nerve case, 144
  histories, 144–145
  history 5-2, 145
  history 5-3, 145
Cranial nerves, 11, 121–145
  I, olfactory, 126
  II, optic, 126–127
  III, oculomotor, 127–131
  IV trochlear, 130, 131
  V – trigeminal, 132–134
  V lesions, 143–144
  VI, abducens, 130, 131

Cranial nerves (*cont.*)
    VII, 188
    VIII, vestibulo-cochlear, 136–138
    IX, glossopharyngeal, 135, 142
    X vagus, 140–141
    XI spinal accessory, 141, 142
    XII hypoglossal, 141, 142
    dysfunction, 141–144
    nerve to each pharyngeal arch, 124, 139
Cranial portion, 141
Cutaneous sensory receptors, 40
Cytoarchitecture
    cerebellum of, 277–278
    hippocampus of, 349
Cytology, 193–196
Cytoskeleton, 28

D

Damage to temporal lobe
    and aphasia, 358
    and aggressive behavior, 359
    effects on hearing, 358
    and effects on memory, 359
    and klüver-bucy syndrome, 358
    and psychiatric disturbances, 359
    to temporal lobe and complex partial seizures, 359
    and unilateral effects on memory, 359
    and visual defects, 358
Decline of functional neurosurgery, 372–373
Decorticate preparation, 228
Degeneration, 47–49
Deglutition, 115
Delayed response test, 371
Dementia, 366
Dendrites, 23–24
Dendritic Spines, 24–25
Dentate Gyrus, 349
    cytoarchitectural, 349–350
    molecular layer, 350
    polymorphic cell, 350
Dentaten gyrus granule cell layer, 350
Descending tracts in the spinal cord, 73
Development of the cerebral cortex, 212–214
Diagram illustrating a chordotomy, 76
Diaphragma sellae, 401
Diencephalon, 12-13, 147–165
Differences between the spinal cord and brain stem, 89–110
Differential diagnosis of parkinson's disease, 267

Disease of
    cochlea, 137
    eighth nerve, 137
Disorders of recent memory; the amnestic confabulatory syndrome of diencephalic origin; wernicke-korsakoff's, 363–364
Disorders of Motor Development, 241–242
Disturbance in concept of body image (neglect) and denial of illness, 304
Dominant hemisphere in the parietal lobules, 303–304
Dopaminergic pathways, 161
Dorsal horn = lamina 1-6, 61
Dorsal lateral PMA, 239
Dorsal longitudinal fasciculus, 172
Dura mater, 401–402
Dysarthria, 377
Dysfunction in the eye due to lesions in the brain stem, 130
Dyslexia, 389–390

E

EEG. alpha rhythm, 217
Effectors, 39–40
Effects
    decreased dopaminergic input on thalamus and cortex = Less excitation, 255
    disease on the cerebellum, 280–294
    lesions
        of amygdala, 347
        in occipital visual areas, 327–328
        spinal, brainstem and cerebral lesions on the motor system, 225–231
    stimulation
        of amygdaloid region, 346–347
        of areas 17, 18, and 19, 326
        of speech areas, 382–383
Embryological considerations, 124–125
Emetic center, 116
Emotional
    brain, 353–355
    response, 371
Endothelial cells, 42
Enteric nervous system, 187
Entorhinal region, 346, 350–351
Entorhinal reverberating circuit/perforant pathway, 354
Ependymal cells, 45–46
Epithalamus, 343
Essential tremor, 294
Etiology of parkinson disease, 260–261
Evoked potentials, 215–216

Excitatory synapses, 37–38
Extracellular space and the CSF, 54
Extrastriate visual cortex Areas 18 (V2 & V3) and 19 (V4 & V5), 323

F
Falx cerebelli, 401
Falx cerebri, 401
Fibrous astrocytes, 41
Filum terminale, 55
Final effect of this system on the thalamus, 255
Fixation system holds the eyes still during intentional gaze on an object, 245
Fixed pupil, 315–316
Flaccid paralysis, 242
Fluent aphasias associated with lesions of the dominant inferior parietal areas: angular and supramarginal gyri, 392
Fluent aphasia-posterior aphasia-Wernicke's, 380–382
Fluent aphasias, 383
Follicle-Stimulating Hormone (FSH), 179
Food intake, 182
Fornix, 352–353
Frontal (Area 8) and parieto-occipital eye fields, 243
Frontal lobe, 199–200
Functional centers in the brain stem, 110–118
Functional localization, 15
  in hypothalamic nuclei, 174
  in Hypothalamus, 182–185
  within the anterior horns, 65
Functional neurosurgery, 371–373
Functional organization
  of cranial nerves, 122–124
  of thalamic nuclei, 149–158
Functions in the lobes of the cerebrum, 17
Functions of the cerebellum, 279–280
Fundamental types of cerebral cortex, 196–198

G
GABA-ergic pathways, 161
Gag reflex, 139
Gamma system, 69
Generalized chorea, 268
Genu of the internal capsule, 159
Gerstmann's syndrome, 303–304
Gic pathways, 161
Glands associated with the brain, 22
Glial response to injury, 52
Golgi tendon organs, Ib, 69–70

Golgi type I and II neurons, 24
Gonadotropin (luteinizing hormone [LH]), 179
Grasp reflex, 229
Gray matter, 59–60
Gross landmarks in the medulla, 91
Growth hormone (Somatotropin, or STH), 179
Guidelines for localizing disease to and within the brain stem, 118–120

H
Hemiballismus, 164
Hemichorea and hemiballismus, 268
Herpes zoster involvement of the Gasserian ganglion, 134
Heterogenetic, 197
Hierarchy of function in the limbic system, 373
Higher cortical functions, 377
Hippocampal commissure, 377–397
Hippocampal formation, 346–351
Hippocampal- molecular *layer*, 349
Hippocampal polymorphic layer (stratum oriens), 349
Hippocampal pyramidal cell layer (stratum pyramidal), 349
Hippocampal sectors. CA1, CA2, CA3, CA4, 349
Homogenetic, 197
Homonymous hemianopsia
  lesions behind the optic chiasm, 329
  with macular sparing is seen with lesions in the visual cortex, 329
Hormones produced
  by hypothalamus, 176–177
  in adenohypophysis, 177–181, 189
How do we confirm the location of the pathology, 216–217
How do we study function, 215
How the brodmann areas got their numbers, 199
How the cranial nerves got their numbers, 126
Human studies on hippocampus, 347
Huntington's Chorea, 269–271
Huntington Disease-specific mutation repeat in the CAG series coding for poly glutamine tracts at the 4p16.3 locus on this chromosome, 269
Hyper complex cell, 325
Hypophysiotrophic Area, 176
Hypophysis cerebri, 175
Hypothalamic
  nuclei, 167–170
  hypophyseal portal system, 175–176
  hypophyseal tracts- neurosecretory system, 173

Hypothalamus, 374
  and emotions, 184
  and light levels, 184–185
  and the autonomic nervous system, 181–182, 186
  neuroendocrine system, and autonomic nervous system, 167–189

I

Immediate or short term working memory, 361–362
Inferior cerebral veins, 408
Inferior longitudinal fasciculus, 209, 211
Inferior parietal lobule and language, 205
Inferior radiations, 160
inferior temporal 20, 207
Inhibitory synapses, 26, 38
Injury to nerve VII at the stylomastoid foramen, 136
Injury to the chorda tympani, 136
Inner granule cell layer, 278
Innervation in the pelvis and perineum, 189
Input to reticular formation, 112
Instinctive grasp, 229, 230, 232
Instinctive tactile avoiding reaction, 229–230
Intermediate region lamina = 7, 62
Internal capsule, 158–159
Internal carotid, 406
  cavernous portion, 406
  cerebral/supraclinoid/intradural, 406
  cervical portion, 406
  meningeal branch, 406
  petrous portion, 406
Internal cerebral veins (of Galen), 408
Interneurons, 63–64
Interpeduncular nucleus, 342
Interruption of anterior thalamic radiation or destruction of dorsomedial nucleus, 372
Intracortical associations between striate and nonstriate cortex, 323
Intralaminar nuclei, 157

K

Kinetic tremors, 294
Kluver and Bucy (1937), 340
Klüver-Bucy Syndrome, 358

L

Lactogenic hormone (Prolactin), 179
Lamina, 3
Laminar organization of central gray, 61–64
Language functions in the nondominant parietal hemisphere, 396
Lateral Geniculate Nucleus (LGN), 320–322
Lateral nuclear, 64
Lateral Premotor Area (PMA), 239
Lateral ventricles, 403–404
Layer I: molecular or plexiform layers, 198
Layer II: external granular layer, 198
Layer III: external pyramidal layer, 198
Layer IV: internal granular layer, 198
Layer V: internal or large and giant pyramidal cell layer, 198
Layer VI: fusiform or spindle cell multiform layer, 198
Left common carotid, 406
Left subclavian, 406
Lens, 314–316
Leocortex, 3-layers, 348
Lesion, 209
  of diencephalon and adjacent regions producing the amnestic confabulatory disorder seen in the Korsakoff .syndrome, 365
  in extrastriate areas 18 and 19 produce deficits in visual association, including defects in visual recognition and reading, 327
  of inferior cerebellar peduncle, 291
  of middle cerebellar peduncle, 291
  in occipital cortex, 336–337
  in optic nerve before the chiasm-result monocular blindness, 329–331
  in optic radiation; result noncongruous homonymous hemianopsia or quadrantanopia, 334–336
  of the superior cerebellar peduncle, 291
Light reflexes, 315, 322
Limbic cortical regions, 375
Limbic nuclei-anterior, medial, lateral dorsal, midline and intralaminar nuclei, 153, 154
Limbic system, 339–355
Lack of awareness of hemiplegia, 396
Lesion at optic chiasm, result bitemporal hemianopsia, 331–333
Limbic system
  and corticospinal and corticobulbar pathway, 355
  temporal lobe and prefrontal cortex, 339–375

Index 493

Lingual taste buds, 116–117
Local circuits within the striatum, 252
Location of Postganglionic Autonomic Neurons, 187
Location of Preganglionic Autonomic Neurons, 186–187
Location of the corticospinal tracts as shown by degeneration caused, 74
Long term memory, labile stage, 362
Long-term Memory stage of Remote memory, 362
Lower Motor Neuron Lesion, 77, 79
LP, 153
Lumbar Sympathetic Ganglia, 189

M

Major gyri in the parietal lobe, 205
Methods for the study of functional localization in cerebral cortex, 215–219
Motor systems III: cerebellum and movement, 275–294
Major cerebellar syndromes, 282–294
Major indirect outflow pathway, 254
Major input into the basal ganglia, 252
Major voluntary motor pathways, 246–250
Mammillothalamic tract, 172
Management of parkinson's disease, 263–267
Manganese poisoning, 262
Mass reflex, 226
Mechanicoreceptor, 40, 308
Meckel's cave, 401
Medial forebrain bundle, 172
Medial nuclear complex, 154
Medial nuclear division, 64
Medulla, 89–97
Meniere's disease, 138
Meninges, ventricular system and vascular system, 401–409
Meninges-coverings
  of brain, 401–403
  of spinal cord, 56
Mesencephalic locomotion pattern generator:, 224
Mesocortex-a transitional type of 6-layer cortex, 348
Mesocortical system, 257
Mesolimbic system, 257
Microanatomy of the Striatum, 255–256
Microglial cells, 44–45
Microscopic changes in Alzheimer's, 367–368
Microtubules, 31–32
Mid line cerebellar tumor in the adult, 284–285

Midbrain, 103–110
Midbrain preparation, 227–228
Middle Cerebral Artery (MCA), 407
Middle cerebral veins, 408
Middle Cranial Fossa (CN III, IV, V, VI), 122
Middle group,169
Middle temporal 21, 207
Middle tunic (vascular and pigmented), 314–316
Midline cerebellar tumor in a child, 283–284
Midline nuclei, 156–157
Mitochondria, 28–29
Modern concepts of the plasticity of the primary motor cortex, 235–237
Modulation of pain transmission, 71–72
Modulators of neurotransmission, 39
Monoamine nuclei, 113
Monocular blindness. Lesion in the retina or the optic nerve, 329
Mononuclear Cells, 43–45
Motor Area 4, 200
Motor control of the foot from the motor-sensory cortex, 19
Motor cranial nerve lesion, 141–143
Motor system II. Basal ganglia, 251–273
Motor system I movement and motor pathways, 223–250
Motor/Ventral Horn Cells, 64
MPTP Toxicity, 262
Muscle spindle, 67–69
Myelin, 33–35
Myelin Sheath-the insulator in an aqueous media, 33
Myelination, 34–35, 214

N

Nasal visual field, 321
Nausea and vomiting, 138
Necrosis, 263
Neglect syndrome, 396
Neocerebellum, 277
Neocortex, 197–198, 348
Neocortex- 6 layers, 348
Nerve growth factors, 52
Nerve roots, 56, 58
Neural crest cells, 47
Neurochemically defined nuclei in the reticular formation affecting consciousness, 113
Neurofibrillar tangles, 32
Neuroendocrine system, the hypothalamus and its relation to the hypophysis, 173–185

Neurohypophysis, 169, 170, 174–177
Neuroleptic agents, 261
Neuronal cytoskeleton, 30–31
Neurophysiology correlates of cortical cytoarchitecture and basis of EEG, 217–219
Neurosecretory granules, 30, 169, 170
Neurotransmitters, 38–39
Nigral-striatal pathways, 256
Nociception and pain, 70–77
Nociceptive stimulus, 70
Non-cortical system, 245
Non-declarative memory (implicit or reflexive), 362
Non-dominant hemisphere in the parietal lobules, 304–308
Nonfluent aphasia, 380, 393–396
Nonfluent aphasia-anterior aphasia-Broca's, 380
Non-specific associational, 156–158
Non-thalamic sources of input efferent projections, 212
Noradrenergic (norepinephrine) pathway, 161
Nuclei, 5
Nuclei of the thalamus, 148–149
Nucleus, 26
Nystagmus, a jerk of the eyes, 138

O
Occipital lobe, 208–209, 356
Occipital lobe and eye movements, 328–329
Occipital somites, 124, 125
Ocular dominance columns, 324
Oligodendrocytes, 40
Ophthalmic artery, 407
Optic nerve, 330
    disease, 127
    termination in LGN, 321
Optic Nerve>Optic Chiasm>Optic Tract>LGN>Visual cortex, 320, 321
Optical righting reflex, 228–229
Opticokinetic movements, 245–246
Organization
    of neurons in ventral horn, 64
    sensory receptors, 66–69
    post central gyrus, 300
Origins of cranial nerves and associated muscles, 124
Other causes of cerebellar atrophy, 287
Other causes of trigeminal symptoms, 133–134
Other motor pattern centers, 225

Other movement disorders associated with diseases of the basal ganglia, 272
Other pathological processes, 262–263
Other possible inputs to thalamus, 160–163
Other reflexes associated with the cerebral cortex, 230–231
Outer fibrous tunic, 313
Outer molecular layer, 278
Output of reticular system, 112–114
Overlap with the cerebellar system, 257
Overview
    eye movements, 264
    localized lesions in the visual system, 329, 330
    dopaminergic systems, 256–257
    role of the prefrontal area in motor and cognitive function, 261
    tremors, 293

P
Pain and temperature, 75
    spinal/descending nucleus of V, 312
Pain receptors, 70–71
Paleocerebellum, 277
Papez, J. W., 379
Papez circuit, 163, 353–354
Parahippocampal gyrus of the hippocampal formation, 344
Parallel fibers, 279, 280
Parallel processing in the visual cortex, 323
Paralysis of the intrinsic muscles, 131
Parasympathetic (Craniosacral), 186
Parasympathetic fibers originate from segments S2 to S4, 189
Pariaxial mesenchyme, 125
Parietal lobe, 205–206
Parietal lobe and tactile sensation from the body, 308
Parietal lobules-superior and inferior parietal lobules, 302
Parkinson's disease and the parkinsonian syndrome, 258
Pars optica of the retina, 316
Partial lesions in area 17, 327
Pathology of parkinsonian lesions, 259–260
Pattern of reflex recovery, 225
Perceptual pathways, 325–326
Pericytes, 41
Peripheral nerve regeneration, 49–51
Peripheral nervous system, 4
Perivascular cells, 40
Periventricular system, 172

Index 495

Photoreceptor layer of the retina, 314
Physiologic tremor, 293, 294
Pia mater, 401, 402
Pineal, 22
Pineal body, 185
Pituitary, 22
Placing reactions, 229
Placodes, origin of special sensory
    nerves, 125
Pleasure/punishment areas, 374–375
Pons, 97
Postcentral gyrus
    lesions, 297–298
    stimulation, 296–297
Posterior cerebral arteries (PCA), 408
Posterior circulation, 406, 407
Posterior columns–tactile sensation from the
    neck, trunk and extremeties
    (fasciculus gracilis and cuneatus),
    309–310350
Posterior communicating arteries, 434
Posterior cranial fossa (CN VIII-XII), 122
Posterior inferior cerebellar artery
    (PICA), 291
Posterior limb of the internal capsule, 159
Posterior radiation, 160
Posterior root-sensory nerves, 58
Posterior temporal 37, 207
Postnatal development of motor reflexes, 231
Postural tremors, 293
Prefrontal
    areas, 232–233
    cortex Areas 9, 10, 11, 12, 13, 14, and 46,
        204, 240
    granular areas and emotions, 369–373
    lobe anatomy and functional localization,
        369–370
    lobotomy and prefrontal leucotomy,
        371–372
    non-motor areas, 221
Preganglionic autonomic nuclei, 66
Premotor cortex, 237
Primary motor cortex, 231
Primary motor cortex Area 4, 233, 234, 238
Primary sensory neuron, 74
Primary sulci, 212–214
Principal pathways of the limbic system,
    352–355
Progressive dementing processes, 366
Projection fibers, 71, 210
Proprioception from the head, 312
Protoplasmic astrocytes, 41
Pupillary muscles, 315
Pupillary reflexes, 315–316

Pure word deafness, 391
Purkinje cell layer, 278
Pyramidal cells, 194–195
Pyramidal tract, 233

Q

Quadrantanopia, due to partial lesions of the
    geniculocalcarine radiations, 329

R

Reactions dependent on cerebral cortex, 228
Readiness potential (Bereitschafts potential),
    239
Reading, 382
Recovery, 47
Reflex response to stretch, 66–67
Regeneration in the central nervous
    system, 51
Regions in the brain stem, 88
Relationship between the thalamus and the
    cerebral cortex, 159–163
Relationship of primary motor, premotor and
    prefrontal cortex, 231–241
Reorganization of gray and white matter from
    spinal cord gray to tegmentum of
    brain stem, 10–11
Repetition, 382
Respiration centers, 114
Response of nervous system to injury,
    47–52
Rest tremor, 293
Reticular formation, 374
Reticular formation of the brain stem and
    spinal cord, 341–342
Reticular nucleus of thalamus, 156
Retina and visual fields, 319–321
Retinal disease, 127
Retinal representation in the occipital
    cortex, 324
Retrograde changes in the cell body-
    chromatolysis, 47–48
Return of deep tendon reflexes/stretch
    reflexes, 242–243
Return of distal hand movement, 243
Return of selective ability to grasp, 43
Return of the instinctive tactile grasp reaction
    with the capacity for projectile
    movement, 243
Return of traction response, 243
Rods and cones, 316
Rods:vision in dim light and night vision,
    316–318

Role(s)
  ascending reticular system of, 113
  astrocytes in the central nervous system of, 41
  corpus callosum in transfer of information of, 397
  descending systems of, 113
  hypothalamus of, 228
  limbic system in memory of, 361–369
  limbic system in psychiatric disorders of, 373
Rough endoplasmic reticulum –Nissl body, 27–28

S

Saccadic eye movements, 243–244
Satellite cells, 46
Schwann cells, 46
Second order neuron(s), 74, 117–118
Segmental function, 64–70
Selective vulnerability of hippocampus, 350–351
Sensory
  and motor relay nuclei-the ventrobasal complex & lateral nucleus, 149–153
  cranial nerve lesion, 143–144
  ganglia, 5
  information from the foot to the sensori-motor cortex, 19
  receptors, 66
Septum, 343
Serotoninergic pathway, 161
Simple cell, 324–325
Sleep cycle, 182
Slow waves, 218
Smooth neurons, 195–196
Smooth pursuit in contrast to saccade, 245
Soma, 24
Somatosensory function and the parietal lobe, 295–312
Special somatic sensory nuclei- vision and audition, the lateral geniculate and medial geniculate nuclei of the metathalamus, 154–156
Special visceral afferent taste buds (chorda tympani) from ant 2/3 of tongue via petrotympanic fissure, 135
Specific associational- polymodal/ somatic nuclei-the pulvinar nuclei, 154, 155
Spinal cord, 8–9, 55–83
  parasymapthetic segments S2-S4, 188
structure and function, 56–60

Spinal
  pathways, 83
  pattern generator, 224
  portion, 141
  shock, 225
Spinocerebellar degenerations, 293
Spinothalamics/anterolateral column, 91–92
Spiny neurons, 194
Spiny stellate neurons, 195
Stimulation, 215
Stretch receptors, 67–69
Stria terminalis, 354
Striate cortex area 17(V1), 323
Striatum also receives a major dopaminergic input from the substantia nigra compacta, 252–253
Structure of the eye, 313–319
Studies of Jacobsen and Nissen, 371
Studies of recovery of motor function in the human, 242–243
Subcortical fibers, 16
Subcortical structures, 341–343
Subcortical white matter afferents and efferents, 209–212
Subjective taste, 117
Sublentiform portion, 159
Subthalamus, 150, 163–165
Summary of cortical circuitry, 198
Summer's sector, 349
Superior and inferior parietal lobules, 302–308
Superficial cerebral veins, 408
Superior cerebellar artery, 292
Superior colliculus, 106, 245
Superior longitudinal fasciculus, 211
Superior radiations, 160
Supplemental motor cortex and language, 380
Supplementary Motor Area (SMA), 238–239
Supporting cells in the peripheral nervous system, 46–47
Supporting cells of the central nervous system, 40–46
Suppressor areas for motor activity (Negative Motor Response), 240
Sympathetic, 188
Sympathetic (thoracolumbar), 187
Sympathetic system, 186, 188–189
Symptoms
  following stimulation of the temporal lobe, 356–357
  from ablation of or damage to the temporal lobe, 358–359
  of disease involving the temporal lobe, 356–361
Synapse, 35–40

Synaptic
  structure, 36
  transmission, 38–40
  types, 36–37
  vesicles, 37–38
Syndrome
  of anterior lobe, 285–287
  of cerebellar peduncles, 291
  of floccular nodular lobe and other midline cerebellar tumors, 283–285
  oflateral cerebellar hemispheres (Neocerebellar or Middle-Posterior Lobe Syndrome), 288–291

T
Tactile sensation
  from the body–medial lemnsicus, 308–310
  from the head, 310–312
  chief/main nucleus of V, 312
Tardive dyskinesia and other tardive reactions, 272–273
Task specific kinetic tremor, 294
Taste and the VPM, 153
Taste, 116–118
Tectum, 103–104, 106
Tegmentum, 85, 98, 105–106
Temporal lobe, 206–208, 355–361
Temporal visual field, 329, 331
Tentorium cerebelli, 401
Termination of optic radiation, 324
Thalamic
  borders, 160
  input onto the cortical layers, 160
  radiations and the internal capsule, 160
  syndrome (of Dejerine), 163
Thalamolenticular portion, 159
Thalamus, 159, 374
The Babinski response. Upper, 75
The cranial nerves. *See* cranial nerves
The decerebrate preparation, 226
The following case history provides an example of epidural metastatic tumor compressing the spinal cord: Case 3-2, 81–83
The inner tunic, 316
The intracortical association fiber system, 355
The limbic brain as a functional system, 373–375
The major direct outflow pathway, 254
The motor-sensory cortex, 14, 18–19
The neuron, 3–5, 23
The pulvinar nuclei, 155
The senses, 5–6

These two types of fibers (short U and long), 15–16
Third order neurons, 74
Third ventricle, 403
Thoracic Sympathetic Ganglia, 189
Thyrotrophic Hormone (TSH), 181
Tics, 273
Tinnitus, 137
Topographic patterns of representation in cerebellar cortex, 279–280
Toxic agents, 261–262
Transection of the brain stem- the decerebrate preparation, 226–227
Transmitter replacement, 264–265
Tremor
  at rest, 263
  during target directed movements: intention tremor, 294
Trigeminal neuralgia (tic, douloureux), 133
Types of aphasia, 380–383
Types of microglia cells, 44

U
Umbar puncture, 60
Uncinate fasciculus, 211
Unilateral lesion of extrastriate areas 18 & 19, 328
Upper and lower motor neurons lesions, 77–83
Upper motor neuron lesion (UMN), 77–79
Useful facts on the cranial nerves, 122

V
V1 (area 17) and V2 (area 18)-columnar organization, 325
VA, 152
Vascular lesions in Area 17, 27
Vascular lesions within the calcarine cortex, 338
Vascular syndromes of the cerebellum- vertebral basilar, 291–294
Venous
  circulation of the brain, 408
  sinuses, 409
Ventral
  amygdalofugal pathway, 355, 365
  basal nuclear complex, 52
  Horn = Lamina 8 and 9, 62–63
  lateral PMA, 239
  posterior nuclear complex, 152
  roots-motor, 58
Ventricular system, 403–405

Vergence movements, 245
Vertebrals, 425
Vertigo, 138
Vestibulo-ocular movements hold, 245
Visceral sensory receptors, 40
Vision, 55–156
Visual
   acuity, 331
   field deficits produced by lesions in the optic pathway, 329–338
   fields, 320, 360
   pathway, 320–323
   perceptions, 356
   system & occipital lobe, 334–338
VL, 151
Voluntary control of lower motor neurons in the spinal cord via the corticospinal tracts, 73–74
Vomiting, 115–116

Von Economo's encephalitis, 263
VPL, 153
VPM, 160

W

Wallerian degeneration, 48–49
Water balance and neurosecretion, 183–184
Wernicke's aphasia and Wernicke's area, 383–385
Wernicke's receptive aphasia area, 380
Wernicke-Korsakoff's syndrome, 287
White matter, 59–60
   of diencephalon, 158–159
   tracts, 73
Word finding and selection, 382
Working memory, 241
Writing, 418

Neuroanatomy for the Neuroscientist

This textbook provides a single text for the undergraduate, and graduate student and for the first and second year medical students learning Neuroanatomy and Neurosciences.

Key Features of this volume include:

*More then 250 illustrations with 4 color illustrations including micrographs, diagrams, photos of gross CNS structures, and images from CT and MRI scans.

*Case Studies illustrative of disease at each level of the CNS.

*List of Movies that provide examples of dysfunction in the CNS and are an invaluable adjunct to teaching.

TABLE OF CONTENTS

PART I. INTRODUCTION TO THE CENTRAL NERVOUS SYSTEM

Chapter One	Introduction to the Central Nervous System
Chapter Two	Neurocytology
Chapter Three	Spinal Cord
Chapter Four	Brain Stem
Chapter Five	Cranial Nerves
Chapter Six	Diencephalon
Chapter Seven	Hypothalamus
Chapter Eight	Cerebral Cortex Functional Localization

PART II. THE SYSTEMS WITHIN THE CENTRAL NERVOUS SYSTEM

Chapter Nine	Motor System I Movement and Motor Pathways
Chapter Ten	Motor Functions II Basal Ganglia
Chapter Eleven	Motor Functions III Cerebellum
Chapter Twelve	Somatosensory Function and the Parietal Lobe
Chapter Thirteen	Visual System and Occipital Lobe
Chapter Fourteen	Limbic System and the Temporal Lobe
Chapter Fifteen	Higher Cortical Functions

PART III. THE NONNERVOUS ELEMENTS WITHIN THE CNS

Chapter Sixteen	Meninges, Ventricular System, Vascular System
Chapter Seventeen	Cerebral Vascular Disease
Chapter Eighteen	Movies on the Brain

Stanley Jacobson, Ph.D. is Professor of Anatomy and Cellular Biology, Tufts University Health Sciences Campus, Boston, MA

Elliott M. Marcus M.D. is Professor Emeritus of Neurology University of Massachusetts School of Medicine; Chairman Emeritus, Department of Neurology, St. Vincent Hospital and Fallon Clinics, Worcester MA; Lecturer in Neurology Tufts University School of Medicine.

Printed in the United States of America